Ferdinand von Mueller

Select Extra-Tropical Plants

Readily Eligible for Victorian Industrial Culture or Naturalisation

Ferdinand von Mueller

Select Extra-Tropical Plants
Readily Eligible for Victorian Industrial Culture or Naturalisation

ISBN/EAN: 9783337024604

Printed in Europe, USA, Canada, Australia, Japan

Cover: Foto ©berggeist007 / pixelio.de

More available books at **www.hansebooks.com**

SELECT EXTRA-TROPICAL PLANTS,

READILY ELIGIBLE FOR

Industrial Culture or Naturalization,

WITH INDICATIONS OF THEIR NATIVE COUNTRIES
AND SOME OF THEIR USES;

BY

BARON FERD. VON MUELLER,

K.C.M.G., M.D., PH.D., F.R.S., F.L.S., F.G.S., F.C.S., F.R.G.S., C.M.Z.S., H.M.R.H.S., H.C.M.S.A.,
H.F.R.B.S., H.M.G.S.E., H.M.B.S.E., ETC.,

GOVERNMENT BOTANIST FOR VICTORIA.

"Omnia enim in usus suos creata sunt."—SYRACH, XXXIX., 21, 26.

New Victorian Edition, Revised and Enlarged.

MELBOURNE: JOHN FERRES, GOVERNMENT PRINTER.
1885.

TO

THE HONORABLE GRAHAM BERRY, M.L.A.,

CHIEF SECRETARY OF THE COLONY OF VICTORIA,

AN ENLIGHTENED PROMOTER OF RURAL INDUSTRIES AND A GENEROUS
SUPPORTER OF THE AUTHOR'S RESEARCHES,

THIS VOLUME
IS RESPECTFULLY DEDICATED.

PREFACE.

In the volumes, issued by the Victorian Acclimatisation-Society from 1871 to 1878, five contributions have appeared concerning such industrial plants, as are available for culture in extra-tropical countries, or in high mountain-regions within the tropics. These writings were mainly offered with a view of promoting the introduction and diffusion of the very many kinds of plants, which may be extensively reared in the forests, fields or pastures of temperate geographic latitudes. But the work thus originated became accessible merely to the members of the Society, while frequent calls arose for these or some similar data, not only throughout the Australian communities, but also abroad. The whole was therefore re-arranged and largely supplemented, first for re-issue in Victoria, and lately also in India, under the auspices of the Central Government at Calcutta. Subsequently the work was likewise honored by being reprinted, with numerous additions, for the use of New South Wales; and at nearly the same time it went through a German translation, by Dr. Goeze, in Herr Th. Fischer's publishing establishment in Cassel; while last year it appeared revised and still further augmented, more particularly for North-American use, through the generous interest of one of the most enterprising scientific publishers in the United States, Mr. George Davis of Detroit. The early Victorian edition having become exhausted, the present one is offered now, still further enlarged by such notes as could be made very recently. As stated in the preface to the original essays, they did not claim completeness, either as a specific index to, or as a series of notes on the respective rural or technologic applicability of the plants enumerated. But what these writings may perhaps aspire to, is to bring together some condensed data in popular language on all the principal utilitarian plants, hitherto known to prosper in extra-tropical zones. Information of this kind is widely scattered through many and often voluminous works in several languages; yet such volumes apply generally to countries with a climatic zone far narrower than that, for which these pages were written.

Most, but not all the books, which it was desirable to consult, were at the author's command; thus the necessity of further successive supplements will be apparent, even irrespective of needful references to future discoveries; because in the progress of geographic, medical, technologic and chemical inquiries many new plants are likely to be disclosed, and additional uses of known plants to be elucidated. Thus, for instance, among the trees and shrubs, or herbs and grasses, occurring in the middle and higher altitudinal zones of Africa, or, nearer to us, of New Guinea and the Sunda-Islands, many specific forms may be expected to occur, which we could transfer to extra-tropical countries or to mountains in equinoctial regions. Moreover the writer would modestly hope, that his local efforts may prove to be useful in other parts of the globe for extending rural pursuits; indeed, through the generous action of an enlightened American, Capt. Ellwood Cooper, President of the State Board of Horticulture of California, the first fragmentary publications, then offered for Australian use, were deemed worthy of re-issue in San Francisco. Gradual or partial reprints had also previously appeared in weekly journals of Sydney and San Francisco and in some other periodicals.

As already intimated, the rapid progress of tillage almost throughout all colonial dominions and in other new States is causing a growing desire for general and particular indications of such plants, which a colder clime excludes from the northern countries, in which many of the colonists spent their youth; and it must be clear to any reflecting mind, that in all warmer latitudes, as compared with the Middle-European zone, is existing a vastly enlarged scope for cultural choice of plants. Thus, indicative as these notes merely are, they may yet facilitate the selection. More extensive information can then be sought for in larger, though less comprehensive works already extant, or likely still to be called forth by local requirements in other countries. The writer should even not be disinclined, under fair support and encouragement, to issue, collateral to the present volume, also another, exclusively devoted to the industrial plants of the hotter zones, for the promotion of tropical culture, particularly in our Australian continent.

Considerable difficulty was experienced in fixing the limits of such remarks, as are admissible into the present pages, because certain plants may be important only under particular climatic conditions and cultural applications, or they may have been overrated in regard to the copiousness and relative value of their yield. Thus it was not always easy to sift the chaff from the grain, when these notes were gathered; the remarks, offered in these pages, might indeed under less rigorous restrictions have been indefinitely extended; and although the author has for more than twenty years been watching for industrial tests the plants, introduced by him into the Melbourne Botanic Garden, he had still to a very large extent to rely implicitly on the experience of other observers elsewhere. It may also be at once here stated, that when calculations of measurements and weights were quoted, such always represent the maximum as far as hitherto on record. It was not always found easy, to fix with accuracy the geographic range of the species for this work in concise terms, as even some of the best and newest taxologic works relate not with sufficient distinctness; what is truly indigenous and what merely naturalized in any particular part of the globe. Furthermore schematic indices, to facilitate general views over the geographic distribution of plants, such as given for Australia in " a systematic census of plants with geographic and literary annotations " have not been yet forthcoming for any of the other great divisions of the earth with completeness. To draw prominent attention to the primarily important among the very many hundreds of plants, referred to in these pages, the leading species have been designated with an asterisk. It has not been easy in numerous instances, to trace the original source of that information on utilitarian plants, which we fiud recorded in the various volumes of phytologic or rural or technologic literature; many original observations are however contained in the writings of Bernardin, Bentham, Bentley, Brandis, Brockhaus, Candolle, Chambers, Collins, Dyer, Drury, Engelmann, Engler, Flueckiger, Fraas, Freyn, Asa Gray, Grisebach, Hanbury, Hooker, King, Koch, Langethal, Lawson, Lindley, Lorentz, Loudon, Martius, Masters, Meehan, Meyer, Michaux, Naudin. Nuttall, Oliver, Pereira, Philippi, Porcher, Rosenthal, Roxburgh, Sargent, Seemann, Simmonds, Stewart, Trimen, Wittstein and others, to whose names reference is cursorily made in the text. The volumes of the Agricultural Department at Washington, of the Austrian Apotheker-Verein, of the Journal of Applied Science, of the Bulletin de la Société d'Acclimatation de France and of several other periodicals have likewise afforded data, utilized on this occasion.

In selecting notes from general rural literature great caution had to be exercised, to guard against being misled by perhaps sometimes faulty nomenclature. Furthermore in choosing or elaborating the data for entries into this work, it had constantly to be kept in view, that the information is intended for the bread-winning portion of communities in young colonies mainly if not exclusively; nothing beyond this is aimed at.

In grouping together at the close of this volume all the genera, enumerated according to the products, which they yield, facility is afforded for tracing out any series of plants, regarding which special economic information may be sought, or which may at any time prominently engage the attention of the cultivator, the manufacturer or the artisan. Again, the placing together in index-form of the respective industrial plants according to their geographic distribution, as has likewise been done in the concluding pages, has rendered it easy, to order or obtain from abroad the plants of such other countries, with which any settlers or colonists may be in relation, through commercial, literary or other intercourse. Lists like the present may also aid in naming the plants and their products with scientific correctness in establishments of economic horticulture or in technologic or other educational collections. If the line of demarcation between the plants, admissible into this list and those which should have been excluded, has occasionally been extended in favor of the latter, then it must be pleaded, that the final value of any particular species for a peculiar want, locality or treatment cannot always be fully foretold. Doubtless, many plants of primary importance for rural requirements, here again alluded to, have long since been secured by intelligent early pioneers of immigration, who timely strove to enrich the cultural resources of their adopted country. In these efforts the writer, so far as his public or private means would permit, has endeavored for more than a quarter of a century to take an honorable share. But although such plants are introduced, they are not in all instance as yet widely diffused, nor tested in all desirable localities. For the sake of completeness even the most ordinary cultural plants have not been passed, as the opportunity seemed an apt one, to offer a few cursory remarks on their value also.

The writer entertains a hope, that a copy of this plain volume may find a place in the library of every educational establishment for occasional and perhaps frequent reference to its pages. The increased ease of communication, which has latterly arisen between nearly all parts of the globe, places us now also in a fair position for independent efforts, to suggest or promote introductions of new vegetable treasures from unexplored regions, or to submit neglected plants of promising value to unbiassed original tests. It may merely be instanced, that after the lapse of more than three centuries since the conquest of Mexico, only the most scanty information is extant on the timber of that empire, and that of several thousand tropical grasses not many dozen have been tried with rural or chemical exactitude for pasture-purposes, not to speak of many prominently utilitarian trees, shrubs and herbs, restricted to cool mountain-regions elsewhere within the tropics, but never yet carried to the lowlands of higher latitudes. For inquiries of such kind every civilized State is striving to afford in well-planned, thoughtfully directed and generously supported special scientific establishments the needful aid, not merely for adding to the prosperity, comfort and enjoyment of the present generation, but also with an anticipation of earning the gratitude of posterity; and this, as a rule, is done with a sensitive jealousy, to maintain also thereby the fair fame of the country for scientific dignity and industrial development. Friendly consideration will recognize the fact, that a desire to arouse more and more such a spirit of emulation has much inspired the writer, to offer these pages, he trusting that enlightened statesmanship far and wide will foster the aims, which he has had in view, through liberal and circumspect support.

Melbourne, August 1885.

SELECT PLANTS,

READILY ELIGIBLE FOR INDUSTRIAL CULTURE

IN

EXTRA-TROPICAL COUNTRIES.

Aberia Caffra, J. Hooker and Harvey.

The "Kai-Apple" of Natal and Caffraria. This tall shrub serves for hedges; it bears only slight frost. The rather large fruits are edible, and can be converted into preserves. Allied South-African species are A. Zeyheri and A. tristis (Sonder).

Acacia acuminata, Bentham.

A kind of "Myall" from Western Australia, attaining a height of 40 feet. The scent of the wood comparable to that of raspberries. It is the best of West-Australian woods for charcoal. The stems much sought for fence-posts, very lasting for this purpose, even when selected young. A similar tree with hard and scented wood is A. Doratoxylon (A. Cunn.), of the dry regions of South-Eastern Australia.

Acacia aneura, F. v. Mueller.

Arid desert-interior of extra-tropic Australia. A tree never more than 25 feet high. The principal "Mulga" tree. Mr. S. Dixon praises it particularly as valuable for fodder of pastoral animals; hence it might locally serve for ensilage. Mr. W. Johnson found in the foliage a considerable quantity of starch and gum, rendering it nutritious. Cattle and sheep browse on the twigs of this and some allied species, even in the presence of plentiful grass, and are much sustained by such Acacias in seasons of protracted drought. Dromedaries in Australia crave for the Mulga as food. Wood excessively hard, dark-brown, used preferentially by the natives for boomerangs, sticks to lift edible roots, end-shafts of Phragmites-spears, woomerangs, nulla-nullas and jagged spear-ends.

Acacia Arabica, Willdenow.

The "Kikar" or "Babur." Northern and Central Africa, also in South-Western Asia, growing in dry, calcareous soil. This small tree can be utilized for thorny hedges. It furnishes one of the best kinds of gum arabic for medicinal and technical purposes. The lac-insect lives also on the foliage, and thus in Sind the lac is mainly yielded by this tree. The stem attains a circumference of 10 feet. The astringent pods are valuable for tanning, also the bark, which is known as "Baboot" bark; the wood is very durable if water-seasoned, extensively used for wheels, well-curbs and many kinds of imple-ments, also for the knees and planks of boats. This species is of com-paratively quick growth. A. Ehrenbergiana (Hayne) is among the species, which yield gum arabic in North-Africa. A. latronum (Willdenow) and A. modesta (Wallich) form thorny hedges in India according to Dr. Brandis.

Acacia armata, R. Brown.

Extra-tropical Australia. The Kangaroo-Thorn. Much grown for hedges, though less manageable than various other hedge-plants, and not so fire-proof. More important for covering coast-sand with an unapproachable prickly vegetation.

Acacia binervata, De Candolle.

Extra-tropical East-Australia. A tree attaining a height of 40 feet. The bark used by tanners, but not quite so valuable as that of A. decurrens (W. Dovegrove).

Acacia Catechu, Willdenow.

India, East-Africa, up to 5,000 feet. A tree attaining 40 feet in height. Will bear some frost (Brandis). Wood hard, heavy, ex-tremely durable, locally chosen for underground posts, particularly also mill-work. The extract prepared from the bark and heartwood is the catechu of medicine or cutch of tannery. Pure cutch is worth about £25 per ton; 4 tons of bark will produce 1 ton of cutch. A. Suma (Kurz) is closely allied.

Acacia Cavenia, Hooker and Arnott.

The "Espino" of the present inhabitants of Chili, the "Cavan" of the former population. A small tree with exceedingly hard wood, resisting underground-moisture. The plant is well adapted for hedges. The husks contain 32 per cent. tannin (Sievers), particularly valuable as a dye-material.

Acacia Cebil, Grisebach. (*Piptadenia Cebil*, Grisebach.)

La Plata-States. This is one of the most useful of all trees there, on account of its bark, which is exceedingly rich in tannic acid; a species well worthy of introduction elsewhere, even as an ornamental tree. Numerous other Acaciæ, particularly the Australian species, deserve yet tests for tannin.

Acacia concinna, De Candolle.

India, China. Praised by Dr. Cleghorn as a valuable hedge-shrub. The pod contains saponin.

Acacia dealbata, Link.

South-Eastern Australia and Tasmania. This tree is generally known amongst Australian colonists as Silver-Wattle. It prefers for its habitation humid river-banks, and sometimes attains there a height of 150 feet, supplying a clear and tough timber used by coopers and other artisans, but principally serving as select fuel of great heating power. The bark of this variety is much thinner and greatly inferior in quality to that of the Black Wattle, yielding only about half the quantity of tannin-principle. It is chiefly employed for lighter leather. This tree is distinguished from the Black Wattle by the silvery or rather ashy hue of its young foliage: it flowers early in spring, ripening its seeds in about 5 months, while the Black Wattle occurs chiefly on drier ridges, blossoms late in spring or at the beginning of summer, and its seeds do not mature in less than about 14 months.

Acacia decurrens, Willdenow.*

The Black Wattle. ' From the eastern part of South-Australia, through Victoria and New South Wales, to the southern part of Queensland, also in Tasmania. A small or middle-sized tree. Its wood is used for staves, for turners' work, occasionally also for axe- and pick-handles and many other purposes; it supplies an excellent firewood; a chief use of the tree would be also, to afford the first shelter in treeless localities for raising forests. Its bark, rich in tannin, and its gum, not dissimilar to gum arabic, render this tree highly important. The English price of the bark ranges generally from £8 to £11. In Melbourne it averages about £5 to £8 per ton. It varies, so far as experiments made in my laboratory have shown, in its contents of tannic principle from 30 to 40 per cent. in bark completely dried. In the mercantile bark the percentage is somewhat less, according to the state of its dryness—it retaining about 10 per cent. moisture. $1\frac{1}{2}$ lbs. of Black Wattle-bark give 1 lb. of leather, whereas 5 lbs. of English Oak-bark are requisite for the same results, but the tannic principle of both is not absolutely identical. Melbourne tanners consider a ton of Black Wattle-bark sufficient to tan 25 to 30 hides; it is best adapted for sole-leather and other so-called heavy goods. The leather is fully as durable as that tanned with oak-bark, and nearly as good in color. Bark carefully stored for a season improves in tanning power considerably. From experiments made under the author's direction it appears, that no appreciable difference exists in the percentage of tannin in Wattle-bark, whether obtained in the dry or in the wet season. The tannin of this Acacia yields a gray precipitate with ferric, and a violet color with ferrous salts; it is completely precipitated from a strong aqueous solution by means of concentrated sulphuric acid. The bark improves by age and desiccation, and yields about 40 per cent. of catechu, rather more than half of which is tannic acid. Bichromate of potash added in a minute quantity to the boiling solution of mimosa-tannin produces a ruby-red liquid, fit for dye-purposes; and this solution gives with the salts of sub-oxide of iron black pigments, and with the salts of the full oxide of iron red-brown dyes. As far back as 1823 a fluid extract of Wattle-bark was shipped to London, fetching then the extraordinary price of £50 per ton, one ton of bark yielding 4 cwt. of extract of tar-consistence (Simmonds), thus saving much freight and cartage.

Tan extract is best obtained from the bark by hydraulic pressure and evaporation of the strong liquid thus obtained in wide pans under steam-heat, or better still, to avoid any decomposition of the tannic acid, by evaporation under a strong current of cold air. For cutch or terra japonica the infusion is carefully evaporated by gentle heat. The estimation of tannic acid in Acacia barks is effected most expeditiously by filtering the aqueous decoction of the bark after cooling, evaporating the solution and then re-dissolving the residue in alcohol and determining the weight of the tannic principle obtained by evaporating the filtered alcoholic solution to perfect dryness.

The cultivation of the Black Wattle is extremely easy, being effected by sowing either broadcast or in rows. Seeds can be obtained in Melbourne at about 5s. per lb., which contains from 30,000 to 50,000 grains; they are known to retain their vitality for several years. For discrimination in mercantile transactions it may be noted, that the seeds of the genuine A. decurrens are somewhat smaller, comparatively shorter, rounder and not so flat as those of A. dealbata, while the funicular appendage does not extend so far along the seeds, nor is the pod quite so broad; from those of A. pycnantha they differ in being shorter, thus more ovate than oblong.

Seeds should be soaked in warm water before sowing. Any bare, sterile, unutilized place might most remuneratively be sown with this Wattle-Acacia; the return could be expected in from five to ten years. Full-grown trees, which supply also the best quality, yield as much as 1 cwt. of bark. Mr. J. Dickinson states, that he has seen 10 cwt. of bark obtained from a single tree of gigantic dimensions at Southport. A quarter of a ton of bark was obtained from one tree at Tambo without stripping all the limbs. The height of this tree was 60 feet, and the stem 2 feet in diameter. The rate of growth of the tree is about 1 inch in diameter of stem annually. It is content with the poorest and driest soil, although in more fertile ground it shows greater celerity of growth. This Acacia is perhaps the most important of all tan-yielding trees of the warm temperate zones, for its strength in tannic acid, its rapidity of growth, its contentedness with almost any soil, the ease with which it can be reared and its early yield of tanner's bark, and indeed also gum and stave-wood. This tree is to be recommended for poor land, affected with sorrel. It is hardier than Eucalyptus globulus, thus enduring the climate of South England, although it hardly extends to sub-alpine elevations.

The wood of this Acacia, particularly when the trees are in an unhealthy state, is sometimes bored by the large larva of a moth, namely that of Eudoxyla Eucalypti.

For fuller information the "Report on Wattle-bark," presented in 1878 to the Parliament of Victoria by a special commission, may be referred to.

Acacia estrophiolata, F. v. Mueller.

Central Australia. A tree, attaining a height of 30 feet and a stem-diameter of 1 foot, enduring the extremest of dry heat; suitable for cemeteries on account of its pendent branches. It flowers almost constantly, and accommodates itself to all sorts of soil, even sand. Wood very durable, locally much used for implements and especially wheelwright's work (Rev. H. Kempe). Bark rich in tannin.

Acacia excelsa, Bentham.

The Ironbark-Acacia of Queensland, extending into New South Wales. Attains a height of 80 feet. Branches pendent. The wood is dark-colored, hard, heavy and durable, well adapted for furniture and implements; towards the centre it is of a deep pinkish color. The tree exudes a large quantity of clear gum (O'Shanesy), and sprouts again from the root after the stem is cut. Also particularly eligible for cemeteries.

Acacia falcata, Willdenow.

Extra-tropical East-Australia. A small tree. Important for its bark in tanneries.

Acacia Farnesiana, Willdenow.

Dioscorides' small Acacia. Indigenous to Southern Asia; found eastward as far as Japan; a native also of the warmer parts of Australia, as far south as the Darling-River; found spontaneous in tropical and sub-tropical America, but apparently not in tropical Africa. Professor Fraas has recognized in this Acacia the ancient plant. The scented flowers, inappropriately called "cassie" flowers, are much sought for perfumery, and develop successively. This species may well be utilized as a hedge-plant; a kind of gum arabic may also be obtained from it. The scent perhaps obtainable from the fresh and slightly moist flowers by gentle dry distillation under mere steam-heat. Ordinarily the odorous essential oil is withdrawn from the flowers by the enfleurage-process; many Australian Acacias might be thus treated for perfumery.

Acacia fasciculifera, F. v. Mueller.

South-Queensland. Tree, sometimes to 70 feet high; branches pendent. Desirable for culture on account of the excellence of its easily worked dark wood. Eligible also for cemeteries.

Acacia giraffae, Willdenow.

South-Africa. The Camel-Thorn. This tree attains a great age and a height of 40 feet. The trunk assumes a large size, and supplies a wood of great hardness. The tree will grow on the driest soil and in the hottest clime.

Acacia glaucescens, Willdenow.

Queensland and New South Wales. Extreme height about 60 feet. A kind of "Myall," with hard, dark, prettily grained wood, which is however less fragrant than that of some other species.

Acacia gummifera, Willdenow.

This tree yields principally the Gum Arabic of Morocco. (Sir Joseph Hooker and John Ball.) The principal collecting time extends over six weeks in midsummer.

Acacia harpophylla, F. v. Mueller.

Southern Queensland. A tree, sometimes attaining a height of 90 feet, furnishing locally a considerable share of the mercantile wattle-bark for tanning purposes. Wood, according to Mr. P. O'Shanesy, brown, hard, heavy and elastic, of violet odor; it splits freely and is thus also well adapted for fancy lathe-work; used by the natives for spears. The tree grows naturally on sand-lands, almost to the exclusion of other trees and shrubs. Saplings used as stakes in vineyards have lasted 20 years and more. The tree yields also considerable quantities of gum. It is one of the principal " Brigalows " in the scrubs of that designation.

Acacia heteroclita, Meissner.

South-Western Australia. This or an allied species furnishes a particular sort of edible gum, called by the autochthones " Quannot." (Hon. John Forrest.)

Acacia homalophylla, Cunningham.

The Victorian " Myall," extending into the deserts of South-Australia and New South Wales. Never a tall tree. The dark-brown wood is much sought for turners' work on account of its solidity and fragrance; perhaps its most extensive use is in the manufacture of tobacco-pipes. Allied species pass under the aboriginal appellation " Boree."

Acacia horrida, Willdenow.

The " Doornboom" or " Karra-Doorn" of South-Africa. A formidable hedge-bush with thorns often 3 inches long, readily available for impenetrable hedge-rows. It exudes also a gum of good quality, but often of amber-color. This is the principal species used for tanners' bark in South-Africa, where Leucospermum conocarpum (R. Br.) is also extensively employed for the same purpose (Mc Gibbon). It imparts however an unpleasant odor to the leather made with it (McOwan).

Acacia implexa, Bentham.

Victoria, New South Wales, Queensland. A tree of middle size, content with poor soil. Wood firm and close, dark-brown with yellowish stripes; much in demand for turnery, cogwheels and other purposes, which need tenacity and strength (Dickinson). Bark available for tanneries.

Acacia Koa, A. Gray.

Hawaii; there one of the most valuable of timber-trees. Stem reaching a height of 60 feet, topped by wide-spreading phyllodinous foliage. Wood easy to work, particularly in a fresh state; formerly much used for boat-building and for building purposes generally; also suitable for cabinet-work. Species of Metrosideros, some ascending to 8,000 feet, one overtopping all other trees, furnish a large share of hard, tough and very durable timber in the Hawaian islands. Their wood varies from a light red to a purplish hue. (Hon. Judge McCully.)

Acacia leiophylla, Bentham.* (*A. saligna,* Bentham not Wendland.)

South-Western Australia, where it is the principal tree chosen for tanners' bark. It is a wide-spreading small tree, fit for avenues ; emitting suckers. The weeping habit fits it particularly also for cemeteries. The bark contains nearly 30 per cent. of mimosa-tannin, and is extensively used by tanners in West-Australia. Perfectly dried leaves yield from 7 to 8 per cent. mimosa-tannic acid, giving a lead-precipitate of a light yellow color; the leaves contain also a considerable quantity of sulphate of lime. The London price of fair West-Australian gum arabic from this species was from 46s. to 49s. per cwt. in 1879. The tree has proved in Algeria to resist the sirocco better than most species (Dr. Bonand). A. cyanophylla (Lindley) is a closely allied species, serving the same purposes.

Acacia longifolia, Willdenow.

South-Eastern Australia. This tree is introduced into this list, inasmuch as the very bushy variety known as A. Sophoræ (R. Brown) renders most important service in subduing loose coast-sand, the lower branches striking root into the soil; it should therefore be disseminated on extensively bare sand-shores in regions, where no severe frosts occur. The bark of A. longifolia is only half as good as that of A. decurrens for tanning, and used chiefly for sheep-skins. The tree is of quick growth—20 to 30 feet in 5 to 6 years (Hartmann).

Acacia macrantha, Bentham.

From Mexico to Argentina, also in the Galapagos-Group. This tree, usually small, provides the "Cuji-pods" for tanning (Simmonds).

Acacia Melanoxylon, R. Brown.*

South-Eastern Australia. Generally known as Blackwood-tree, passing also under the inappropriate name of Lightwood-tree. In irrigated glens of deep soil the tree will attain a height of 80 feet, with a stem several feet in diameter. The wood is most valuable for furniture, railroad-cars and carriages, boat-building (stem and stern-post, ribs, rudder), for tool-handles, crutches, some portions of the work of organ-builders, casks, billiard-tables, pianofortes (for sound-boards and actions) and numerous other purposes. The fine-grained wood is cut into veneers; it takes a fine polish, and is considered almost equal to walnut. The best wood in Victoria for bending under steam; it does not warp and twist. Local experiments gave the strength in transverse strain of Blackwood equal to Eucalyptus-wood of middling strength, approaching that of the American White

Oak, and surpassing that of the Kauri. The bark contains about 20
per cent. mimosa-tannin. The tree has proved, with A. decurrens
and A. dealbata, hardy in the Isle of Arran (Rev. D. Landsborough).

Acacia microbotrya, Bentham.

South-Western Australia. "The Badjong." A comparatively tall
species, the stem attaining a diameter of 1 to 1½ feet. It prefers
river-valleys, and lines brooks naturally. According to Mr. Geo.
Whitfield, a single tree may yield 50 lbs. of gum in a season. The
aborigines store the gum in hollow trees for winter use; it is of
a pleasant, sweetish taste.

Acacia moniliformis, Grisebach.

Argentina. The "Tusca." The young pods are used for feeding
horses and cattle (Dr. Lorentz), like those of Acacia Cavenia in
South-Western America.

Acacia pendula, Cunningham.

New South Wales and Queensland, generally in marshy tracts
of the interior. The "Weeping Myall." Reaching 35 feet in height.
Wood violet-scented, hard, close-grained, beautifully marked; used by
cabinet-makers and turners, in high repute for tobacco-pipes (W.
Hill). The tree is desirable for cemeteries.

Acacia penninervis, Sieber.

Victoria, New South Wales and Queensland. A small tree, so
hardy as to occupy sub-alpine localities. The bark contains about 18
per cent. of tannin.

Acacia pycnantha, Bentham.*

Victoria and South-Australia. The "Golden Wattle" of the
colonists. This tree, which attains a maximum height of about 30
feet, is second only to A. decurrens in importance for its yield of
tanners' bark; the quality of the latter is even sometimes superior to
that of the Black Wattle, but the yield is less, as the tree is smaller
and the bark thinner. It is a tree of rapid growth, content with
almost any soil, but is generally found in poor sandy ground, particularly near the sea-coast, where A. decurrens would not succeed, and
thus also important for binding rolling sand. Experiments instituted
by me have proved the absolutely dried bark to contain about 30
per cent. tanning principle, full-grown sound trees supplying the best
quality. The aqueous infusion of the bark can be reduced by boiling to a dry extract, which in medicinal and other respects is equal to
the best Indian catechu, as derived from Acacia Catechu and A.
Suma. It yields nearly 30 per cent., about half of which or more is
mimosa-tannic acid. This catechu is also of great use for preserving
against decay articles subject to exposure in water, such as ropes,
nets and fishing-lines. The fresh leaves yield 6 per cent. and dry
leaves 15 to 16 per cent. of mimosa-tannin. While, according to
Mr. Simmonds, the import of the bark of oaks and hemlock-spruce

into England becomes every year less, and while the import of sumach and gambir does not increase, the annual demand for tanning substance has within the last twenty years been doubled. A. pycnantha is also important for its copious yield of gum, which is in some localities advantageously collected for home-consumption and also for export. The wood, though not of large dimensions, is well adapted for staves, handles of various implements and articles of turnery, especially bobbins (Dickinson). By improved methods, the fragrant oil of the flowers could doubtless be fixed, though its absolute isolation might be difficult and unremunerative. The tree as a rule seeds well.

Acacia retinodes, Schlechtendal.

South-Eastern Australia. Ascertained so early as 1846 by Dr. Hermann Behr to yield a good tanners' bark and much gum. This Acacia is ever-flowering, and in this respect almost exceptional. It likes river-banks, but never grows beyond the height of a small tree. A. neriifolia (A. Cunningham) of New South Wales is a closely allied species.

Acacia Sentis, F. v. Mueller.

Interior of Australia. This shrub or small tree is suitable for hedges. The seeds of this species and also of A. Kempeana, A. cibaria and some others are eaten by the natives. Horses, cattle, sheep and camels browse in Central Australia with avidity on the foliage (Rev. H. Kempe). This species will endure most protracted drought and a shade-temperature of 118° F.

Acacia Seyal, Delile.

In the Libyan and Nubian Deserts. This thorny tree exudes a brownish kind of gum arabic. It is adapted for the most arid desert-country. In any oasis it forms a large and shady tree. Native name "Soffar." Can be utilized for thorny hedges as well as A. tortilis (Forskael), the latter also yielding gum arabic.

Acacia stenocarpa, Hochstetter.

Abyssinia and Nubia. A large tree, which yields the brownish "Suak-" or "Talha"-Gum, a kind of gum arabic. (Hanbury and Flueckiger.)

Acacia stenophylla, Cunningham.

On banks of water-courses in the interior of Australia, as far south as the Murray-River. A tree with exquisite, hard, dark wood, serving the same purposes as Myall-wood, and also known as Iron-wood. Attains a height of 60 feet and a stem-diameter of 2 feet.

Acacia supporosa, F. v. Mueller.

South-Eastern Australia. Straight stems over 50 feet long are formed by this tree; the wood is tough and elastic, fit for carriage-shafts, gunstocks, various select tools. (L. Morton.)

Acacia Verek, Guillemin and Perrottet.

From Senegambia to Nubia. Affords the best white gum arabic of the Nile-region, and a large quantity of this on a commercial scale. A. Etbaica (Schweinfurth) from the same region produces also a good mercantile gum.

Acanthophoenix rubra, H. Wendland.

Mauritius and Réunion. This palm has proved hardy in Florida, also as far south as Sydney (C. Moore). Height reaching 60 feet. The upper rings of the stem are of a bright red. In gardens usually passing as an Areca.

Acanthosicyos horrida, Welwitsch.

In the deserts of Angola, Benguela and Damarland. This thorny erect, cucurbitaceous shrub bears fruit of the size and color of oranges and of pleasant acidulous taste. The seeds are also edible. No rain occurs in the Acanthosicyos- and Welwitschia-region, but the mean heat does not exceed 70° F. and the soil is kept somewhat moist through capillarity from beneath.

Acer Campbellii, J. Hooker and Thomson.

The chief Maple of the North-Eastern Himalayas. A large tree. Freely reproduced by seed or coppice. Wood pale, close-grained, particularly valuable for planking (Gamble).

Acer campestre, Linné.

The British Maple. Extends from South- and Middle-Europe to Northern Africa, also to many parts of Asia. Height reaching 40 feet, in shelter and deep soil; the yellow and purple tints of its foliage in autumn render the tree then particularly beautiful. Occurs in Norway to 63° 26′ N. L. (Prof. Schuebeler). The wood is compact and fine-grained, and sought for choice furniture, machinery and musical instruments. The tree can be trimmed into hedges. Comparatively quick in growth, and easily raised from seed. These remarks apply to many kinds of maples.

Acer circinatum, Pursh.

The Vine-Maple of North-Western America, forming in some parts of Oregon impenétrable forests on account of its long branches bending to the ground and striking root; its autumnal tint gives quite a picture to the landscape. The stem is sometimes 40 feet long, but slender. Found to be hardy as far north as Christiania or even Nyborg, in lat. 70° 10′, where the mean annual temperature is 29° F., the highest 95° and the lowest—40° F. (Professor Schuebeler). The wood is heavier and of closer grain than that of A. macrophyllum (Dr. Gibbons); very tough, used for helves.

Acer dasycarpum, Ehrhart. (*A. saccharinum,* Linné.)

The Silver-Maple of North-America. Requires a rather warmer climate than the other American maples, but has proved hardy in Norway as far as 59° 55′ N. (Schuebeler). Height reaching 50 feet; stem sometimes 9 feet in diameter. Much praised for street-planting;

growth comparatively rapid. It produces no suckers, nor is the tree subject to disease. A most beautiful tree, with a stout stem and a magnificent crown, growing best on the banks of rivers with limpid water and a gravelly bed, but never in swampy ground, where the Red Maple takes its place. The wood is pale and soft, of less strength and durability than that of many of its congeners, but makes excellent charcoal. It may be cut into extreme thinness for wood-paperhangings (Simmonds). The tree also yields maple-sugar, though not in such quantity as A. saccharinum. With other maples, an early yielder of honey to bees.

Acer macrophyllum, Pursh.

Large Oregon-Maple. From British Columbia to Northern Mexico. A fine shade-tree of quick growth; sometimes reaching a height of 90 feet; stem attaining 16 feet in circumference; wood whitish, beautifully veined; delights on banks of streams. The inner bark can be utilized for baskets, hats and superior mats; the hard and close wood is a substitute for hickory. The wood when curled is splendid for ornamental work. Maple-sugar is also manufactured from the sap of this species (Sargent).

Acer Negundo, Linné. (*Negundo aceroides*, Moench.)

The Box-elder of North-America. Hardy in Norway to 59° 55' N. (Schuebeler). A tree, deciduous like the rest of the maples; may attain a height of about 50 feet, and is rich in saccharine sap; according to Vasey it contains almost as much as the Sugar-maple. In California it is used extensively as a shade-tree. Cultivated, the stem attains about 8 inches in diameter in 8 years (Brewer). The wood is yellow, marked with violet and rosy veins (Simmonds). Rate of stem-girth in Nebraska about 2 feet in fourteen years (Governor Furnas).

Acer niveum, Blume.

Continental and Insular India, up on the forest-ranges. This is the tallest of the maples, attaining a height of 150 feet. Several other large maples, worthy of cultivation particularly in parks, occur on the mountains of India.·

Acer palmatum, Thunberg.

Japan. A beautiful tree, with deeply cleft leaves; various varieties with red- and yellow-tinged leaves occur. Unhurt by frost at 0° F. (Gorlie). Should it be an aim, to bring together all the kinds of maples, which could be easily grown in appropriate spots, then Japan alone would furnish 22 species.

Acer pictum, Thunberg.

From Persia to Japan. Ascending the Himalayas to 9,000 feet. Foliage turning yellow and red in autumn. Wood close-grained and elastic, particularly sought for load-poles, ploughs, chairs and various implements and utensils (Brandis); twigs lopped off for fodder (Gamble).

Acer platanoides, Linné.

The Norway-Maple, extending south to Switzerland. Up to 80 feet high. Found hardy in Norway (cultivated) to 67° 56' N.; attained in 59° 46' a stem-diameter of 3½ feet (Schuebeler). The pale wood much used by cabinet-makers. Tint of the autumn-foliage golden-yellow. The tree is of imposing appearance, and much recommended for ornamental gardening; it gives a denser shade than most of the other maples. Sap of this species also saccharine.

Acer Pseudo-Platanus, Linné.

The Sycamore-Maple, or Spurious Plane. Middle and Southern Europe, Western Asia. Hardy to 67° 56' N. in Norway (Schuebeler). The celebrated maple at Trons, under which the Grisons swore the oath of union in 1424, exists still (Langethal). Attains a height of over 100 feet. The wood is compact and firm, valuable for various implements, instruments and cabinet-work; thus mangles, presses, dishes, printing and bleaching works, beetling-beams, and in foundries the patterns are often made of this wood (Simmonds); for the back, neck, sides and circle of violins, for pianofortes (portion of the mechanism) and harps it is utilized, it being free-cutting and clean on the end-grain. This like some other maples furnishes a superior charcoal for intense and continuous heat (Hartig). Will admit of exposure to sea-air. The sap also saccharine.

Acer rubrum, Linné.

The Red Maple of North-America. Hardy in Norway to 63° 26' N. (Schuebeler). A tree, attaining over 100 feet in height, 5 feet in stem-diameter. This species grows well with several other maples even in dry, open localities, although the foliage may somewhat suffer from hot winds, but thrives most luxuriantly in swampy, fertile soil. It is valued for street-planting. The foliage turns red in autumn, the flaming tints being indescribable (A. J. Cook). The wood is of handsome appearance, used in considerable quantity for saddle-trees, yokes, turnery and various furniture; that of old trees is somewhat cross-grained, and thus furnishes a portion of the curled Maple-wood, which is so beautiful and much in request for gun-stocks and inlaying. The tree yields also Maple-sugar, but like A. dasycarpum, only in about half the quantity obtained from A. saccharinum (Porcher). The flowers of some, if not all, maples are early frequented by bees for honey.

Acer saccharinum, Wangenheim.* (*A. nigrum*, Michaux.)

The Sugar- or Rock-Maple; one of the largest of the genus. Eastern North-America, extending to Arizona. It is the national emblem of Canada. In the cooler latitudes often 80 or rarely 120 feet high, with a stem 3 to 4 feet in diameter. Hardy to 59° 55' N. in Norway (Schuebeler). The wood is strong, tough, hard, close-grained, of rosy tinge, and when well seasoned is used for axle-trees, spokes, shafts, poles and furniture, exteriors of pianos, saddle-trees, wheel-wrights' work, wooden dishes, founders' patterns and flooring; not apt to warp; preferred for shoe-lasts; when knotty or curly it furnishes the Bird's-eye and Curly Maple-wood. From the end of February till the early part of April the trees, when tapped, will yield the saccharine fluid, which is so extensively converted into Maple-sugar, each tree yielding 12 to 24 gallons of sap in a season, 3 to 6 gallons giving 1 lb. of sugar; but exceptionally the yield may rise to 100 and more gallons. The tapping process commences at the age of 20 years, and may be continued for 40 years or more without destruction of the tree (G. Maw). According to Porcher, instances are on record of 33 lbs. of sugar having been obtained from a single tree in one season. The Sugar-Maple is rich in potash, furnishing a large proportion of this article in the United States. The bark is important for the manufacture of several American dyes. The tree is particularly recommendable in Australia for sub-alpine regions. It bears a massive head of foliage on a slender stem. The autumnal coloring is superb. In the Eastern States of North-America the Sugar-Maple is regarded as the best tree for shade-avenues. Numerous other maples exist, among which may be mentioned Acer Creticum (Linné) of South- Europe, 40 feet; A. lævigatum, A. sterculiaceum and A. villosum (Wallich) of Nepal, 40 feet.

Achillea Millefolium, Linné.

Yarrow or Millfoil. Europe, Northern Asia and North-America. A perennial medicinal herb of considerable astringency, pervaded with essential oil, containing also a bitter principle (achillein) and a peculiar acid, which takes its name from the generic appellation of the plant. Fitted for warrens and light sandy soil. Recommended by many for sheep-pastures, but disregarded by Langethal. Found indigenous in Norway as far as 71° 10′ N. (Schuebeler).

Achillea moschata, Wulfen.

Alps of Europe. The "Genipi" or "Iva" of the Swiss. This perennial herb ought to bear transferring to any other alpine mountains. With the allied A. nana (Linné) and A. atrata (Linné) it enters as a component into the aromatic medicinal Swiss tea. A. fragrantissima (Reichenbach) is a shrubby species from the deserts of Egypt, Turkey and Persia, valuable for its medicinal flowers.

Achras Sapota, Linné. *(Sapota Achras, Miller.)*

The "Sapodilla-Plum" of the West-Indies and Central America. A fine evergreen tree, producing delicious fruit. Yields also gutta-percha. The bark possesses tonic properties. Achras Australis (R. Brown; Sideroxylon australe, J. Hooker), a tree yielding also tolerably good fruit, occurs in New South Wales and Queensland. Other sapotaceous trees, producing table-fruit, such as the Lucuma mammosa (the Marmalade-tree), Lucuma Bonplandi, Chrysophyllum Cainito (the Star-Apple), all from West India, and Lucuma Cainito of Peru, might also be subjected to trial-culture in sub-tropical forest-valleys;

so furthermore many of the trees of this order, from which gutta-
percha is obtained (species of Dichopsis, Isonandra, Sideroxylon,
Cacosmanthus, Illipe, Mimusops, Imbricaria and Payeuia) might prove
hardy in sheltered woodlands, as they seem to need rather an equable
humid and mild climate, than the heat of the torrid zone.

Aconitum Napellus, Linné.

The "Monk's Hood." In the colder parts of Europe and Asia,
extending to the Himalayas and also to arctic America, especially in
mountainous regions. A powerful medicinal plant of perennial growth,
but sometimes only of biennial duration, variable in its forms. It
was first introduced into Australia, together with a number of other
Aconites, by the writer. All the species possess more or less modi-
fied medicinal qualities, as well in their herbage as in their roots; but
so dangerously powerful are they, that the plants should never be administered except as prescribed by a qualified physician. Napellus-
root contains three alkaloids: aconitin, napellin and narcotin. The
foliage contains also a highly acrid volatile principle, perhaps chemically not unlike that of many other Ranunculaceæ. Aconitin, one of
the most potent poisons in existence, can likewise be obtained from
the highly powerful Nepalese and Himalayan Aconitum ferox
(Wallich) and probably from several other species of the genus.

Acorus Calamus, Linné.

The "Sweet Flag." Europe, Middle and Northern Asia, North-
America. In Norway indigenous to 61° N., cultivated up to 63° 26′
(Schuebeler). A perennial pond- or marsh-plant. The aromatic
root is used as a stomachic and also in the preparation of confectionery, in the distillation of gin and liqueurs, and in the brewing of
some kinds of beer. The flavor of the root depends mainly on a
peculiar volatile oil.

Acrocomia Mexicana, Karwinski.

Mexico, in the cooler regions up to 3,000 feet, with a mean-temperature of 65° F. (Drude). A prickly palm, reaching 20 feet in height,
accompanied by very splendid Chamædora-Palms in the shade of oak-
forests. A. Totai (Martius) of Argentina yields sweet fruit.

Actæa spicata, Linné.

The "Baneberry." On wooded mountains, mainly on limestone-
soil in Europe, Northern Asia and North-America. A perennial
medicinal herb. Its virtue depends on peculiar acrid and bitter as
well as tonic principles. In North-America this species and likewise
A. alba (Bigelow) are also praised as efficacious antidotes against
ophidian poisons.

Adenostemum nitidum, Persoon.

Southern Chili, where this stately tree passes by the appellations
"Queule, Nuble and Aracua." Wood durable and beautifully
veined. Fruit edible.

Adesmia balsamica, Bertero.

The " Jarilla " of Chili. A small shrub, remarkable for exuding a fragrant balsam of some technic value (Philippi).

Ægiceras majus, Gaertner.

Southern Asia, Polynesia, Northern and Eastern Australia. This spurious Mangrove-tree extends far south into New South Wales. It may be employed for preventing the washing away of mud by the tide, and for thus consolidating shores subject to inundation by sea-floods.

Æschynomene aspera. Linné.

The " Solah " of tropical Asia and Africa. A large perennial erect or floating swamp-plant. Introduced from the Botanic Gardens of Melbourne early into the tropical parts of Australia. Pith-hats are made from the young stems of this plant; this pith is also a substitute for cork in some of its uses. The Solah is of less importance for cultivation than for naturalization.

Æsculus Californica, Nuttall.

California. This beautiful tree attains a height of 50 feet, with a stem 6 feet in diameter, the crown spreading out exceptionally over a width of 60 feet, the upper branches touching the ground. In full bloom it is a magnificent ornament, with its crowded snow-white flowers, visible for a long distance. The wood is light and porous, and used for the yokes of oxen and for various other implements (Dr. Gibbons).

Æsculus Hippocastanum, Linné.

The " Horse-Chestnut Tree." Indigenous to North-Greece, Thessaly and Epirus, on high ranges (Heldreich), where it is associated with the Walnut, several Oaks and Pines, at an altitude of 3–4,000 feet, occurring likewise in Imeretia, the Caucasus (Eichwald), and possibly wild also in Central Asia, ascending the Himalayas to 10,000 feet. One of the most showy of deciduous trees, more particularly when during spring " it has reached the meridian of its glory, and stands forth in all the gorgeousness of leaves and blossoms." Height reaching 60 feet, circumference of stem sometimes 16 feet. In cool climates one of the choicest of trees for street-planting. Flowers sought by bees in preference to those of any other northern tree except the Linden. Even in Norway, in latitude 67° 56′ N., a cultivated tree attained a height of 60 feet and a stem-circumference of 11 feet (Schuebeler). It will succeed even in sandy soil, but likes sheltered spots. The wood adapted for furniture, also particularly for mould-patterns in casting, the slips of pianofortes and a variety of other purposes ; it remains free from insects. The seeds yield starch copiously, and supply also food for various domestic animals ; the bark serves as a good tanning material. A variety is known with thornless fruits. Three species occur in Japan and several in North-America and South-Asia, mostly not of great height.

Æsculus Indica, Colebrooke.

In the Himalayas, from 3,500 to 9,000 feet. Height finally 50 feet; trunk comparatively short, occasionally with a girth of 25 feet. Never quite without leaves. Can be used like the Horse-Chestnut as an ornamental shade-tree. Twigs lopped off for fodder in India. Wood whitish, soft, available for various purposes, particularly liked for water-troughs, drinking-vessels, platters (Gamble). Other Asiatic species are A. Punduana (Wallich), A. Chinensis (Bunge), A. dissimilis (Blume).

Æsculus lutea, Wangenheim. (*A. flava*, Aiton.)

The "Buck-eye." North-America. This showy tree rises occasionally to a height of 80 feet. The wood is light, soft and porous, not inclined to split or crack in drying. It is valuable for troughs, bread-trays, wooden bowls and shuttles (Simmonds); also for ceiling and wainscoting (Mohr).

Æsculus turbinata, Blume.

Japan. Allied to A. Chinensis. The seeds are there used for human food.

Agaricus cæsareus, Schaeffer.

In the spruce-forests of Middle and Southern Europe. Trials might be made, to naturalize this long famed and highly delicious mushroom in our woodlands. It attains a width of nearly one foot, and is of a magnificent orange-color. Numerous other edible Agarics could doubtless be brought into this country by the mere dispersion of the spores in fit localities. As large or otherwise specially eligible may here be mentioned, on the authority of Dr. Rosenthal, who alludes to many more, A. extinctorius L., A. melleus Vahl, A. deliciosus L., A. giganteus Sowerby, A. Cardarella Fr., A. Marzuolus Fr., A. Eryngii Cand., A. splendens Pers., A. odorus Bulliard, A. auricula Cand., A. oreades Bolt., A. esculentus Wulf., A. mouceron Tratt., A. socialis Cand., A. laccatus Scop., all from Europe, besides numerous other highly valuable species from other parts of the globe. Professor Goeppert adds as edible species, sold in Silesia and other parts of Germany: A. decorus Fries, A. fusipes Bull., A. gambosus Fries, A. procerus Scop., A. scorodonius Fries, A. silvaticus Schaeff., A. virgineus Wulf., A. volemus Fries, besides the almost cosmopolitan A. campestris Linné and A. arvensis Schaeffer. Dr. M. C. Cooke mentions of Agarics besides as European (mostly British) kinds, fit for the kitchen: A. rachodes Vitt., A. personatus Fr., A. nebularis Batsch, A. dealbatus Sow., A. geotropus Bull., A. salignus Tratt., A. prunulus Scop., A. mutabilis Schaeff., A. squarrosus O. Muell., A. pudicus Viv. Dr. L. Planchon noted the following among the French edible species additionally: A. vaginatus Bull., A. ovoideus Bull., A. rubescens Fr., A. caligatus Viv., A. terreus Schaeff., A. albellus Cand., A. nudus Bull., A. crassipes Desm., A. piperatus L., A. oreades Bolt., A. cylindrius

Cand., A. pluteus Fr., A. bombycinus Schaeff. Several of these extend spontaneously to Australia. Mushroom-beds are best made from horse-manure, mixed with one-eighth loam, the scattering of the mushroom-fragments to be effected, when the temperature of the hot-bed has become reduced to 85° F., this sowing to be made 2–3 inches deep and 4 inches apart; 1 inch sifted loam over the damp bed and some hay to cover the whole. After two months mushrooms can be gathered from the bed. Mushroom-beds can also be prepared in spare places of cellars, stables, sheds and other places, where equability of mild temperature and some humidity can be secured. According to Mr. C. F. Heinemann, of Erfurt, the needful hot-beds can best be made one above another, inclined forward, generating a temperature of from 60° to 90° F., a surface-layer of cut straw being applied subsequently, to be removed after about two weeks, then to be replaced by a stratum of rich loam as a matrix for the root-like organs of the pushing fungs. In Japan mushrooms are reared on decayed split logs, and largely consumed. In France mushrooms are grown in caves to an enormous extent. Puff-balls when young are also edible, and some of them delicious (Meehan).

Agaricus flammeus, Scopoli.

Europe, Asia. In Cashmere particularly noticed as a large and excellent edible mushroom (Dr. Atchison). Some of the noxious mushrooms become edible by drying. Professor Morren mentions among edible Belgian species Agaricus laccatus, Scop., Russula integra, Fr. Any kind of cavern might be turned into a mushroom field; the spawn is spread on fermented manure, and kept moist by water, to which some saltpetre is added. They all afford a highly nutritious nitrogenous food, but some require particular cooking. See also the agaric-like mushrooms noticed under Cantharellus, Coprinus, Cortinarius, Russula.

Agaricus ostreatus, Jacquin.

On trunks chiefly of deciduous trees throughout Europe. The delicious oyster-mushroom, renowned from antiquity, hence prominently on this occasion mentioned. For fuller information on fungs for the table consult as very accessible works Badham's "Esculent Funguses of England," and Cooke's "British Fungi"; for systematic characteristics see the works of Fries and of Berkeley.

Agave Americana, Linné.

One of the gigantic Aloes of Central America, quite hardy at Port Phillip. In the open air it comes into flower in about ten years or later. The flowering stem may shoot up to the extent of ten feet in a week, and may finally rise to 40 feet; mellaginous sap flows also from the flower-stem. Mr. Fred. Hickox at Clunes saw the young offshoots producing also small flower-bunches, while the maternal plant was in bloom. The pithy stem can be utilized for some of the purposes, for which cork is usually employed—for instance, to form the bottoms of insect-cases. The honey-sucking birds and bees are very fond of the flowers of this prodigious plant. The leaves of this and some other Agaves, such as A. Mexicana, furnish the strong Pita-fibre, which is adapted for ropes, and even for beautiful textile frabrics. The strength of ropes of this fibre is considerably greater than that of hemp-ropes, as well in as out of water. The leaves contain saponin. The sap can be converted into alcohol, and thus the "Pulque" beverage is prepared from the young flower-stem. Where space and circumstances admit of it, impenetrable hedges may be raised in the course of some years from Agaves. One kind rose in Fiji also to 38 feet (Thurston).

Agave inaequidens, K. Koch.

A species closely allied to A. Americana; it seems to include A. Hookeri and A. Fenzliana, Jacobi, according to Baker (in Bot. Mag., 6589 and Gardener's Chron., 1871, p. 718).

Agave rigida. Miller. (*A Ixtli*, Karwinski.)

Yucatan. The Chelem, Henequen and Sacci of the Mexicans, furnishing the Sisal-hemp. Drs. Perrine, Scott and Engelmann indicate several varieties of this stately plant, the fibre being therefore also variable, both in quantity and quality. The leaves of the Sacci or Sacqui giving the largest return. The yield of fibre begins in four or five years, and lasts for half a century or more, the plant being prevented from flowering by cutting away its flower-stalk when very young. The leaves are from 2 to 6 feet long and 2 to 6 inches wide; the flower-stem attains a height of 25 feet; the panicle of flowers is about eight feet long, bearing in abundance bulb-like buds. Other large species of Agave, all fibre-yielding, are A. antillarum (Descourtil) from Hayti; A. Parryi (Engelmann) from New Mexico; A. Palmeri (Engelmann) from South-Arizona, up to a cool elevation of 6,000 feet. Concerning the uses of Agaves refer also to Dr. Fr. P. Porcher's "Southern Fields and Forests," p. 596–599 (1869).

Agonis flexuosa, De Candolle.

The Willow-Myrtle of South-Western Australia. A tree, attaining finally a height of 60 feet, with pendent branches. One of the best of evergreen trees for cemeteries in a climate free from frost. The foliage is rich in antiseptic oil.

Agriophyllum Gobicum, Bunge.

Eastern Asia. The "Soulchir" of the Mongols. Prevalsky says, that the seeds of this plant, wild as well as cultivated, afford a great part of the vegetable food of the Ala-Shan nomads. Several other annual salsolaceous herbs belong to the genus Agriophyllum, among them A. arenarium (Bieberstein) being closely cognate to A. Gobicum.

Agrostis alba, Linné.

The Fiorin or White Bent-Grass. Europe, Northern and Middle Asia, North-Africa, North-America. Perennial, showing a predilection for moisture ; can be grown on peat-soil. It is the herd-grass of the

United States and valuable as an admixture to many other grasses, as it becomes available at the season, when some of them fail. Sinclair regards it as a pasture-grass inferior to Festuca pratensis and Dactylis glomerata, but superior to Alopecurus pratensis The variety with long suckers (A. stolonifera) is best adapted for sandy pastures, and helps to bind shifting sand on the sea-coast, or broken soil on river-banks. It luxuriates even on saline wet soil or periodically inundated places, as well observed by Langethal. It is more a grass for cattle-runs than for sheep-pastures, but wherever it is to grow, the soil must be penetrable. Its turf on coast-meadows is particularly dense and of remarkable fineness. For sowing, only one-sixth of the weight of the seeds as compared with those of the rye-grass is needed. The creeping variety is also valuable for fine and enduring lawns.

Agrostis rubra, Linné.

Northern Enrope, Asia and America. A perennial grass, called red-top and also herd-grass in the United States of North-America. Professor Meehan places it for its value on pasture-land among grasses cultivated there next after Phleum pratense and Poa pratensis (the latter there called blue grass), and before Dactylis glomerata, the orchard-grass of the United States.

Agrostis scabra, Willdenow.*

The hair-grass of North-America. Recently recommended as one of the best lawn-grasses, forming a dense turf. It will grow even on poor gravelly soil, and endure drought as well as extreme cold. Its fine roots and suckers spread rapidly, forming soon dense matted sods (Dr. Channing). It starts into new growth immediately after being cut, is selected for its sweetness by pasture-animals, has proved one of the best grasses for dairyground, and suppresses weeds like Hordeum secalinum. One bushel of seed to an acre suffices for pastures; two bushels are used for lawns. A. perennans (Tukermann) is an allied species of similar value.

Agrostis Solandri, F. v. Mueller.

Extra-tropical Australia and New Zealand. Produces a large quantity of sweet fodder in damp localities (Bailey). Valuable as a meadow-grass (W. Hill). In Australia it is essentially a winter-grass, but available also in our sub-alpine regions (J. Stirling). Chemical analysis in spring gave the following results: Albumen, 4·08; Gluten, 8·81; Starch, 1·34; Gum, 2·50; Sugar, 9·75 per cent. (F. v. Mueller and L. Rummel.) Under the name A. Forsteri (Roemer and Schultes) only some forms of this very variable grass are comprehended.

Agrostis vulgaris, Withering.

Europe, Northern Africa, Middle Asia, North-America. One of the perennial grasses, which disseminate themselves with celerity, even over the worst of sandy soils. Though not a tall grass, it may be destined to contribute perhaps with others largely to the grazing capabilities of desert-lands; yet it will thrive also even in moist soil and alpine regions, and is essentially a grass for sheep-pastures; counted by Hein and many others among valuable lawn-grasses.

Ailantus glandulosa, Linné.

South-Eastern Asia. A hardy, deciduous tree, reaching 60 feet in height, of rather rapid growth and of very imposing aspect in any landscape. Particularly valuable on account of its leaves, which afford food to a silk-worm (*Attacus Cynthia*) peculiar to this tree. Wood extremelv durable, pale-yellow, of silky lustre when planed, and therefore valued for joiners' work; it is tougher than that of oak or elm, easily worked, and not liable to split or warp. In Southern Europe this tree is much planted for avenues. Valuable also for reclaiming coast-sands, and to this end easily propagated by suckers and fragments of roots, according to Professor Sargent. The growth of the tree is quick even in poor soil, but more so in somewhat calcareous bottoms. Thrives on chalk (Vasey). Professor Meehan states, that it checks the spread of the rose-bug, to which the tree is destructive. In Norway hardy to latitude 63° 26′ N. (Schuebeler), but suffers from frost in youth.

Aira cæspitosa, Linné.

Widely dispersed over the globe, reaching as indigenous also Australia. A rough fodder-grass, best utilized for laying dry any moist meadows; affords fair pasturage, if periodically burnt down (J. Stirling). Extends to 71° 7′ N. in Norway (Schuebeler).

Albizzia basaltica, Bentham.

Eastern sub-tropic Australia. A small tree. The wood praised by Mr. P. O'Shanesy for its beautiful reddish color and silky lustre. Cattle like the foliage. As a genus Pithecolobium differs no more from Albizzia than Vachelia from Acacia or Cathartocarpus from Cassia. The oldest generic name is Zygia, but no species was early described under that name.

Albizzia bigemina, F. v. Mueller. (*Pithecolobium bigeminum*, Martius.)

India, up to Sikkim and Nepal, ascending in Ceylon to 4,000 feet. Desirable for Australian forestry on account of its peculiar dark and hard wood. Another congener, A. subcoriacea (Pithecolobium subcoriaceum, Thwaites), from the mountains of India is deserving of utilitarian cultivation with numerous other tall species.

Albizzia dulcis, F. v. Mueller. (*Pithecolobium dulce*, Bentham.)

Mexico. A valuable hedge-plant. The sweet pulp of the pod is regarded as wholesome.

Albizzia Julibrissin, Durazzini.

From the Caucasus to Japan. A favorite ornamental red-flowering Shade-Acacia in Southern Europe. Naturally it seeks river-banks.

Albizzia latisiliqua, F. v. Mueller. (*Lysiloma latisiliqua,* Bentham.)

Tropical America. A large spreading tree; trunk attaining a diameter of 3 feet; wood excellent for select cabinet-work, excelling according to Nuttall the Mahogany in its variable shining tints, which appear like watered satin; it is hard and close-grained.

Albizzia Lebbek, Bentham.

The Siris-Acacia of Southern and Middle Asia and Northern Africa. Available as a shade-tree. It produces also a good deal of gum; the flowers much sought for honey by bees. An allied species is the North-Eastern Australian A. canescens (Bentham).

Albizzia lophantha, Bentham. (*Acacia lophantha,* Willdenow.)

South-Western Australia. One of the most rapidly-growing plants for copses and first temporary shelter in exposed localities, but never attaining the size of a real tree. It produces seeds abundantly, which germinate most easily. For the most desolate places, especially in desert-tracts, it is of great importance, quickly affording shade, shelter and a copious vegetation. Cattle browse on the leaves, The bark contains only about 8 per cent. mimosa-tannin; but Mr. Rummel found in the dry root about 10 per cent. of saponin, so valuable in silk- and wool-factories. Saponin also occurs in Xylia dolabriformis of Southern Asia. In Australia this plant is found better even than the Broom-bush for sheltering new forest-plantations in open sand-lands; in rich soil known to have grown 14 feet in a year.

Albizzia micrantha, Boivin. (*A. odoratissima,* Bentham.)

India; ascending to 5,000 feet ; growing in almost any kind of soil; hardy in sub-tropical countries. A middle-sized tree; timber particularly hard, dark-colored, durable and strong ; well adapted for naves and felloes (Drury and Brandis). Regarded by Roxburgh as one of the most valuable jungle-timbers.

Albizzia Saman, F. v. Mueller. (*Pithecolobium Saman,* Bentham.)

The " Rain-tree " or " Guango," extending from Mexico to Brazil and Peru. It attains a heighth of 70 feet, with a trunk 6 feet in diameter, the colossal branches expanding to 150 feet ; it is of quick growth, and in outline not unlike an oak ; it is content with light soil, and forms a magnificent feature in a landscape. In India it attained in ten years a stem-girth of about 6 feet at 5 feet from the ground, its ramifications by that time spreading out to 90 feet (Blechyndon). It thrives particularly in the dry salt-pond districts of the West Indies, and likes the vicinity of the sea, "its foliage possessing the power, to an unusual extent, to attract, absorb and condense aerial humidity " (Consul for France at Laroto). Not ascending to above 1,000 feet altitude in Jamaica ; resisting drought. The pods mature for fodder at a time, when grass and herbage on pastures become parched. Rain and dew fall through the foliage, which is shut up at night, thus allowing grass to grow underneath. The tree thrives best where the rainfall fluctuates between 30 and 60 inches a year.

One of the best trees in mild climates for shade by the roadsides. The wood is hard and ornamental; but the principal utility of the tree lies in its pulpy pods, which are produced in great abundance and constitute a very fattening fodder for all kinds of pastoral animals, which eat them with relish (Jenman, J. H. Stephens).

Albizzia stipulata, Bentham.

Continental and insular South-Asia, extending to the Himalayas and China, ascending to 4,000 feet. An umbrageous tree of easy culture.

Alchemilla vulgaris, Brunfels.

Europe, West-Asia, Arctic North-America, Alpine Australia; extending in Norway to 71° 10′ N. (Schuebeler). This perennial herb is important for moist dairy-pastures. The same can be said of other congeners; for instance, A. alpina (L.) from the coldest parts of Europe, North-Asia and North-America; A. Capensis (Thunberg) and A. elongata (Ecklon and Zeyher) of South-Africa, some Abyssinian species, as well as A. pinnata (Ruiz and Pavon) and other congeners of the Andes.

Aletris farinosa, Linné.

The "Colic-root" of the woodlands of North-America. This pretty herb is of extreme bitterness, and is employed medicinally as a tonic; inaptly called also "Star-Grass."

Aleurites cordata, R. Brown.

From Japan to Nepal, also in Bourbon. This tree deserves cultivation for its beauty and durable wood in warm humid districts. The oil of the seeds serves as a varnish. Perhaps in localities, quite free from frost, it would be of sufficiently quick growth.

Aleurites triloba, R. and G. Forster.

The "Candlenut-tree," a native of some of the tropical regions of both hemispheres; it furnishes a valuable dye from its fruits and copious oil from its seeds, the yield being about one-third. I found the tree barely able to endure the winters of Melbourne.

Alibertia edulis, A. Richard.

Guiana and Brazil, southward to extra-tropic latitudes, widely dispersed through the drier regions. The fruit of this shrub is edible and known as "Marmeladinha." A. Melloana (J. Hooker), of Southern Brazil, seems to serve the same purpose.

Alkanna tinctoria, Tausch.

On sandy and calcareous places around and near the Mediterranean Sea, extending to Hungary. Cultivated in the open air to perfection up to 59° 55′ N., by Professor Schuebeler. This perennial herb yields the "alkanna-root," used for dying oleaginous and other substances. It might be naturalized. Can be grown in almost pure coast-sand.

Allium Ampeloprasum, Linné.

The British Leek. Middle and Southern Europe, Western Asia and North-Africa. Called in culture the Summer-Leek, a variety of which is the Pearl-Leek. The specific name arose already from Dioscorides' writings.

Allium Ascalonicum, Linné.

The Shallot. South-Western Asia. Specific name already used by Theophrastos and Plinius.

Allium Canadense, Kalm.

Eastern North-America. This Garlic could be cultivated or naturalized on moist meadows for the sake of the tops of its bulbs, which are sought for pickles of superior flavor.

Allium Cepa, Linné.

Wild in Turkestan (Dr. A. Regel). The ordinary Onion. At Aschersleben and Quedlinburg alone about 130,000 lbs. of onion-seeds are raised annually (B. Stein). The specific name in use already by Plinius and Columua. Raw onions quench excessive thirst. One of the staple-products of Bermuda (Sir J. Lefroy).

Allium fistulosum, Linné.

Central Asia. The Welsh Onion.

Allium Porrum, Linné.

The Leek. Regarded by Gay and Regel as a cultural variety of A. Ampeloprasum.

Allium roseum, Linné.

Countries on the Mediterranean Sea. This, with Allium Neapolitanum (Cyrillo), one of its companions, yields edible bulbs, according to Heldreich.

Allium rubellum, Bieberstein. (*A. leptophyllum*, Wallich.)

The Himalayan Onion. Captain Pogson regards the bulbs as sudorific; they are of stronger pungency than ordinary onions; the leaves form a good condiment.

Allium sativum, C. Bauhin.

The Garlic. Soongarei and in the farthest N.W. of India, as shown by Dr. von Regel. Nearest allied to A. Scorodoprasum. The "Allium" of Plinius.

Allium Schœnoprasum, Linné,

The Chives. Europe, Northern Asia and North-America. Cultivated in Norway to latitude 70° 22′ (Schuebeler). Available for salads and condiments. This species of Allium seems here not yet so generally adopted in our culinary cultivation as other congeners.

C

Allium Scorodoprasum, Linné.

The Sand-Leek. Europe and North-Africa. Resembles in some respects Garlic, in others Shallot. The Scorodoprason of Dioscorides according to Fraas seems A. descendens (Linné), indigenous to South-Europe.

Alnus glutinosa, Gaertner.

The common Alder. Throughout Europe and extra-tropical Asia; indigenous to 64° 10' N. lat. in Norway (Schuebeler). Reaches a height of 90 feet; attaining even in lat. 61° 47' a stem-diameter of 10 feet. Easily clipped, when young, into hedges; well adapted for river-banks; recommended by Wessely for wet valleys in coast-sand; wood soft and light, turning red, furnishing one of the best charcoals for gunpowder; it is also durable under water, and adapted for turners' and joiners' work; the wood is also well suited for pump-trees and other underground-work, as it will harden almost like stone. The tree is further valuable for the utilization of bog-land. A. incana (Willd.) extends to North-America; it is of smaller size; was found to grow over 60 feet high in lat. 70° of Norway by Professor Schuebeler. The bark of several alders is of medicinal value, and a decoction will give to cloth saturated with lye an indelible orange-color (Porcher); it contains a peculiar tannic principle to the extent of 36 per cent. (Muspratt). American alder-extract has come into use for tanning; it renders skins particularly firm, mellow and well-colored (Eaton). A. Oregana (Nuttall), of California and Oregon, rises to a height of 80 feet; its wood is extensively used for bent-work (Meehan). A. Japonica and A. firma (Siebold and Zuccarini), of Japan, furnish wood there for carvers and turners, and bark for black dye (Dupont). A. rubra (Bongard), the Red or Tag Alder of California and British Columbia, rises to fully 100 feet, and gets finally a stem-diameter of 6 feet. The timber is excellent for piles and bridge-bottoms, also for pumps, and proved exceedingly durable; the wood serves further for carving, turnery, furniture, machinery (Dr. Kellogg).

Alnus Nepalensis, D. Don.

Himalayas, between 3,000 and 9,000 feet. Reaches a height of 60 feet. With another Himalayan alder, A. nitida (Endlicher), it can be grown along streams for the sake of its wood.

Aloe dichotoma, Linné fil.

Damara and Namaqua-land. This species attains a height of 30 feet, and occasionally an expanse of 40 feet. The stem is remarkably smooth, with a girth sometimes of 12 feet. It is a yellow-flowering species. A. Zeyheri is still more gigantic than the foregoing, it attaining exceptionally a height of 60 feet (Dyer); it occurs in Caffraria and Natal, with a stem 16 feet in circumference at 3 feet from the ground. A. Bainesii and A. Barberæ are identical, according to Mr. J. G. Baker. A. speciosa (Baker) rises also to a height of 25 feet. All grand scenic plants.

Aloe ferox, Miller.

South-Africa. This species yields the best Cape-aloes, as observed by Dr. Pappe. The simply inspissated juice of the leaves of the various species of the genus constitutes the aloe-drug. It is best obtained by using neither heat nor pressure for extracting the sap. By re-dissolving the aqueous part of Aloes in cold water, and reducing the liquid through boiling or other processes of exsiccation to dryness, the extract of aloes is prepared. The bitter sap, used for dressing wounds, keeps off flies very effectually. It deserves introduction particularly in veterinary practice. All species are highly valuable, and can be used, irrespective of their medicinal importance, to easily and inexpensively beautify any rocky or otherwise inarable spot. '

Aloe linguiformis, Miller.

South-Africa. According to Thunberg, the purest gum-resin is obtained from this species.

Aloe Perryi, Baker.

Socotra. It is now known, that it was this species, which furnished the genuine "Aloes," renowned in antiquity (Baker, Balfour). It grows best in limestone-soil, and ascends to 3,000 feet. Flowers turning from scarlet to yellow. Closely allied to A. vulgaris.

Aloe plicatilis, Miller.

South-Africa. The drug of this species acts more mildly than that of A. ferox.

Aloe purpurascens, Haworth.

South-Africa. Another of the plants, which furnish the Cape-aloes of commerce. The South-African Aloe arborescens (Miller) and A. Commelyni (Willdenow) are also utilized for aloes, according to Baillon, Saunders and Hanbury.

Aloe spicata, Thunberg.

South-Africa. This also furnishes Cape-aloes, and is an exceedingly handsome plant.

Aloe vera, Miller. *(A. succotrina,* Lamarck.)

South-Africa. A purplish flowered species, figured already by Commelyn in 1697 (Baker). Yields the common Socotrine-aloes and Moka-aloes.

Aloe vulgaris, Bauhin. (*A. vera,* Linné; *A. Barbadensis,* Miller.)

The Yellow-flowered Aloe. Countries around the Mediterranean Sea, also Canary-Islands, on the sandy or rocky sea-coast. Such places could also be readily utilized elsewhere for this and allied plants. Dr. Sibthorp has identified this species with the Aλόη of Dioscorides; thus it is also the real Aloe of Plinius; hence it is not probable, that A. vulgaris is also simultaneously of American origin, although it is long cultivated in the Antilles, and furnishes from thence the main supply of the Barbadoes-aloes, also Curaçoa-aloes; likewise in East-India this species seemingly only exists in a cultivated state. Haworth found the leaves of this and of A. striata softer and more succulent, than these of any other aloe. It is said to be the only species with yellow flowers among those early known, and it is also the only one, which Professors Wilkomm and Parlatore record as truly wild in Spain and Italy.

Aloexylon Agallochum, Loureiro.

Cochinchina, on the highest mountains. The precious aloe-wood, so famed from antiquity for its balsamic fragrance and medicinal properties, is derived from this tree.

Alopecurus bulbosus, Linné.

Middle- and South-Europe. An important rural grass for salt-marshes.

Alopecurus geniculatus, Linné.

Europe, Asia, North-Africa. A perennial fodder-grass, valuable for swampy ground; easily naturalized.

Alopecurus pratensis, Linné.

Meadow " Fox-tail" grass. Europe, Northern Africa, Northern and Middle Asia. In Norway indigenous to lat. 69° 11' (Schuebeler). One of the best of perennial pasture-grasses. It reaches its full perfection only after a few years of growth, as noticed by Sinclair. For this reason it is not equal to Dactylis glomerata for crop-rotation, but it is more nutritious than the latter, although the annual return in Britain has proved less. Langethal places it next to Timothy-grass for artificial pastures. Sheep thrive well on it. Sinclair and others have found, that this grass, when exclusively combined with white clover, will support after the second season five ewes and five lambs on an acre of sandy loam; but to thrive well it needs land not altogether dry. In all permanent artificial pastures this Alopecurus should form one of the principal ingredients, because it is so lasting and so nutritive. It is also one of the best grasses for maritime or alluvial tracts of country. In alpine regions it would also prove prolific, and might gradually convert many places there into summer-pastures. It does not altogether dislike shade, is early flowering and likes the presence of lime in the soil.

Alstonia constricta, F. v. Mueller.

Warmer parts of East-Australia, particularly in the dry inland-districts. The bark of this small tree is aromatic-bitter, and regarded as valuable in ague, also as a general tonic. It is allied to the Dita-bark of India and North-Eastern Australia, procured from Alstonia scholaris (R. Brown), and from this bark a peculiar alkaloid, the Porphyrin of Hesse, is prepared. The sap of all Alstonias should be tried for caoutchouc, that of A. plumosa and another species yielding Fiji-rubber (Hooker).

Alstrœmeria pallida, Graham.

Chili. Palatable starch can be obtained from the root of this plant, which for its loveliness alone deserves a place in any garden. The tubers of others of the numerous Alstrœmerias can doubtless be practically utilized in a similar manner.

Althæa officinalis, Linné.

The real "Marsh-Mallow." Europe, Northern Africa, Northern and Middle Asia. Hardy to lat. 59° 55' in Norway (Schuebeler). A tall perennial herb, with handsome flowers. The mucilaginous root and also the foliage are used for medicinal purposes. The plant succeeds best on damp, somewhat saline soil.

Amarantus Blitum, Linné.

Southern Europe, Northern Africa, South-Western Asia. This annual herb is a favorite plant amongst allied ones for spinage, but not the only species of this genus, as also many other Amarantaceæ serve for culinary purposes. The dried plant contains 10 to 12 per cent. nitrate of potash. It arrives at maturity in two or three months, producing on favorable soil about 4 tons per acre, calculated to contain about 400 lbs. saltpetre. A. cruentus L., A. hypochondriacus L. and A. caudatus L. are cultivated in Ceylon, though not all of the agreeable taste of real spinage. A. Mangostanus, A. Gangeticus, A. melancholicus, A. tristis, L. and A. polystachyus, Willdenow, likewise furnish in Southern Asia either foliage for spinage or seeds for porridge. Amarants for spinage must be well boiled and the water repeatedly changed.

Amarantus paniculatus, Linné. (*A. frumentaceus,* Royle.)

In tropical countries of Asia and also America. An annual herb, attaining a height of 6 feet, yielding half a pound of floury nutritious seeds on a square yard of ground in three months, according to Roxburgh. Extensively cultivated in India for food-grain; the leaves serve as vegetable.

Amelanchier Botryapium, De Candolle.

The "Grape-pear" of North-America; also called "Shadbush." Cultivated in Norway as far north as 59° 55' (Schuebeler). This handsome fruit-tree attains a height of 30 feet. Its purplish or almost black fruits are small, but of pleasant subacid taste, and ripen early in the season. It bears abundantly; and Mr. Adams, of Ohio, has calculated the yield at 300 bushels per acre annually, if the variety oblongifolia is chosen; it is the Dwarf "June-berry" of North-America. This bush or tree will live on sandy soil; but it is one of those hardy kinds, particularly eligible for alpine ground; it is remarkably variable in its forms.

Anacylus Pyrethrum, De Candolle.

Countries near the Mediterranean Sea. The root of this perennial herb is used medicinally.

Ananas sativa, Schultes.

South-America. The famous "Pine-apple" plant. Mr. Th. Mead remarks, that in Florida the fruit is produced by merely piling a little brush over the plant during the three cool months of the year. Fruits even to 10 lbs. weight have there been produced. The leaves yield an excellent fibre.

Andropogon annulatus, Forskael.

Intra- and Sub-tropical Africa, Asia and Australia. Recommended by Mr. Walter Hill as a meadow-grass. Dr. Curl observes, that in New Zealand it is both a summer- and autumn-grass, that it does not grow fast in winter, but at the period of its greatest growth sends up an abundance of foliage.

Andropogon argenteus, De Candolle.

Pronounced by Leybold to be one of the best pasture-grasses of the Cordilleras of Chili.

Andropogon australis, Sprengel. (*Sorghum plumosum,* Beauvois.)

Tropical and also Eastern Extra-tropic Australia as far south as Gippsland. Brought under notice by Mr. Ch. Moore as an admirable perennial pasture-grass. The allied A. serratus (Thunberg) of tropical Australia, Southern Asia, China and Japan, serves similar purposes. General Sir H. Macpherson proved some Indian Andropogons well adapted for Silos. ·

Andropogon avenaceus, Michaux. (*Sorghum avenaceum,* Willdenow.)

North- and Central-America. This tall perennial grass lives in dry sandy soil, and should be tried for growth of fodder.

Andropogon bombycinus, R. Brown.

Australia. This robust grass, which is generally well-spoken of by graziers, seems to like a somewhat strong soil, and is often found among the rocks on hill-sides. The bases of the stems of this species, like several others of the genus, are highly aromatic (Bailey). It will live in shifting sand and endure the hottest desert-clime. The Australian A. procerus (R. Br.) and the Mediterranean A. laniger (Desf.) are closely allied congeners.

Andropogon Calamus, Royle.

Central India. The " Sweet Cane " and " Calamus " of Scripture, according to Calcott; it is regarded as the aromatic reed of Dioscorides by Royle. From this species the gingergrass-oil of Nemaur, an article much used in perfumery, is distilled. It is the "Cusha" of India. Gibson and Dalzell regard it identical with A. nardoides (Nees) of South-Africa.

Andropogon cernuus, Roxburgh.* (*Sorghum cernuum,* Willd.)

One of the Guinea-corns. India, where it is much cultivated, as in other tropical countries. It is annual (according to Hackel); but

Roxburgh distinctly asserts that " the plant is of two or more years' duration, if suffered to remain." It forms the " staff of life" of the mountaineers beyond Bengal. It reaches a height of 15 feet, with leaves over 3 feet long. The thick stems root at the lower joints, and cattle are very fond of them. The grain is white. The specific limits of the various sorghums are not well ascertained. This belongs to the series of A. Halepensis.

Andropogon erianthoides, F. v. Mueller.

Eastern Sub-tropical Australia. Mr. Bailey observes of this perennial species, that " it would be difficult to find a grass superior for fodder to this; it produces a heavy crop of rich, sweet, succulent foliage; it spreads freely from roots and seeds, and shoots again when fed down."

Andropogon falcatus, Steudel.

India and Queensland. Considered by Mr. Bailey a good lawn-grass, as it is of dwarf compact growth, and of bright verdure.

Andropogon involutus, Steudel.

. From Nepal to China. The " Bhaib-Grass," used for the manu-facture of ropes, string-matting and other textile articles; approaches Esparto in the quality of its fibres; grows readily on dry soil. (Dr. Hance).

Andropogon Gryllus, Linné.

In the warm temperate and the hot zone of the eastern hemisphere. A perennial pasture-grass, of easy dissemination, particularly useful in arid climes.

Andropogon Halepensis, Sibthorp. (*Sorghum Halepense*, Persoon.) .

Southern Europe, warmer parts of Asia, Northern Africa. Praised already by Theophrastus more than 2,000 years ago. Not easily repressed in moist ground. A rich perennial grass, cultivated often under the name of Cuba-grass. All the vernacular names given to this grass should be discontinued in maintaining the very appropriate original appellation " Haleppo-grass." It keeps green in the heat of summer (J. L. Dow.), is not eaten out by pastoral animals (Hollings-worth); the roots resist some frost; three tons can be cut from one acre in a single season; it yields so large a hay-crop, that it may be cut half a dozen times in a season, should the land be rich. All kinds of stock have a predilection for this grass. It will mat the soil with its deep and spreading roots; hence it should be kept from cultivated fields. Detrimental to Lucerne on meadows (Rev. Dr. Woolls). In Victoria hardy up to 2,000 feet elevation.

Andropogon Ischaemum, Linné.

Southern Europe, Southern Asia, Africa. One of the fittest of grasses for hot dry sand-regions, and of most ready spontaneous dispersion. Perennial. Succeeds well on lime-soil and that contain-ing gypsum. In its new annual upgrowth it is particularly liked by sheep. It needs burning off in autumn.

Andropogon montanus, Roxburgh.

Southern Asia, Northern and Eastern Australia. Rapid in growth and valuable for fodder when young; resists fire better than many other grasses (Holmes). Perennial, like most other species of this large genus.

Andropogon muricatus, Retzius.

India. The " Vitivert " or " Kus-kus." A grass with delightfully fragrant roots. According to Surgeon-Major Dr. G. King, the fragrant Indian mats are made of this grass, and according to Prof. Lindley, awnings, tatties, covers for palanquins and screens are manufactured from this species; also an essence.

Andropogon Nardus, Linné. (*A. Ivarancusa,* Blane.)

Southern Asia. Perennial. One of the lemon-scented species. Prof. Hackel suggests, that A. Calamus may be referable to this species. Kunth unites with this A. citriodorus (De Candolle) the A. citratus of many botanic gardens, while Link referred this grass to A. Schœnanthus. It yields an essential oil for condiment and perfumery, and is occasionally used for tea. Simmonds gives the export value of this oil as from Ceylon alone at £7,000. " Citrionella" Oil to the extent of 40,000 lbs. annually is in Ceylon distilled from this grass (Piesse).

Andropogon nutans, Linné. (*Sorghum nutans,* Gray.)

North-America. A tall, nutritious, perennial grass, content with dry and barren soil.

Andropogon pertusus, Willdenow.

Southern Asia, Tropical and Sub-tropical Australia. Perennial. Mr. Nixon, of Benalla, regards it as one of the best grasses to withstand long droughts, while it will bear any amount of feeding. It endures cold better than some other Andropogons of Queensland, according to Mr. Bailey's observations.

Andropogon provincialis, Lamarck.

Southern Europe. Strongly recommended by Bouché for fixing loose maritime sand. Attains a height of 5 feet. A. furcatus (Muehlenberg) is the same species, according to Hackel.

Andropogon refractus, R. Brown.

Northern and Eastern Australia, Polynesia. Mr. Bailey observes of this perennial grass, that it is equally excellent for pastures and hay, and that it produces a heavy crop during summer; the root is fragrant. According to Mr. Holmes, it is easily inflammable, of inferior fodder-value, but is particularly useful for mattresses.

Andropogon saccharatus, Roxburgh.* (*Sorghum saccharatum,* Persoon.)

Tropical Asia, or perhaps only indigenous in equatorial Africa. The Broom-corn or Sugar-Millet; also passing as " Durra," " Dochna " and " Battari." A tall annual species. Produces of all grasses, except the Teosinté and maize, the heaviest of all fodder-crops in warm climates. From the saccharine juice sugar is obtainable. A sample of such, prepared from plants of the Melbourne Botanic Garden, was shown already at the Exhibition of 1862. This Sorghum furnishes also material for a well-known kind of brooms. Mr. Simmonds relates, that as many as 150,000 doz. of these brooms have been made in one single factory during a year. To pigs this plant is very fattening also. The plant can be advantageously utilized for preparing syrup. For this purpose the sap is expressed at the time of flowering, and simply evaporated; the yield is from 100–200 gallons from the acre. Already in 1860 nearly seven millions of gallons of sorghum-treacle were produced in the United States. General Le Duc, then commissioner for agriculture at Washington, states that Mr. Seth Kenny, of Minnesota, obtained from the "Early Amber" variety up to 250 gallons of heavy syrup from one acre of this sorghum. Machinery for the manufacture of sorghum-sugar on plantations can be erected at a cost of £50 to £100. Sorghum juice can be reduced to treacle and sugar without the use of chemicals, beyond clearing with lime and neutralizing the lime remaining in the juice by sulphurous acid. Raw sorghum-sugar is nearly white. By an improved method Mr. F. L. Stewart obtained 10 lbs. of sugar from a gallon of dense syrup. At the State-University's experimental farm, in Wisconsin, Professors Swenson and Henry have proved, that sorghum-sugar, equal to the best cane-sugar, can be produced at $4\frac{1}{2}$ cents per pound. The seeds are very valuable for stable-fodder as well as for poultry-feed, and may even be utilized for bread and cakes. The stem can be used as a culinary vegetable. See also the elaborate chemical reports by Dr. P. Collier, Washington, 1880–1882; further, the essay by Commissioner Hon. G. B. Loring, 1883.

An able report from a committee of the National Academy of Science, submitted to the Senate of the United States on the sugar-producing capacity of the Sorghum, gives the following conclusions: That from about 4,500 most accurate analyses, instituted by Dr. Collier in the Department of Agriculture in Washington, the presence of cane-sugar in the juice of the best varieties of Sorghum (Amber, Honduras and Orange) in such quantity is established, as to equal the yield of that of sugar-cane, the average quantity of crystallisable sugar in the juice being 16 per cent., the average percentage of juice in stripped stalks from plants grown at Washington being 58 per cent., the sugar really obtainable by ordinary processes of manufacture from the juice being 11·30 per cent. It is further demonstrated, that the Sorghum-stalks should be cut only, when the seeds are already of a doughy consistence or still harder, and the stalks should be worked up immediately after cutting; further it is shown, that the sugar from Sorghum is not inferior to that from cane or beet, and that excellent sugar can likewise be obtained from maize; that Sorghum when advanced to maturity will resist some frost; that no more than $12\frac{1}{2}$ per cent. of the sugar is lost by the evaporation of the juice to syrup, quite as great a loss taking place in the juice of sugar-cane by defecation, skimming and inversion (change into glucose or grape-sugar). The committee considered this new Sorghum-industry placed already on a safe and profitable footing. Sorghum-stubbles are of surprising value as pastoral feed. Sugar-Sorghum was introduced into the United States for rural purposes only 30 years ago ; but its culture spread with unexampled rapidity there, having only its counterpart in the tea- and cinchona-culture of the last decennia in India. The process of manufacture of sugar from Sorghum is not more difficult than that of cane, and less complicated than that of beet. Some varieties mature in 80 days, others require twice that time, the Orange-variety becoming perennial, and is with the Amber-variety most saccharine, while the Honduras-variety gives a very heavy crop; the broomcorn-variety is poor in sugar. Sorghum will thrive well in sandy loam too light for maize and in a clime too dry for that corn, and can be grown closer. The variety A. bicolor (Roxburgh) ripens its seeds in Lower India within three or four months of the time of sowing, the produce being often upwards of one-hundredfold, and the grain particularly wholesome for human sustenance. Sorghum giganteum (Edgeworth) represents a form of very tall growth.

Andropogon Schœnanthus, Linné. (*A. Martini*, Roxburgh.)

Southern Asia and Tropical Australia, extending to Japan. One of the " Lemon-grasses." It will live in arid places. The medicinal Siri-Oil is prepared from the root. The Australian A. bombycinus (R. Brown) approaches in affinity this species.

Andropogon scoparius, Michaux.

North-America. Takes permanent possession of sandy or otherwise poor land, and is regarded as one of the best forage-resources of the prairies.

Andropogon sericeus, R. Brown.

Hotter regions of Australia, even in desert-tracts, also extending to New Caledonia and the Philippine-Islands. A fattening perennial pasture-grass, worthy of praise.

Andropogon Sorghum, Brotero.* (*Sorghum vulgare*, Persoon.)

The large "Indian millet" or "Guinea-corn" or the "Durra." Warmer parts of Asia, but according to Alph. De Candolle perhaps indigenous only in Tropical Africa. It matures seed even at Christiania in Norway (Schuebeler). A tall annual plant. The grains can be converted into bread, porridge and other preparations of food. It is a very prolific corn; Sir John Hearsay counted 12,700 seeds on one plant; it is particularly valuable for green fodder. The panicles are used for carpet-brooms, the fibrous roots for velvet-brushes. The grain of this millet is in value superior to that of A. saccharatus. A variety (A. caffrorum, Kunth) yields the "Kaffir-corn." The " Imphee " affords a superior white food-grain. Even A.

saccharatus may be only an extreme form of A. Sorghum. In Central Australia it ripens within three months (Rev. H. Kempe). A kind of beer called "Merisa" is prepared from the seed. Many others of the numerous species of Andropogon, from both hemispheres, deserve our attention.

Anemone Pulsatilla, Linné.

Europe and Northern Asia. On limestone-soil. This pretty perennial herb is of some medicinal importance.

Angophora intermedia, De Candolle.

South-Eastern Australia. This is one of the best of the Angophoras, attaining a large size, and growing with the rapidity of an Eucalyptus, but being more close and shady in its foliage. It would be a good tree for lining public roads and for sheltering plantations. Exudes much kino. The Rev. J. Tennison-Woods states, that it is not rarely over 150 feet high, that the wood is hard and very tough, bearing dampness well, but that the many kino-veins lessen its usefulness. Carefully selected, it can be employed in carpenters' and wheelwrights' work. Mr. Kirton observes, that a single tree of this species or of A. lanceolata, will yield as much as two gallons of liquid kino. Timber useful, when extra-toughness is to be combined with lightness (Reader). A lanceolata (Cavanilles) of the same geographic region is a closely allied species. Flowers of all Angophoras much frequented by the honey-bee (Ch. French).

Angophora subvelutina, F. v. Mueller.

Queensland and New South Wales. Attains a height of 100 feet. The wood is light and tough, soft while green, very hard when dry, used for wheel-naves, yokes, handles and various implements; it burns well and contains a large proportion of potash (C. Hartmann).

Anona Cherimolia, Miller.

Ecuador to Peru. One of the "Custard-Apples." This shrub or tree might be tried in frostless forest-valleys, where humidity and rich soil will prove favorable to its growth. It is hardy in the mildest coast-regions of Spain, also in Chili. It yields the Cherimoyer-fruit. The flowers are very fragrant. A. muricata L. (the Sour Sop), A. squamosa L. and A. sericea Dunal (the Sweet Sop) and A. reticulata (the Custard-Apple), all natives of the Antilles, can probably only under exceptionally favorable conditions be grown in any extra-tropic countries, though they produce fruits still in Florida.

Anthemis nobilis, Linné.

The true "Chamomile." Middle and Southern Europe, Northern Africa. A well-known medicinal plant, frequently used as edgings for garden-plots. Flowers in their normal state are preferable for medicinal use to those, in which the ray florets are produced in increased numbers. They contain a peculiar volatile oil and two acids, similar to angelic and valerianic acid. Hardy in Norway to lat. 63° 52' (Schuebeler).

Anthemis tinctoria, Linné.

Middle and Southern Europe, Orient. An annual herb. The flowers contain a yellow dye.

Anthistiria avenacea, F. v. Mueller.

Extra-tropical and Central Australia. A nutritious, perennial pasture-grass. Called by Mr. Bailey " one of the most productive grasses of Australia"; it produces a large amount of bottom-fodder, and it has also the advantage of being a prolific seeder, while it endures a hot, dry clime.

Anthistiria ciliata, Linné fil. *(Anthistiria Australis,* R. Brown.)

The well-known "Kangaroo-grass," not confined to Australia, but stretching through Southern Asia also and through the whole of Africa; perennial, nutritious, comparatively hardy, ascending to sub-alpine elevations. Chemical analysis of this grass during its spring-growth gave the following result:—Albumen, 2·05; gluten, 4·67; starch, 0·69; gum, 1·67; sugar, 3·06 per cent. (F. v. Mueller and L. Rummel.) Several species of Anthistiria occur variously dispersed from South-Africa to Japan, deserving introduction and naturalization into countries of warm-temperate or tropical climates.

Anthistiria membranacea, Lindley.

Interior of Australia. Esteemed as fattening ; seeds freely (Bailey). Particularly fitted for dry, hot pastures, even of desert-regions.

Anthoxanthum odoratum, Linné.

The "Scented Vernal-grass." Europe, Northern and Middle Asia, North-Africa. Found wild in Norway to lat. 71° 7′ (Schuebeler). Perennial, and not of great value as a fattening grass, yet always desired for the flavor, which it imparts to hay. Perhaps for this purpose the scented Andropogons might also serve. On deep and moist soils it attains its greatest perfection. It is much used for mixing among permanent grasses in pastures, where it will continue long in season. Discarded as a pasture- or hay-grass by Professor Hackel, liked by many as an admixture to lawn-grasses. It would live well in any alpine region. Dr. Curl observes, that in New Zealand it grows all the winter, spring and autumn, and is a good feeding-grass. The lamellar crystalline cumarin is the principle, on which the odor of Anthoxanthum depends.

Anthriscus Cerefolium, Hoffmann.

Europe and Western Asia. The Chervil. An annual culinary plant; its herbage used as an aromatic condiment, but the root is seemingly deleterious. The plant requires in hot countries a shady situation (Vilmorin); the foliage forms the principal ingredient of of what in France is called "fines herbes."

Anthyllis vulneraria, Linné.

The Kidney-vetch. All Europe, Northern Africa, Western Asia. This perennial herb serves as sheep-fodder, and is particularly recommended for calcareous soils. It would also live in any alpine region. Indigenous in Norway as far north as lat. 70° (Schuebeler).

Apios tuberosa, Moench.

North-America. A climber with somewhat milky juice. The mealy tubers are edible.

Apium Chilense, Hooker and Arnott.

Western Extra-tropic temperate America. A stouter plant than the ordinary celery, but of similar culinary use.

Apium graveolens, Linne.

The " Celery." Europe, Northern Africa, Northern and Middle Asia. Grows in Norway to lat. 70° (Schuebeler). It is here merely inserted with a view of pointing out, that it might be readily naturalized anywhere on sea-shores. The wild plant is however by some considered unwholesome in a raw state. The fruitlets are occasionally utilized for condiment. Seeds will keep for several years.

Apium prostratum, La Billardière.

The Australian Celery. Extra-tropical Australia, New Zealand, Extra-tropical South-America. This also can be utilized as a culinary vegetable.

Apocynum cannabinum, Linné.

North-America. Locally known as "Indian Hemp." A perennial herb. This is· recorded among plants yielding a textile fibre. A. androsæmifolium (Linné) yields medicinally the "Bitter Root" of the United States, that of A. cannabinum being likewise medicinally valuable.

Aponogeton crispus, Thunberg.

From India to New South Wales. The tuberous roots of this water-herb are amylaceous and of excellent taste, though not large. The same remarks apply to A. monostachyos (Linné fil.) and several other species, all from the warmer regions of the eastern hemisphere.

Aponogeton distachyos, Thunberg.

South-Africa. This curious water-plant might be naturalized in ditches, swamps and lakes, for the sake of its edible tubers. The scented flowering portion of this plant affords spinage.

Aquilaria Agallocha, Roxburgh.

Silhet and Assam. A tree of immense size. It furnishes the fragrant calambac or agallochum-wood, known also as aggur or tuggur or the aloe-wood of commerce, famed since ancient times. The odorous portion is only partially distributed through the stem. This wood is also of medicinal value.

Arachis hypogæa, Linné.*

Brazil. The "Earth-nut, Pea-nut or Ground-nut." The seeds of this annual herb are consumed in a roasted state, or used for the expression of a palatable oil, which is not readily becoming rancid. The plant is a very productive one, and yields a very quick return. It ranks also as a valuable fodder-herb; the hay is very nutritious, much increasing the milk of cows. A light somewhat calcareous soil is best fitted for its growth. On such soil 50 bushels may be obtained from the acre, but Mr. Bernays notes the return up to 120 bushels, the weight of which ranging from 25 to 32 lbs. In 1880 the pea-nut-crop in the Southern United States came to 2,820,000 bushels, representing a value of £517,000.

Aralia cordata, Thunberg.

China. The young shoots provide an excellent culinary vegetable.

Aralia Ginseng, Decaisne and Planchon. (*Panax Ginseng*, Meyer.)

China and Upper India, ascending to 12,000 feet. This herb furnishes the celebrated Ginseng-root, so much esteemed as a stimulant by the Chinese, the value of which however may be overrated. The root, to be particularly powerful, needs probably to be obtained from high mountain-elevations. The species is closely related to the North-American A. quinquefolia (Decaisne and Planchon).

Araucaria Bidwilli, Hooker.*

Southern Queensland. Bunya-Bunya. A tree, attaining 250 feet in height, with a fine-grained, hard and durable wood, particularly valuable for furniture; it shows its beautiful veins best when polished. The seeds are large and edible. Growth in height at Port Phillip, 30–40 feet in 20 years, the big strobiles ripening there.

Araucaria Brasiliensis, A. Richard.*

Southern Brazil. A tree to 180 feet high, producing edible seeds. Dr. Saldanhada Gama reports, that the wood makes spendid boards, masts and spars, and that the tree also yields a good deal of turpentine. Except a few palms (Mauritia, Attalea, Copernicia), this seems the only tree, which in Tropical South-America forms forests by itself. (Martius.)

Araucaria Cookii, R. Brown.

New Caledonia, where it forms large forests. Height of tree to 200 feet. Habit and technical value similar to that of A. excelsa. Growth at Port Phillip not quite 30 feet in 20 years.

Araucaria Cunninghami, Aiton.*

"Moreton-Bay Pine." Eastern Australia, between 14° and 32° south latitude, extending also to New Guinea, according to Dr. Beccari. The tree attains a height of 200 feet, with a trunk 6 feet in diameter. Growth in height at Port Phillip 30 to 40 feet in 20 years. The timber is fine-grained, strong and durable, if not exposed to alternately dry and wet influences ; it is susceptible of a high polish, and thus competes with satin-wood, and in some respects with birdseye-maple. Value in Brisbane, £2 15s. to £3 10s. per 1,000 superficial feet. The tree grows on alluvial banks as well as on rugged mountains, overtopping all other trees. The resin, which exudes from it, has almost the transparency and whiteness of crystal, and is often pendent in the shape of icicles, which are sometimes 3 feet long and 6 to 12 inches broad (W. Hill). Araucarias should be planted by the million in fever-regions of tropical countries for hygienic purposes, on account of their antiseptic exhalations.

Araucaria excelsa, R. Brown.*

"Norfolk-Island Pine." A magnificent tree of remarkable symmetry, sometimes to 220 feet high, with a stem attaining 10 feet in diameter, and with regular tiers of absolutely horizontal branches. The timber is useful for ship-building and many other purposes. Growth in height at Port Phillip about 40 feet in 20 years.

Araucaria imbricata, Pavon.*

Chili and Patagonia, The male tree attains generally a lesser height than the female, which reaches 150 feet. At Brest it attained 72 feet in 55 years (de Kersanton). This species furnishes a hard and durable timber, as well as an abundance of edible seeds, which constitute a main-article of food of the natives. Eighteen good trees will yield enough of vegetable food for a man's sustenance all the year round. The wood is yellowish-white, full of beautiful veins, capable of being polished and worked with facility. It is admirably adapted for ship-building. The resin is pale and smells like frankincense (Lawson). The tree is most frequently found on rocky eminences almost destitute of water (J. Hoopes). It is hardier than any other congener, having withstood the frosts of Norway up to latitude 61° 15' (Schuebeler). Endures also the clime in many parts of Scotland. Grows more slowly in Australia than the native species.

Araucaria Rulei, F. v. Mueller.

New Caledonia. A magnificent tree, with large shining leaves; doubtless not merely of decorative but also of utilitarian value. A closely allied species, A. Muelleri (Brogniart), comes with A. Balansæ and A. montana from the same island.

Arbutus Menziesii, Pursh.

North-Western America. An evergreen tree, attaining a height of 150 feet, with a stem reaching 8 feet in diameter. It is of comparatively quick growth (Dr. Gibbons), and it belongs to the coast-tract exclusively. Wood exceedingly hard, not apt to rend, splendid for furniture. The tree requires a deep loamy soil (Bolander), and is fit only for shady, irrigated woodlands; likes the company of Pinus Douglasii and of Sequoias. It would be valuable at least as a highly ornamental garden- or park-plant, being the tallest among about a thousand Ericeæ of the world.

Archangelica officinalis, Hoffmann.

Arctic zone and mountain-regions of many parts of Europe. The young shoots and leaf-stalks of this biennial herb are used for Angelica-confectionery; the roots are of medicinal use, and likewise they are chosen as well as the fruitlets for the distillation of some cordials. Hardy in Norway to lat. 71° 10' (Shcuebeler). In any sub-alpine regions this plant would particularly establish its value. The surprisingly gigantic Angelica ursina (Regel; Angelophyllum ursinum, Ruprecht) forms a conspicuous feature in the landscape of Kamtschatka and Sachalin. It is delineated on plate XVI. of Lindley and Moore's Treasury of Botany; it ought to become an important plant for annual scenic culture.

Arctostaphylos uva ursi, Sprengel.

Europe, Northern Asia and North-America, in colder regions, extending to the arctics. A medicinal small shrub, which could best be reared in the heath-moors of alpine tracts. Valuable also as a honey-yielding plant (Cook).

Arenga saccharifera, La Billardière.

India, Cochinchina, Philippines and also most Southern Japan (Doederlein). This Palm attains a height of 40 feet. The black fibres of the leaf-stalks are adapted for cables and ropes, intended to resist wet very long. The juice convertible into toddy or sugar; the young kernels made with syrup into preserves. This Palm dies off as soon as it has produced its fruit; the stem then becomes hollow, and can be used for spouts and troughs of great durability. The pith supplies sago, about 150 lbs. from a tree, according to Roxburgh.

Argania Sideroxylon, Roemer and Schultes.

Western Barbary, on dry hills. "The Argan-tree." Its growth is generally slow, but it is a long-lived tree. Though comparatively low in stature, its foliage occasionally spreads to a circumference of 220 feet. It sends out suckers from the root. The fruit serves as food for cattle in Morocco; but in Australia the kernels would be more likely to be utilized by pressing an oil from them. Height of tree exceptionally 70 feet.

Aristida prodigiosa, Welwitsch.*

Angola, on the driest sand-hills. A perennial fodder-grass, of which the discoverer speaks in glowing terms of praise. In the West-African desert-country, in places devoid of almost all other vegetation, the zebras, antelopes and hares resort with avidity to this grass; it also affords there in the dry season almost the only fodder for domestic grazing animals. Moreover, this seems to indicate, that the closely cognate A. plumosa, L. and A. ciliata, Desf., of the countries at or near the Mediterranean Sea,. might likewise be encouraged in their natural growth or be cultivated. All feathery

· grasses are among the most lovely for minor decorative purposes or designs, and this may also be said of the Australian plumous Stipa elegantissima (La Billardière) and S. Tuckeri (F. v. M.).

Aristolochia Indica, Linné.

Tropical Asia, North-Eastern Australia and Polynesia. A perennial climber; the leaves famed as an alexipharmic. Can only be grown in places free from frost.

Aristolochia recurvilabra, Hance.

The green "Putchuck" of China. A medicinal plant, largely obtained at Ningpo. The present value of its export is from £20,000 to £30,000 annually.

Aristolochia serpentaria, Linné.

The "Snake-root" of North-America. The root of this trailing herb is valuable in medicine; it contains a peculiar volatile oil. Several other Aristolochiæ deserve culture for medicinal purposes, —for instance, Aristolochia ovalifolia (the Guaco) and A. anguicida, from the mountains of Central America.

Aristotelia Macqui, L'Héritier.

Chili. The berries of this shrub, though small, have the pleasant taste of bilberries, and are largely consumed in Chili (Philippi). The plant would thrive in mild forest-valleys.

Arnica montana, Linné.

Colder parts of Europe and Western Asia. This pretty herb is perennial and of medicinal value. It is particularly eligible for subalpine regions. Hardy in Norway to lat. 62° 47' (Schuebeler). The active principles are arnicin, a volatile oil, caproic and caprylic acids.

Arracacha xanthorrhiza, Bancroft.

Mountain-regions of Central America. A perennial umbelliferous herb. The root is nutritious and palatable. There are yellow, purple and pale varieties, which are kept up by division of the roots.

Artemisia Abrotanum, Linné.

Countries at the Mediterranean Sea. A very odorous shrub, known as "Southern Wood." The foliage used in domestic medicine, also as condiment. The plant is easily grown from cuttings.

Artemisia Absinthium, Linné.

Europe, North- and Middle-Asia and North-Africa. The "Wormwood." A perennial herb, valuable as a tonic and anthelminthic. Should be avoided where bees are kept (Muenter). Recommended for cultivation as a preventative of various insect-plagues, even the Phylloxera. Several other species of Artemisia deserve cultivation for medicinal purposes. Active principles: Absinthin, an oily substance indurating to a crystalline mass; also a volatile oil peculiar to the species.

D

Artemisia Cina, Berg.

Kurdistan. This herb furnishes the genuine santonica-seeds (or rather flowers and fruits), a vermifuge of long-established use. Some other Asiatic species yield a similar drug.

Artemisia Dracunculus, Linné.

Northern Asia. The " Tarragon or Estragon." A perennial herb, used as a condiment. Its flavor depends on two volatile oils, one of them peculiar to the plant. Hardy in Norway to lat. 63° 52' (Schuebeler). Propagation by division of root; the wild plant has but little flavor (Vilmorin).

Artemisia Mutellina, Villars.

Alps of Europe. This aromatic, somewhat woody plant deserves to be established in any alpine region. This plant and A. glacialis, L., A. rupestris, L. and A. spicata, Wulf. comprised under the name of " Genippi," serve for the preparation of the Extrait d'Absinthe (Brockhaus).

Artemisia Pontica, Linné.

Middle and Southern Europe, Western Asia. More aromatic and less bitter than the ordinary wormwood. Hardy to lat. 63° 45' in Norway (Schuebeler). Many other species of this genus deserve attention of the culturist.

Artocarpus communis, R. and G. Forster. (*A. incisa*, Linné fil.)

South Sea-Islands, Moluccas and Sunda-Islands.

The Tahiti "Bread-fruit tree." It stretches in the Sandwich-Islands through cultivation almost beyond the tropics. According to Dr. Seemann's excellent account seedless varieties exist, and others with entire leaves and with smooth and variously shaped and sized fruits; others again ripening earlier, others later, so that ripe bread-fruit is obtainable more or less abundantly throughout the year. The fruit is simply boiled or baked or converted into more complicated kinds of food. Starch is obtainable from the bread-fruit very copiously. The very fibrous bark can be beaten into a sort of rough cloth. The light wood serves for canoes. The exudation, issuing from cuts, made into the stem, is in use for closing the seams of canoes, and could be turned to technic account.

Artocarpus integrifolia, Linné fil.

India. The famous " Jack-Tree," ascending like the allied A. Lakoocha (Roxburgh) to 4,000 feet ; only fit for places free of frost. The fruit attains exceptionally a weight of 80 lbs.; it is eaten raw or variously prepared; the seeds when roasted are not inferior to chestnuts (Dr. Roxburgh).

Arundinaria falcata, Nees.

The Nirgal- or Ningala-Bamboo of the Middle Himalayan zone. The canes attain a diameter of only 4 inches, are durable and applied to manifold useful purposes. This bamboo does not necessarily require moisture. In reference to various bamboos see the Gardeners' Chronicle of December, 1876, also the Bulletin de la Société d'Acclimation de Paris, 1878. The closely-allied Jurboota-Bamboo of Nepal, which occurs only in the cold altitudes of from 7,000 to 10,000 feet, differs in its solitary stems, not growing in clumps. The Tham-or Kaptur-Bamboo is from a still colder zone, at from 8,500 to 11,500 feet, only 500 feet or less below the inferior limits of perpetual glaciers (Major Madden). The wide and easy cultural distribution of bamboos by means of seed has been first urged and to some extent initiated by the writer of the present work.

Arundinaria Falconeri, Munro. (*Thamnocalamus Falconeri*, J. Hooker.)

Himalaya, at about 8,000 feet elevation. A tall species with a panicle of several feet in length. Allied to the foregoing species.

Arundinaria Hookeriana, Munro.

Himalaya, up to nearly 7,000 feet. Grows to a height of about 15 feet. Vernacularly known as " Yoksun and Praong." The seeds are edible, and also used for a kind of beer (Sir Jos. Hooker).

Arundinaria Japonica, Siebold and Zuccarini.

The "Metake" of Japan. Attains a height of from 6 to 12 feet. Uninjured by even the severest winters at Edinburgh, with 0° F. (Gorlie).

Arundinaria macrosperma, Michaux and Richard.

Southern States of North-America, particularly on the Mississippi. This bamboo-like reed forms there the cane-brakes. Fit for low borders of watercourses and swamps. According to C. Mohr it affords throughout all seasons of the year an abundance of nutritious fodder. It requires to be replanted after flowering, in the course of years. Height reaching 20 feet.

Arundinaria spathiflora, Trinius.

Himalaya, at elevations of 8,000 to 10,000 feet, growing among firs and oaks in a climate almost as severe as that of England, snow being on the ground from 2 to 3 months.

Arundinaria tecta, Muehlenberg.

Southern States of North-America. A cane, growing 10 feet high. Prefers good soil, not subject to inundations ; ripens its large mealy seeds early in the season, throwing out subsequently new branches with rich foliage. Fire destroys this plant readily (C. Mohr).

Arundinella Nepalensis, Trinius.

Middle and Southern Africa, Southern Asia, Northern and Eastern Australia. This grass commences its growth in the spring-weather, and continues to increase during the whole summer, forming a dense mat of foliage, which grows as fast as it is fed off or cut. In New Zealand it is only a summer-grass, but valuable for its rapid growth at that season, and for thriving on high dry land (Dr. Curl).

Arundo Ampelodesmos, Cyrillo.

Southern Europe, Northern Africa. Almost as large as a Gynerium. The tough flower-stems and leaves readily available for tying.

Arundo Bengalensis, Roxburgh.

China, India. Closely allied to A. Donax. The long panicle beautifully variegated with white and violet (Hauce).

Arundo conspicua, G. Forster.

New Zealand and Chatham-Islands. Although not strictly an industrial plant, it is mentioned here as important for scenic effect, flowering before the still grander A. Sellowiana comes in bloom, but not quite so hardy as that species, still bearing considerable frost.

Arundo Donax, Linné.

The tall, evergreen, lasting Bamboo-reed of Southern Europe and Northern Africa. It is one of the most important plants of its class for quickly producing a peculiar scenic effect in picturesque plantations, also for intercepting almost at once the view of unsightly objects, and for giving early shelter. The canes can be used for fishing-rods, for light props, rustic pipes, distaffs, baskets and various utensils. Readily flowering when strongly manured. The root is used medicinally in France; easily transplanted at any season. Cross-sections of the canes are very convenient, placed closely and erect, for sowing into them seeds of pines, eucalypts and many other trees, seedlings of which are to be forwarded on a large scale to long distances (J. E. Brown), in the same manner as bamboo-joints are used in India.

Arundo Karka, Roxburgh.

India, China, Japan. The Durma-mats are made of the split stems of this tall reed.

Arundo Pliniana, Turra.

On the Mediterranean and Adriatic Seas. A smaller plant than A. Donax, with more slender stems and narrower leaves, but similarly evergreen, and resembling the Donax-reed also in its roots.

Arundo saccharoides, Grisebach. (*Gynerium saccharoides*, Humboldt.)

Northern parts of South-America. Attaining a height of 20 feet. Like the following, it is conspicuously magnificent.

Arundo Sellowiana, Schultes. (*Arundo dioica*, Sprengel *non* Loureiro, *Gynerium argenteum*, Nees.)

The "Pampas-grass" of Uruguay, Paraguay and La Plata-States. A grand autumnal-flowering reed, with gorgeous feathery panicles. As an industrial plant it deserves here a place, because paper can be prepared from its leaves, as first shown by the author.

Asparagus acutifolius, Linné.

In all the countries around the Mediterranean Sea, also in the Canary-Islands. Although a shrubby Asparagus, yet the root-shoots, according to Dr. Heldreich, are collected in Greece, and are tender and of excellent taste, though somewhat thinner than those of the ordinary herbaceous species ; in Istria and Dalmatia they are consumed as a favorite salad. The shrub grows on stony rises, and the shoots are obtained without cultivation. A. aphyllus, L. and A. horridus, L., according to Dr. Reinhold, are utilized in the same manner, and all may probably yield an improved product by regular and careful culture.

Asparagus albus, Linné.

Countries around the Mediterranean Sea, also in the Canary-Islands. Serves for garden-hedges.

Asparagus laricinus, Burchell.

South-Africa. Dr. Pappe observes of this shrubby species, that with some other kinds of that country it produces shoots of excellent tenderness and aromatic taste.

Asparagus officinalis, Linné.

Europe, North-Africa, North-Asia. The well-known Asparagus-plant, which, if naturalized on any coast, would aid in binding the sand. Hardy in Norway to lat. 64° 12' (Schuebeler). The foliage contains inosit-sugar; the shoots yield asparagin. Sea-weeds are a good additional material for forcing asparagus.

Asperula odorata, Brunfels.

The "Woodruff." Europe, Western and Northern Asia, Northern Africa, there as in Southern Europe only on mountains, always a forest-plant. Indigenous in Norway to lat. 66° 59' (Schuebeler). A perennial herb with highly fragrant flowers; it deserves naturalization in forests; it contains much cumarin in its flowers, and serves in Germany for preparing the "Maitrank."

Aspidosperma Quebracho, Grisebach.

Argentina. Shrub or tree, even tall, with wood fit for xylography. The bitter bark is astringent and febrifugal (Lorentz), being almost as rich in tannin as that of Acacia Cebil. The leaves even contain 27½ per cent.; both have the advantage of producing an almost colorless leather (Sievert). F. Jean states, that even the Quebracho-wood contains 14 to 16 per cent. of tannic and 2 to 3 per cent. of gallic acid.

Astartea fascicularis, De Candolle.

South-Western Australia. A tall shrub or small tree, with a predilection for swampy ground, rather remarkable for quick growth, and evidently destined to take its place in antimalarian plantations. The foliage is locally used for tea.

Astragalus adscendens, Boissier and Haussknecht.

Persia, in alpine elevations of 9,000 to 10,000 feet. A shrub, attaining a height of 4 feet. Yields gum-tragacanth in abundance (Haussknecht). Many species of this genus, numerous in various parts of Europe and Asia, in California and in some other parts of the globe, deserve attention for pasture and other agronomic purposes.

Astragalus arenarius, Linné.

Europe and Western Asia. A perennial fodder-herb for any sandy desert-country.

Astragalus brachycalyx, Fischer.

Kurdistan. A low shrub, affording gum-tragacanth (Flueckiger).

Astragalus Cephalonicus, Fischer. (*A. aristatus*, Sibthorp.)

Cephalonia. A small shrub, yielding a good tragacanth; and so probably also does the true A. aristatus of l'Héritier.

Astragalus Cicer, Linné.

Middle and Southern Europe and Middle Asia. A nutritious and well flavored perennial herb, much sought by grazing animals. It requires, according to Langethal, deep friable grounds and, like most leguminous herbs, calcareous ingredients in the soil.

Astragalus Creticus, Lamarck.

Candia and Greece. A small bush, exuding the ordinary vermicular tragacanth; the pale is preferable to the brown sort.

Astragalus glycyphyllos, Linné.

Europe and Northern Asia. Succeeds on light soil, also in forest-regions. It has been recommended as a perennial, substantial fodder-plant.

Astragalus gummifer, La Billardière.

Syria and Persia. This shrub also yields a good kind of tragacanth.

Astragalus hypoglottis, Linné. (*A. Danicus*, Retzius.)

Colder regions of Europe, Asia and North-America. This perennial plant is regarded as a good fodder-herb on calcareous and gravelly soil, and would likely be of importance in any alpine region. Of the enormous number of supposed species of this genus (according to Boissier, not less than 750 merely in Asia Minor and the adjoining countries) many must be of value for pasture, like some of the closely-allied Australian Swainsonas, though they also may include deleterious species; thus A. Hornii and A. lentiginosus (Gray) of California, and A. mollissimus (Nuttall) of Texas, are known as loco-weeds, and are poisonous to cattle and horses, just as Swainsona lessertifolia (Candolle) and S. Greyana (Lindley) have shown themselves highly injurious to pastoral animals in Australia; the active principle of these plants may however become of importance in medicine.

Astragalus microcephalus, Willdenow.

From Turkey to Russian Armenia. Gum-tragacanth is collected largely also from this species (Farnsworth) and from the nearly allied A. pycnocladus (Boissier and Haussknecht).

Astragalus Parnassi, Boissier. (*A. Cylleneus,* Heldreich.)

Greece. This small shrub furnishes there almost exclusively the commercial tragacanth. It ascends to elevations of 7,000 feet, becoming therefore alpine.

Astragalus strobiliferus, Royle.

Asiatic Turkey. A brown tragacanth is collected from this species.

Astragalus stromatodes, Bunge.

Syria, at elevations of approximately 5,000 feet. Exudes Aintab-tragacanth, which is also obtained from A. Kurdicus (Boissier).

Astragalus venosus, Hochstetter.

From Abyssinia to Central Africa. This perennial herb is subjected to regular cultivation for fodder, known as "Hamat-Kochata" (D. Oliver).

Astragalus verus, Olivier.

Asiatic Turkey and Persia. This shrub furnishes the Takalor- or Smyrna-tragacanth, or it is derived from an allied species.

Astrebla pectinata, F. v. Mueller.* (*Danthonia pectinata,* Lindley.)

New South Wales, Queensland, Northern and Central Australia, in arid regions, always inland. A perennial desert-grass, resisting drought; sought with avidity by sheep, and very fattening to them and other pasture-animals.

Astrebla triticoides, F. v. Mueller.* (*Danthonia triticoides,* Lindley.)

The "Mitchell-grass." Of nearly the same natural distribution as the preceding, and equalling that species in value. Both so important as to deserve artificial rearing even in their native country.

Atalantia glauca, J. Hooker.

New South Wales and Queensland. This Desert-lemon is mentioned here, to draw attention to the likelihood of its improving in culture, and to its fitness for being grown in arid climes.

Atriplex albicans, Aiton.

South-Africa. A good salt-bush for pastures there (McOwan).

Atriplex crystallinum, J. Hooker.

South-Eastern Australia and Tasmania, on the brink of the ocean and exposed to its spray. This herb vegetates solely in salty coast-sands, which it helps to bind, like Cakile.

Atriplex halimoides, Lindley.

Gregarious over the greater part of the saline desert-interior of Australia, reaching the South- and West-coasts. A dwarf bush, with its frequent companion, A. holocarpum (F. v. M.), among the very best for saltbush-pasture. All readily raised from seeds.

Atriplex hortense, Dodoens.

Northern and Middle Asia. The "Arroche." An annual spinage-plant. Hardy in Norway to lat. 70° (Schuebeler).

Atriplex Muelleri, Bentham.

Interior of Australia, reaching the South- and West-coasts. Cattle and especially sheep are so fond of it, that they often browse it to the root. This species approaches in its characteristics closely to A. roseum (Linné) from Europe, Northern Africa and Western Asia; which thus perhaps may be of greater rural significance also, than hitherto supposed.

Atriplex nummularium, Lindley.

From Queensland through the desert-tracks to Victoria and South-Australia. One of the tallest and most fattening and wholesome of Australian pastoral salt-bushes. Sheep and cattle, pastured on salt-bush-country, are said not only to remain free 'of fluke, but to recover from this Distoma-disease and other allied ailments.

Atriplex semibaccatum, R. Brown.

Extra-tropic Australia. A perennial herb, very much liked by sheep (R. H. Andrews), thus considered among the best of saline herbage of the saltbush-country.

Atriplex spongiosum, F. v. Mueller.

Through a great part of Central Australia, extending to the South- and West-coast. Available, like the preceding and several other species, for saltbush-culture. Unquestionably some of the shrubby extra-Australian species, particularly those of the Siberian and Californian steppes, could also be transferred advantageously to sub-saline country elsewhere, to increase its value, particularly for sheep-pasture.

Atriplex vesicarium, Hewerd.

In the interior of South-Eastern and Central Australia. One of the most fattening and most relished of all the dwarf pastoral salt-bushes of Australia, holding out in the utmost extremes of drought, and not scorched even by sirocco-like blasts. Its vast abundance over extensive saltbush-plains of the Australian interior, to the exclusion of almost every other bush except A. halimoides, indicates the facility, with which this species disseminates itself. Splendid wool is produced in regions, where A. vesicarium and A. halimoides almost monopolize the ground for enormous stretches. With other woody species easily multiplied from cuttings also.

Atropa Belladonna, Linné.

The " Deadly Nightshade." Southern and Middle Europe and Western Asia. A most important perennial medicinal herb. The highly powerful atropine is derived from it, besides another alkaloid, belladonnine.

Audibertia polystachya, Bentham.

California. A shrub, attaining a height of 10 feet; keeps the bees roaring with activity about its flowers for honey during the whole spring (A. J. Cook). The same can be said of A. Palmeri (Gray) and some other species.

Avena elatior, Linné. (*Arrhenatherum elatius,* Beauvois.)

The tall Meadow-Oat-grass. Europe, Middle Asia, North-Africa. Indigenous in Norway to lat. 68° 11′ (Schuebeler). This grass should not be passed altogether on this occasion, although it becomes easily irrepressible on account of its wide-creeping roots. It should be chosen for dry and barren tracts of country, having proved to resist occasional droughts better than rye-grass. Mr. J. L. Dow regards it as one of the very best of grasses for sandy soil. Yields more green feed in the southern states of North-America during winter than most other grasses (Loring). The bulk yielded by it is great; it submits well to pasturing, and gives two or three crops of hay annually; it is however not so much relished by animals as many other grasses.

Avena fatua, Linné.

Wild Oat. Europe, Northern Africa, Northern and Middle Asia, eastward as far as Japan. The experiments of Professor Buckman seem to indicate, that our ordinary cultivated Oat (Avena sativa, L.) is descended from this plant. Cultivated in California for fodder, but requiring early cutting, as it matures and sheds its seed in July. For this reason it is also hard to exterminate it in grain-fields, where it sometimes proves quite troublesome, except by change of crops.

Avena flavescens, Linné. (*Trisetum flavescens,* Beauvois.)

Yellowish Oat-grass. Europe, Northern Africa, Middle and Northern Asia, eastward as far as Japan. One of the best of perennial meadow-grasses, living on dry soil; fitted also for alpine regions. Lawson observes, that it yields a considerable bulk of fine foliage, and that it is eagerly sought by sheep, but that it thrives best intermixed with other grasses. It likes particularly limestone-soil, where it forms a most valuable undergrass, but is not adapted for poor sand, nor will it stand well the traversing of grazing animals (Langethal).

Avena pratensis, Linné.

Meadow-Oat-grass. Europe, Northern Asia. Indigenous in Norway, to lat. 66° 40′ (Schuebeler). It thrives well on dry clayey soil, is well adapted also for alpine mountains, where it would readily establish itself, even on heathy moors. It produces a sweet fodder, but not in so great quantity as several other less nutritious grasses. It is perennial, and recommended by Langethal for such ground, as contains some lime, being thus as valuable as Festuca ovina. Eligible also for meadows, especially under a system of irrigation.

Avena pubescens, Linné.

Downy Oat-grass. Europe, Northern and Middle Asia. A sweet perennial grass, requiring dry but good soil containing lime; it is nutritious and prolific, and one of the earliest kinds, but not well resisting traffic. Several good Oat-grasses are peculiar to North-America and other parts of the globe. Their relative value for fodder is in many cases not exactly known, nor does the limit assigned to this volume allow of their being enumerated specially.

Avena sativa, Linné.

The Common Oats. In Middle Europe cultivated before the Christian era, and in Switzerland already at the Bronze-Age. A. de Candolle regards it as probably indigenous to Eastern temperate Europe, particularly the Austrian Empire, thence perhaps extending to Siberia. Annual. Important for fodder, green or as grain—for the latter indispensable. Fit for even poor or moory or recently drained land, though not so well adapted for sandy soil as rye, nor well available for calcareous ground; resists wet better than other cereals; best chosen as first crop for inferior land, when newly broken up; middling grassy soil is particularly suited for oats; in rich ground more prolific for green fodder. It succeeds in rotation after every crop, though variously as regards yield, and best after clover. In volcanic soil of the Victoria-colony as much as 75 bushels of Oats have been obtained from an acre in one harvest, and in most favorable places in New Zealand exceptionally even double that quantity. Its culture extends not quite so far towards polar and alpine regions as barley, on account of the longer time required for its maturing; yet it will ripen still at lat. 69° 28′ in Norway (Schuebeler). Varieties with seeds separating spontaneously from the bracts (chaff) are : A. nuda, L. and A. Chinensis, Metzger, the Tatarian and Chinese Oats, which are the sorts preferred for porridge and cakes. Other varieties or closely allied species are: A. orientalis, Schreber, which is very rich in grain, and on account of the rigidity of its stem especially fitted for exposed mountain-localities; A. brevis, Roth, the short-grained oats, which is particularly suitable for stable-fodder; A. strigosa, Schreber, which is a real native of Middle-Europe, and deserves preference for sandy soil. Russian quas-beer is made of oats (Langethal, Brockhaus).

Averrhoa Carambola, Linné.

Continental and insular India. Not hurt by slight frost, except when very young. Sir Jos. Hooker found this small tree on the Upper Indus as far as Lahore. The fruit occurs in a sweet and acid

variety; the former is available for the table raw, the other for preserves. That of A. Bilimbi (Linné) is of similar use, especially for
tarts.

Avicennia officinalis, Linné.

From the coasts of South-Asia to those of South-Africa, all
Australia and New Zealand. It is proposed by Dr. Herm. Behr, to
plant this tree for consolidating muddy tidal shores. The copious
nectar from the flowers eagerly sought by bees.

Azima tetracantha, Lamarck.

From South-India to South-Africa. A hedge-bush, growing freely
in every kind of soil.

Baccharis pilularis, De Candolle.

California and Oregon. This evergreen bush can be grown for
hedges, used also for garlands, wrappers of flower-boquets and other
decorative purposes, as cut branches do not wither for a considerable
time. It attains a height of 15 feet (Professor Bolander), and could
readily be naturalized along sandy sea-shores.

Backhousia citriodora, F. v. Mueller.

Southern Queensland. Though only a small tree, it is well worth
cultivating for the fragrance of its lemon-scented foliage, from which
also a culinary and cosmetic oil can be cheaply distilled.

Bactris Gasipæs, Humboldt. (*Guilielma speciosa*, Martius.)

The "Peach-Palm" of the Amazon-River, ascending to the warm
temperate regions of the Andes. Stems clustered, attaining a height
of 40 feet. The fruit grows in large bunches; Dr. Spruce describes
it as possessing a thick, firm and mealy pericarp, and when cooked
to have a flavor between that of the potato and chestnut, but superior
to either.

Bacularia Arfakiana, Beccari.

In Araucaria-forests of New Guinea up to 6,000 feet. A reed-
like palm, evidently desirable for decorative purposes.

Bacularia monostachya, F. v. Mueller. (*Areca monostachya*, Martius.)

Eastern Australia, extending to extra-tropical latitudes. One of the
best among small Palms for table-decoration. The stems sought for
walking-sticks.

Baloghia lucida, Endlicher. (*Codiæum lucidum*, J. Mueller.)

East-Australia. A middle-sized tree. The sap from the wounded
trunk forms, without any admixture, a beautiful red indelible
pigment.

Balsamodendron Mukul, Hooker.

Scinde and Beluchistan. Yields the Bdellium-resin.

Balsamodendron Myrrha, Nees. (*Commiphora Myrrha,* Engler.)

Deserts of Arabia. This tree yields the commercial " Myrrh," but perhaps some other species may produce the same substance.

Balsamodendron Opobalsamum, Kunth. (*B. Gileadense,* Kunth. *Commiphora Opobalsamum,* Engler.)

Deserts of Arabia, Abyssinia and Nubia. A small tree. This species furnishes Mekka- or Gilead-Balsam. B. Capense (Sonder) is a closely allied congener from Extra-tropical South-Africa. Some other Balsam-shrubs deserve introduction into warm dry regions.

Bambusa arundinacea, Roxburgh.*

The " Thorny Bamboo " of India. It likes rich, moist soil, and delights on river-banks; it is of less height than Bambusa vulgaris, also sends up from the root numerous stems, but with bending branches, thorny at the joints. Used in continental India for hedges. According to Kurz it will thrive in a climate too dry for B. Tulda and B. vulgaris. The seeds of this and some other Bamboos are useful as food for fowls. Whenever seeds of any Bamboos can be obtained fresh and disseminated soon, large masses of these plants could easily be raised in suitable forest-ground of other countries ; Bamboo-seeds moreover, like Palm-seeds, ought to become a very remunerative article of commercial export for horticultural purposes under practical and experienced hands.

Bambusa aspera, Poiret.

Indian Archipelagus. Attains a height of 120 feet. Stems very strong and thick. This species ascends to cool elevations of 4,000 feet.

Bambusa Balcooa, Roxburgh.*

From the Plains of Bengal to Assam. Proved hardy at the Cape of Good Hope. Height reaching 70 feet. With B. Tulda the principal Bamboo used by the natives for constructing large huts or sheds, but, as Roxburgh has pointed out, in order to render the material durable, it needs long previous immersion in water. Mr. Routledge recommends young shoots of Bamboos as paper-material. The seeds of Bambusa Tulda have been found by me to retain their vitality for some time and to germinate readily.

Bambusa Blumeana, Schultes.

Insular India. This Bamboo, with its spiny buds and pendent branchlets, is according to Kurz one of the best for cattle-proof live-hedges among the Asiatic species. In continental India B. nana and B. arundinacea are much used for the same purpose. Periodic trimming is required.

Bambusa Brandisii, Munro.

Tenasserim, Martaban and Pegu, wild up to elevations of 4,000 feet. Height of stems reaching 120 feet, diameter 9 inches. It likes limestone-soil. Locally much used for rural buildings, affording posts, rafters, flooring-material and shingles; also many utensils, among them buckets, are made of this Bamboo (Dr. Brandis).

Bambusa flexuosa, Munro.

China. Only 12 feet high, but very hardy, having resisted in Southern France a temperature of 8° F (Geoffroy de St. Hilaire).

Bambusa Senaensis, Franchet and Savatier.

Japan. A tall and hardy species, distinguished from all other Japanese Bambusaceæ by its large leaves. Young Bamboo-shoots (probably of several species) constitute part of the nourishment of all classes in Japan (Dupont).

Bambusa spinosa, Roxburgh.*

Bengal. A Bamboo, attaining 100 feet in height. The central cavity of the canes is of less diameter than in most other species; thus the strength for many technic purposes is increased.

Bambusa vulgaris, Wendland.

The large unarmed Bamboo of Bengal. It rises to a height of 70 feet, and the stems may attain a length even of 40 feet in one season, though the growth is slower in cooler climes. It has proved to be capable of resisting occasional night-frosts. It is the best for building bamboo-houses. Immersion in water for some time renders the cane still firmer. To the series of large thornless bamboos belong also Bambusa Tulda and Bambusa Balcooa of India, and Bambusa Thouarsii from Madagascar and Bourbon. These bamboos are much used for various kinds of furniture, mats, implements and other articles. Besides these, Kurz enumerates as among the best Asiatic bamboos for building purposes: Gigantochloa aspera, G. maxima, G. attar; while Teysmann notes G. apus for the same purpose. Kurz recommends further, Bambusa arundinacea, B. Balcooa, B. Brandisii, B. polymorpha, Dendrocalamus Hamiltoni and Schizostachyum Blumei. In the Moluccas, according to Costa, Gigantochloa maxima, or an allied species, produces stems thick enough to serve when slit into halves for canoes. Bamboos are utilized for masts and spars of small vessels. Bambusa Balcooa was found by Wallich to grow 12 feet in 23 days. Bambusa Tulda, according to Roxburgh, has grown at first at the rate of from 20 to 70 feet in a month. Fortune noticed the growth of several Chinese Bamboos to be two to two and a half feet a day. There are many other kinds of Bamboo eligible among the species from China, Japan, India, tropical America and perhaps tropical Africa. Two occur in Arnhem's Land, and one at least in North-Queensland. New Guinea is sure to furnish also additional kinds of technical importance or eminent horticultural value.

Baptisia tinctoria, R. Brown.

The "Wild Indigo" of Canada and the United States. A perennial herb. It furnishes a fair pigment, when treated like the best Indigoferas; also used as an antiseptic in medicine.

Barbarea vulgaris, R. Brown.

In the cooler regions of all parts of the globe, ascending to alpine zones. Hardy to lat. 64° 5′ in Norway (Schuebeler). This herb furnishes a wholesome salad. As with other raw vegetables, particularly watercress (Nasturtium aquaticum, Trag.), circumspect care is necessary, to free such salads from possibly adherent Echinococcus-ova or other germs of entozoa, particularly in localities where hydatids prevail. An excellent honey-plant (Muenter), particularly for cold regions. Several allied species exist.

Barosma serratifolia, Willdenow.

South-Africa. This shrub supplies the medicinal Bucco-leaves. B. crenulata, Hooker (Diosma crenulata, L.) is only a variety of this species. Active principles: a peculiar volatile oil, a peculiar resin and a crystalline substance called diosmin. Empleurum serrulatum (Solander), a small South-African shrub, yields also Bucco-leaves as noted by Prof. Eichler.

Basella lucida, Linné.

India. Perennial. This spinage-plant has somewhat the odour of Ocimum Basilicum; other species serve also for culinary purposes.

Basella rubra, Linné.

From Southern Asia to Japan. This annual or biennial herb serves as a spinage of pleasant coloration, but is not possessed of the agreeable flavor of real spinage. It yields also a rich purple dye, not easily fixed however (Johnson).

Bassowia solanacea, Bentham. (*Witheringia solanacea*, L'Heritier.)

South-America. This perennial herb needs trial-culture, on account of its large edible tubers.

Batis maritima, Linné.

. Central America and northward to Florida, also in the Sandwich-Islands. This shrub can be used, to fix tidal sediments for the reclamation of harbor-lands.

Beesha elegantissima, Hasskarl.

Java, on mountains of about 4,000 feet elevation. Very tall and exceedingly slender; the upper branches pendulous. A hardy species of Bamboo.

Belis jaculifolia, Salisbury. (*Cunninghamia Sinensis*, R. Brown.)

Southern China. A tree, attaining 40 feet in height. Though too slow for timber-growth, it should not be passed in this work, as its Araucaria-like habit entitles it to a place in any arboretum, which is not subject to severe frosts; it proved hardy at Arran (Rev. D. Landsborough). The tree furnishes resin; the timber serves building-purposes well.

Benincasa cerifera, Savi.

India, Philippines, China, Japan, perhaps also North-Eastern Australia and Polynesia. This annual plant produces a large edible gourd, which in an unripe state forms part of the composition of many kinds of curry.

Berberis Asiatica, Roxburgh.

Himalaya. A Berberry-shrub. Hardy in Christiania (Schuebeler). One of the best among numerous species with edible berries. Among these may particularly be mentioned B. Lycium (Royle) and B. aristata (De Candolle), which also yield valuable yellow dye-wood (Dr. Rosenthal). All kinds of Berberry-shrubs must be kept away from cereal fields, as they might become the seat of the Aecidium-state of one of the principal rust-fungs, Puccinia graminis.

Berberis buxifolia, Lamarck.

From Magelhaen's Straits to Chili. This bush, according to Dr. Philippi, is the best among the South-American species for berries, which are comparatively large, black, hardly acid, but slightly astringent. In Valdivia and Chiloe they are frequently consumed.

Berberis Darwinii, Hooker.

Chiloe and South-Chili. Considered one of the most handsome of all shrubs for garden-hedges. Hardy in England; even at Christiania. Several other evergreen Berberry-shrubs serve the same purpose.

Berberis Nepalensis, Sprengel.

Himalayas, at elevations between 4,000 and 8,000 feet. Hardy to lat. 59° 55′ in Norway (Schuebeler). The fruit of this evergreen species is edible.

Beschorneria yuccoides, Hooker.

Mexico, at rocky elevations from 8,000 to 10,000 feet (Finck). This handsome plant will bear slight frost. The stem attains a height of 6 feet, and produces flowers annually like yuccas. The leaves yield a remarkable fine and strong fibre (T. Christy). The two other known species, also from Mexico, B. tubiflora (Kunth) and B. Parmentierii (Jacobi), are probably similarly useful.

Beta vulgaris, Linné.*

The "Beet or Mangold-Wurzel." Middle and Southern Europe, Middle Asia, Northern Africa. Hardy in Norway to lat. 70° 4′ (Schuebeler). This well-known perennial or biennial herb ought to engage the general and extensive attention of any farming population. Can be grown for mere foliage even in sandy soil near the sea, and is often chosen for the first crop on heath-lands in Northern Germany. The herbage is most valuable as a palatable and nutritious spinage; the root is of importance not only as a culinary vegetable, but, as is well-known, also for containing crystallizable sugar. The sugar of the beet is indeed now almost exclusively consumed in Russia, Germany, Austria, France, Sweden and Belgium; and these countries not only produce beet-sugar, but also export it largely to the neighboring States. The white Sicilian Beet is mainly used for salads, spinage and soups. The thick-ribbed variety serves like asparagus or sea-kale, dressed like rhubarb. Cereal soil, particularly such as is fit for barley, is generally adapted also for the culture of beets. The rearing of the root and the manufacture of the sugar can be studied from manifold works; one has been compiled by Mr. N. Levy, of Melbourne. A deeply-stirred, drained soil, rich in lime, brings the saccharine variety of beet to the greatest perfection. The Imperial beet yields from 12 to 20 per cent. sugar. The Castlenauderry, the Magdeburg, the Siberian White-rib and the Vilmorin-Beet are other varieties rich in sugar. About 5 lbs. of seed are required for an acre. In rotation of crops the beet takes its place best between barley and oats. In Middle Europe the yield averages 14 tons of sugar-beets to the acre, and as many hundredweight of raw sugar. The mercantile value of the root, at distilleries, has ranged from 20s. to 30s. per ton. In climates not subject to frost the beet-harvest can be extended over a far greater portion of the year than in Middle Europe. The extraction of the sap is effected generally by hydraulic pressure; the juice is purified with lime and animal charcoal; excess of lime is removed by carbonic acid, and the purified and decolorized juice is evaporated in vacuum pans, with a view to prevent the extensive conversion of the crystallizable sugar into treacle. The production of beet-sugar needs far less labor than that of cane-sugar, and the harvest is obtained in so short a time as eight months. The beet has shown itself subject neither to alarming diseases nor to extensive attacks of insects. It is mostly grown in extra-tropical zones, while the sugar-cane is confined to tropical and sub-tropical latitudes. Beet-culture, by directly or indirectly restoring the refuse, ameliorates the soil to such an extent, that in some parts of Europe land so utilized has risen to fourfold its former value. The beet furthermore affords one of the most fattening stable-fodders; and thus again an ample supply of manure. In the beet-districts of Middle Europe about one-sixth of the arable land is devoted to beets, yet the produce of cereals has not been reduced, while the rearing of fattened cattle has increased. Notwithstanding a heavy tax on the beetsugar-factories in Europe, the industry has proved prosperous, and assumes greater and greater dimensions. In 1865 the sugar-consumption of Europe amounted to 1,583,825 tons, one-third of which had been locally supplied by the beet, from over one thousand factories. Treacle obtained from beet is distilled for alcohol. For establishing remunerative factories on a large and paying scale, it has been suggested, that farmers' companies might be formed. For ascertaining the percentage of sugar in the beet, saccharometers are used. In Germany some scientific periodicals are exclusively devoted to the fostering of this industry. In 1875 the total production of beet-sugar amounted to 1,318,000 tons (Boucheraux)

Betula acuminata, Wallich.

Himalaya, between 3,000 and 10,000 feet. Attains a height of 60 feet, and thrives along forest-streams. The wood is hard, strong and durable. Another Himalayan Birch, B. utilis (D. Don), grows on arid ground, and produces good timber of less hardness.

Betula alba, Linné.

"White Birch." The common Birch of Europe and Extra-tropical Asia and North-America. With some Willows approaching nearer to the North-Pole than any other woody vegetation, but ceding in milder regions before the Beech when occurring mixed. It attains a height of 80 feet, and would, when cultivated, thrive best in moist glens of ranges or in the higher regions of mountains, where it would form at the alpine zone also excellent shelter-plantations. The variety B. pubescens (Ehrhart) attains still a height of 60 feet in lat. 70° N. in Norway (Schuebeler). Content with the poorest soil. The variety B. populifolia (Willd.) extends to North-Eastern America, the variety B. occidentalis (Hooker) to North-Western America. The durable bark serves for rough roofing. Wood white, turning red, tough, adapted for spools, shoe-pegs and many other minor purposes, also for some parts of the work of organ-builders; affords like that of other Beeches charcoal for gunpowder. The oil of the bark is used in preparing the Russian leather.

Betula lenta, Willdenow.

The "Cherry-Birch" of North-America. A tree, reaching to 80 feet in height, and 2 feet in stem-diameter, liking moist ground, but also content with dry soil. Hardy at Christiania in Norway (Schuebeler). Wood rose-colored or dark, fine-grained, excellent for furniture. It is so heavy, that when fresh it will not float in water. It is used for ships' keels, machinery, furniture and other purposes, where strength, hardness and durability are required. Bark of a somewhat aromatic odor. Several Birches occur in Japan, which might well be tried elsewhere.

Betula lutea, Michaux.

The "Yellow or Gray Birch" of North-Eastern America. Height sometimes 80 feet. Adapted for moist forest-land. In timber similar to B. lenta. The wood is used for shoe-lasts and various other purposes.

Betula nigra, Linné.

The "Red Birch" or "River Birch" of North-America. One of the tallest of Birches, occasionally more than 3 feet in stem-diameter, If grown on the banks of streams, it will bear intense heat. The wood is compact, of a light color, easily worked, excellent for turning, also in use by cabinet-makers and carriage-builders; well adapted to sustain shocks and friction (Robb). It is also used for

E

shoe-lasts, bowls and trays, and the saplings and branches for hoops. The bark is well adapted for rough roofing. Twigs of the Red Birch furnish one of the best materials for rough brooms. Hardy at Christiania (Schuebeler).

Betula papyracea, Aiton.

The "Paper-Birch" of North-America. A larger tree than B. alba, with a fine-grained wood and a tough bark; the latter much used for portable canoes. It likes a cold situation. Hardy to lat. 63° 55' in Norway (Schuebeler).

Boehmeria nivea, Gaudichaud.*

The "Ramee or Rheea." Southern Asia, as far east as Japan. This bush furnishes the strong and beautiful fibre, woven into fabric, which inappropriately is called grass-cloth. The bark is softened by hot water or steam, and then the bast separable into its tender fibres. The bast is obtained from the young shoots; it is glossy, tough and lasting, combining to some extent the appearance of silk with the strength of flax. The ordinary market-value of the fibre is about £40 per ton; but Dr. Royle mentions, that it has realized at times £120. The seeds are sown on manured or otherwise rich and friable soil. In the third year, or under very favorable circumstances even earlier, it yields its crops, as many as three annually. The produce of an acre has been estimated at two tons of fibre. This latter, since Kaempfer's time, has been known to be extensively used for ropes and cordage in Japan. Rich forest-valleys seem best adapted for the Ramee, as occasional irrigation can be applied there. In the open lands at Port Phillip it suffers from the night-frosts, although not to such an extent as materially to injure the plant, which sends up fresh shoots, fit for fibre, during the hot season. The plant has been cultivated and distributed since 1854 in the Botanic Garden of Melbourne, where it is readily propagated from cuttings, the seeds rarely ripening there. Cordage of this Boehmeria is three times as strong as that of hemp. Numerous shoots spring after cutting from the same root. Fertile humid soil or rich manuring is necessary for productive returns. Dr. Collyer, of Saharumpore, boils the whole branches with soap-water (a process used here since 1866, for separating the Phormium-fibre) for the easy separation of the fibre, of which he obtained 150 lbs. from a ton of Rheea-branches; the cost of separation and final preparation being calculated at £10 per ton (interest on capital for machinery not counted). He also perfected machinery, to render the process easy and more remunerative. Fibre, further prepared by Bonsor's process, can be spun into the finest yarn. Colonel Hannay and Dr. Forbes Watson record, that in Assam four to six crops are cut annually, that obtained in the cool season providing the strongest fibre; the latter is obtainable to the length of 6 feet. Other species require to be tested, among them the one, which was discovered in Lord Howe's Island, namely Boehmeria calophleba, Moore and Mueller.

Boletus bovinus, Linné.

Europe. Besides this species Dr. Goeppert mentions also the following, as sold for food in the markets of Silesia: B. circinaus, Persoon; B. edulis, Bulliard; B. luteus, Linné; B. sapidus, Harzer; B. scaber, Bulliard; B. subtomentosus, Linné; B. variegatus, Swartz. Dr. Cooke lauds B. æstivalis, Fries, for food. Dr. L. Planchon's list of edible fungs of France includes from this generic group: B. æreus, Bulliard; B. granulatus, Linné; B. castaneus, Bulliard. Among the Victorian species one, allied to B. edulis, attains a weight of 3 lbs., and the pileus then is of 12 inches width. The genus is dispersed over the globe in numerous species, many doubtless fair esculents.

Bongardia Rauwolfi, C. A. Meyer.

From Greece through Turkey to the Caucasus. A perennial herb, the leaves of which are utilized like culinary sorrel.

Borassus Aethiopicus, Martius.

Africa, from Zanzibar to Egypt. A palm of gigantic dimensions, its stem attaining 9 feet in diameter at the base, or 7 feet at 4 feet above the ground; sometimes even stems have been measured having a circumference of 37 feet. Leaves occur as much as 12 feet across; they serve for the manufacture of baskets, mats, ropes and sieves. The edible portion of the fruit is yellow, stringy, of a fruity flavor. The sap obtained from incisions in the stem under the leaves yields a kind of palm-wine. In its natural home the tree always denotes water (Colonel Grant). Sir W. Hooker admits only one species, and regards Africa solely as its home.

Borassus flabelliformis, Linné.

The "Palmyra." From the Persian Gulf to India, extending to 30° North. This noble palm attains a height of 100 feet. The pulp of the fruit serves as food. Enormous masses of sugar and toddy are produced in India from the sap, which flows from incisions of the stalk of the unexpanded flowers. This palm, wherever hardy, should be reared for scenic plantations. Assumed to reach, like the Date-Palm, an age of more than 200 years. Many other Palms are notable for longevity; thus Euterpe oleracea has been calculated to attain 130 years; Cocos oleracea, 650 years; Cocos nucifera, 330 years, according to the number of their stem-rings (Langethal), of which however perhaps more than one are formed in a year.

Boronia megastigma, Nees.

In Western Australia, on margins of swamps. This remarkable bush is recorded here as an emblem of mourning, its externally blackish flowers rendering it especially eligible for graves. Industrially it interests us on account of its very fragrant blossoms, for the sake of which this bush well deserves to be cultivated. The perfume could doubtless be extracted and isolated, particularly by absorbents in vacuum. Individual plants of this Boronia will endure in a cultivated state for very many years. B. heterophylla (F. v. M.) from King George's Sound is of similar but not quite so strong a scent.

Borrago officinalis, Linné.

Southern Europe, Orient. An annual herb, rich in nitrate of potash, occasionally used for medicinal purposes or as an admixture to salad. Readily disseminating itself; not to be overlooked as a honeyplant.

Boswellia papyrifera, Hochstetter.

Morocco, Nubia and Abyssinia, forming entire forests about Bertat on the Atlas. This tree exudes a kind of Olibanum-resin, and represents apparently one of the hardiest species of this or allied genera.

Boswellia serrata, Roxburgh. (*B. thurifera*, Colebrooke.)

India. A deciduous tree, living in arid forest-regions. Yields an aromatic resin. The real Olibanum is exuded by B. Carteri (Birdwood) of Arabia and tropical Africa.

Boussingaultia baselloides, Humboldt.

South-America. This hardy climber is well fitted for bowers; the mucilaginous tubers are edible. It is not uncommonly grown as a climber on verandahs, where no severe frosts occur.

Bouteloua polystachya, Bentham.

New Mexico, Texas and Arizona. One of the Gramma-grasses of the prairies, called with some other species also Muskit-grass; gregarious; famed for nutritive value; fit for arid regions. Dr. Vasey recently enumerated fourteen kinds of Bouteloua, of which B. foenea, B. eriopoda and B. oligostachya (Torrey) are mentioned as those best liked by the herds; others proved excellent in Argentina.

Brabejum stellatifolium, Linné.

South-Africa. The nuts of this small tree are edible, resembling those of our Macadamia ternifolia, to which also Brabejum is closely allied in foliage and flowers. The nuts are also similar to those of the Chilian Guevina Avellana. The fruit should be roasted, otherwise it is deleterious. Flowers sweet-scented.

Brachychiton acerifolium, F. v. Mueller.

The East-Australian Flame-tree. An evergreen shade-tree, with magnificent trusses of crimson blossoms. Like B. populneum (R. Br.), eligible for shading promenades, when rapidity of growth is no object. The mucilaginous sap, when exuded, indurates to a kind of Bassorin-Tragacanth.

Brahea dulcis, Martius.

Mexico, as far as its northern parts, and ascending to 4,500 feet. A Brahea-Palm has also been discovered as far north as Arizona, 32° (Drude).

Brahea edulis, Herm. Wendland.

Lower California. Height to 20 feet. The clusters of plumshaped fruits sometimes weigh 40 lbs., and are eaten by domestic animals.

Brassica alba, Visiani. *(Sinapis alba,* Linné.)

White Mustard. Eastern Europe, Northern Africa, Northern and Middle Asia. An annual. The seeds are less pungent than those of the Black Mustard, but used in a similar manner. The young leaves of both are useful as a culinary and also antiscorbutic salad. Can be employed with great advantage as green manure and suppresses weeds simultaneously (W. Emerson McIvor). The cold-pressed oil of mustard-seed serves for table-use. Dr. M. T. Masters enumerates Brassica dichotoma, B. Pekinensis, B. ramosa and B. glauca among the mustards, which undergo cultivation in various parts of Asia, either for the fixed oil of their seeds or for their herbage. From 15 lbs. to 20 lbs. of seed of the White Mustard are required for an acre. In the climate of California 1,400 lbs. of seed have been gathered from an acre. Can be grown in shallow soil, even on land recently reclaimed from swamps. It prefers argillaceous ground. The return is obtained in a few months. The stalks and foliage after the seed-harvest serve as sheep-fodder. In Norway the plant comes still to perfection as far north as lat. 70° (Schuebeler).

Brassica Chinensis, Linné.

China and Japan. Serves like B. oleracea for cabbage, and produced in cultivation new varieties, particularly for use very late in the season. The seeds in Japan extensively pressed for oil. B. Cretica (Lamarck) is a woody Mediterranean species.

Brassica juncea, J. Hooker and Thomson. *(B. Willdenovii,* Boissier; *Sinapis juncea,* Linné.)

From Middle Africa to China. According to Colonel Drury cultivated all over India for Sarepta-Mustard seed ; also a good salad-plant.

Brassica nigra, Koch. *(Sinapis nigra,* Linné.)

The "Black Mustard." Europe, Northern Africa, Middle Asia. An annual. The seeds simply crushed and then sifted constitute the mustard of commerce. For medicinal purposes the seeds of this species are preferable for sinapisin and especially sinapisms. Through aqueous distillation a volatile oil of extreme pungency is obtained from mustard-seeds. In rich soil this plant is very prolific; and in forest-valleys it is likely to remain free from the attacks of aphides. Chemical constituents: a peculiar fixed oil, crystalline sinapin, the fatty sinapisin, myronic acid and myrosin. All mustards can be regarded as honey-plants. Brassica-seeds of various kinds are retaining their vitality for several years.

Brassica oleracea, Linné.*

An annual or biennial coast-plant, indigenous to various parts of Europe. It is mentioned here with a view of showing, that it might be naturalized on any rocky and sandy sea-shores. One of the best plants for newly reclaimed lands. From the wild plant of the coast have originated various kinds of cabbages, broccoli, cauliflower;

Brussels-sprouts, kale, kohl-rabi &c.   Some regard the fattening
qualities of cabbage as superior to those of turnips, particularly for
stable-food during the autumnal season. , The gluten of cabbages on
one acre has been estimated at 1,500 lbs. against 1,000 lbs. of gluten
obtainable from turnips.   Other races of this species are collectively
represented by Brassica Rapa, L. (B. campestris, L.), the wild Navew,
yielding most of the varieties of turnips, some handed down to us from
ancient times with other cultivated forms.   Again, other varieties are
comprehended within Brassica Napus, L., such as the Swedish and
Teltower turnips, while the Rape-seed, so important for its oil (Colza),
is also derived from a form of B. Napus.   The Rape-plant should
be reared extensively for agrarian produce, giving a rapid return,
wherever it remains free from aphides.   Ordinary Rape is a good
admixture to summer-fodder.   Important where bees are kept.   The
hardier turnips can be produced on the highest Alps, as they are
grown even within the Arctic Circle, and according to Sir J. Hooker
at a height of 15,000 feet in the Himalayan mountains.   Oil-Rape
and Turnips are grown as far north as 70° 22′ in Norway (Schuebeler), yet they also succeed as winter-crops in the hottest parts
of Central Australia.   Rhind mentions a very tall variety, introduced
from the Vendee, as cattle-provender.

Bromus asper, Murray.

Europe, Northern and Middle Asia.   A good perennial fodder-grass
for wood-regions; but like Festuca gigantea late in the season.

Bromus ciliatus, Linné.

North-America.   A   perennial fattening grass, resembling the
Prairie-grass, growing in mild temperate climes all the winter and also
during summer, if drought is not too long continued, starting afresh
after the least rain (Dr. Curl).

Bromus erectus, Hudson.

Europe, Northern Africa.   Important as a perennial nutritious
grass for dry limestone-regions; much liked by cattle and sheep.

Bromus unioloides, Humboldt.*   (B. Schraderi, Kunth.)

From Central America to the sub-alpine zone of Northern Argentina.   In Australia called the Prairie-grass.   It has spread over many
parts of the globe.   The writer saw it disseminated on the mountains
of St. Vincent's Gulf as early as 1847.   It is one of the richest of all
grasses, grows continuously and spreads rapidly from seeds, particularly on fertile and somewhat humid soil, and has proved as a lasting
and nutritious fodder-grass or pasture-grass one of our best acquisitions.   Very early out in the season for fodder.   Kept alive in the
hottest and driest parts of Central-Australia, where it was first introduced by the writer of this work.   In Norway it comes to perfection
up to lat. 67° 55′ (Schuebeler).   Chemical analysis in early spring
gave: albumen, 2·80; gluten, 3·80; starch, 3·30; gum, 17·0; sugar,.
2·30 per cent. (F. v. Mueller and L. Rummel).

Broussonetia papyrifera, Ventenat.

The " Paper Mulberry-tree." Islands of the Pacific Ocean, China, Japan, perhaps only truly indigenous in the last-named country. The bark of this tree or shrub can be converted into very strong paper. It can also be used for textile fabrics; furthermore, the cloth made from it can be dressed with linseed-oil for waterproof coverings. In cultivation the plant is kept like an osier. The leaves cannot be used for silkworms. European fabrics have largely superseded the clothing made of this plant in the South-Sea-Islands.

Buchloe dactyloides, Engelmann.*

The true Buffalo-grass of Kansas, also known as one of the mesquite-grasses, naturally extending from Canada to Texas, forming a large proportion of the food of the buffaloes on the prairies (Engelmann). Dioecious, creeping, only rising to half a foot or less, but overpowering the Boutelouas. It is extremely fattening, but apt to be suppressed by coarser grasses on ground, where these are not trampled out or kept down by pasture-animals. One of the best of summer-grasses, resisting also some frost.

Buddlea Asiatica, Loureiro.

Continental and .insular India up to 7,000 feet, thence to China, the Philippines and New Guinea. Shrub, attaining to 15 feet; eligible for ornamental copses; flowers produced in profusion, sweet-scented.

Buddlea Colvillei, J. Hooker and Thomson.

Himalayan mountains at elevations of 9,000 to 12,000 feet. One of the hardiest of all species, attaining a height of 20 feet, but not so quick of growth as some other kinds. Extremely handsome, with its masses of dark-crimson flowers (Gamble).

Buddlea globosa, Lamarck.

Andes of Chili and Peru. Withstands the winter-cold of Arran. The author has in this as in a very few other instances broke through the rule, adopted for this work, not to accept into it any plant on ornamental value alone; but on the present species almost all the praise of B. Madagascariensis can also be bestowed, and it is really useful likewise for screening unsightly fences quickly and also for other kinds of hedgings. A similar species from the same geographic region is B. connata (Ruiz and Pavon).

Buddlea macrostachya, Bentham.

Mountains of India from 3,000 to 7,000 feet. Shrub with fragrant spikes; flowers with yellowish tube, bluish limb and orange-colored throat. Highly desirable for æsthetic culture with several congeners.

Buddlea Madagascariensis, Lamarck.

Madagascar, Bourbon and Rodriguez. Of the numerous species of Buddlea the most eligible one for shelter-copses on account of its great size and always tidy appearance, as well as vigor and celerity of growth. It is ever-flowering, highly elegant, tolerant to many kinds of soil, bears some frost and like most congeners is easily propagated from cuttings in the open air.

Buddlea paniculata, Wallich. (*B. crispa*, Bentham.)

Himalayan mountains at altitudes between 4,000 and 7,300 feet, extending to Ava, Cabool and Beloochistan. Attaining the size of a small tree, resisting English winters unharmed, merely protected by a wall; flowering there already from the beginning of February till May, scenting the atmosphere around with fragrance (Sir Will. Hooker). The trusses of flowers resemble those of the Lilac in shape and color.

Bursera elemifera, J. Hooker.

Mexico, up to the temperate plateau. This tree furnishes the Mexican Copal or Elemi.

Butea frondosa, Roxburgh.

The "Dhak' or Pulas" of India. This magnificent tree extends to the Himalayan mountains, ascending to elevations of 4,000 feet, and bears a few degrees of frost. It is very rich in a peculiar kind of kino, which according to Muspratt contains up to 73 per cent. of tannin. The Lac-insect is also nourished by this tree.

Butomus umbellatus, Linné.

The "Flowering Rush." Europe, Northern and Middle Asia. This elegant perennial water-plant is mentioned here more for its value in embellishing our lakes and water-courses than for the sake of its roots. The latter, when roasted, are edible. The plant would live in sub-alpine rivulets. In Norway it is hardy to lat. 59° 55′ (Schuebeler).

Buxus microphylla, Siebold and Zuccarini.

Japan. There used for the best of wood-engravings and turnery; considered as good as ordinary box-wood. Native name, Tsougné (E. Dupont).

Buxus sempervirens, Linné.*

The Turkey Box-tree. England, Southern Europe, Northern Africa, South-Western Asia, extending to Upper India and Japan. This slow-growing tree should be planted, to provide the indispensable box-wood for wood-engravers and instrument-makers, no good substitute for it having been discovered as yet. It is also employed for shuttles, rollers and various other select implements, clarionets, flutes, flageolets. Box-wood on account of its extreme density can best be used as an unit in comparative scales of the closeness of various kinds of wood. The box-tree needs calcareous soil for its best development. In Norway it is hardy to lat. 63° 26', according to Prof. Schuebeler, who saw a plant 11 feet high and 6 inches in stem-diameter in lat. 58° 58'. Among allied species B. Balearica attains a height of 80 feet. Other congeners are B. subcolumnaris, B. Cubana, B. Purdieana, B. citrifolia, B. acuminata, B. lævigata, B. Vahlii, B. gonoclada, B. retusa, B. glomerata, B. Wrightii, all from the West Indies; B. Madagascarica, also B. longifolia from Turkey and B. Wallichiana from the Himalayas. It does not however appear to be known, how the wood of any of these, nor of the various species of the Indian genus Sarcococca or the genus Styloceras of the Andes compares with true box-wood; nor is it known, whether or not they are of much more rapid growth.

Cæsalpinia Bonduc, Roxburgh. (*Guilandina Bonduc,* Linné.)

Widely dispersed through the inter-tropical regions of both hemispheres with G. Bonducella, L. These two species would be well adapted for hedges in the warmer parts of the temperate zone.

Cæsalpinia brevifolia, Bentham. (*Balsamocarpon brevifolium,* Clos.)

Chili, the "Algorobillo." The pods of this shrub are extraordinarily rich in tannic acid. Godeffroy found in the husks 68½ per cent. The process of tanning with these pods is accomplished in one-third of the time, required for leather from oak-bark; this material is also especially valuable as giving a bloom to the leather.

Cæsalpinia coriaria, Willdenow.

Wet sea-shores of Central America. Might be naturalized in salt-marshes elsewhere. Colonel Drury states, that each full-grown tree produces annually about 100 lbs. of pods, the husk of which, commercially known as Divi-Divi, is regarded as the most powerful and quickly acting tanning material in India. The mercantile price of the pods is from £8 to £13 per ton.

Cæsalpinia crista, Linné.

West-Indies and Carolina. This shrub or tree furnishes a yellow dye-wood.

Cæsalpinia echinata, Lamarck.

Brazil. The Fernambuc-wood or Red Brazil-wood is obtained from this tree and allied species; they also furnish the dye-principle brazilin.

Cæsalpinia Gilliesii, Wallich. (*Poinciana Gilliesii,* Hooker.)

La Plata-States. This beautiful hardy bush can be utilized for hedges.

Cæsalpinia Sappan, Linné.

South-Asia. The wood furnishes a red dye. This shrub can also be adopted as a hedge-plant.

Cæsalpinia sepiaria, Roxburgh.

Southern Asia, east to Japan. There often utilized as a hedge-bush. It can advantageously be mixed for hedge-growth with *Pterolobium lacerans* (R. Br.), according to Dr. Cleghorn. It furnishes a red dye-wood.

Cæsalpinia tinctoria, Humboldt.

Chili. The bark yields a red dye.

Cæsalpinia vesicaria, Linné. (*C. bijuga*, Swartz.)

West-Indies, on dry savannas and limestone-rocks. This tree furnishes part of the red Fernambuc-wood of commerce, for dye-purposes and select implements.

Cajanus Indicus, Sprengel.*

The Catjang ; in Assam called Gelooa-mah, often also Arhar. A shrubby plant of tropical Africa and perhaps Asia, but ascending to 6,000 feet in the extra-tropical latitudes of the Himalayas. One of the upland-varieties will endure a few degrees of frost (C. B. Clarke). It sustains itself on dry ground, and yields the pulse known as Dhal, Urhur and Congo-Pea. The plant lasts for about three years, attains a height of 15 feet, and has yielded in the richest soil of Egypt 4,000 lbs. of peas to the acre. A crop is obtained in the first year. The seeds can be used as peas in the green state as well as when ripe. Even more utilized in India than Phaseolus radiatus and Cicer arietinum. Some of the tribes of Central Africa use the stem of this shrub in friction with reeds, to strike fire, according to Speke. Several species of Cajanus of the Atylosia-section, partly indigenous to the warmer parts of Australia, might be tested for the sake of the economic value of their seeds. The insect, active in the formation of Lac, lives extensively on the Cajanus, according to Mr. T. D. Brewster of Assam. Silkworms also live on it.

Cakile maritima, Scopoli.

Europe, North-Africa, North- and South-America, extra-tropical Australia. Not unimportant for aiding to cover drift-sand, cast up on low sea-shores; not hurt by the spray. Regarded as antiscorbutic. In Norway hardy to lat. 71° 7′ (Schuebeler).

Calamagrostis longifolia, Hooker.

North-America. Excellent for fixing drift-sand. C. Epigeios (Roth) and C. Halleriana (De Candolle) serve the same purpose according to Wessely.

Calamintha Nepeta, Hoffmansegg.

From England to the countries around the Mediterranean Sea; fond of limestone-soil. It is strongest in odor among several species, but not of so pleasant a scent as C. incana, Boissier, and C. grandiflora, Moench.

Calamintha officinalis, Moench.

Middle and Southern Europe and Middle Asia, Northern Africa. A perennial herb, used like Melissa as a condiment.

Calamus montanus, T. Anderson.

Himalaya, up to 6,000 feet. A hardy climbing palm. The old canes are naked. The light but strong suspension-bridges, by which the large rivers of Sikkim are crossed, are constructed of this palm. It supplies material for the strongest ropes, to drag logs of wood from the forest. Most durable baskets and cane-work of chairs are manufactured from the slit stems. Walking sticks and riding canes made of this species are exported from Sikkim in considerable quantity. Many other Calami serve similar purposes, but probably few, or perhaps none, are equally hardy.

Callitris arborea, Schrader. (*Widdringtonia juniperoides,* Endlicher.)

Southern Africa, 3,000 to 4,000 feet above sea-level. A middle-sized tree, rich in resin.

Callitris calcarata, R. Brown. (*Frenela Endlicheri,* Parlatore.)

Eastern Australia. A tree to 70 feet in height and 3 feet in stem-diameter. Spreads readily over sterile, particularly sandy land. Wood of local importance, strong, durable, and of agreeable smell. It is used for piles, and furnishes planks for boats; it is repugnant to termites. The knotty portion of the stem valuable to cabinet-makers for inlaying. Market price about £5 per 1,000 feet.

Callitris columellaris, F. v. Mueller.

Eastern Australia, on bare and sandy coast-tracts. Height reaching 100 feet. Timber durable, fine-grained, fragrant, capable of a high polish; used for piles of wharves and sheeting of punts and boats; it resists the attacks of chelura and white ants; the knotty portion valued for veneers. The young stems are liked for telegraph-poles according to Mr. Thozet. Present market value of timber £6 per 1,000 superficial feet. (Queensland Exhibition, 1878.)

Callitris Macleayana, F. v. Mueller. (*Frenela Macleayana,* Parlatore.)

New South Wales. A handsome tree, of regular pyramidal growth, attaining a height of 70 feet; the timber is valuable. C. actinostrobus and C. acuminata from South-West Australia are too small for timber-purposes, but the first-mentioned is one of the very few conifers fit for saline soil.

Callitris Parlatorei, F. v. Mueller.

Southern Queensland. Recommended by Mr. F. M. Bailey as a shade-tree. It attains a height of 60 feet. The wood is esteemed by cabinet-makers. Several other species of Callitris are worthy of forest-culture.

Callitris quadrivalvis, Richard.

. North-Africa. A middle-sized tree, yielding the true sandarac-resin. Tables, made of the mottled butt-wood, fetched fabulous prices already at Plinius's time. (J. St. Gardner.)

Callitris verrucosa, R. Brown. (*Frenela verrucosa,* A. Cunningham.)

Through the greater part of Australia. Stems used for telegraph-posts (C. Moore). Wood obnoxious to insects, hard, heavy, light-colored, pleasantly scented, fit for furniture and flooring. This tree disseminates itself with great ease in sandy soil, and will succeed in the driest clime. Planks 2 feet wide can be obtained. Dromedaries, according to Giles, browse on the foliage. Some of the other species are also among the trees, which may be utilized for binding the coast- and desert-sand. They all exude Sandarac.

Calodendron Capense, Thunberg.

Eastern South-Africa. A large and handsome tree, called the Wild Chestnut-tree by the colonists. Particularly fit for promenades. Rate of growth in height at Port Phillip, where it was first brought by the writer, about $1\frac{1}{2}$ foot in a year. Fresh seeds readily germinate.

Calyptranthes aromatica, Saint Hilaire.

South-Brazil. The flower-buds of this spice-shrub can be used almost like cloves, the berries like allspice. Several other aromatic species are eligible for test-culture.

Calyptronoma Swartzii, Grisebach.

West-Indies. A palm, reaching a height of 60 feet. Ascends on tropical mountains to over 3,000 feet elevation. It yields the " long thatch " of Jamaica, the foliage furnishing an amber-colored roofing material, neater and more durable than any other used on that island, lasting twenty years or more without requiring repairs (Jenman). The generic name Calyptrogyne takes precedence.

Camelina sativa, Crantz.

Middle and Southern Europe, temperate Asia. An annual herb, cultivated for the sake of its fibre and the oil of its seeds. It is readily grown after cereals, yields richly even on poor soil, and is not attacked by aphides. Mr. W. Taylor obtained 32 bushels of seed from an acre, and from this as much as 540 lbs. of oil. The return is obtained within a few months. Hardy in Norway to lat. 70° (Schuebeler).

Camellia Japonica, Linné.

This renowned horticultural plant attains a height of 30 feet in Japan. It is planted there on roadsides for shelter, shade and orna-ment (Christie). The wood is used for superior xylography (Dupont). The seeds, like those of C. Sasanqua (Thunberg), are available for pressing oil. C. reticulata (Lindley) from China is conspicuous for its very large flowers, attaining sometimes 20 inches in circumference. Like C. Japonica it is hardy at Arran, flowering there far more freely (Rev. D. Landsborough). In England a very large plant of C. reticulata, reared in Mr. Byam Martin's conservatory, had in October 1848 removed from it 2,600 flower buds, to allow for April 1849 about 2,000 flowers to come to perfection (Sir W. Hooker).

Camellia Thea, Link.* (*Thea Chinensis*, Linné.)

The Tea-shrub of South-Eastern Asia, said to be indigenous also to some localities of Japan, for instance Suruga, traced as spontaneous as far as Manschuria (Fontanier). This evergreen and ornamental bush has proved hardy in the lowlands at Melbourne, where in exposed positions it endures quite unharmed light night-frosts as well as the free access of scorching summer-winds. But it is in humid valleys, with rich alluvial soil and access to springs for irrigation, that the most productive tea-fields can be formed. The plant comes into plentiful bearing of its product as early as the Vine and earlier than the Olive. Its culture is not difficult, and it is singularly exempt from fungus-diseases, if planted in proper localities. Pruning is effected in the cool season, in order to obtain a large quantity of small tender leaves from young branches. Both the Chinese and Assam tea are produced by varieties of a single species, the tea-shrub being indigenous in the forest-country of Assam also. Declivities are best adapted and usually chosen for tea-culture, particularly for Congo, Pekoe and Souchong, while Bohea is often grown in flat countries. In Japan the tea-cultivation extends to 43° north latitude, where the thermometer occasionally sinks to 16° F. (Simmonds), and where in winter time the ground is frozen several inches deep for weeks (General W. G. Le Duc). The Chinese-variety has withstood the winter of Washington in sheltered positions without protection (W. Saunders). The Assam-variety succumbs to frost. For fuller details Fortune's work, "The Tea-Districts of China" might be consulted. The very troublesome Tea-bug of Asia is Helopeltis theivora. Fumigation and the application of birdlime are among the remedies to cope with this insect. The third volume of the Journal of the Agricultural and Horticultural Society of India is mainly occupied by Lieut.-Colonel Edw. Money's and Mr. Watson's elaborate essays on the cultivation and manufacture of tea in India. For more advice on the culture and preparation of tea consult also the writer's printed lecture, delivered in 1875 at the Farmers' Club of Ballarat, further the Report of the Commissioner for Agriculture, Washington, 1877, pp. 349–367, with illustrations; also Bernays's Cultural Industries for Queensland, pp. 181–190.

The tea of commerce consists of the young leaves, heated, curled and sweated. The process of preparing the leaves can be effected by steam-machinery. Already in 1866 three machines for dressing tea were patented in England—one by Messrs. Campbell and Burgess, one by Mr. Thomson, and one by Mr. Tayse. To give an idea of the quantity

of tea, which is consumed at the present time, it may be stated, that
from June to September 1871, 11,000,000 lbs. of tea were shipped
from China alone to Australia, and that the produce of tea in India
from January to June of 1872 was 18,500,000 lbs. In 1840 India sent
its first small sample of tea to the European market, but in 1877
exported to England forty million pounds, that is, as much as the
whole English importation thirty years ago (Burrell). Ceylon alone
exported already in the commercial year 1882-3 one and a half million lbs. of tea. Dr. Scherzer estimates the Chinese home-consumption
at 400,000,000 lbs., others much higher. In 1873 China exported
242,000,000 lbs., Japan, 12,000,000 lbs. Simmonds calculates the
area under tea cultivation in China at 25,000,000 acres. In 1884
Great Britain imported 215,000,000 lbs. of tea, valued at ten and a
half million pounds sterling; of this quantity 66,000,000 lbs. came
from India, after such a comparatively short time of culture. 100 lbs.
of prepared tea is the average yield per acre. Seeds of the tea-bush
are now locally to be gathered in many parts of Australia from plants
distributed by the writer since 1859; and for years to come the cultivation of the tea-bush, merely to secure local supplies of fresh seeds,
ready to germinate, will in all likelihood prove highly lucrative. Tea
contains an alkaloid, caffein, a peculiar essential oil and Bohcic acid,
along with other substances.

Canavalia gladiata, De Candolle.*

Within the tropics of Asia, Africa and America. This perennial
climber grows to an enormous height, and bears an abundant crop of
large edible beans, which can be used green (Sir Walter Elliott). It
varies with red and white seeds, and in the size of the latter, which
are said to be wholesome. C. ensiformis (D.C.) is another variety.
C. obtusifolia is deleterious.

Canna Achiras, Gillies.

Mendoza. One of the few extra-tropic Cannas, eligible for arrowroot culture.

Canna coccinea, Roscoe.

West-Indies. Yields, with some other Cannas, the particular
arrowroot called Tous Les Mois.

Canna edulis, Edwards.*

The Adeira of Peru. One of the hardiest of arrowroot-plants.
Seeds will germinate even when many years old. Plants, supplied at
the Botanic Garden of Melbourne, have yielded excellent starch at
Melbourne, Western Port, Lake Wellington, Ballarat and other
localities in the colony of Victoria. The Rev. Mr. Hagenauer, of the
Gippsland Aboriginal Mission-station, obtained over one ton from an
acre. The Rev. Mr. Bulmer found this root to yield 28 per cent. of
starch. The gathering of the roots is effected there about April.
The plants can be set out in ordinary ploughed land. Starch grains

remarkably large. This Canna resembles a banana in miniature, hence it is eligible for scenic plantations; the local production in Gippsland is already large enough to admit of extensive sale.

Canna flaccida, Roscoe.

Carolina. Probably also available for arrowroot, though in the first instance, like many congeners, chosen only for ornamental culture.

Canna glauca, Linné.

One of the West-Indian Arrowroot-Cannas.

Cannabis sativa, C. Bauhin.*

The Hemp-plant, seemingly indigenous to various parts of Asia, as far west as Turkey and as far east as Japan, recorded recently by Dr. A. v. Regel as naturally also wild in Turkestan; A. de Candolle gives Dahuria and Siberia as the native country. Long cultivated for its fibre. It exudes the churras or hasheesh, a medicinal resinous substance of narcotic properties, particularly in hot climates. The foliage also contains a volatile oil, while the seeds yield by pressure the well-known fixed hemp-oil. The staminate plant is pulled for obtaining fibre in its best state immediately after flowering; the seeding plant is gathered for fibre at a later stage of growth. Good soil, well-drained, never absolutely dry, is needed for successful hemp-culture. Hemp is one of the plants yielding a full and quick return within the season. The average summer-temperatures of St. Petersburg (67° F.) and of Moscow (62° F.) admit still of the cultivation of this plant. The Hemp-plant serves as a protection against insects on cultivated fields, if sown along their boundaries. The seeds are sometimes used in medicine, and are a favorite fodder for various cagebirds. The importations of Hemp into the United Kingdom in 1884 were 1,335,000 cwt., worth over two million pounds sterling.

Canella alba, Murray.

West-India and Florida. An evergreen tree, to 50 feet high, aromatic in all its parts; the bark particularly used, less in medicine than as a condiment.

Cantharellus edulis, Persoon. (*C. cibarius,* Fries.)

The Chantarelle. Various parts of Europe, occurring also in South-Eastern Australia and some other parts of the globe. Dr. Goeppert mentions this among the many mushrooms, admitted under Government-supervision for sale in Silesia.

Capparis sepiaria, Linné.

From India to the Philippine-Islands, ascending to cool elevations and living in arid soil. A prickly bush, excellent for hedges. Dr. Cleghorn mentions also as hedge-plants C. horrida (L. fil.), C. aphylla (Roth), C. Roxburghii (D.C.), some of which also yield capers.

Capparis spinosa, Linné.

The Caper-Bush. Southern Europe and Northern Africa, Southern Asia and Northern Australia. A somewhat shrubby and trailing plant, deserving already for the sake of its handsome flowers a place in any garden. It·sustains its life even in arid deserts. Light frosts do not destroy this plant; the soil requisite for greatest productiveness should be of calcareous clay. The flower-buds and young berries, preserved in vinegar with some salt, form the capers of commerce. Samples of capers, prepared from plants of the Botanic Garden of Melbourne, were placed already twenty years ago in our Industrial Museum, together with many other products, emanating from the writer's laboratory. The Caper-plant is propagated either from seeds, or suckers, or cuttings; it is well able to withstand either heat or drought. The buds, after their first immersion in slightly salted vinegar, are strained and afterwards preserved in bottles with fresh vinegar. In sheltered plains of Provence annually about 1,760,000 lbs., worth at an average 7d. per pound, are collected. The shrub comes into full bearing at the fifth year, the harvests continuing well for many years afterwards (Masters). ·Chemical principle: Rutin.

Capsicum annuum, Linné.

Central America. An annual herb, which yields the Chillies, and thus also the material for cayenne-pepper. Chemical principle: the acrid, soft, resinous capsicin. Comes to seeding in Christiania still.

Capsicum baccatum, Linné.

The Cherry-Capsicum. A perennial plant. Brought from Brazil to tropical Africa and Asia, where other pepper-capsicums are likewise now naturalized.

Capsicum frutescens, Linné. (*C. fastigiatum,* Blume.)

Tropical South-America. The berries of this shrubby species are likewise converted into cayenne-pepper.

Capsicum longum, De Candolle.

Some of the hottest parts of America. An annual herb, also yielding cayenne-pepper. C. grossum (Willd.) is also mentioned by Colonel Drury as a very pungent species. The summers of the warm temperate zone admit of the successful growth of at least the annual species of Capsicum in all the lowlands. C. humile also binds sand even when brackish.

Capsicum microcarpum, De Candolle.

South-America. It is this species, which is used by preference in Argentina. There are annual and perennial varieties.

Caragana arborescens, Lamarck.

The Pea-tree of Siberia, reaching to 70° North. The seeds are of culinary value, but particularly used for feeding fowls. The leaves yield a blue dye (Dr. Rosenthal).

Carex arenaria, Linné.

Western Europe and Northern Asia. Hardy to lat. 62° 30' in Norway (Schuebeler). One of the most powerful of sedges for subduing rolling sand, its rigid foliage not attracting grazing animals. The roots are of medicinal value.

Carex Moorcroftiana, Falconer.

The Loongmur of the Alps of Thibet. One of the best of sedges for fixing the shifting sand by its deeply penetrating and creeping roots. It forms an intricate net-work on the surface and beneath. Outliving most other fodder-plants at its native places, it becomes available for cattle- and horse-food—particularly in the cold of winter, and is held to be singularly invigorating to pasture-animals.

Carica Candamarcensis, Morren.

Andes of Ecuador up to an elevation of about 9,000 feet (Prof. Jameson). A small, slender tree. Fruit to nine inches long and sometimes nearly as broad, edible and wholesome, of delicious scent and grateful taste (Sir Jos. Hooker). Other large-fruited Caricas occur in andine regions, comprised under the vernacular name Camburu (Spruce). Their cultural rearing seems possible in frostless regions only. Used raw or cooked.

Carica Papaya, Linné.

West-Indies and Mexico to Peru. Cultivated northward still in some parts of Florida, elsewhere to 32° N. (A. de Candolle). The Papaw-tree. A small branchless tree of short vitality, only fit for regions not subject to frost. Fruit generally of the size of a small melon; eaten boiled or preserved in sugar or pickled in vinegar (Sir James Smith). Fresh seeds germinate readily. The acrid milky juice of the tree, much diluted with water, renders any tough meat, washed with it, tender for cooking purposes by separating the muscular fibres (Dr. Holder). Medicinally the juice has been administered as a vegetable pepsin and as an anthelmintic. Fruits ripen successively.

Carissa Arduina, Lamarck.

South-Africa. A shrub with formidable thorns, well adapted for boundary-lines of gardens, where rapidity of growth is not an object. Quite hardy at Melbourne. C. ferox (E. Meyer) and C. grandiflora (A. de Cand.) are allied plants of equal value. The fruit of the latter largely used for jam. The East Australian C. Brownii (F. von Mueller) can be similarly utilized. The flowers of all are very fragrant. C. Carandas (Linné) extends from India to China; its berries are edible; it is also a strong hedge-plant.

Carpinus Americana, Michaux.

The Water-Beech or Ironwood of North-America, thriving best on the margins of streams. The wood is fine-grained, tough and compact, used for cogs of wheels and any purpose, where extreme hardness is

F

required, such as yokes (Robb). It is often speckled and somewhat curled, thus fitted for superior furniture (Simmonds). C. Caroliniana (Walter) is the oldest name. Very closely allied to the following.

Carpinus Betulus, Linné.

The Hornbeam. Middle and Southern Europe and Western Asia. A tree to 80 feet high. Wood pale, of a horny toughness and hardness, close-grained, but not elastic. It is used for wheel-wrights' work, for cogs in machinery and for turnery (Laslett). It furnishes a good coal for gunpowder. This tree would serve, to arrest the progress of bush-fires, if planted in copses or hedges, like willows and poplars, around forest-plantations. In Norway it is hardy to lat. 63° 26' (Schuebeler). Four species occur in Japan : C. cordata, C. erosa, C. laxiflora and C. Japonica (Blume). Carpinus viminea (Wallich) is a species with durable wood from the middle regions of Nepal.

Carthamus tinctorius, Linné.

From Egypt to India. The Safflower. In Norway grown to lat. 70° 22' North. A tall, annual, rather handsome herb. The florets produce yellow, rosy, ponceau and other red shades of dye, according to various admixtures. Pigment principles : carthamin and carthamus-yellow. For domestic purposes it yields a dye ready at hand from any garden. In India the Carthamus is also cultivated for the sake of the oil, which can be pressed from the seeds.

Carum Ajowan, Bentham. (*C. Copticum*, Bentham.)

From the countries around the Mediterranean Sea to India. The fruits of this annual herb form an excellent culinary condiment with the flavor of thyme. Its peculiar oil is accompanied by cymol and thymol.

Carum Bulbocastanum, Koch.

Middle- and South-Europe, North-Africa, Middle Asia, on lime-stone soil, extending in Cashmere to 9,000 feet elevation. The tuberous roots and also the leaves serve as a culinary vegetable ; the fruits as a condiment.

Carum Capense, Sonder.

South-Africa, where the edible, somewhat aromatic root is called Fenkelwortel.

Carum Carui, Linné.

The Caraway-Plant. Perennial. Europe, Northern and Middle Asia. Extends in Norway to. lat. 71° 7'. A wholesome adjunct, if interspersed among the herbs of sheep-pastures. It might be naturalized even on our Alps, and also along the sea-shores. The Caraway-oil is accompanied by two chemical principles : carven and carvol. Among the many other purposes, for which it is employed, is that of entering into the scents of soaps and cheap essences of per-fumery (Piesse). The seeds will keep three years (Vilmorin). On rich soil, in Essex, as much as 20 cwt. seeds on an acre have been produced (G. Don). Royle mentions two varieties or allied plants from Upper India.

Carum ferulifolium, Koch. (*Bunium ferulifolium*, Desfontaines.)

A perennial herb of the Mediterranean regions. The small tubers are edible.

Carum Gairdneri, Bentham.

Western North-America, particularly in the Sierra Nevada. A biennial herb, the tuberous root of which furnishes an article of food as well as the root of the allied Californian C. Kelloggii (A. Gray). Geyer probably had this plant in view, when he mentions the tubers of an umbelliferous plant, which are among the dainty dishes of the nomadic Oregon-natives. The truly delicious root bursts on being boiled, showing its snowy white farinaceous substance, which has a sweet cream-like taste and somewhat the aroma of parsley-leaves (Lindley).

Carum Petroselinum, Bentham. (*Apium Petroselinum*, Linné.)

The Parsley. South-Europe, North-Africa and Orient. This biennial, well-known culinary herb is always desirable on pastures as a preventive or curative of some kidney- and liver-diseases of sheep, horses and cattle. In Norway it is hardy to lat. 70° (Schuebeler). The root is also valuable for the table. The essential oil of the fruits contains a peculiar stearopten. Mr. J. W. Fedarb had individual plants of a very curly variety growing uninterruptedly for nineteen years without seeding; pieces taken from them grow readily, continuing the variety.

Carum Roxburghianum, Bentham.

Southern Asia, where it is extensively cultivated for curries, particularly in North-Western India (Atkinson).

Carum segetum, Bentham. (*Anethum segetum*, Linné.)

Around the Mediterranean Sea, extending to Western Europe. An aromatic annual herb, available for culinary purposes.

Carya alba, Nuttall.*

The Shagbark-Hickory and Shellbark-Hickory also. Eastern North-America, extending to Canada and Carolina. Professor Schuebeler found it to be hardy in Norway to lat. 63° 52'. A deciduous tree, reaching a height of 90 feet, which delights in rich forest-soil. Wood heavy, strong, elastic and tenacious, but not very durable; used for chairs, agricultural implements, carriages, baskets (Sargent), whip-handles and a variety of other purposes. Yields the main supply of hickory-nuts. All the hickories are extensively used in North-America for hoops. Circumference of stem 2 feet above ground 30 inches after 24 years at Nebraska (Furnas).

Carya amara, Nuttall.

The Bitternut-Tree or Swamp-Hickory. Eastern North-America, extending to Georgia and Texas. A tree, sometimes 80 feet high. Wood less valuable than that of other hickories. Richest of all North-American trees in potash, in which most hickories abound. Hardy at Christiania. The flowers of all the Caryas yield much honey (Damkoehler).

Carya glabra, Torrey.* (*Carya porcina*, Nuttall.)

The Hog-nut-Tree. Eastern North-America, reaching Canada and Florida. Often to 80 feet high. Wood very tough; the heart-wood reddish or dark-colored; much used for axletrees and axehandles. Rate of stem-growth in Nebraska, 38 inches circumference in 24 years (Furnas).

Carya microcarpa, Nuttall.

The Balsam-Hickory. Eastern North-America. A fine lofty tree, attaining a height of 80 feet, with a stem 2 feet in diameter. The wood is pale and tough, and possessed of most of the good qualities of C. tomentosa, to which this species is also in other respects allied. Also very closely related to C. alba. The nut is of pleasant taste, but small (Nuttall).

Carya oliviformis, Nuttall.*

The Pacan or Pecannut-Tree of Eastern North-America, extending to .Texas. A handsome tree, reaching 70 feet in height, with a straight trunk. The most rapid growing of all the hickories (Meehan). Its wood is coarse-grained, heavy and compact, possessing great durability ; in strength and elasticity it surpasses even that of the White Ash (Harrison). The nuts are usually abundant, and the most delicious of all walnuts ; they form an article of considerable commerce in the Southern States. Texas annually exports nuts to the value of over £10,000 (Dr. C. Mohr). The tree matures fruits as far north as Philadelphia. It commences to bear in about eight years. The fresh nuts should be packed in dry moss or sand into casks for distant transmission. Although the wood of all the hickories is not well adapted for building purposes, as it is subject to the attacks of insects and soon decays if exposed to the weather, yet its great strength and elasticity render it extremely useful for implements, articles of furniture, hoops and many minor purposes, besides supplying locally the very best of fuel. Hickories, even when very young, do not well bear transplanting, C. amara perhaps excepted. C. alba and C. glabra would be particularly desirable for the sake of their timber, and C. oliviformis on account of its fruit. The bark of all the hickories contains yellow dye-principles ; by the addition of copperas an olive color is produced ; by the addition of alum, a green color. Hickory-stems are known to attain 12 feet in girth.

Carya sulcata, Nuttall.*

The Furrowed-Hickory and the Shellbark-Hickory of some districts ; also one of the Shagbark-Hickories. North-America, in the

Eastern States. A tree, to 80 feet high in damp woods. Its rate of growth is about 18 inches in a year, while young. Heart-wood pale-colored. Seed of sweet pleasant taste. Wood similar to that of C. alba, but paler. The tree is still hardy in Christiania.

Carya tomentosa, Nuttall.*

The Mockernut-Tree or White-Heart Hickory. Eastern North-America, extending to Canada. A large tree, likes forest-soil, not moist. Heart-wood pale-colored, remarkable for strength, elasticity, heaviness and durability, yet fissile; used for axles, spokes, felloes, handles, chairs, screws, sieves and the best of mallets; the saplings for hoops and wythes. Hickory is the most heat-giving amongst all North-American woods. Nut small, but sweet; very oily. A variety produces nuts as large as a small apple, which are called King-Nuts.

Caryota urens, Linné.

India. One of the hardier Palms, ascending the Himalayas to an altitude of 5,000 feet, according to Dr. Thomas Anderson, yet even there attaining a considerable height, though the temperature sinks in · the cooler season to 40° F. Drude mentions, that species of this genus ascend to an elevation of 7,500 feet, where the temperature occasionally approaches the freezing point. The trunk furnishes a sago-like starch. This palm flowers only at an advanced age, and after having produced a succession of flowers dies away.. From the sap of the flower-stem, just as from that of the Cocos- and Borassus-Palm, toddy and palm-sugar are prepared, occasionally as much as 12 gallons of liquid being obtained from one tree in a day. The fibre of the leaf-stalks can be manufactured into very strong ropes, also into baskets, brushes and brooms. It also serves the Indian races as tinder. The outer wood of the stem answers for turnery. Several allied species exist, one extending to North-Eastern Australia.

Casimiroa edulis, Llav and Levarz.

Mexico, up to the cool heights of 7,000 feet. This tree comes into bearing in about ten years. The kernel of its fruit is deleterious (Hernandez), but the pulp of a delicious, melting, peach-like taste (Garner), partaking of which is said to induce sleep. The tree thrives well in a clime like that of Santa Barbara, California. The fruit is about an inch in diameter, pale-yellow, of a rich subacid taste, and most palatable when near decay. Efforts to propagate it from cuttings were not successful, and seeds do not seem to reach perfection in California. The Spanish inhabitants call the tree Zapote (Calif. Hortic. Magaz. 1880).

Cassia acutifolia, Delile.

Indigenous or now spontaneous in Northern and Tropical Africa and South-Western Asia. Perennial. The leaflets merely dried constitute part of the Alexandrian- and also Tinnevelly-senna. The active principle of senna—namely, cathartic acid—occurs also in the Coluteas and in Coronilla varia, according to C. Koch.

Cassia angustifolia, Vahl.

Northern and Tropical Africa and South-Western Asia, indigenous or cultivated. Perennial. Yields Mecca-senna, also the Bombay- and some of the Tinnevelly-senna.

Cassia artemisioides, Gaudichaud.

Sub-tropical and extra-tropical Australia. The species of this series are shrubby and considered valuable for arid and sandy sheep-runs as affording feed. They brave intense heat, and are adapted for rainless regions.

Cassia fistula, Linné.

Southern Asia. The long pods of this ornamental tree contain an aperient pulp of pleasant taste and of medicinal value. It is also used in the manufacture of cake-tobacco. Traced by Sir Jos. Hooker to the dry slopes of the Central Himalayas.

Cassia Marilandica, Linné

An indigenous Senna-plant of the South-Eastern United States of North-America. Perennial.

Cassia obovata, Colladon.

South-Western Asia; widely dispersed through Africa as a native or disseminated plant. Perennial. Part of the Alexandrian Senna and also Aleppo-senna is derived from this plant; less esteemed and less collected than the other species. It furnishes also Tripolis, Italian, Senegal and Tanacca Senna.

Castanea sativa, Miller.* (*C. vulgaris,* Lamarck; *C. vesca,* Gaertner.)

The Sweet Chestnut-tree. South-Europe and Temperate Asia, as far as Japan; a variety with smaller fruit extending to North-America. Professor Schuebeler records, that even in Norway at latitude 58° 15′ a chestnut-tree attained a height of 33 feet with a stem 4 feet in circumference; in a shrubby state it is found as far north as 63°. It reaches an enormous age; at Mount Etna a tree occurs with a stem 204 feet in circumference. At other places trees are found 10 feet in diameter, solid to the centre. The tree does not readily admit of transplantation. The wood is light, cross-grained, strong, elastic and exceedingly durable, well adapted for staves and wheel-cogs, the young wood for hoops and mast-rings. The wood is comparatively rich in tannic acid (about 4 to 6 per cent.), and hence used for preparing a liquid extract; the bark contains 12 per cent. tannin (Wiesner). The leaves furnish food for the Bombyx Jamamai (Dupont). The greatest importance of the tree rests on its adaptability for shade-plantations, its nutritious nuts and timber-value. The American wood is slightly lighter in color than that of the Red Oak, and available for shingles and rails; chestnut-rails in North-America have lasted for half a century. The wood is beautifully laminated (Simmonds), and largely employed for furniture, for the inside finish of railroad-cars and steamboats (Vasey). The American nuts are smaller, but sweeter than the European; they are largely used for fattening hogs (Robb). Rate of stem-growth in Nebraska, 24 inches in 14 years, diametrically (Furnas).

Castanopsis argentea, A. de Candolle.

A lofty tree in the mountains of India, produces also edible chestnuts. Other species of the genus Castanopsis are valuable, thus according to the Rev. B. C. Henry the nuts of the Chinese C. jucunda (Hance) are edible.

Castanopsis chrysophylla, A. de Candolle.

The Oak-Chestnut of California and Oregon. A tree, attaining a height of 150 feet and 8 feet in stem-diameter. Either for beauty or utility worthy of cultivation (Dr. Gibbons). The leaves are golden-yellow underneath. Wood durable, highly prized by joiners, wheelwrights and even shipbuilders (Dr. Kellogg).

Castanopsis Indica, A. de Candolle.

Mountains of India, at about 4,000 feet. This Oak-Chestnut produces seeds with the taste of filberts.

Casuarina Decaisneana, F. v. Mueller.

Central Australia, where it is the only species of the genus. The tree is one of the largest among its congeners, and particularly valuable for arid sandy regions. The wood is exceedingly hard, and resists the attacks of termites and also decay; the stem-wood is straight and easily fissile (Rev. H. Kempe). Dromedaries delight in getting the branchlets of this tree for food (E. Giles).

Casuarina distyla, Ventenat.

Extra-tropical Australia. A shrubby species, well adapted for fixing the sand-drifts of sea-coasts. All Casuarinas can be pollarded for cattle-fodder.

Casuarina equisetifolia, Forster. (*C. litorea,* Rumph.)

Eastern Africa, Southern Asia, tropical and sub-tropical Australia, Polynesia. Attains a maximum height of 150 feet. Splendid for fuel, giving great heat and leaving little ashes. The timber is tough, nicely marked. The tree will live in somewhat saline soil at the edge of the sea. Colonel Campbell-Walker estimates the yield of firewood from this tree as four times as great as the return from any tree of the forests of France. Known to have grown in 10 years to a height of 80 feet, but then only with a comparatively slender stem (Blechyndon). In India the wood is much used as fuel for railway-locomotives ; the tree is there also extensively employed, to reclaim sand-land of the coast, it succeeding in growth down to highwater-mark, throwing often out decumbent branches, which develop roots, further to fix the sand and to throw up independent shoots (Dr. Bidie). It yields a lasting wood for piles of jetties and for underground-work, and is much used for knees of boats and for tool-handles (Wilcox). The cost of raising Casuarinas in India has been from £4 to £10 per acre, and the return, after only eight years, £13 to £32.

Casuarina Fraseriana, Miquel.

South-Western Australia. A middle-sized tree; the wood easily split into shingles. The best furniture-wood of South-Western Australia, as it does not rend. This tree is adapted even for sterile heath-land.

Casuarina glauca, Sieber.

Widely distributed through South-Eastern Australia, even in desert-country, but nowhere forming forest-like masses. This species attains in favorable places a height of 80 feet. Its hard durable wood is valuable; used for staves, shingles and various utensils (Woolls). Important for its rapid growth, for its resistance to exposure, for shelter plantation and its speedy supply of fuel,—a remark which applies to the following species also.

Casuarina quadrivalvis, La Billiardière.

The Coast-Sheoak of South-Eastern Australia. Not living merely in coast-sand, but also on other barren places, reaching the inland-hills. Height attaining 60 feet. The foliage of this species is drooping. The male tree is very eligible for avenues, but the female less slightly. Cattle are fond of the foliage; indeed it is a "stay-by to all kinds of stock" in drought, branches then being lopped from the trees for feed. For arresting the ingress of coast-sand by belts of timber this is one of the most important trees. It produces seed early and copiously like other Casuarinas and is easily raised. The foliage, like that of the other species, is acidulous from a crystallizable substance allied to bicitrate of lime.

Casuarina suberosa, Willdenow.

The erect-branched Sheoak of South-Eastern Australia. Height reaching 40 feet. A beautiful shady species. Casuarina trichodon (Miq.) and C. Huegeliana (Miq.) are arboreous species of South-Western Australia, valuable for their wood.

Casuarina torulosa, Aiton.

New South Wales and Queensland. Attains a height of 70 feet. The tough wood of this handsome tree is in demand for durable shingles and furniture-work, as well as for staves and veneers; it is also one of the best for oven-fuel.

Catalpa bignonioides, Walter.*

Southern States of North-America, extending to Illinois. A tree of rapid growth in warm humid climates, attaining a height of about 20 feet in four years. Professor Meehan observed the stem to attain a diameter of 4 feet in twenty years, even in the latitude of New York. Rate of growth in the clime of Nebraska, as recorded by Governor Furnas, considerably less. In many parts of the United States it is a favorite tree for shade-lines. When closely planted it will grow tall and straight, with a stem fully 50 feet to the first branch. It prefers bottom-lands, but will succeed in almost any soil and position, according to Mr. Barney. It is hardier than most Eucalypts, but will not stand severe frosts. According to Professor Burrill, it is not liable to be destroyed by insects; bears seeds when quite young. Professor Meehan considers the wood to be as durable as that of the best Chestnut-trees; indeed, it lasts for an almost indefinite period. General Harrison insists, that there is nothing like it for posts. Catalpa-pickets of the old French stockade are still sound. Logs thrown across water-courses for crossing have lasted for three generations; railway-posts and platforms of this wood are almost indestructible. Logs a century old, and posts half a century old, were not in the least decayed (Barney). Railway cross-ties made of this wood are also very durable, a tree twenty years old furnishing sufficient timber for four ties. Canoes of Catalpa-wood never crack or decay.

Catalpa Kaempferi, Siebold and Zuccarini.

Japan. Grows in eight years to about 25 feet in height, with a trunk of 2 feet circumference; bunches of flowers very large and fragrant (Hovey). Proved hardy at Christiania (Schuebeler). C. Bungei (Meyer) from North-China, or a closely allied species, can be grown from hedges. Flowers of all Catalpas sought by bees.

Catalpa speciosa, Warder.

In the Mississippi-states. Hardier and taller than C. bignonioides; blooming earlier; leaves inodorous, flowers larger, growth as rapid and wood as durable; also only with a very thin layer of destructible sapwood (Dr. Engelmann). Found to have attained in 40 years a stem-circumference of 40 feet at 4 feet from the ground (Letterman).

Catha edulis, Forskael.

Arabia and Eastern Africa. The leaves of this shrub, under the designation of Kafta or Cat, are used for a tea of a very stimulating effect, to some extent to be compared to that of Erythroxylon Coca. To us the plant would be mainly valuable for medicinal purposes.

Ceanothus rigidus, Nuttall.

California. One of the best of hedge-shrubs, available for dry situations. Evergreen; to 12 feet high; the branches becoming densely intricate. In the coast-tracts it is replaced by C thyrsiflorus (Escholtz), which can also be used for hedges and copses, and will live in mere coast-sand. C. prostratus (Bentham) forms natural mats on slopes made by roads and slides, which it gradually covers, and with its pretty blue flowers soon decorates (Professor Bolander). Irrespective of their beauty, the different species are worthy of cultivation as forming excellent wind-breaks. A fair tea is made from the leaves of C. velutinus (Dr. Gibbons). Some species are relied on as forage-plants.

Cedrela australis, F. v. Mueller.

Eastern Australia, as far south as 36°. The Australian Red Cedar. Foliage deciduous in cool regions. Attains a height of 200

feet and a stem-girth of 18 feet towards the base. Messrs. Danger
and Name measured a tree on the Macleay-River, 48 feet in stem-
circumference at 10 feet from the ground; it yielded 80,000 feet of
sound timber. The Rev. Dr. Woolls noted in New South Wales
trees so large as to yield 30,000 feet (superficial) of timber. Market-
value in Brisbane £7 10s. to £8 10s. per 1,000 superficial feet. The
light, beautiful wood is easily worked and susceptible of high polish;
it is very much in request for furniture, for turnery including
stethoscopes, for the manufacture of pianofortes, for boat-building,
frames of window-blinds and a variety of other joiners' work; thus
it is highly prized for building racing boats, which weigh little over
30 lbs., though 30 feet long, and yet prove durable (S. Edwards).
The timber from the junction of the branches with the stem furnishes
choice veneers. The bark contains a considerable quantity of tannin,
which produces a purplish leather (Fawcett). This tree is hardy
at Melbourne, but of slow growth in open exposed gardens and poor
soil. C. glabra (Cas. de Cand.) and C. microcarpa (C. de Cand.)
yield Cedar-wood in Sikkim, according to Dr. Geo. King. C. serrata
(Royle) grows at higher altitudes, and furnishes a different but also
good timber (G. King).

Cedrela Brasiliensis, A. de Jussieu.*   (*C. fissilis*, Vellozo.)

From Argentina extending to Mexico. The timber is soft,
fragrant and easily worked; it is known as Acajou-wood. The
wood of C. odorata (Linné) from Central America furnishes the
principal material for cigar-boxes there (Laslett). The Surinam
Cedar-wood is furnished by C. Guianensis (A. de Jussieu).

Cedrela febrifuga, Blume.

Java, Sumatra, Timor, in cooler mountain-regions. More closely
allied to C. australis than to C. Toona. A tree, rising finally to a
height of 200 feet. Bark of tonic property. Hasskarl further notes
from Java C. Teysmanni and C. inodora. Cedrelas occur also in
New Guinea.

Cedrela Sinensis, A. de Jussieu.*

China and Japan. An elegant tree, hardy in South-Europe.
It furnishes a wood not unlike that of the Singapore-cedar, reddish
in color, particularly sought for cigar-boxes and similar articles.

Cedrela Toona, Roxburgh.*

The Singapore-cedar. Southern India, ascending the Himalayas
to 8,000 feet. Foliage deciduous. One of the most important of all
timber-trees for furniture-wood, which is easily worked, light,
seasons readily, takes polish well and is applicable for a multitude of
purposes in joinery. Dr. Brandis gives the stem-girth of trees 35
years old as 7 feet, when the tree grew on rich and moist soil; trees
with 30 feet stem-circumference are known.

Cedrela Velloziana, Roemer.

Brazil. A magnificent tree, with odorous wood of a red hue.

Cedronella cordata, Bentham.

Southern States of North-America. A perennial herb, fragrant like the following.

Cedronella triphylla, Moench,

Madeira and Canary-Islands. A shrubby plant with highly scented foliage. The volatile oil obtainable from it resembles that of Melissa, but is somewhat camphoric.

Celtis australis, Linné.

The Lotus-tree of South-Europe, North-Africa and South-Asia,. ascending the Himalayas to 9,000 feet. Attains a height of about 50 feet. Though of rather slow growth, this tree can be used for avenues, as its stem finally reaches to 6 feet in diameter. It is supposed, that this Celtis reaches the age of fully 1,000 years. Berries edible. Wood hard and dense, eligible particularly for turners' and carvers' work. Used also by instrument-makers for flutes and pipes. The stem-wood is fine-grained, easily cleft and of a splendid yellow tinge; the branch-wood is one of the best for whip-sticks.

Celtis occidentalis, Linné.

The Hackberry-tree. Eastern States of North-America. Height reaching to 80 feet. Hardy as far north as Christiana. The sweet fruit edible. Wood elastic and fissile.

Celtis Sellowiana, Miquel.

Argentina. Tree to 40 feet high. Wood strong, used for wagons, posts, turnery. Fruit edible (Hieronymus).

Celtis Sinensis, Persoon. (*C. Japonica,* Planchon.)

China and Japan. The "Henoki." A tree bearing extreme cold. Wood useful for carpenters' and turners' work. Fruit edible, but small.

Celtis Tala, Gillies.

From Texas to the La Plata-States. A thorny shrub, or under favorable circumstances a good-sized tree. This plant can be used for forming impenetrable hedges but also shade-avenues. One or two other Argentine species serve the same purpose.

Cephaelis Ipecacuanha, Richard.

Brazil, in mountain-woods, consociated with Palms and Tree-ferns. It is not unlikely, that this herb, which is perennial and yields the important medicinal ipecacuanha-root, would live in warm extra-tropic forest-regions. Active principles: emetin and ipecacuanha-acid.

Cephalantus occidentalis, Linné.

North-America, extending to Canada and Mexico. A tree, attaining a height of about 50 feet on streams, easily disseminated, flowering profusely already in a shrubby state, more important for

ornamental than for industrial growth, admitted here however as a
rich yielder of honey from its fragrant flowers, which last through
several months. The bitter bark, particularly that of the root, used
therapeutically (Dr. Kellogg). The plant was introduced first by the
writer into Victoria, where it thrives to perfection.

Cephalotaxus drupacea, Siebold and Zuccarini.

China and Japan. This splendid Yew attains a height of 60 feet
and is very hardy. According to Dr. Masters, the C. Fortunei
(Hooker) is merely a variety.

Ceratonia Siliqua, Linné.*

The Carob-Tree, indigenous to the Eastern Mediterranean regions.
It attains a height of 50 feet, and resists drought well; succeeds best on a
calcareous subsoil. Wood pale-reddish. The saccharine pods, Algaroba
or St. John's Bread, of value for domestic animals. In some parts of
South-Europe even used for human food. The frequent unsexuality
of the flowers accounts to some extent for the want of productiveness
in fruit of this plant, where but few plants exist and no bees are
kept. The seeds germinate readily. The exportation of the pods
for cattle-food from Creta is very large. The fruit is used for a
medicinal syrup, an imitation of chocolate and a liqueur (Wittmack).
In some of the Mediterranean countries horses and stable-cattle are
almost exclusively fed upon the pods. The meat of sheep and pigs
is greatly improved in flavor by this food, while its fattening properties are twice those of oil-cake. The pods contain about 66 per
cent. of sugar and gum. To horses and cattle 6 lbs. a day are given
of the crushed pods, raw or boiled, with or without chaff. The
Spanish conquerors took this plant early to Central- and South-
America. The seeds should by geographic explorers be carried
through the central regions of Australia, and be sown on humid spots
particularly in the limestone-formation. Instances are on record of a
tree having yielded nearly half a ton of pods in a season (Chambers).

Ceratopetalum apetalum, Don.

Extra-tropic Eastern Australia. A beautiful tree with long cylindrical stem. Height reaching 90 feet, diameter 3 feet. Wood soft,
light, tough, close-grained, fragrant, good for joiners' and cabinet-
makers' work, locally in request for coachbuilding and therefore
called coach-wood by the colonists.

Cercocarpus ledifolius, Nuttall.

California. Becomes in favorable spots a tree 40 feet in height,
with a stem-diameter of 2½ feet. The wood is the hardest known in
California. It is of dark color, very dense, used for bearings in
machinery (Dr. Gibbons). C. parvifolius is of lesser dimensions.

Cereus Engelmanni, Parry.

Utah. A dwarf species, with large scarlet flowers and fruits of
strawberry-flavor and refreshing taste. C. Lecomtei attains there
the size of a flour-barrel.

Cereus Quixo, Gay.

Chili. This stately Cactus attains a height of 15 feet, and is one of the hardiest species. The charming snow-white flowers are followed by sweetish mucilaginous fruits, available for the table (Philippi). C. giganteus (Engelmann), from New Mexico, which attains the stupendous height of 60 feet, with a proportionate columnar thick-ness, also yields edible fruit, and lives unprotected at Port Phillip, withstanding the sea-air close to the shores, and growing at the rate of nearly a foot a year. It was introduced by the writer many years ago. Columnar species of Cereus rising to a height of 40 feet occur also in Argentina. C. repandus and C. triangularis (Haworth), of the West-Indies and Mexico, together with several other species, are available as hedge-plants in places free from frost. Née speaks of a Mexican Cactus (probably an Echinocactus) five feet in diameter by 3 feet in height.

Cereus Thurberi, Engelmann.

North-Western Mexico and Arizona in arid regions. Attains a height of 20 feet; the fruits vary in size from that of a hen's egg to that of an orange; it is of delicious flavor, pleasant taste and very nutritious.

Cerinthe major, Linné.

Countries around the Mediterranean Sea. A handsome but annual herb, particularly alluded to by G. Don as a honey-plant of superior value. A few congeners exist in the same regions, among which C. minor (Bauhin) is biennial or pauciennial, and C. alpina (Kitaibel) perennial.

Ceroxylon andicola, Humboldt. *

The Wax-palm of New Granada, ascending the Andes to 11,000 feet. One of the most majestic and at the same time one of the most hardy of all Palms, attaining occasionally a height of 180 feet. The trunk exudes a kind of resinous wax, about 25 lbs. being obtainable at a time from each stem; this, after the admixture of tallow, is used for candles. There are several other andine palms, which could be reared in Australian forests or in sheltered positions about our dwellings.

Ceroxylon australe, Martius. (*Juania australis,* Drude.)

Juan Fernandez, latitude 34° South, on the higher mountains.

Ceroxylon Klopstockia, Martius.

Venezuela. This very tall Wax-palm reaches elevations of 6,000 feet.

Cervantesia tomentosa, Ruiz and Pavon.

Forest-mountains of Peru. This tree yields edible seeds. It is likely to prove hardy in lower forest-regions of the warmer extra-tropic countries.

Cestrum nocturnum, Linné.

West-Indies, Southern Mexico. Praised above almost all other
plants for its fragrance in Mexico, its flowers lasting through the
summer and autumn, and their scent being particularly powerful at
night (Dr. Barroeta)'.

Cetraria Islandica, Acharius.

Colder regions of Europe, Asia and North-America. This
renowned lichen, inappropriately called "Iceland-Moss," deserves
translocation to other cold parts of the globe; it yields on boiling a
nutritious jelly, pleasant after removal of the bitter principle (Cet-
rarin), the latter rendering this lichen additionally valuable in medicine.

Chærophyllum bulbosum, Linné.

Middle Europe and Western Asia. The Parsnip-chervil. A biennial
herb. The root a very palatable culinary esculent, three times as
rich in starch as potatoes; to be kept some time before consumed
(Vilmorin).

Chamædora elatior, Martius,

Mexico, at an elevation of 4–5,000 feet. This graceful palm attains
only a height of about 12 feet and bears some frost. With many of
its congeners available for table-decoration. The oldest generic
name is Morenia.

Chamærops excelsa, Thunberg.* (*Trachycarpus excelsus,* Wendland.)

Southern China, as far north as Napong, also in Japan. This
Fan-palm is highly desirable, although not very tall, as the name
would indicate. The hardiest of all palms; has stood 3° F. with
only a slight litter (Count de Saporta). Hardy in the mild middle
coast-regions of England. Cordage prepared from the leaves does
not decay in water (Dupont). Rate of growth while young at Melbourne about 1 foot a year. The hairy covering of the stem. of this
palm and of Livistona Chinensis is utilized for fixing lime-plaster to
buildings in Japan (Christie). C. Fortunei (Hooker), the Chusan-
palm from North-China, is a variety or closely allied species. It
attains a height of about 30 feet, and endures considerable frost. The
leaves can be employed for plaiting palm-hats, the fibrous leaf-sheaths
for making brushes, brooms and cordage. Other hardy palms might
be naturalized and used for various purposes, irrespective of their
ornamental features.

Chamærops humilis, Linné.

The Dwarf Fan-palm of South-Europe, North-Africa and the
most south-western parts of Asia. Height to 20 feet. It is very
ornamental for gardens and plantations, and particularly eligible for
scenic effect. Hats, mats, baskets, fans and brushes are made from
the leaves.

Chamærops Khasyana, Griffith. (*Trachycarpus Khasyanus,* H. Wendland.)

In the Himalayas at elevations of from 4,000 to 8,000 feet, also, according to Kurz, in dry pine-forests of Martaban and Ava.

Chamærops Martiana, Wallich. (*Trachycarpus Martianus,* H. Wendland.)

Ascends the mountains of Nepal to 8,000 feet. This Fan-palm attains a height of 50 feet, and is altogether a noble object. Reaches higher altitudes in the Himalayas than any other species, indeed where snow occurs or covers the soil four or five months during the year.

Chamærops Ritchieana, Griffith. (*Nannorhops Ritchieana,* H. Wendland.)

Arid mountains of Afghanistan; seemingly the only native palm there. Extensively used for cordage; leaves also made into baskets and mats; fruit locally used like dates (Aitkinson). Has proved hardy even in England.

Chelidonium majus, Fuchs.

The Celandine. Europe and Western Asia, wild to latitude 63° N. in Norway. A perennial herb of medicinal value. Chemical principles: chelerythrin and chelidonin; also a yellow pigment, chelidoxanthin.

Chelone glabra, Linné.

North-America. The "Balmony." A perennial herb, which has come into therapeutic use.

Chenopodium ambrosioides, Linné.

Tropical and sub-tropical America. "Mexican Tea" and "Worm-seed." An annual medicinal herb. Chenopodium anthelminticum seems to be a perennial variety of this spécies. Easily naturalized.

Chenopodium auricomum, Lindley.

Australia, from the Darling-River to Carpentaria and Arnhem's Land. A tall perennial herb, furnishing a nutritious and palatable spinage. It will live in arid desert-regions. It is one of the "Blue Bushes" of the squatters, who value it as a nutritive and wholesome pastoral plant. Several other species of Chenopodium, among them the European C. bonus Henricus (Linné), afford fair spinage, but they are annual.

Chenopodium Blitum, F. v. Mueller. (*Blitum virgatum,* Linné.)

From South-Europe to Middle Asia. An annual herb, in use there as a cultivated spinage-plant. The fruits furnish a red dye. The genus Blitum was reduced to Chenopodium by the writer in Caruel's *Nuovo Gionale Botanico* many years ago, and in 1864 by Dr. Ascherson, who gave to B..virgatum the name Chenopodium foliosum. C. capitatum, Ascherson (*Blitum capitatum,* Linné) may not be really a distinct spécies. Nyman regards its nativity unascertained. Some of this group of plants are useful to anglers, attracting fish, when thrown into rivers or lakes.

Chenopodium nitrariaceum, F. v. Mueller.

Interior of Australia, especially in localities occasionally humid, reaching in some places the south-coast. A rather tall " Salt-bush," liked particularly by sheep.

Chenopodium Quinoa, Willdenow.

New Granada, Peru, Chili. An annual herb. Admitted here as a savory and wholesome spinage-plant, which can be grown so quickly, as to become available during the short summers of even the highest habitable alpine altitudes. In Peru the seeds are used for a nutritious porridge (Tschudi, Markham).

Chionachne cyathopoda, F. v. Mueller.

Tropical and Eastern sub-tropical Australia. With C. barbata (R. Brown) and C. Wightii (Munro) of India and Queensland a valuable fodder-grass, yielding a large return. Scleraclne punctata (R. Brown) from Java is closely allied.

Chloris scariosa, F. v. Mueller.

Tropical Australia. Particularly recommended by Mr. Walter Hill as a pasture-grass. Dr. Curl mentions, besides this, C. divaricata (R. Brown), from North- and East-Australia, as useful summer- and autumn-grasses, even in the cooler clime of New Zealand.

Chloris truncata, R. Brown.

The Windmill-grass. South-Eastern Australia, as far south as Port Phillip. This perennial and showy grass is regarded by Mr. Walter Bissill as an excellent summer- and autumn-grass, of ready growth and relish to grazing animals. C. ventricosa (R. Br.) is another valuable East-Australian species. Several other congeners from the eastern or western world deserve the attention of graziers. Chemical analysis will determine their nutritive value, though the degree of liking of such grasses by pasture-animals can only be found out by rural tests.

Chlorogalum pomeridianum, Kunth.

California, frequent on mountains. This lily-like plant attains a height of 8 feet. The heavy bulb is covered with many coatings, consisting of fibres, which are used for cushions and mattresses; contracts are entered into for the supply of this material on a very extensive ·scale (Professor Bolander). The inner part of the bulb serves as a substitute for soap, and the possibility of utilizing it for technological purposes, like the root of Saponaria, might be tested, as it contains saponin.

Chloroxylon Swietenia, De Candolle.

The Satin-wood. Mountains of India. Like the allied Flindersias, possibly this tree would prove hardy in sheltered places of milder extra-tropic latitudes, the cognate Cedrela australis advancing in East-Australia southward to the 36th degree. A resin, valuable for varnishes, exudes from the stem and branches.

Chondrus crispus, Lyngbye.

Shores of the Northern Atlantic Ocean. "Caragaheen." This well-known alg yields a nutritious and palatable gelatine on boiling, and has thus become even of some therapeutic importance. The ready steam-communication all over the world affords doubtless now the opportunity of carrying also highly useful algs widely from shore to shore in portable aquaria. In Australia the Eucheuma speciosum (J. Agardh) and Gelidium glandulifolium (Harvey) are marine jelly-weeds, well deserving of wide translocation.

Chrysanthemum cinerarifolium, Boccone. (*Pyrethrum cinerarifolium*, Trevisan.)

Austria. Furnishes the Dalmatian Insect-powder. It is superior even to the Persian powder as an insecticide; it will keep for years. It is prepared from half-opened flowers during dry weather, and ex-siccated under cover. Best applied in puffs from a tube. To be used also against aphides (W. Saunders). [See further U. S. Agricultural Report for 1881-2.]

Chrysanthemum parthenium, Persoon. (*Pyrethrum parthenium*, Smith.)

Middle and Southern Europe. " Feverfew." The root, foliage and flowers of this perennial herb are in request for medicinal purposes since ancient times; the variety with yellow foliage serves for edging of garden-plots, ribbon- and carpet-culture.

Chrysanthemum roseum, Adam. (*Pyrethrum roseum*, Bieberstein.)

Sub-alpine South-Western Asia. This perennial herb, with C. coronopifolium (Willdenow) yields the Persian Insect-powder.

Chusquea Culeou, E. Desvaux.

Chili, Valdivia, Argentina. This Bamboo exceeds not often 20 feet in height; the autochthones on the La Plata-River use it for lances. C. heterophylla and C. Cumingii (Nees) serve in the same region for thatch-roofing (Hieronymus). C. andina (Philippi) grows in Chili near the snow-line.

Cicer arietinum, Dodoens.

South-Europe and South-Western Asia. The Gram or Chick-Pea. An annual herb, valuable as a pulse for stable-food, but an extensive article also of human diet in India. Colonel Sykes counted as many as 170 seeds on one plant. In Spain, next to wheat, the most ex-tensively used plant for human food (Honorable Caleb Cushing). The seeds can be converted into pea-meal or can be used in various other ways for culinary purposes.

Cichorium Endivia, Linné.

South-Europe, North-Africa, Orient, Middle Asia. A biennial plant, used even in ancient times as a culinary vegetable. In Nor-way it grows to lat. 70° (Schuebeler). The inner leaves are bleached for food by tying the outer leaves together (Vilmorin).

G

Cichorium Intybus, Linné.

Chicory. A well-known perennial plant, indigenous to Europe, Northern Africa and South-Western Asia. The roots much used as a substitute for coffee; 5,000 tons of Chicory valued at £68,000 were imported into the United Kingdom in 1884. This plant requires a rich deep loamy soil, but fresh manure is detrimental to the value of the root. It is also a good fodder-plant, especially for sheep, and can be kept growing for several years, if cut always before flowering. The root can be dressed and boiled for culinary purposes; the leaves are useful for salad, particularly when the plants are removed to dark warm places for bleaching (Kuehnel); seeds will keep for several years (Vilmorin); medicinal use can be made also of the fresh root. Indigenous in Norway to lat. 63° 30′ (Schuebeler).

Cimicifuga racemosa, Elliott.

The "Black Snake-Root" and also "Cohosh" of North-America. A perennial herb of medicinal value, the root possessing emetic properties.

Cinchona Calisaya, Ruiz and Pavon.*

Andes of Peru, New Granada, Brazil and Bolivia, 5,000 to 6,000 feet above the ocean. This tree attains a height of 40 feet; it yields the Yellow Bark and also part of the Crown-Bark. It is one of the richest yielders of quinin, and also produces cinchonidin, but little of other alkaloids. The most valuable species in Bengal, braving occasional night-frost. This has flowered at Berwick (Victoria) already ten years ago under the care of Mr. G. W. Robinson, from plants supplied by the author, therefore as far south as Port Phillip, where also good seeds mature. It grows under conditions more limited than those of C. succirubra, nor is it so easily propagated. All of its varieties do not furnish bark of equal value. The Santa Fé variety ascends the Andes of New Granada 10,000 feet, and produces the highly valuable soft Columbia-bark. The variety Ledgeriana comes from Brazil, south-east of Lake Titicaca. Its bark yielded in Java 11 to 12 per cent. of quinin.

Renewed bark, obtained by covering the stem, where the bark has been removed, with moss or matting, according to Mr. McIvor's method, realized double the ordinary market-price, and in C. succirubra even more (Woodhouse). Young Cinchona-plants are subject to the attacks of Helopeltis Antonii, which insect preys also on the Tea-bush.

Cinchona cordifolia, Mutis.*

Peru and New Granada on the Andes, at between 6,000 and 8,000 feet elevation, and according to Mr. Willis Weaver at Bogota (probably under the shelter of forests) up to the frosty region of 9,500 feet. Provides the hard Cartagena-bark or West Pitaya-bark, one extremely rich in alkaloids. It is a species of robust constitution, grows with rapidity and vigor. The thickest bark is obtained in the highest altitudes, which are often involved in ·misty humidity by passing clouds (Cross).

Cinchona micrantha, Ruiz and Pavon.

Cordilleras of Bolivia and Peru. This tree attains a height of 60 feet, and from it part of the Grey and Huanuco-Bark as well as Lima-Bark are obtained. It is comparatively rich in cinchonin and quinidin, contains however also quinin.

Cinchona nitida, Ruiz and Pavon.

Andes of Peru and Ecuador. This tree rises to 80 feet under favorable circumstances. It also yields Grey Bark and Huanuco-Bark, besides Loxa-Bark. It will probably prove one of the hardiest species. It contains predominantly cinchonin and quinidin.

Cinchona officinalis, Linné (partly).* (*Cinchona Condaminea*, Humboldt.)

Andes of New Granada, Ecuador and Peru, at a height of 6,000 to 10,000 feet. Yields Crown- or Brown Peru-bark, besides part of the Loxa-bark. Comparatively rich in quinin and cinchonidin. The temperature of the middle regions of the Andes, where this tree grows, is almost the same as that of the Canary-Islands. Super-abundance of moisture is particularly pernicious to this species. The hardiest of all cultivated kinds. The Crispilla-variety endures a temperature occasionally as low as 27° F.

Cinchona lancifolia (Mutis) is considered by Weddell a variety of C. officinalis. This grows, where the mean annual temperature is that of Rome, with however less extremes of heat and cold. It yields part of the Pitaya-Bark.

Cinchona Pitayensis must also be referred to C. officinalis as a variety. This attains a height of 60 feet and furnishes also a portion of the Pitaya-bark. It is this particular cinchona, which in Upper India yielded in some instances the unprecedented quantity of 11 per cent. alkaloids, nearly 6 per cent. quinin, the rest quinidin and cinchonin; this plant is now annihilated for bark-purposes in its native forests.

Cinchonas raised from seeds, provided by the writer of this work, have withstood the slight frosts at San Francisco (G. P. Rixford).

The Uritusinga- or Loxa-variety grows in its native forests to a height of 60 feet and more (Pavon), and attained in Ceylon in fifteen years a height of 28 feet with a stem-girth of nearly 2 feet. The price of its bark in 1879 was about 7s. per pound, and of renewed bark 11s. Mr. McIvor obtained 6,850 cuttings from one imported plant in twenty months; but all Cinchonæ produce seeds copiously, so that the raising of great numbers of plants can be effected with remarkable facility. The bark has yielded 7·4 to 10·0 per cent. sulphate of quinin (Howard).

Iu Java some of the best results were obtained with Cinchona Hasskarliana, Miq., a species seemingly as yet not critically identified. Cinchona-seeds do not long retain their vitality; but as they are so very light, no difficulty exists in sending them speedily even to widely distant places.

Cinchona succirubra, Pavon.*

Middle Andine regions of Peru and Ecuador. A tree, attaining a height of 40 feet, yielding the Red Peru-Bark, rich in cinchonin and cinchonidin. It is this species, which is predominantly cultivated on the mountains of Bengal. In India it thrives at lower elevations than other Cinchonas, proves of quicker growth, and there the mixed cheap Cinchona-alkaloids forming the "Quinettum" are largely derived from this plant. (G. King, J. S. Gamble.) It has been found hardy in Lower Gippsland and the Westernport-District of Victoria. It grew in Madeira at an elevation of 500 feet, after having been planted two and a half years, to a height of 20 feet, flowering freely.

All these Cinchonas promise to become of importance for culture in the warmest regions of extra-tropical countries, on places not readily accessible or eligible for cereal culture. The Peruvian proverb, that Cinchona-trees like to be " within sight of snow," gives some clue to the conditions, under which they thrive best. They delight in the shelter of forests, where there is an equable temperature, no frost, some humidity at all times both in air and soil, where the ground is deep and largely consists of the remnants of decayed vegetable substances, and where the subsoil is open. Drippage from shelter-trees too near will be hurtful to the plants. Closed valleys and deep gorges, into which cold air will sink, are also not well adapted for cinchona-culture. The cinchona-region may be considered as interjacent between the coffee- and the tea-region, or nearly coinciding with that of the Assam-tea. Cross found the temperature of some of the best natural Cinchona-regions to fluctuate between 35° and 60° F. We here ought to consociate the Peru-bark plants with naturally growing fern-trees, but only in the warmest valleys and richest soil. The best temperature for Cinchonas is from 53° to 66° F.; but for the most part they will endure in open places a minimum of 32° F.; in the brush-shades of the Botanic Garden of Melbourne, where already many years ago Cinchonas were raised by the thousand, they have even resisted uninjured a temperature of a few degrees less, wherever the wind had no access, while under such very slight cover the Cinchonas withstood also a heat of a few degrees over 100° F.

The plants are most easily raised from seed, best under some cover such as mats; they produce seeds copiously a few years after planting. C. succirubra, first introduced into California by the writer of this work, together with the principal other species, thrives well in the lower coast-ranges as far north as San Francisco; better indeed than C. Calisaya, according to Dr. Herman Behr. The quantity of alkaloids in the bark can be much increased by artificial treatment, if the bark is only removed to about one-third on one side of the stem and the denuded part covered with moss or straw matting (kept moist), under which in one year as much bark is formed as otherwise requires three years' growth—such forced bark moreover containing the astounding quantity of 25 per cent. alkaloids, because no loss of these precious substances takes place by gradual disintegration through age. The root-bark of some cinchonas has proved to contain as much as 8 per cent. of alkaloids (see Gardeners' Chronicle, 1877, p. 212). The income from Java-plantations is considerably over double the cost of the expenses of culture and transit. Mr. Howard's opinion, that cinchonas in lowland-plantations produce a far less quantity of alkaloids, needs further confirmation, particularly regarding the valuable quinin and cinchonidin ; probably however geologic conditions have in all instances to be taken into account.

Young Cinchona-plants are set out at distances of about 6 feet. The harvest of bark begins in the fourth or fifth year. The price varies in Europe from 2s. to 9s. per lb., according to quality. The limits assigned to this literary compilation do not admit of entering further into details on this occasion ; but I may add, that in the Darjeeling-district over three millions of Cinchona-plants were already in cultivation during 1869 in Government-plantations. Cultivation of Cinchona for commercial purposes was first initiated in Java through Dr. Hasskarl in 1851. In 1880, 240,000 lbs. of bark were already exported from this island. The British harvest in the Madras Presidency alone amounted to 150,000 lbs. in 1875. Surgeon-Major Dr. G. King reports in 1880, that four million trees of Cinchona succirubra are now under his control in the Sikkim-plantations. This has proved the hardiest species ; it grows under a wide range of conditions, and seeds freely; thus it is the most valuable Cinchona in the elevations of Sikkim. In the Neilgherries more than 600,000 Cinchona plants were distributed from the Government-plantations in 1879, and 1,322 lbs. of seed (Barlow); from 80,000 to 250,000 seedlings being obtainable from one pound of sound seed, as almost every grain will grow. All its varieties produce bark of great value. The total amount of alkaloids is at an average 4 per cent. If the trees were cut every seven or eight years and simultaneous re-planting should take place, Dr. King could keep up an annual supply of 366,000 lbs. of bark. In 1883 there were as many as 128 millions of plants under cultivation in British India, of which 22 millions were two years old. The importations of Cinchona-bark into the United Kingdom in 1884 amounted to 106,000 cwt., of the value of £907,000; in 1882 the quantity was 139,000 cwt. and the cost £1,781,000. The total number of deaths of the Indian population from fever is considered to approach a million and a half annually.

Cinna arundinacea, Linné.

North-America. There recorded as a good fodder-grass ; perennial, somewhat sweet-scented. Particularly adapted for forest-meadows. Blyttia suaveolens (Fries) is, according to Dr. Asa Gray, a variety with pendent flowers.

Cinnamomum Camphora, Fr. Nees.[*]

The Camphor-tree of China and Japan, north to Kinsin, attaining a height of about 40 feet. It endures the occasional frosts of a clime like that of Port Phillip, though the foliage will suffer. The wood, like all other parts of the tree, is pervaded by camphor, hence resists the attacks of insects. The well known camphor is obtained by distilling or boiling the chopped wood or root; the subsequently condensed camphoric mass is subjected to a purifying sublimation-process.

Cinnamomum Cassia, Blume.

Southern China. It is not unlikely, that this tree, which produces the Chinese cinnamon or the so-called Cassia lignea, may prove hardy outside the tropics. Sir Joseph Hooker found on the Khasya-mountains up to 6,000 feet three cinnamons producing similar bark —namely: C. obtusifolium, C. pauciflorum and C. Tamala (Nees), the latter species extending to Queensland. Dr. Thwaites notes the true Cinnamon-tree (C. Zeilanicum, Breyn) even up to 8,000 feet in Ceylon, but the most aromatic bark comes from lower altitudes. Cinnamon-leaves yield a fragrant oil and the root gives a sort of camphor. Mr. Ch. Ford has ascertained, that the Chinese cut Cinnamomum Cassia when 6 years old, the time chosen being from March to May, after which season the bark loses much of its aroma. The branches are cut to near the root. The bark on distillation affords the Cassia-oil, 1 cwt. of bark yielding nearly 1 lb. of oil, which is much in use for confectionery and culinary purposes and the preparation of scented soaps. Oil can also be obtained from the foliage.

Cinnamomum Loureiroi, Nees.

Cochin-China, and Japan. A middle-sized tree. The leaves locally in use as a condiment and for perfumery.

Cistus Creticus, Linné.

Countries on the Mediterranean Sea, particularly the eastern. This shrub, with C. Cyprius (Lamarck) furnishes the best ladanum-resin. Other species yield a less fragrant produce.

Citrus Aurantium, Linné.

The Orange-tree (in the widest sense of the word). A native of South-Eastern Asia. A plant of longevity; thus a tree at Versailles, known as the "Grand Bourbon," is still in existence, though planted in 1421. Stems of very good Orange-trees have gained such a size, as to require two men to clasp them. If intervening spaces exist in orangeries, they might be used for raising herbaceous honey-plants. Any specific differences, to distinguish C. Aurantium from C. Medica, if they once existed, are obliterated now through hybridization, at least in the cultivated forms. In Central India a peculiar variety is under culture, producing two crops a year; the blossoms of February and March yield their ripe fruit in November and December, whereas from the flowers of July mature fruits are obtained in March and April. To prevent exhaustion only alternate fruiting is allowed. Nearly 5 million bushels of oranges and lemons, representing a value of £1,782,000, were imported into the United Kingdom during 1884. It is not unusual for orange-trees to continue in full bearing for 60 or 70 years, and after that the wood is still valued for its durability, fragrance and beauty. . The Sorrento-honey derives its delicious perfume from orange-flowers, and it has become classical as the best, and analogous to that of Hymethus (Laura Redden). As prominent varieties of C. Aurantium the following may be distinguished :—

> *Citrus Bergamium*, Risso. From the fruit-rind of this variety Bergamotte-oil is obtained; the flowers also yield oil. The Mellarosa-variety furnishes a superior oil and exquisite confitures.
>
> *Citrus Bigaradia*, Duhamel. The Bitter Orange. This furnishes from its flowers the Neroli-oil, so delicious and costly as a perfume. It is stated, that orange-flowers to the value of £50 might be gathered from the plants of an acre within a year. The rind of the fruit is used for candied orange-peel. Bitter principle: hesperidin in the rind, limonin in the seeds.
>
> *Citrus decumana*, Linné. The Shaddock or Pompelmos. The fruit will exceptionally attain a weight of 20 pounds. The pulp and thick rind can both be used for preserves.
>
> *Citrus dulcis*, Volkamer. The Sweet Orange, of which many kinds occur. The St. Michael Orange has been known to bear in the Azores on sheltered places 20,000 fruits on one tree in a year. Navel-oranges, weighing 19 ounces, have been obtained at Rockhampton; other varieties have been known to reach 3 pounds (Thozet). Neroli-oil is also obtained from the flowers of this and closely allied varieties. The oil of orange-peel might be used as a cheap and pleasant one in the distillation of costly odorous substances.
>
> *Citrus nobilis*, Loureiro. The Mandarin-Orange. The thin peel separates most readily from the deliciously flavored sweet pulp. There are large and small fruited Mandarin-oranges; the Tangerine-variety is one of them. Some varieties are excellent for hedges, for which they are much used in Japan. Burnt earth is valuable as an admixture to soil in orangeries. On the high authority of Dr. Piesse it may be stated, that recently rather more than 1¼ millions pounds weight of orange-flowers were gathered annually for perfumery-purposes merely at Nice and Cannes.

Citrus Australasica, F. v. Mueller.

Coast-forests of extra-tropical Eastern Australia. A shrubby species, with oblong or almost cylindrical fruits of lemon-like taste, measuring 2 to 4 inches in length. They are thus very much larger than those of Atalantia glauca of the eastern desert-interior of tropic Australia, but both are of similar taste. These plants are entered on this list, together with C. Planchoni, merely to draw attention to them as probably capable of improvement in their fruit through culture.

Citrus Japonica, Thunberg.*

The Kumquat of Japan. A shrubby Citrus with fruits of the size of a gooseberry, from which on account of their sweet peel and acid pulp an excellent preserve can be prepared.

Citrus Medica, Linné.

The Citron-tree (in the widest sense of the word). Indigenous to Southern Asia. For the sake of convenience it is placed here as distinct from C: Aurantium. As prominent varieties of the Citrus Medica may be distinguished :—

Citrus Cedra, Gallesio. The true Citron. From the acid tubercular fruit essential oil and citric acid can be obtained, irrespective of the ordinary culinary use of the fruit. A large variety with thick rind furnishes candied the citrionate or succade. The Cedra-oil comes from a particular variety.

Citrus Limonium, Risso. The true Lemon. Lemon-juice is largely pressed from the fruit of this variety, while the thin, smooth, aromatic peel serves for the production of volatile oil or for condiments. The juice of this fruit is particularly rich in citric acid. A large variety is the Rosaline-Lemon.

Citrus Limetta, Risso. The true Lime. The best lime-juice is obtained from this variety, of which the Perette constitutes a form. Less hardy than most other varieties. The Lime is one of the best and most enduring hedge-plants for warmer countries (H. A. Wickham).

Citrus Lumia, Risso. The Sweet Lemon, including the Pearlemon with large pear-shaped fruit. Rind thick and pale; pulp not acid. This variety serves for particular condiments.

Citrus trifoliata, Linné. Japan. Much grown as a hedge-shrub in its native country; used often as stock for grafting oranges on.

Coal-oil proved the most effectual remedy in Florida to dislodge scale from any kind of citrus-plant; half a pint of oil is to be mixed with sifted wood-ashes and then with 6 gallons of water, this fluid to be syringed over the trees. The import of lemons and oranges into the United Kingdom during 1884 was valued £1,782,686.

Citrus Planchoni, F. v. Mueller. (*C. Australis,* Planchon, partly.)

Forests near the coasts of sub-tropic Eastern Australia. A noble tree, fully 40 feet high, or according to C. Hartmann even 60 feet high, with globular fruit about the size of walnuts, called in Australia Native Oranges. The species first appeared under the above name in the "Report on the Vegetable Products of the Intercolonial Exhibition of 1867." Its beautiful wood takes a high polish; hence it is made use of for the finest cabinet-work. Through regular culture doubtless the fruit could be enlarged and improved.

Cladrastis tinctoria, Rafinesque.

Eastern States of North-America. Yellow-wood. The wood of this tree produces a saffron-yellow dye.

Clavaria botrytis, Persoon.

Europe. This and the following are species, admitted for sale among Silesian mushrooms, according to Dr. Goeppert: C. brevipes (Krombholz), C. flava, C. formosa, C. grisea (Persoon), C. muscoides (L.), C.

aurea (Schæffer), C. palmata (Scop.), C. crispa (Wulfen). Morren mentions as much consumed in Belgium C. fastigiata (L.). Bergner and Trog illustrate C. botrytis (Persoon). Several of these extend naturally to Australia, where also various other species exist. The puff-balls, comprising species of Bovista, Lycoperdon and Scleroderma, are not specifically admitted as recommendable culinary fungs into this work, as they are only convertible into food when very young, and many become soon noxious. All Clavarias seem adapted for human sustenance; their growth should therefore be encouraged.

Claytonia perfoliata, Donn.

From Mexico to California and Cuba. An annual succulent herb, serving for salad and also spinage. The genus could be reduced to Montia.

Clinostigma Mooreanum, F. v. Mueller. (*Kentia Mooreana*, F. v. M.)

Dwarf-palm of Lord Howe's Island, where it occurs only on the summits of the mountains, at about 3,000 feet elevation. Likely to prove one of the hardiests of all palms.

Coccoloba uvifera, Jacquin.

Central America, northward to Florida. A tree, attaining a large size, fit for sandy sea-shores. Sir J. Lefroy noticed in Bermuda stems 6 feet in girth. The dark-blue sweet or acidulous berries are edible. A kind of kino is obtained from the bark; the wood yields a red dye. Dr. Rosenthal notes as likewise producing edible fruits:— C. nivea (Jacq.), C. pubescens (L.), C. excoriata (L.), C. flavescens (Jacq.), C. diversifolia (Jacq.).—C. Leoganensis (Jacq.) is also a coast-tree; other species belong to forest-regions of mountains. They are all natives of the warmer zones of America.

Cochlearia Armoracia, Linné. (*Nasturtium Armoracia*, Fries.)

The Horse-radish. Middle Europe and Western Asia. Perennial. Grown in Norway to lat. 70° 22′ N. The volatile oil of the root allied to that of mustard.

Cochlearia officinalis, Linné.

Water-cress. Shores of Middle and Northern Europe, Northern Asia and North-America, also on saline places inland, even on the Pyrenees. A biennial herb, like the allied C. Angelica and C. Danica (Linné.), valuable as an antiscorbutic, hence deserving naturalization. It contains a peculiar volatile oil.

Cocos australis, Martius.

From Brazil to Uruguay and the La Plata-States. One of the hardiest of all palms, hardier than even the Date-palm, withstanding unprotected a cold, at which oranges and almonds are injured or destroyed. It remained perfectly uninjured at Antibes at a temperature of 15° F. (Naudin). C. pityrophylla ascends the Andes to 7,800 feet (de Dentérghem).

Cocos flexuosa, Martius.

' Brazil, extending far south. This slender and rather tall decorative Palm belongs to the dry Cactus-region with C. coronata, C. capitata, Astrocaryum campestre, Diplothemium campestre and Acrocomia sclerocarpa (Martius). Cocos coronata withstood at Hyères a temperature of 22° F. (Bonnet).

Cocos plumosa, Loddiges.

South-Brazil. This splendid Feather-palm attains a height of 60 feet. It is one of the hardiest of all palms, requiring no protection at Port Phillip. Stem comparatively slender.

Cocos regia, Liebmann.

Mexico, up to 2,500 feet. A Palm of enormous height; almost sure to prove hardy in the mildest extra-tropic latitudes.

Cocos Romanzoffiana, Chamisso.

Extra-tropic Brazil. This noble Palm attains a height of 40 feet.

Cocos Yatay, Martius.*

Rio Grande do Sul, Uruguay and Argentina. Forms distinct forests mainly with C. australis and C. Datil (Drude). The last mentioned bears date-like fruits, according to Dr. Lorentz. The kernels of the nuts of C. Yatay are edible. The incomparably valuable strictly tropical Cocoanut-palm Cocos nucifera (Linné) has fruited at the verge of the tropics in Queensland at Rockhampton under the care of Mr. J. S. Edgar.

Coffea Arabica, Linné.

Mountains of South-Western Abyssinia, extending as indigenous, according to Welwitsch and Peters, to Mozambique and Guinea. The Coffee-plant. This shrub or small tree has been admitted into this list, not without great hesitation, merely to avoid passing it. The cultivation within extra-tropical boundaries can only be tried with any prospect of success in the warmest and at the same time moistest regions, frost being detrimental to the Coffee-plant. In Ceylon the coffee-regions are between 1,000 and 5,000 feet above the ocean; but Dr. Thwaites observes, that the plant succeeds best at an elevation of from 3,000 to 4,500 feet, in places, where there is a rainfall of about 100 inches a year. The temperature there hardly ever rises above 80° F., and almost never sinks below 45° F. Coffee requires moist weather, whilst it ripens its fruit, and a season of drier weather to form its wood. Average-yield in Ceylon 4 to 5 cwt. per acre. An extraordinarily prolific variety of coffee was introduced twenty years ago by the writer of this work into Fiji, where it now forms the main plantations. The Coffee-plant has been found hardy as far north as Florida. For many particulars see the papers of the Planters' Association of Kandy. The importations of Coffee into the United Kingdom in 1884 amounted to 1,134,000 cwt. (about one-quarter being for home-consumption) valued at 3¾ million pounds sterling. Chemical principles: caffein, a peculiar tannic acid and quinic acid. The loss sustained in 1878 alone by the ravages of parasitic fungus-growth on Coffee-plants in Ceylon amounted to £2,000,000, the total loss since 1869 from this source reaching £15,000,000 (Abbay). The destruction of this Coffee-leaf Fungus (Hemileia vastatrix) is effected by applying flower of sulphur, particularly in dewy weather, and by dressing the ground with quicklime (Morris). See also essay by Mr. T. Dyer, in Journal of Microsc. Soc. New series, vol. XX. In America coffee-plantations have suffered not only from the attacks of an erysiphoid fungus, but also the Cemiostoma-fly. Elsewhere a beetle (Xylotrogus quadrupes) and a brown scaly bug (Lecanium coffeæ) have attacked the plant. Coffee-leaves have recently come into use as a substitute for tea.

Coffea Liberica, Bull.

Guinea. The Liberian Coffee-plant, distinguished already by Afzelius. According to Dr. Imray this species has shown immunity from the Cemiostoma-fly, and it is less affected by the Hemileia-mould. It grows to the size of a real tree, is a rich bearer, and the berries are larger than those of the ordinary coffee-bush; but the (useless) pulp is about twice as large in proportion to the seeds. The fruit requires a longer time to ripen (a year), but this species can be grown in hot tropical countries down to the coast (Lietze, Regel).

Colchicum autumnale, Linné.

The Meadow-Saffron. Middle and Southern Europe, Western Asia, The seeds and roots of this pretty bulbous-tuberous herb are important for medicinal use. The plant has been introduced into Australia by the writer with a view to its naturalization on moist meadows in our ranges. Active principle: colchicin. The plant proves hardy in Norway to lat. 67° 56′ (Schuebeler).

Colocasia antiquorum, Schott.*

The Taro. From Egypt through Southern Asia to the South-Sea Islands; apparently also indigenous in the warmer parts of East-Australia. The stem-like, tuberous, starchy roots lose their acridity by the processes of boiling, roasting or baking. It is the Kolkas of the Arabs and Egyptians, and one of their most esteemed and abundant vegetables. Immense quantities are harvested and kept during the winter. A splendid starch is obtainable from the tubers of this and the following species. The plant proves hardy as far south as Melbourne, and is also cultivated in New Zealand. The tops of the tubers are replanted for a new crop. Taro requires a rich, moist soil, and would grow well on banks of rivers. For scenic culture it is a very decorative plant. Colocasia esculenta is a variety of this species.

Colocasia Indica, Kunth. (*Alocasia Indica*, Schott.)

South-Asia, South-Sea Islands and Eastern Australia. Cultivated for its stem and tubers on swamps or rivulets. This stately plant will rise in favorable localities to a height of 12 feet, the edible trunk attaining a considerable thickness, the leaves sometimes measuring 3 feet in length. In using the stem and root for food, great care is needed, to expel all acridity by some heating process. Colocasia odora and C. macrorrhiza seem varieties of this species. Several other aroid plants deserve attention for test-culture on account of their edible roots, among them Cyrtosperma edule (Seemann) from the Fiji-Islands.

Combretum butyraceum, Caruel.

The Butter-tree of Caffraria and other parts of South-Eastern Africa. The Caffirs call the fatty substance, obtained from this tree, Chiquito. It is largely used by them as an admixture to their food, and is also exported. It contains about one-quarter olein and three-quarters margarin. This butter-like fat is extracted from the fruit, and is of an aromatic flavor. The tree should be hardy in the warmer and milder parts of extra-tropical countries.

Comptonia asplenifolia, Solander.

The Sweet-Fern of North-America. This dwarf shrub is perhaps quite worthy of dissemination on sterile hills, as the foliage contains nearly 10 per cent. of tannin; an extract of the leaves has come into the tanning trade. The plant is also not without medicinal value.

Condalia microphylla, Cavanilles.

The Piquillin. Chili and Argentina. A bush, yielding sweet, edible, succulent fruit.

Conium maculatum, Linné.

The Poison-Hemlock. Europe, Northern Africa, Northern and Western Asia. A biennial herb, important for medicinal purposes. It should however not be allowed to stray from its plantations, as it is apt to be confounded with culinary species of Anthriscus, Chæro-phyllum and Myrrhis, and may thus cause, as a most dangerous plant, disastrous mistakes. Active principles: coniin in the fruit, also conhydrin. The wild or naturalized plant best for therapeutic use.

Conopodium denudatum, Koch.

Western Europe. The small tuberous roots of this herb, when boiled or roasted, are available for food, and known as Earth-Chestnuts. The plant is allied to Carum Bulbocastanum.

Conospermum Stœchadis, Endlicher.

West-Australia. The question has arisen, whether this shrub, with C. triplinervium (R. Brown), ought to be introduced into any desert-country. All kinds of pasture-animals browse with avidity on the long, tender and downy flower-stalks and spikes, without touching the foliage, thus not destroying the plant by close cropping.

Convallaria majalis, Linné.

Europe, Northern and Middle Asia to Japan. The "Lily of the Valley." Far famed as a lovely fragrant spring-plant, desirable for naturalization in any temperate forest-regions, quite a trade-plant for bouquet-sellers, reintroduced into medicine also latterly, particularly in the treatment of cardial affections and dropsy. Both root and flowers have also sternutatorian properties.

Convolvulus floridus, Linné fil.

Canary-Islands. A shrubby species, not climbing or winding. With C. scoparius it yields the Atlantic Rosewood from stem and root.

Convolvulus Scammonia, Linné.

Mediterranean regions and Asia Minor. A perennial herb. The purgative drug, Scammonia-resin, is obtained from the root, which will grow to 2 feet in length. Plants readily raised from seeds. To obtain the drug, a portion of the root is laid bare, and into incisions made some shells are inserted, to collect the juice, which is daily removed (Maw).

Copernicia cerifera, Martius.*

Brazil, extending into Bolivia and Argentina. This magnificent Fan-palm has been proved to be hardy as far south as Sydney, by Mr. Charles Moore. It resists drought in a remarkable degree, and prospers also on a somewhat saline soil. The stem furnishes starch; the sap yields sugar; the fibres of the leaf-sheets are converted into ropes, which resist decay in water; the leaves can be used for mats, hats, baskets and brooms, and many other articles are prepared from them. The inner part of the leaf-stalks serves as a substitute for cork. This palm however is mainly valued for the Carnauba-wax, with which its young leaves are coated, and which can be detached by shaking. This is harder than bees' wax, and is used in the manufacture of candles. Each tree furnishes about 4 lbs. annually. In 1862 no less than 2,500,000 lbs. were imported into Great Britain, realizing about £100,000.

Coprinus comatus, Fries.

Europe, Asia. Included by Dr. L. Planchon among the Champignons for French kitchens. Other species elsewhere are probably quite as good, but they all can only be used for food in a very young state. More important are the deliquiscent species of Coprinus, such as C. atramentarius, C. ovatus, C. cylindraceus (Fries), for the preparation of a black water-color and also ink, both indelible (Wilson's Rural Cyclop.); the black fluid emitted needs the addition of some antiseptic to preserve it. Various Coprini are also common in Victoria.

Corchorus acutangulus, Lamarck.

Tropical Africa, South-Asia and North-Australia. This plant is specially mentioned by some writers as a jute-plant. A particular machine has been constructed by Mr. Le Franc, of New Orleans, for separating the jute-fibre. With it a ton of fibre is produced in a day by four men's work. This apparatus can also be used for other fibre-plants. The seeds of the Corchorus, which drop spontaneously, will reproduce the crop.

Corchorus capsularis, Linné.*

From India to Japan. One of the principal jute-plants. An annual, attaining a height of about twelve feet, when closely grown, with almost branchless stem. A nearly allied but lower plant, Corchorus Cunninghami (F. v. Mueller) occurs in tropical and sub-tropical Eastern Australia. Jute can be grown where cotton and rice ripen, be it even in localities comparatively cold in the winter, if the summer's warmth is long and continuous. The fibre is separated by steeping the full-grown plant in water from five to eight days; it is largely used for rice- wool- and cotton-bags, carpets and other similar textile fabrics, and also for ropes. In 1884 Great Britain imported 5,111,000 cwt. of jute, valued at £3,600,000. In 1883 the quantity amounted even to 7,372,000 cwt. of the value of £4,520,000, and a large quantity is also sent to the United States. Jute is sown on good land, well ploughed and drained, but requires no irrigation, although it likes humidity. The crop is obtained in the course of four or five months, and is ripe when the flowers are replaced by fruit-capsules. Good paper is made from the refuse of the fibre. Jute has been found, like hemp, to protect cotton from caterpillars, when planted around fields (Hon. T. Watts). In India jute often alternates with rice and sugar-cane ; as a crop it requires damp soil. It does not require drained land, according to Mr. C. B. Clarke. Unlike cotton, it will bear a slight frost. Under favorable circumstances 2,000 to 7,000 lbs. may be obtained from an acre. It is best grown on temporarily flooded ground, as otherwise it proves an exhaustive crop. Two hundred million pounds of jute were woven in 1876 in Dundee, and fifty million gunny-bags were exported from Britain in one single year, according to S. Waterhouse. Jute does not decay so easily as hemp, when exposed to moisture.

Corchorus olitorius, Linné.*

South-Asia and North-Australia. Furnishes, with the foregoing species, the principal supply of jute-fibre. As it also is an annual, it can be brought to perfection in the summers of the warm temperate zone. The foliage can be used for spinage. The fibre is not so strong as hemp, but very easily prepared. It will not endure long exposure to water. The seeds will keep for several years. The allied Corchorus trilocularis (Linné), of Indian origin, is likewise wild in eastern tropical and sub-tropical Australia.

Cordyline Banksii, J. Hooker.

New Zealand. This lax- and long-leaved Palm-Lily attains a height of 10 feet; its stem is usually undivided. This and the following species have been admitted into this list for a double reason, not only because they are by far the hardiest, quickest growing and largest of the genus, and thus most sought in horticultural trade for scenic planting, but also because their leaves furnish a fair fibre for textile purposes. The small seeds are produced in great abundance and germinate with extreme readiness. The same may be said of the three following species; their seeds can with the greatest ease be sent to the remotest distances. These Palm-Lilies ought to be naturalized copiously in forest-ranges by mere dissemination.

Cordyline Baueri, J. Hooker. (*C. Australis*, Endlicher *non* J. Hooker.)

Norfolk-Island. The stem of this stately species attains a height of 40 feet, and becomes ramified in age. It is very intimately allied to the following.

Cordyline indivisa, Kunth.

New Zealand. The stem of this thick and rigid-leaved palm-like species rises to a height of 20 feet, and remains undivided. Leaves finally 5 inches broad; yield the toi-fibre. Aged leaves persistent in a perfectly downward position for many years. Panicle at first erect. Berries white. Grows without protection in Arran (Capt. Brown).

Cordyline superbiens, C. Koch. (*C. Australis*, J. Hooker *non* Endlicher.)

New Zealand. The stem of this noble thin-leaved plant attains a height of 40 feet, and is branched. Aged leaves readily seceding; berries blue. Hardy at Torquay (W. Wood), Power's Court, Limerick, and in others of the milder localities of South-England and Ireland, also in the Island of Arran, where it grows luxuriously and flowers (Rev. D. Landsborough). It will stand a minimum temperature of 20° F. (Gorlie).

Cordyline terminalis, Kunth.

South-Asia, Polynesia, East-Australia. The roots are edible, when roasted. The leaves, like those of other species, can be utilized for textile fibre. The splendid decorative Cordylines with red or variegated foliage belong to this species.

Coriandrum sativum, Linné.

Orient and Middle Asia. An annual or biennial herb, its fruits much in use for condiments. The essential oil peculiar. Ripens seeds in Norway to lat. 68° 40' (Schuebeler). The seeds will keep for several years; 20 lbs. will be sufficient for one acre, returning 10–14 cwt. (G. Don).

Cornus florida, Linné.

The Dogwood of Eastern North-America. A showy tree, sometimes 30 feet high. The wood in great demand for shuttles, handles, harrow-teeth, horse-collars and sledge-runners. The root-bark is of therapeutic value. The tree is hardy still at Christiania (Schuebeler).

Cornus mas, Linné.

Europe, also Asia quite to Japan. This deciduous shrub or small tree is deserving of attention, as from the fruits a very palatable preserve can be prepared (Freyn). It answers also for hedge-growth.

Cornus Nuttalli, Audubon.

North-Western America. This is the largest of the genus, attaining a height of 80 feet, with a stem 2 feet in diameter. One of the most showy of Californian forest-trees. The wood is hard and close-grained, similar to that of the preceding species. The natives use the small twigs for making baskets (Dr. Gibbons). The white spring-inflorescence is visible for miles; in autumn again the scarlet fruit-coloration becomes an ornament to the landscape (Prof. Bolander).

Cortinarius cinnamomeus, Fries.

Europe and Asia. This mushroom, together with *C. violaceus* (Fries), is mentioned among numerous congeners by Drs. Badham and Cooke as particularly eligible for the table. Dr. Planchon recommends also *C. turbinatus* (Fries).

Corylus Americana, Walter.

Eastern North-America. Not tall; easily naturalized by dissemination, but fruit small and hard-shelled (A. Gray).

Corylus Avellana, Linné.

Europe, Northern Africa, Northern and Middle Asia. The ordinary Hazel, so well known for its filberts or cob-nut, one variety yielding the Barcelona-nut. A tree attaining a height of 30 feet; wood elastic; young shoots serving for hoops. The earliest flowering tree in northern countries. Loudon's account also of this tree is extensive and excellent. Chambers says, that generally about £100,000 worth of hazel-nuts are annually imported into Britain.

Corylus Colurna, Linné. (*C. Bizantina,* l'Ecluse.)

From Hungary to Greece and the Himalayas, there at from 5,500 to 10,000 feet elevation. The Constantinople-Nut Tree, the tallest of hazels, attaining 60 feet in height, of rather quick growth. Hardy at Christiania in Norway (Schuebeler). This, as well as the Nepal-Hazel (Corylus ferox, Wallich) and the Japan-Hazel (C. heterophylla, Fischer) might be naturalized in forest-gullies for their filberts.

Corylus maxima, Miller. (*C. rubra,* Borkhausen ; *C. tubulosa,* Willdenow ; *C. Lambertii,* Loddiges.)

Recorded as indigenous to Hungary and Istria by A. de Candolle, who however places the species nearer to C. Americana. Prof. C. Koch thinks, that it may have sprung from C. Avellana. It yields the red filbert or Lambert-nut.

Corylus Pontica, C. Koch.

Caucasus. Taller than C. maxima. Fruit similar to the Barcelona-nut; much consumed in Constantinople (C. Koch); called also Pontinian-nut.

Corylus rostrata, Aiton.

North-America, both east and west. Never tall. Nut small, but kernel sweet. C. Mandschurica (Maximowicz), from the Amur-region is a closely allied species, or perhaps only a variety.

Corynocarpus lævigata, Forster.

The Karaka of New Zealand and the principal forest-tree of the Chatham-Islands, attaining a height of 60 feet. The wood is light, and used by the natives for canoes. The pulp of the fruit is edible. Cattle browse on the foliage. In rich irrigated soil the tree can be adopted for very shady avenues.

Corynosicyos edulis. (*Cladosicyos edulis,* J. Hooker.)

Guinea. A new cucumber-like plant, with edible fruits about 1 foot long and 3 inches in diameter. Referred recently by Cogniaux to the genus Cucumeropsis.

Crambe cordifolia, Steven.

From Persia and the Caucasus to Thibet and the Himalayas, up to 14,000 feet. The root and foliage of this Kale afford an esculent. Flower-stems reaching 10 feet in height; the long-stalked leaves measure more than 2 feet in width. The root bears severe frost (Gorlie). C. Kotschyana (Boissier) is an allied plant.

Crambe maritima, Linné.

Sea-Kale. Sandy coasts of Europe and North-Africa, in Norway to nearly 60° N. A perennial herb; the young shoots used as a wholesome and agreeable vegetable. Should be naturalized.

Crambe Tataria, Wulfen.

From Eastern Europe to Middle Asia. Perennial. Leaves likewise used for culinary purposes. According to Simmonds the large fleshy roots also form an esculent. Can be grown still at Christiania.

Cratægus æstivalis, Torrey and Gray.

The Apple-Haw. South-Eastern States of North-America. The small juicy fruit of an agreeable acid taste.

Cratægus apiifolia, Michaux.

Eastern North-America. Highly serviceable for hedges.

Cratægus Azarolus, Linné.

Welsh Medlar. South-Eastern Europe and South-Western Asia. Hardy still in Christiania, Norway (Schuebeler). The pleasantly acidulous fruits can be used for preserves.

Cratægus coccinea, Linné.

Eastern North-America, there called White Thorn. A valuable hedge-plant; also very handsome. Spines strong. It braves the winters of Norway as far north as lat. 67° 56′ (Schuebeler).

H

Cratægus cordata, Aiton.

South-Eastern States of North-America. Also much employed for hedges.

Cratægus Crus Galli, Linné.

The Cockspur-Thorn. Eastern North-America. Regarded as one of the best species for hedges. Spines long and stout. Hardy to lat. 63° 26′ (Schuebeler). Fruit edible.

Cratægus oxyacantha, Linné.

The ordinary Hawthorn or White Thorn or Quick. Europe, North-Africa, North- and West-Asia. In Norway it grows to lat. 67° 56′; Professor Schuebeler found the plant to gain still a height of 20 feet in lat. 63° 35′. Recorded here as one of the most eligible among deciduous hedge-plants, safe against pastoral animals. The wood is considered one of the best substitutes for boxwood by engravers. The flowers are much frequented by bees for honey. C. monogyna (Jacquin) is a variety.

Cratægus Mexicana, Mocino and Sesse. (*Mespilus Mexicana,* C. Koch.)

A shrub, hardy in England. The fruit is of about one inch size and edible.

Cratægus parvifolia, Aiton.

Eastern North-America. For dwarf hedges. Spines long, slender, sharp and numerous.

Cratægus pyracantha, Persoon.

The Fire-Thorn. Southern Europe, South-Western Asia. This species is evergreen. It is likewise adapted for hedges, though slower in growth than the Hawthorn, but altogether not difficult to rear. Hardy in Norway to lat. 59° 55′ (Schuebeler). Referred by Boissier to Cotoneaster.

Cratægus tomentosa, Linné.

South-Eastern States of North-America. Reaching a height of 20 feet. Fruit edible. The list of American Hedge-thorns is probably not yet exhausted by the species mentioned; all afford honey. Two species, C. rivularis (Nuttall) and C. Douglasii (Lindley) occur in California and Oregon.

Crepis biennis, Linné.

Europe, Western Asia. Bosc regards this plant as useful for winter-pastures, in cool climes it keeping well green. The flowers afford food for bees.

Crithmum maritimum, Linné.

The real Samphire. Sea-shores of Western and Southern Europe, North-Africa and the Orient. A perennial herb. Settlers on the coast might readily disseminate and naturalize it. It is held to be one of the best plants for pickles, the young leaves being selected for that purpose.

Crocus sativus, C. Bauhin.

The Dye-Saffron. South-Eastern'Europe and the Orient. The stigmata of this particular autumnal flowering crocus constitute the costly dye-substance. The best is collected from the flowers as they daily open in succession. At any early stage of colonization it would not be profitable, to grow saffron commercially; but as the plant is well adapted for many extra-tropical countries or for high elevations within the tropics, it might be planted out into various unoccupied mountain-localities with a final view to naturalize it, and to thus render it available from native sources at a later period. It has additional claims on account of its prettiness. Noted as a bee-plant even by the ancients (Muenter). In Norway it is grown as far north as lat. 67° 56'. Likes calcareous light soil.

Crocus serotinus, Salisbury. (*C. odorus*, Bivona.)

South-Europe. This species also produces saffron rich in pigment. The bulbs of several species are edible.

Crotalaria Burhia, Hamilton.

Beloochistan, Afghanistan, Scinde. This perennial herb grows in arid places, and like the following yields Sunn-fibre.

Crotalaria juncea, Linné.

The Sunn-Hemp. Indigenous to Southern Asia and also widely dispersed through tropical Australia. An annual herb, rising under favorable circumstances to a height of 10 feet. In the colony of Victoria, Sunn can only be cultivated in the warmest and moistest localities. It comes to maturity in four or five months. The plant can also be grown as a fodder-herb for cattle. It requires rich, friable soil. If a superior soft fibre is desired, the plant is pulled while in flower; if strength is the object, the plant is left standing until it has almost ripened its seeds. The steeping process occupies about three days. For the purpose of obtaining branchless stems it is sown closely. Cultivated in the Circars, according to Roxburgh, to feed milch-cows.

Crotalaria retusa, Linné.

Asia, America and Australia within and near the tropics. A perennial herb. Its fibre resembles that of C. juncea, and is chiefly used for ropes and canvas. Others of the multitudinous species of Crotalaria deserve to be tested for their fibres.

Croton lacciferus, Linné.

Ceylon, up to 3,000 feet. Valuable for the warmer forest-regions of temperate climes, on account of its peculiar exuding lac-resin.

Crozophora tinctoria, Necker.

South-Europe, North-Africa and the Orient. An annual herb. The turnsole-dye is prepared by exposing the juice to the air, or by treating it with ammonia.

Cryptomeria Japonica, D. Don.

The Sugi or Japanese Cedar. Japan and Northern China. The largest tree in Japan, the trunk attaining 35 feet in circumference (Rein) and 120 feet in height. Stem long, clear, of perfect straightness; the plant is also grown for hedges; in Japan it yields the most esteemed timber, scented like that of Cedrela (Christie). It requires forest-valleys for successful growth. The wood is durable, compact, soft and easy to work; more extensively utilized in Japan than any other. In the Azores the tree is preferred even to the Pinus Haleppensis for timber-culture, on account of its still more rapid growth in that insular climate. Several garden-varieties exist. Lives unprotected still at Christiania.

Cucumis Anguria, Linné.

Wild in tropical America, but according to Sir Jos. Hooker and Prof. Naudin perhaps of African origin; all other species belonging to the eastern hemisphere. Annual. The fruit serves for pickles.

Cucumis cicatratus, Stocks.

Scinde, where it is called "Wungee." The edible ovate fruit is about 6 inches long. Deemed a wild form of C. Melo by Cogniaux.

Cucumis Citrullus, Seringe. (*Citrullus vulgaris,* Schrader.)

Indigenous probably only in Eastern Africa. The Water-Melon. It is simply mentioned here, to indicate the desirability of naturalizing it in any desert. In those of South-Africa it has become spontaneously established, and retained the characters of the cultivated fruit.

Cucumis Colocynthis, Linné. (*Citrullus Colocynthis,* Schrader.)

From the Mediterranean regions to India. An annual herb. The medicinal extract of colocynth is prepared from the small gourd of this species. Active principle: colocynthin.

Cucumis Melo, Linné.

The Melon. Originally from the country about the Caspian Sea, but some forms indigenous to India, northern and tropical Africa and tropical Australia, if really all the forms united by Cogniaux are conspecific. The best varieties might also be naturalized in sand-deserts, particularly in places where some moisture collects. In seasons of drought the Muscat-Melon, introduced by the author into Central Australia, has borne fruit there more amply than any other variety. Some of the Bokhara-varieties are remarkably luscious and large. Apparently remunerative results have been gained in Belgium from experiments, to cultivate melons for sugar and treacle. The seeds thus obtained in quantity become available for oil-pressing. The root contains melonemetin. The Japan C. conomon (Thunberg) belongs to this species. Prof. Naudin investigated extensively the variability of this and allied plants. Some varieties of melons and pumpkins ripen in Scandinavia during the long summers there in the open air far north; all are annual.

Cucumis Momordica, Roxburgh.

Cultivated in India. It produces cucumbers 2 feet long, bursting slowly when ripe into several divisions. Young, the fruit is used like cucumbers, older like melons. Referred by Cogniaux to the varieties of C. Melo.

Cucumis sativus, Linné.

The Cucumber. North-Western India. Indicated here merely for completeness' sake, also because gherkin-pickling ought to become a more extended local industry. Dr. G. King brought under notice and Indian culture the Chinese Cucumber "Solly-Qua," which attains a length of 7 feet. It must be trained on walls or trellises, to afford to the fruit sufficient scope for suspension. For definitions of numerous varieties of Melons, Cucumbers and Gourds, as well as for full notes on their cultivation, see, irrespective of other references, G. Don's Dichlamydeous Plants III, 1–42. Seeds will retain their vitality for ten years or more (Vilmorin).

Cucurbita maxima, Duchesne.

Large Gourd or Pompion. Indigenous probably in South-Western Asia. Yields some sorts of pumpkins. Instances are on record of fruits having weighed over 2 cwt. This species also is eligible for naturalization in the interior. Amongst other purposes it serves for calabashes. The seeds will keep about six years.

Cucurbita Melopepo, Linné.

The Squash. May be regarded as a variety of C. Pepo. It will endure storage for months.

Cucurbita moschata, Duchesne.

The Musky Gourd. Doubtless also from the Orient, but its exact nativity never traced (A. de Candolle). A variety, much cultivated in Italy, produces fruits so large as occasionally to weigh fully 40 lbs. (Vilmorin).

Cucurbita Pepo, Linné.

The Pumpkin and Vegetable Marrow, as well as the Succade-Gourd. Countries on the Caspian Sea, but A. de Candolle believes it to be of North-American origin, where some other though not culinary species of this genus occur. Its naturalization in hot deserts would be a boon. The seeds on pressure yield a fixed oil; they are also anthelmintic. Most of the ornamental gourds are varieties of this species. This, with many other Cucurbitaceæ, yields much honey for bees. The fruit of the perennial C. melanosperma (A. Braun) is not edible.

Cudrania Javensis, Trecul.

East-Australia, Southern and Eastern Asia to Japan, East-Africa. This climbing thorny shrub can be utilized for hedges. Fruit edible, of a pleasant taste; the root furnishes a yellow dye.

Cudrania triloba, Hance.

China. The leaves of this shrub serve as food for silkworms according to Mr. F. B. Forbes.

Cuminum Cyminum, Linné.

North-Africa. The fruits of this annual herb are known as Cumin, and used for certain condiments, as also in medicine. Cuminum Hispanicum (Merat) is similar. Essential oil peculiar.

Cupania sapida, Cambessedes. (*Blighia sapida,* Koenig.)

Western tropical Africa. A tree, to 30 feet high, if not sometimes higher. Flowers so fragrant as to be worth distilling. Succulent portion of the fruit eatable, improved by frying. This Cupania may endure slight frost as some of its congeners.

Cupressus Benthami, Endlicher.

Mexico, at elevations from 5,000 to 7,000 feet. A beautiful tree, reaching 60 feet in height. The wood is fine-grained and exceedingly durable. Rate of growth at Port Phillip as much as 30 feet in height within 15 years. Professor C. Koch deems it identical with C. thurifera.

Cupressus fragrans, Kellogg.

The Ginger-Pine or Oregon-Cedar. California. A tree, reaching 150 feet in height, with a clear trunk for 70 feet and a stem-diameter reaching 6 feet. Wood abounding in aromatic oil (J. Hoopes).

Cupressus funebris, Endlicher.

Thibet. The Weeping Cypress. Attains a height of 90 feet. One of the most eligible trees for cemeteries; can be grown from the lowlands of India to 7,000 feet or even higher.

Cupressus Lawsoniana, Murray. (*Chamæcyparis Lawsoniana,* Parlatore.)

Northern California. This splendid red-flowered Cypress grows to 100 feet in height, with a stem to 2 feet in diameter, and furnishes a valuable timber for building purposes, it being clear, easily worked, free from knots, elastic and very durable (Sargent); it is however to be avoided for cabinet work on account of the soft and coloring resin permeating it (Dr. Kellogg). Hardy to lat. 61° 15' in Norway (Schuebeler).

Cupressus Lindleyi, Klotzch.

On the mountains of Mexico. A stately Cypress, reaching a height of 120 feet. It supplies an excellent timber. Prof. C. Koch points out the very close affinity of this species to C. thurifera, and restores its older name C. Coulteri (Forbes), suggesting that this cypress may be derived from C. pendula (l'Heritier), which so long was termed inaptly C. Lusitanica.

Cupressus macrocarpa, Hartweg.* (*C. Lambertiana,* Gordon.)

California, from Monterey to Noyo, in the granite- as well as sandstone-formation; sometimes in Sphagnum-moors. This beautiful and shady tree attains to a height of 150 feet, with a stem of 9 feet in circumference, and is one of the quickest growing of all conifers, even in poor dry soil. One of the best shelter-trees on sea-sands, naturally following the coast-line, never extending many miles from the shore, and occurring in localities, where the temperature does not rise above 80° F. nor sink below the freezing point (Bolander); nevertheless it proved even hardy in Christiania. Richer in its yields of tar than the Scotch Fir, according to American writers. Not to be planted on places where stagnant humidity exists under ground.

Cupressus Nutkaensis, Lambert. (*Chamæcyparis Nutkaensis,* Spach; *Thuja excelsa,* Bongard.)

The Yellow Cedar or Cypress of Alaska and the neighboring States. Height of tree reaches 100 feet. Timber soft, pale, clear, durable, tough and close, also scented; worked with ease; used for boat-building and many other purposes; the bast for mats and ropes. Can be trimmed for hedge-growth. The Cypresses of the sections Chamæcyparis and Retinospora are now regarded by Sir Joseph Hooker and Mr. George Bentham as species of Thuja. Prof. C. Koch placed them, as did previously the author of this work, in the genus Cupressus.

Cupressus obtusa, F. v. Mueller. (*Retinospora obtusa,* Siebold and Zuccarini.)

The Hinoki of Japan. Attains a height of 100 feet; stem to 5 feet in circumference. It forms a great part of the forests at Nipon. Growing naturally between 1,200 and 4,200 feet elevation on the transition of the compact alluvial clays to eruptive granite (Dupont). The bark is used for thatching, also for cordage and tow. The wood is white-veined and compact, assuming when planed a silky lustre. According to Mr. Christie, it is durable, close-grained and easily worked. It is selected in Japan for temples. There are varieties of this species with foliage of a golden- and of a silvery-white hue. Hardy at New York, even in exposed localities. One of the finest of evergreen trees for the vicinity of dwellings. It resembles C. Lawsoniana, but excels it; it is also hardier and of more rapid growth (Rev. H. W. Beecher). Easily multiplied from layers of the lower branches.

Two other Japanese Cypresses deserve introduction—namely C. breviramea (Chamæcyparis breviramea, Maximowicz) and C. pendens (Chamæcyparis pendula, Maximowicz).

Cupressus pisifera, F. v. Mueller. (*Chamæcyparis pisifera,* Sieb. and Zucc.)

The Savara of Japan. It attains a height of 30 feet. Stem occasionally 3 feet in diameter (Rein). Very hardy like the foregoing, bearing the frosts of Norway at least to lat. 59° 55' (Schuebeler); also of beautiful aspect and quick growth. There is also a variety with golden-yellow foliage. Less esteemed than C. obtusa; grows in about the same localities, but is content with poorer soil, and bears more heat (Dupont).

Cupressus sempervirens, Linné.

The Common Cypress. South-Europe and South-Eastern Asia, on Mount Lebanon up to 5,000 feet. It is famous for the great age it attains, and for the durability of its timber, which is next to imperishable. Doors from this wood in St. Paul's Church in Rome have lasted over 600 years. Both varieties, namely C. pyramidalis (Targioni) and C. horizontalis (Miller) widely under culture; attains in warm countries occasionally a height of 100 feet and a stem-girth of 9 feet. Hardy in England. Near Somma a cypress is still shown, which—so it is said—was renowned already at Cæsar's time on account of its great size. The wood is prized for trunks and boxes, as rendering the contents proof against most kinds of insects (Dr. Brandis). At present its wood is much sought for the manufacture of musical instruments. Young records the stem-circumference of a Cypress at Lago Maggiore at 54 feet, and this was known even 600 years ago as a venerable tree, thus far one of the few most favored trees in the whole creation.

Cupressus thurifera, Humboldt, Bonpland and Kunth.

Mexican White Cedar; 3,000 to 4,500 feet above sea-level. A handsome pyramidal tree, upwards of 40 feet high.

Cupressus thuyoides, Linné. (*Chamæcyparis sphæroidea*, Spach; *Thuja sphæroidalis*, Cl. Richard.)

White Cedar of North-America; in moist and swampy ground. Height of tree reaching 80 feet; diameter of stem 3 feet. The wood is reddish, light, clear, easy to split, soft and fragrant; it turns red when exposed to the air. Extensively used for a great variety of purposes—for boat-building, cooperage, railway-ties, particularly also shingles; it is fine-grained and easily worked. Mohr says, that the wood when well seasoned offers the finest material for hollow-ware. For furniture, it admits of a high finish and has a pleasing hue. The old wood resists the successions of dryness and moisture better than any other American Cypress hitherto tried. Circumferential rate of stem-growth in Nebraska 22 inches at 2 feet from the ground in 12 years (Furnas).

Cupressus torulosa, Don.*

Nepal-Cypress. Northern India; 4,500 to 8,000 feet above the sea-level. Average ordinary height 40 feet, but much larger dimensions are on record; thus Dr. Stewart and Major Madden mention a tree 150 feet in height and 17 feet in stem-girth. The reddish fragrant wood is as durable as that of the Deodar-Cedar and highly valued for furniture. The tree prefers limestone-soil. Splendid for wind-breaks and tall hedges. Dr. Brandis thinks, that it may attain an age of 1,000 years.

Cyamopsis psoraloides, De Candolle.

Southern Asia. This annual is mentioned by Dr. Forbes Watson among the plants, which furnish throughout the year table-beans to a portion of the population of India.

Cycas Normanbyana, F. v. Mueller.

A noble Queensland-species, deserving introduction, and capable of being shipped to long distances in an upgrown state without emballage.

Cycas revoluta, Thunberg.

The Japan Pine-Palm. The trunk attains in age a height of about 6 feet, and is rich in sago-like starch. The slow growth of this plant renders it only valuable for scenic decorative culture; it endures the climate of Melbourne without protection. Cycas media, R. Br., may also prove hardy, and would be a noble horticultural acquisition, as it is the most gigantic of all Cycadeæ, attaining a height of 70 feet in tropical East-Australia. C. Siamensis (Miquel) will endure a temperature occasionally as low as the freezing point. Like the Zamia-stems, the trunks of any Cycas admit of translocation, even at an advanced age; and like the stems of many kinds of tree-ferns, they can be shipped on very long voyages packed as dead goods in closed wood-cases, deprived of leaves and soil, for subsequent revival in conservatories, as shown many years ago by the writer of this work. The Macrozamias can be associated with the hardier palms in gardens, M. spiralis advancing naturally southward to the 37th degree. One genuine Zamia occurs as indigenous in Florida, several in Mexico are extra-tropical, while Z. Chiqua (Seemann), or a closely allied species, ascends to 7,000 feet in Central America. The South-African species of Encephalartos also endure the night-frosts of Melbourne perfectly well.

Cymopterus glomeratus, De Candolle.

North-America, Missouri-region. Root edible (Dr. Rosenthal).

Cynara Cardunculus, Linné.

The Cardoon. Mediterranean regions; extending to the Canary-Islands. A perennial herb. The bleached leaf-stalks serve as esculents. The foliage employed also as a substitute for rennet. This as well as the following will come to perfection in Norway to lat. 63° 52′ (Schuebeler). Readily raised from seeds. The root also edible (Vilmorin).

Cynara Scolymus, Linné.

The Artichoke. South-Europe and North-Africa. The receptacles and the base of the flower-scales well known as a vegetable. The plant is perennial, and here merely mentioned as entitled to extended culture, grouped with other stately plants. Several other species are worthy of cultivation. In Italy Artichokes are much grown under olive-trees, to utilize spare-ground. The plant is greatly benefited in cultivation by a dressing with sea-weed or any other manure containing sea-salt (G. W. Johnson). The leaves serve instead of rennet. Seeds will keep for several years. To preserve good varieties sprouts are replanted, from which all the buds except two or three of the strongest are removed. Low-lying ground and somewhat peaty soil are well adapted for this plant (Vilmorin).

Cynodon Dactylon, Cl. Richard.*

Widely dispersed over the warmer parts of the globe, thus as indigenous reaching the northern parts of the colony of Victoria; stretching also into Middle Europe and West-England. Hardy in Norway to lat. 63° 52' (Schuebeler). Passes under the names of Bermuda-Grass, Indian Couch-Grass, Doab, Doorba or Doorva or Bahama-Grass. An important grass for covering bare, barren land, or binding drift-sand, or keeping together the soil of abrupt declivities, or consolidating earth-banks against floods. It is not without value as a pasture-grass; resists extreme drought, and may become of great importance to many desert-tracts, as it keeps alive even in the hottest and driest parts of Central Australia; also one of the best of all grasses in tropical countries for hay (Eggers). Placed likewise above all other grasses for pasture- and stable-value in Louisiana (Seiss). Difficult to eradicate, but for permanent pastures on exhausted land in mild climes not surpassed. The dispersion is best effected by the creeping rooting stems cut into short pieces; each of these takes root readily, but it can also be disseminated and grains are now always in the seed-markets. In arable land this grass, when once established, cannot easily be subdued. The stems and roots are used in Italy for preparing the Mellago graminis. Roxburgh already declared this grass to be by far the most common and useful for pastures of India, particularly in the drier regions; that it flowers all the year, and that it forms three-fourths of the food of the cows and horses there. Excellent also as a lawn-grass in mild climates, on account of its dwarf and creeping growth and as enduring trampling pertinaciously. Chemical analysis, made very early in spring, gave the following results:—Albumen 1·60, gluten 6·45, starch 4·00, gum 3·10, sugar 3·60 per cent. (F. v. Mueller and L. Rummel).

Cynosurus cristatus, Linné.

The Crested Dogstail-Grass. Europe, Northern Africa, Western Asia. A perennial grass, particularly valuable as withstanding drought, the root penetrating to considerable depth. The stems can also be used for bonnet-plaiting. Though inferior in value for hay, this grass is well adapted for permanent pasture, as it forms dense tufts without suffocating other grasses or fodder-herbs. Recommended also as an admixture to lawn-grasses by Hein and others.

Cyperus corymbosus, Rottboell.

India, North-Australia, Madagascar. This stately perennial species may be chosen to fringe our lakes and ponds. It is extensively used for mats in India.

Cyperus esculentus, Linné.

Southern Europe, Western Asia, various parts of Africa. Produces the "Chufa" or Ground-Almond, an edible root, which contains about 27 per cent. of starch, 17 per cent. of oil, and 12 per cent. of saccharine substance; other (French) analyses give 28 per cent. oil, 29 starch, 14 sugar, 7 gum, 14 cellulose. This plant does not injuriously spread like the C. rotundus, and can be reared on sand-land, though in rich loose soil the harvest is far more plentiful. The tubers, of which as many as 100 to 150 may be obtained from each plant, are consumed either raw or cooked. Hogs root them up for food. The oil surpasses in excellence of taste all other oils used for culinary purposes. The tubers are a fair substitute for coffee, when properly roasted; the root-crop is available in from four to six months. The plant may become important in the most dreary and arid desert-countries through naturalization. In Norway it can be grown to lat. 67° 56′ (Schuebeler). The root of the North-American C. phymato-des (Muehlenberg) is also nutty.

Cyperus Papyrus, Linné.

The Nile-Papyrus, wild in various regions of Africa. Attains a height of 16 feet. Though no longer strictly a utilitarian plant, as in ancient times, it could scarcely be passed on this occasion, as it ought to become valuable in the horticultural trade. Its grand aspect recommends it as very decorative for aquatic plantations.

Cyperus Syriacus, Parlatore.

The Syrian or Sicilian Papyrus. This is the Papyrus-plant usual in garden-cultivation. It found its way to Australia first through the action of the writer of this work. The plants in the Melbourne Botanic Garden attain a height of 8 feet, but suffer somewhat from frost. Other tall decorative Cyperi deserve introduction, for instance: C. giganteus (Rottboell) from the West-Indies and Guiana; these kinds of plants being hardier than the generality of others from the tropics.

Cyperus tegetum, Roxburgh.

North-Eastern Africa, India, China and North Australia. This Galingale-Rush might be naturalized on river-banks to obtain material for the superior mats made of it in Bengal. The fresh stems are slit longitudinally into three or four pieces, each of which curls round while drying, and can then be worked into durable and elegant mats. In China it is cultivated like rice, but in brackish ground only, where narrow channels will allow the water to flow in and out with the rising and receding tide (Hance and Dilthey).

Cyperus textilis, Thunberg. (*Cyperus vaginatus*, R. Brown.)

Widely dispersed over the Australian continent, also occurring in Southern Africa. It is restricted to swampy localities, and thus is not likely to stray into ordinary fields. In the colony of Victoria it is one of the best indigenous fibre-plants, and it is likewise valuable as being with ease converted into pulp for good writing paper, as shown by the author some years ago. Its perennial growth allows of regular annual cutting. The natives of the Murray-River use this as well as Carex tereticaulis (F. v. M.) for nets.

Cytisus prolifereus, Linné fil.

Canary-Islands. The "Tagasaste." A fodder-shrub for light dry soil; rather intolerant to frost (Dyer).

Cytisus scoparius, Link. (*Spartium scoparium,* Linné.)

The Broom-Bush. Europe, North-Asia; wild in Norway to 58° N. Of less significance as a broom-plant than as one of medicinal value. It can also be used for tanning purposes. Most valuable for arresting drift-sand. Easily raised from seeds. An alkaloid (spartein) and a yellow dye (scoparin) are obtainable from this shrub.

Cytisus spinosus, Lamarck.

Countries around the Mediterranean Sea. This bush forms a strong prickly garden-hedge, handsome when closely clipped (W. Elliott).

Dacrydium Colensoi, Hooker.

New Zealand. A beautiful tree, growing to 50 feet in height and producing hard and incorruptible timber. Chiefly eligible for cool humid forest-regions.

Dacrydium cupressinum, Solander.

New Zealand. Native name, Rimu; the Red Pine of the colonists. This stately tree attains the height of 200 feet, and furnishes a hard and valuable wood, very lasting for fences, but readily decaying in water-works. Professor Kirk recommends the timber on account of its great strength for girders and heavy beams anywhere under cover. With other New Zealand conifers particularly eligible for forest-valleys. A most suitable tree for cemeteries, on account of its pendulous branches. The bark possesses fair tan-properties.

Dacrydium Franklini, J. Hooker.

Huon-pine of Tasmania, where it is endemic; only found in moist forest-recesses, and thus might be planted in ferntree-gullies of South-Eastern Australia also. Height of tree sometimes 100 feet; stem-circumference reaching 20 feet. The wood is highly esteemed for boat-building and various artisans' work. It is the best of Australian woods for carving, also extensively used for the rougher kinds of xylography and in the manufacture of pianos.

Dacrydium Kirkii, F. v. Mueller.

New Zealand. The "Manoao." A pyramidal tree, attaining 80 feet in height; stem-diameter to 4 feet. Timber of a reddish color and extreme durability (Professor Kirk). Bears seeds abundantly.

Dactylis glomerata, Linné.*

Europe, North-Africa, Northern and Middle Asia. The Cocksfoot-grass. One of the best of perennial tall pasture-grasses, adapted as well for dry as moist soil, thus even available for wet clays. It will live under the shade of trees in forests; fit also for coast-sands. It is indigenous in Norway to lat. 68° 50' (Schuebeler). Its yield of fodder is rich and continuous, but its stems are hard. It is generally liked by cattle, unless when by understocking or neglect it has been allowed to become rank. Langethal observes : " What the Timothy-grass is for the more dry sandy ground, that is the Cocksfoot-grass for more binding soil, and no other (European) grass can be compared to it for copiousness of yield, particularly if the soil contains a fair quantity of lime. It grows quickly again after the first cutting, and comes early on in the season. The nutritive power of this grass is of first class." The chemical analysis, made very late in spring, gave the following results: Albumen 1·87, gluten 7·11, starch 1·05, gum 4·47, sugar 3·19 per cent. (Von Mueller and Rummel).

Dactylis litoralis, Willdenow. *(Aeluropus laevis,* Trinius.)

From the Mediterranean countries to Siberia. This stoloniferous grass can be utilized for binding coast-sands; but it is of greater importance still in sustaining a Kermes-insect (Porphyrophora Hamelii), which produces a beautiful purple dye (Simmonds).

Dalbergia latifolia, Roxburgh.

India, up to cool but not cold regions. A deciduous tree, attaining a height of 80 feet. The wood tough and heavy, in local request for ornamental furniture, yokes, wheels, ploughs, knees of boats; its color from nut-brown to dark-purplish, streaked and spotted with lighter hues (Brandis, Gamble).

Dalbergia melanoxylon, Guillemin and Perrottet.

Tropical Africa, extending to Southern Egypt. A small tree with spiny branches; the wood described variously as blackish and purplish; according to Colonel Grant used for arrow-tips, wooden hammers and other select implements.

Dalbergia Miscolobium, Bentham.

Southern Brazil. This tree supplies a portion of the Jacaranda-wood (Tschudi).

Dalbergia nigra, Allemao.

Brazil, down to the Southern Provinces. A tall tree, likely to prove hardy in warmer extra-tropic regions. It yields a portion of the Jacaranda- or Palisander-Wood, also Caviuna-Wood, which for rich furniture have come into European use. Several Brazilian species of Machærium afford, according to Saldanha da Gama, a similar precious wood, also timber for water-works and railway-sleepers, particularly M. incorruptibile (Allemao), M. legale and M. Allemai (Bentham).

Dalbergia Sissoo, Roxburgh.

The Indian Sissoo-tree, extending to Afghanistan, ascending to elevations of 5,000 feet, attaining a height of 80 feet. It may be worthy of test, whether in localities almost free of frost, particularly along sandy river-banks, this important timber-tree could be natural-ized, the Sissoo bearing occasional frosty cold better than the Sâl.

Brandis found the transverse strength of the wood greater than that of teak and of sâl; it is very elastic, seasons well, does not warp or split, is easily worked, and takes a fine polish. It is also durable as a wood for boats. The tree is easily raised from seeds or cuttings, and is of quick growth. The supply of its wood has fallen short of the demand in India. Colonel Campbell-Walker states, that in the Panjàb artificial rearing of Sissoo is remunerative at only 15 inches annual rainfall, with great heat in summer and occasional sharp frosts in winter; but irrigation is resorted to at an annual expense there of four shillings per acre. Sterile land is by the Sissoo-planting greatly ameliorated.

Dammara alba, Rumph. (*D. orientalis*, Lambert.)

Agath-Dammar. Indian Archipelagus and mainland. A splendid tree, up to 100 feet high, with a stem to 8 feet in diameter, straight and branchless for two-thirds in length. It is of great importance on account of its yield of the transparent Dammar-resin, extensively used for varnish.

Dammara Australis, Lambert.*

Kauri-Pine. North-Island of New Zealand. This magnificent tree measures, under favorable circumstances, 180 feet in height and exceptionally 17 feet in diameter of stem; the estimated but perhaps overrated age of such a tree being 700 to 800 years. It furnishes an excellent, remarkably durable timber, straight-grained, and much in use for masts, boats, superior furniture, casks, rims of sieves, and is particularly sought for decks of ships, lasting for the latter purpose twice as long as the deal of many other pines. It is also available for railway-break-blocks and for carriages, and regarded as one of the most durable among timbers of the Coniferæ. Braces, stringers and tie-beams of wharves remained, according to Professor Kirk, for very many years in good order under much traffic. In bridge-building also the Kauri-timber gave excellent results; it can likewise be used advantageously for the sounding-boards of pianofortes. Kauri-wood is also used for light handles of many implements and for various instruments, including stethoscopes, for wool-presses, the bodywork of waggons, butter-casks, brewers' vats; further, in ship-building for bulwarks and for the sides of boats. In strength it is considerably superior to Baltic Deal. Kauri ought to be extensively introduced into our denser forests. Auckland alone exports about £20,000 worth of Kauri-timber annually. It is easily worked, and takes a high polish. This tree yields besides the Kauri-resin of commerce, which is also largely obtained from under the stem. The greatest part is gathered by the Maoris in localities, formerly covered with Kauri-forests; pieces weighing 100 lbs. have been found in such places.

Dammara macrophylla, Lindley.

Santa-Cruz Archipelagos. A beautiful tree, often 100 feet high, resembling D. alba.

Dammara Moorei, Lindley.

New Caledonia. Height of tree about 50 feet.

Dammara obtusa, Lindley.

New Hebrides. A fine tree, resembling D. Australis, reaching 200 feet in height, with a long, clear trunk.

Dammara ovata, C. Moore.

New Caledonia. This tree is rich in Dammar-resin.

Dammara robusta, C. Moore.

Queensland-Kauri. A tall tree, known only from Rockingham's Bay, Fraser's Island and Wide Bay. It thrives well, even in open, exposed, dry localities at Melbourne. Height attaining 180 feet; largest diameter of stem 6 feet; wood free from knots and easily worked. Market value £3 10s. for 1,000 superficial feet of timber. As much as 12,000 feet (superficial) of good timber have been cut from one tree, that not being the largest. The species is closely allied to the Indian D. alba, and yields likewise Dammar-resin.

Dammara Vitiensis, Seemann.

In Fiji. Tree to 100 feet high; probably identical with Lindley's D. longifolia.

Danthonia bipartita, F. v. Mueller.

From the interior of New South Wales and Queensland to West-Australia. Available as a tender-leaved and productive perennial grass, particularly for any desert-regions.

Danthonia Cunninghami, J. Hooker.

New Zealand. A splendid alpine fodder-grass with large panicles; it attains a height of 5 feet, and forms tussocks. Pasture-animals relish the young foliage and the flower-masses (J. Buchanan).

Danthonia nervosa, J. Hooker. (*Amphibromus Neesii*, Steudel.)

Extra-tropical Australia. One of the best of perennial nutritious swamp-grasses.

Danthonia penicillata, F. v. Mueller.

Extra-tropical Australia and New Zealand, ascending to alpine elevations. Mr. A. N. Grant mentions this as the most gregarious of grasses in Riverina, though after seeding early in summer it becomes parched, until it pushes afresh after the first autumnal rains. It is most easily disseminated. Dr. Curl found this perennial grass useful for artificial mixed pasture. Its principal value is in spring. Noted as very valuable in its native localities.

Danthonia robusta, F. v. Mueller.

Australian Alps. Forms large patches of rich forage near or at the very edge of glaciers. The tall D. rigida (Raoul) of New Zealand is closely allied.

Datisca cannabina, Linné.

From Greece to Upper India. A perennial herb of medicinal value; the stems furnish a strong textile fibre; the leaves and roots yield a superior yellow dye.

Daucus Carota, Linné.

The Carrot. Europe, North-Africa, extra-tropical Asia, east to Japan. Biennial. Admits of naturalization along shores. In Norway it is grown to lat. 70° 22′ (Schuebeler). Beyond the ordinary culinary utilization it serves for the distillation of a peculiar oil. Large-rooted varieties as well as the herb give a good admixture to stable-fodder. Carrot-treacle can also be prepared from the root. Requires lime in the soil for its prolific culture. The chemical substances carotin and hydrocarotin are derived from it. Mess. Dippe in Ouedlinburg keep about 130 acres under culture merely for carrot-seeds. They will retain their vitality for a few years ordinarily preserved.

Debregeasia edulis, Weddell.

The Janatsi or Teon-itsigo of Japan. Berries of this bush edible; fibre valuable for textile fabrics. A few Indian species, with fibre resembling that of Boehmeria, ascend the Himalayas for several thousand feet, and may therefore be very hardy—namely: D. velutina, D. Wallichiana, D. hypoleuca. The latter extends to Abyssinia, where it has been noticed at elevations of 8,000 feet. D. dichotoma occurs on mountains in Java.

Decaisnea insignis, J. Hooker and Thomson. (*Slackea insignis*, Griffith.)

Himalaya at 6,000 to 10,000 feet elevation. This showy shrub or miniature-tree produces fruit full of juicy pulp of pleasant sweetness.

Dendrocalamus giganteus, Munro.

Malacca and the adjacent islands. Habit of Gigantochloa maxima; therefore one of the mightiest of all Bamboos. It continues constantly to add stems from its root, several hundred sometimes belonging to the same tuft. Stems reach a height of 100 feet and a circumference of 33 inches; the joints are occasionally as much as 18 inches wide and the walls an inch thick (Dr. Trimen). Locally much used for rural buildings, furnishing posts, rafters, flooring material and shingles (Brandis). Buckets and many other domestic utensils are readily made of this Bamboo. The equally gigantic Dendrocalamus Brandisii (Bambusa Brandisii) of British Burmah has internodes sometimes over 1 foot long, and ligneous substance of over 1 inch thickness. Deciticus of Burmah attains a height of 30 feet, and ascends to 3,000 feet (Kurz).

Dendrocalamus longispathus, Kurz.

British Burmah, where with D. calostachyus (Kurz) it ascends to about 3,500 feet; the former rises to a height of 60 feet. D. membranaceus (Kurz) attains there nearly the same height.

Dendrocalamus Hamiltoni, Nees.

Himalayas, between 2,000 and 6,000 feet. Height reaching 60 feet. The young shoots of this stately Bamboo are edible in a boiled state (Hooker). It endures great cold as well as dry heat (Kurz).

Dendrocalamus strictus, Nees.*

India, extending to Burmah. Grows on drier ground than Bamboos generally. Its strength and solidity render it fit for many select technic purposes. It attains a height of 100 feet, and occasionally forms forests of it own. It endures great cold as well as dry heat (Kurz). Readily raised from seed.

Desmodium acuminatum, De Candolle.

North-America. With D. nudiflorum (D.C.) mentioned by C. Mohr as a nutritive plant for stock, and particularly adapted for forest-soil.

Desmodium triflorum, De Candolle.

In tropical regions of Asia, Africa and America. A densely matted perennial herb, alluded to on this occasion as recommendable for places too hot for ordinary clover, and as representing a large genus of plants, many of which may prove of value for pasture. Dr. Roxburgh already stated, that it helps to form the most beautiful turf in India, and that cattle are very fond of this herb. Colonel Drury informs us, that it is springing up on all soils and situations, supplying the place of Trifolium and Medicago there. D. Canadense (D.C.) is also an excellent fodder-herb (Rosenthal).

Dichopsis Gutta, Bentham.* (*Isonandra Gutta*, Hooker.)

The "Gutta-Percha" or the "Gutta-Taban" Tree. Malayan Peninsula and Sunda-Islands. Attains a height of 150 feet. It seems not altogether hopeless to render this highly important tree a denizen of the mildest wood-regions in temperate climes, Murton having traced it to elevations of 3,500 feet. The milky sap, obtained by ringing the bark at 5 to 15 inches interstices, is boiled for an hour before gradual exsiccation, otherwise the product becomes brittle; 5 to 20 catties yielded by one tree. Genuine Gutta-Percha is only got from plants of the sapotaceous order, as far as hitherto known. Besides Dichopsis Gutta, which yields the best red Gutta-Percha of Borneo, but is slow of growth, the following are actually drawn into use for obtaining this gum-resin: Imbricaria coriacea, A. de Cand.; Mimusops Elengi, L.; M. Manilkara, G. Don; Sideroxylon attenuatum, D.C.; Illipe (Bassia) sericea, Blume; Payenia macrophylla, P. Leerii (which affords the Sundek-Gutta in brackish coast-lands, as shown by Dr. Trimen) and P. Maingayi, Clarke ; Dichopsis obovata, D. polyantha, D. Krantziana, Benth.; Cocosmanthus macrophyllus, Hassk., all from tropical Asia ; Chrysophyllum Africanum, A. de Cand., from tropical Africa; Achras Sapota, L., Mimusops globosa, Gaertner, from Central America; but many of these often at cool elevations. Possibly other

I

sapotaceous trees, including some Australian, could be worked for
Gutta-Percha. Of this article 62,000 cwt. were introduced into Great
Britain in 1884, valued at £462,000, of which quantity this Dichopsis
must have supplied a large proportion. Pierre, after the indications
of Bentham, adopts the generic name Palaguium, and adds as Gutta-
Percha yielding Dichopsis Malaccensis, D. Oxleyana, D. formosa, D.
princeps and D. Borneensis, all previously undescribed species, either
from Malacca or Sumatra or Borneo (see Bulletin mens. de la Soc.
Linn. de Paris, Juin 1885).

Dicksonia Billardierii, F. v. Mueller.   (*D. antarctica,* La Billardière ;
Cibotium Billardierii, Kaulfuss.)

South-Eastern Australia, New Zealand. This tree-fern is mentioned here, as it is the very best for distant transmission, and endures
some frost. It attains a height of 40 feet. Hardy in the Island
of Arran with D. squarrosa and Cyathea medullaris (Rev. D. Landsborough). This species above all others should be disseminated in warmer extra-tropical countries, thus with us in West-
Australia. Important also as commercial plants among fern-trees are
Cyathea medullaris, of South-Eastern Australia and New Zealand;
Cyathea dealbata, the Silvery Tree-fern and C. Smithii, from New
Zealand only; because when upgrown their shipment is not attended
with the same difficulty as that of the tall Alsophila Australis
(which attains 60 feet) and numerous other tree-ferns, about 200
species of which are now known. Those mentioned are among the
hardiest of this noble kind of plants. Anthelmintic properties, which
may exist in these and many other ferns, have not yet been searched
for. The dust-like spores should be scattered through moist forest-
valleys, to ensure new supplies of these superb forms of vegetation for
the next century. D. Billardierii is nowhere antarctic.

Digitalis purpurea, Dodoens.

The Foxglove. Western Europe. A biennial and exceedingly
beautiful herb of great medicinal value, easily raised. In Norway it
grows to lat. 63° 52' (Schuebeler). Chemical principles : digitalin,
digitaletin and three peculiar acids (Wittstein).

Dimochloa Andamanica, Kurz.

Andamans. A scandent Bamboo, rising to fully 100 feet. Should
be of particular value for scenic culture. D. Tjankorreh (Buese)
extends from Java to the Philippines, ascends to 4,000 feet elevation,
but is not so tall as the other species.

Dioscorea aculeata, Linné.*

The Kaawi-Yam. India, Cochin-China, South-Sea Islands. Stem
prickly, as the name implies, not angular. Leaves alternate, undivided. It ripens later than the following species, and requires no
reeds for staking. It is propagated from small tubers. This yam is
of a sweetish taste, and the late Dr. Seemann regarded it as one of

the finest esculent roots of the globe. A variety of a bluish hue, cultivated in Central America (for instance at Caracas), is of very delicious taste.

Dioscorea alata, Linné.*

The Uvi-Yam. India and South-Sea Islands. The stems are four-angled and not prickly. The tubers, of which there are many varieties, will attain under favorable circumstances a length of 8 feet, and the prodigious weight of 100 pounds. This species and the preceding are the two principal kinds cultivated in tropical countries. D. alata is in culture supported by reeds. It is propagated from pieces of the old root, and in warm climes comes to perfection in about seven months. The tubers may be baked or boiled. It is this species, which has been successfully cultivated in New Zealand and also in the Southern States of North-America.

Dioscorea glabra, Roxburgh.* (*D. Batatas*, Decaisne.)

The Chinese Yam. From India to China. Not prickly. The root is known to attain a length of 4 feet, with a circumference of 1¼ inches, and a weight of about 14 lbs. The inner portion of the tuber is of snowy whiteness, of a flaky consistence and of a delicious flavor; preferred by many to potatoes, and obtainable in climes too hot for potato-crops. The bulblets from the axils of the leaf-stalks, as in other Dioscoreas, serve as sets for planting, but the tubers from them attain full size only in the second year. The upper end of the tubers offers ready sets, but there are dormant eyes on any portion of the surface of the tubers (Sir Samuel Wilson, General Noble). First grown in Australia by the author in 1858. A remarkably hardy species ; its yam-root keeps well (Vilmorin).

Dioscorea globosa, Roxburgh.

India. Roxburgh states this to be the most esteemed yam in Bengal.

Dioscorea hastifolia, Nees.

Extra-tropical Western Australia, as far south as 32°. It is evidently one of the hardiest of the yams, and on that account deserves particularly to be drawn into culture. The tubers are largely consumed by the local aborigines for food. This the only plant, on which they bestow any kind of cultivation, crude as it is. Fit for arid situations, but fond of lime.

Dioscorea Japonica, Thunberg.

The hardy Japan-Yam. Not prickly. The material here for comparison is not complete, but seems to indicate, that D. transversa and D. punctata (R. Br.) are both referable to D. Japonica. If this assumption should prove correct, then we have this yam along the coast-tracts of North- and East-Australia, as far south as latitude 33°. In Australia we find the wild root of good taste and large size; the tubers are eaten by the savages raw when young, roasted when aged (E. Palmer).

Dioscorea nummularia, Lamarck.

The Tivoli-Yam. Continental and insular India, also South-Sea Islands. A high-climbing, prickly species, with opposite leaves. Roots cylindrical, as thick as a man's arm; their taste exceedingly good.

Dioscorea oppositifolia, Linné.

India and China. Not prickly. One of the edible yams.

Dioscorea pentaphylla, Linné.

Continental and insular India, also South-Sea Islands. Likewise a good yam. A prickly species, with alternate divided leaves.

Dioscorea purpurea, Roxburgh.

India. In Bengal considered next best to D. alata and D. globosa.

Dioscorea quinqueloba, Thunberg.

Japan, and there one of several yam-plants with edible tubers. Among numerous congeners are mentioned as providing likewise root-vegetables : D. piperifolia (Humboldt) from Quito, D. esurientum (Fenzl) from Guatemala, D. tuberosa and D. conferta (Vellozo) from South-Brazil, D. Cayennensis (Lamarck) from tropical South-America, D. triphylla (Linné) from tropical Asia, D. deltoidea (Wallich) from Nepal. Of these and many other species the relative quality of the roots and their adaptability to field-cultivation, require to be more fully ascertained.

Dioscorea sativa, Linné.

Southern Asia, east as far as Japan, also in the South Sea-Islands, North- and tropical East-Australia, likewise recorded from tropical Africa. Stem cylindrical, not prickly. The acrid root requires soaking before boiling. The plant has proved hardy in the Southern States of North-America. Starch is very profitably obtainable from the tubers.

Dioscorea spicata, Roth.

India. Roots used like those of other species.

Dioscorea tomentosa, Koenig.

Ooyala-Yam. India. The nomenclature of some of the Asiatic species requires further revision.

Dioscorea trifida, Linné fil.

Central America. One of the yams there cultivated. Various other tuberous Dioscoreæ occur in tropical countries, but their respective degrees of hardiness, taste and yield are not recorded or ascertained. The length of the warm season in many extra-tropical countries is probably sufficient for ripening all these yams.

Diospyros Ebenum, Koenig.*

Ceylon, where it furnishes the best kind of Ebony-wood. It is not uncommon up to an elevation of 5,000 feet in that island, according to Dr. Thwaites; hence I would recommend this large and valuable tree for test-plantations in warm extra-tropical lowland forest-regions, where also D. quæsita and D. oppositifolia, the best Calamander-trees, and D. melanoxylon should be tried. Many other species of Diospyros could probably be introduced from the mountains of various tropical regions either for the sake of their ebony-like wood or their fruit. Black Ebony-wood sinks in water. The price in England ranges from £8 to £10 per ton, from 700 to 1,000 tons being imported into Britain annually for pianoforte-keys, the string-holders of musical instruments, the fingerboard and tail-piece of violins, sharp note-pieces of pianos, harmoniums and cabinet-organs, and other select purposes. The following species, some of which may prove hardy, yield Ebony-wood, according to Hiern : *India*—D. Ebenum, Koen, D. melanoxylon, Roxb., D. silvatica, Roxb., D. Gardneri, Thw., D. hirsuta, L. fil., D. discolor, Willd., D. Embropteris, Thw., D. Ebenaster, Retz., D. montana, Roxb., D. insignis, Pers., D. Tupru, Hamilt., D. truncata, Zoll., D. ramiflora, Wall; *Africa*—D. Dendo., Welw., D. mespiliformis, Hochst.; *Mauritius*—D. tesselaria, Poiret; *Madagascar*—D. haplostylis, Boivin, D. microrhombus, Hiern.

Diospyros Kaki, Linné fil.

The Date-plum of China and Japan. A rather slow-growing not very productive tree, hardy at Port Phillip, comes into bearing when only five years old. The fruit is yellow, pink or dark-purple, variable in size, but seldom larger than an ordinary apple; it can readily be dried on strings. A hard and soft variety occur. It has ripened as far north as Philadelphia (Saunders). The most famed varieties are, according to the Rev. Mr. Loorins : Ronosan, Nihon, Micado, Daimio, Taikoon, Yamato, the latter particularly large and saccharine, and with the Jogen-variety mostly used for drying. In Japan this is thought to be the best native fruit (Christie); attains one pound in weight. There is also a small seedless variety. Dried Kaki-fruit is considered superior to figs. For drying the fruit is peeled; it requires a month to exsiccate. The Hyakuma-variety when shrivelled measures as much as 4 by 3 inches (Jarmain). The green fruits serve as medicinal astringents (Dupont). Fruits weighing nearly a pound have been obtained at Melbourne.

Diospyros Lotus, Linné.

From Northern China to the Caucasus. The ordinary Date-plum. The sweet fruits of this tree, resembling black cherries, are edible and also used for the preparation of syrup. The wood, like that of D. chloroxylon, is known in some places as Green Ebony. It resembles Mottled Ebony; it must not however be confounded with other kinds, such as are furnished by some species of Excœcaria, Nectandra and Jacaranda. This tree endures the winters of Northern Germany (C. Koch); in the Crimea it rises to 40 feet (Loudon).

Diospyros Texana, Scheele.

Mexico and Texas. Tree, reaching a height of 30 feet; fruit globose, black, luscious (A. Gray).

Diospyros Virginiana, Linné.

The North-American Ebony or Parsimon or Persimmon, indigenously restricted to the South-Eastern States. A tree, reaching 70 feet in height, sends suckers up from the roots. Wood heavy, very hard, blackish or brownish, valuable for shuttles instead of box-wood. (Jos. Gardner); for turnery, also shoe-lasts (Sargent); for shafts one of the very best (Michaux). The stem exudes a kind of gum. The sweet variety yields a good table-fruit. Ripens fruit to 41° North in Illinois (Bryant). Hot summers promote the early ripening and sweetness of the fruit, the delicious taste not alone depending on early frost. The final sweetness depends upon chemical decomposition. The flowers yield honey. The species is of very close affinity to D. Lotus, as remarked by A. Gray, and endures the clime of Northern Germany also.

Diplothemium campestre, Martius.

Southern Brazil. A dwarf Palm of dry regions. The sweetish pulpy outer covering of the small fruits is edible.

Diposis Bulbocastanum, De Candolle.

Chili. The tubers of this perennial herb are edible (Philippi).

Dipsacus fullonum, Linné.

Fuller's Teasel. Middle Asia. A tall biennal herb. The thorny fruit-heads are used for fulling in cloth-factories. The import into England during one of the last years was valued at £5,000. The plant is most easily reared. The use of these teasels has not yet been superseded by any adequate machinery. The young leaves can be used as food for silkworms (Thorburn). The flowers are rich in nectar for honey (A. J. Cook), which is of excellent quality (Quinby).

Dirca palustris, Linné.

Eastern States of North-America, extending to Canada. An ornamental forest-shrub, the tough bark of which is serviceable for straps and whipcords.

Distichlis maritima, Rafinesque. (*Festuca distichophylla*, J. Hooker.)

North- and South-America, extra-tropical Australia. This dwarf creeping Grass is of great value for binding soil in arid places, forming rough lawns, edging garden-plots and covering coast-sand.

Dolichos gibbosus, Thunberg.

South-Africa. This woody climber is one of the most eligible for covering rustic buildings with a close and almost ever-flowering vegetation.

Dolichos Lablab, Linné.

India, probably from thence only spread widely through the tropics. An annual herb, sometimes lasting through several years, cultivated up to 7,000 feet in the Himalayas ; ripening its fruit at Port Phillip.

The young pods, as well as the ripe seeds, of several varieties available for culinary use. It delights in rich soil, and ripens in hot countries within three months ; its yield is about forty-fold, according to Roxburgh. The whole plant forms excellent stable-feed for cattle.

Dolichos uniflorus, Lamarck.

Tropical and sub-tropical Africa and Asia. An annual herb, the Horse-Gram of South-India, where it is extensively grown. Colonel Sykes got over 300 seeds from a moderate-sized plant. Dr. Stewart saw it cultivated up to 8,000 feet. Content with poor soils ; well adapted for stable pulse.

Dorema Ammoniacum, D. Don.

From Persia extending to Afghanistan and Turkestan, up to 4,000 feet. A tall perennial herb, yielding the gum-resin Ammoniacum, which might be obtained from plants introduced, especially as this plant will grow in dry regions.

Dracæna Draco, Linné.

The Dragon-blood Tree of the Canary-Islands. An imposing feature in scenic horticulture, with D. schizantha (Baker) of eastern tropical Africa ; it yields one kind of Dragon-blood resin. The famed Dragon-tree of Teneriffe, measured in 1831, showed 46 feet stem-circumference, and even at the commencement of the 15th century was celebrated for its age and large proportions.

Dracocephalum Moldavica, Linné.

Northern and Middle Asia. An annual showy scent-herb; also of some medicinal value and particularly for culinary condiment.

Drimys Winteri, R. and G. Forster.

Extra-tropical South-America. The Canelo of Chili, sacred under the name of Boighe to the original inhabitants. Attains in river-valleys a height of 60 feet. The wood never attacked by insects (Dr. Philippi). Bark used for medicinal purposes. The Australian and New Zealand species may be equally valuable.

Duboisia Hopwoodii, F. v. Mueller.

The Pitury. Inland desert-regions from New South Wales and Queensland to near the west-coast of Australia. This shrub deserves cultivation on account of its highly stimulating properties. D. myoporoides (R. Br.) of East-Australia and New Caledonia has come into use for ophthalmic surgery. The alkaloid of the latter, duboisin, is allied to piturin, and important as a mydriatic (Bancroft). The tree attains in deep forest-glens a height of 60 feet (Ralston), but flowers even as a shrub.

Duvaua longifolia, Lindley.

La Plata-States. This shrub and the allied D. latifolia, called Molle there, yield foliage rich in tannin (about 20 per cent.), which, as it does not give any color to leather, is much valued for particular currying (Dr. Lorentz).

Dypsis pinnatifrons, Martius.

Madagascar. This dwarf Palm proved hardy in Sydney, together with Copernicia cerifera (C. Moore).

Ecbalium Elaterium, Richard.

The Squirting Cucumber. Mediterranean regions and Orient. An annual. The powerful purgative Elaterium is prepared from the pulp of the fruit. Chemical principles : elaterid, elaterin, hydroelaterin.

Echinocactus Fendleri, Engelmann.

Mexico. A species attractive for its large rosy flowers and, like the orange-flowered E. gonacanthus and E. Simpsoni, E. conoideus, E. phœniceus, E. viridiflorus, E. viviparus and E. paucispineus, among the most hardy of North-American Cacteæ (E. G. Loder).

Echium candicans, Linné fil.

Madeira. This showy shrub, with the allied E. fastuosum (Jacquin) is an exquisite honey-plant ; other species deserve in this respect also attention.

Ectrosia Gulliveri, F. v. Mueller.

North-Eastern Australia. A pretty but annual grass, useful for covering stony slopes in dry exposed localities (Bailey).

Ehrharta diplax, F. v. Mueller. (*Microlæna avenacea,* J. Hooker.)

New Zealand. This tall perennial grass is fond of woodlands, and deserves introduction elsewhere. It is likely to prove a rich pasture-grass. A few Australian species, particularly of the section Tetrarrhena, are readily accessible, and so indeed also the South-African Ehrhartas, all adapted for a warm temperate clime ; the majority perennial, and several of superior rural value. Ehrharta caudata (Munro) is indigenous to Japan.

Ehrharta longiflora, Smith.

South-Africa. Easily disseminated and, like other perennial species from the same part of the world, fit to grow in sand-land as a pasture-grass. Eagerly consumed by cattle (Cole).

Ehrharta stipoides, La Billardière.

Extra-tropical Australia, also New Zealand. Often called Weeping-Grass. A perennial grass, which keeps beautifully green all through the year. For this reason its growth for pasturage should be encouraged, particularly as it will live on poor soil. Mr. W. H. Bacchus, of Ballarat, considers it nearly as valuable as Kangaroo-Grass, and in the cool season more so. He finds it to bear overstocking better than any other native grass, and to maintain a close turf. Hence it is praised by Mr. O. Tepper as a lawn-grass. High testimony of the value of this grass is also given by Mr. Rankin, of Gippsland, after many years experiments. However, it does not always seed copiously. The chemical analysis, made in spring, gave the following results: Albumen, 1·66, gluten 9·13, starch 1·64, gum 3·25, sugar 5·05 per cent. (F. v. Mueller and L. Rummel).

Elæagnus hortensis, Bieberstein.

From South-Europe and North-Africa to Siberia and China. The fruits of this shrub, known under the name of Trebizonde-dates, are used in Persia for dessert. Flowers highly fragrant (G. W. Johnson).

Elæagnus parvifolius, Royle.

From China to the Himalayas. This bush has been introduced into North-America as a hedge-plant, and, according to Professor Meehan, promises great permanent success, as it has already achieved a high popularity in this respect. In Norway hardy to lat. 59° 55′ (Schuebeler). Sever other species might well be experimented on in the same manner.

Elæagnus umbellatus, Thunberg.

Japan. The fruits of this or an allied species are edible, of a particular and pleasant flavor, and especially adapted for confectionery. This bush resists frost as well as drought, and bears in prodigious abundance throughout the year (Joseph Clarté). It can be struck from cuttings, and comes into bearing in the third year.

Elegia nuda, Kunth.

South-Africa. A rush, able with its long roots to bind moving sand; it also affords good material for thatching (Dr. Pappe). Many of the tall Restiaceæ of South-Africa would prove valuable for scenic effect in gardens and conservatories, and among these may specially be mentioned Cannamois cephalotes (Beauvois).

Elephanthorrhiza Burchelli, Bentham.

South-Africa. The huge club-footed roots of this somewhat shrubby plant are extraordinarily rich in tannin (Prof. Mac Owan). All grazing animals like the foliage much; it starts from the root again after frost (Mrs. Barber). An allied species is E. Burkei.

Eleusine Coracana, Gaertner.

Southern Asia, east to Japan, ascending the Himalayas to 7,000 feet. Though annual, this grass is worthy of cultivation on account of its height and nutritiveness. It is of rapid growth, and the produce of foliage and seeds copious. Horses prefer the hay to any other dry fodder in India, according to Dr. Forbes Watson. The large grains can be used like millet. E. Indica (Gaertner) only differs as a variety. It extends to tropical Australia, and is recorded also from many other tropical countries.

Eleusine stricta, Roxburgh.

India. The increase of grain of this annual grass in rich soil is at times five-hundredfold. E. Tocusso (Fresenius) is a valuable kind from Abyssinia, seemingly allied to E. stricta. The Arabian and Himalayan E. flagellifera (Nees) is perennial. Other species of Eleusine are deserving of trial.

Elymus arenarius, Linné.*

The Sea Lyme-Grass. Europe and North-Asia, on sand-coasts, growing in Norway to lat. 71° 7′. One of the most important and vigorous of grasses for binding drift-sand on the sea-shores. Endures being gradually covered with sand, but not so completely as Psamma. The North-American E. mollis (Trinius) is allied to this species.

Elymus condensatus, Presl.*

The Bunch-Grass of British Columbia and California, extending to lat. 58°. This is favorably known as adapted for sand-land. Prof. Bolander says, that it does excellent service in fixing soil on steep banks. Mr. W. Gorlie noted it to bear severe frost, as much as 0° F. Bunches become fully a yard in diameter and bear stalks up to 10 feet high, so that in annual bulk of weight of produce it surpasses all British pasture-grasses. It is also earlier than any of them, and its young growth never suffers from spring-frosts; moreover it is highly nutritious and greedily eaten in all its stages by stock. This grass should have some claim for adoption in scenic culture.

Elymus Virginicus, Linné.

Eastern North-America. Perennial, easily spreading, but fit for river-banks; of some fodder value (C. Mohr).

Embothrium coccineum, R. and G. Forster.

From Chili to the Straits of Magellan. The Notra or Ciruelillo. A tree of exquisite beauty, but seldom reaching above 30 feet in height. The wood is utilized for furniture. E. lanceolatum is merely a variety (Dr. Philippi). The equally gorgeous E. emarginatum of the Peruvian Andes and E. Wickhami (F. v. M.) from Mount Bellenden-Ker of North-Queensland, deserve, with the East-Australian allied Stenocarpus sinuatus (Endlicher), a place in any sheltered gardens or parks of the warm temperate zone.

Encephalartos Denisonii, F. v. Mueller. (*Macrozamia Denisonii*, Moore and Mueller.)

New South Wales and Queensland, in the litoral forest-tracts. This noble Pine-palm is hardy as far south as Melbourne, and with E. spiralis, E. Preissii and the South-African species to be regarded as a most desirable acquisition to any garden-scenery in mild zones. All admit of translocation even when of large size and when many years old. The lifted stems, with an unusual tenacity of life, sometimes remain dormant for several years. After removal they can be shipped in close cases as dead goods, the leaves being previously cut away, but such shipments should not be exposed to severe frost on transit. Where naturally these Pine-palms abound, an excellent starch may profitably be got from their stems.

Engelhardtia spicata, Blume.

The Spurious Walnut-tree of the mountains of Java, Burmah and the Himalayas. It reaches a height of 200 feet. Wood pale-red, hard and heavy, manufactured into the solid cart-wheels and large troughs, which are in use throughout the Sunda-Islands (Brandis). The bark is rich in tan-substance (Roxburgh).

Eremophila longifolia, F. v. Mueller.

Desert-regions throughout Australia. In the hot season this tall bush or small tree affords food to sheep in desert-tracts, when grass and herbage fail (A. N. Grant). Sheep browse on many other species of this highly ornamental genus. All resist drought and great climatic heat.

Eremurus aurantiacus, Baker.

Afghanistan, 7,000 to 9,000 feet. The leaves of this liliaceous plant form for two months in the year almost the sole vegetable, on which the natives of Hariab depend; it is an agreeable food, crisp and somewhat hard, but neither tough nor fibrous (Dr. Aitchison). Likely to become valuable as a spring-vegetable.

Erianthus fulvus, Kunth.

Interior of Australia. A sweet perennial grass, of which cattle are so fond, as to eat it closely down, and thus cause it to die out (Bailey). Readily raised by re-dissemination.

Erianthus Japonicus, Beauvois.

Japan. Bears frosts of 0° F. (Gorlie). Stems woody at the base, reaching a height of 6 feet with spikes nearly a foot long. The striped-leaved variety is particularly decorative.

Eriochloa annulata, Kunth.

In tropical and sub-tropical regions around the globe. Perennial. Endures moderate cold in South-Queensland, and affords fodder all the year round (Bailey). It resists drought, and is fattening and much relished by stock (Dr. Curl). E. punctata (Hamilton) has a similarly wide range, and is of equal pastoral utility.

Erodium cygnorum, Nees.

Extra-tropical Australia. This herb yields a large amount of feed even in the sandy desert-tracts of Central Australia, and is relished by all kinds of pasture-animals.

Eruca sativa, Lamarck.

From Southern Europe to Central Asia, ascending the Himalayas to 10,000 feet. An annual herb, not unimportant as an oil-plant; much cultivated in some parts of Siberia for its seed (Dr. A. Regel).

Ervum Lens, Linné. (*Lens esculenta,* Moench.)

The Lentil. South-Europe, South-Western Asia. Cultivated up to an elevation of 11,500 feet in India. Annual, affording in its seeds a palatable and very nutritious food. A calcareous soil is essential for the prolific growth of this plant. The leafy stalks, after the removal of the seeds, remain a good stable-fodder. The variety called the Winter-Lentil is more prolific than the Summer-Lentil. Valuable as honey-yielding for· bees. Seeds will retain their vitality for about four years (Vilmorin). The "Revalenta Arabica" consists mainly of lentil-flour.

Eryngium maritimum, Linné.

Europe, North-Africa, South-Western Asia. This perennial herb deserves dissemination on sandy oceanic shores. Root of medicinal value. Young shoots serve as a substitute for asparagus (Dr. Rosenthal), so those of E. campestre (Linné), a plant of similar geographic range.

Eryngium pandanifolium, Chamisso.

South-Brazil, Paraguay, Misiones and Chaco. This or an allied species, called "Caraguata" with bromeliaceous habit, yields there textile fibre, which is long and silky (Kew Report, 1877, p. 37; Gard. Chron., 1882, p. 431, E. H. Egerton).

Erythroxylon Coca, Lamarck.*

Peru and Eastern Bolivia. This shrub is famed for the extraordinary stimulating property of its leaves, which pass under the names of Spadic and Coca. They contain two alkaloids, cocain and hygrin; also a peculiar tannic acid. The cocain has become of great importance in medicine, as a topical anæsthetic, particularly in ophthalmic surgery. In the native country of the plant its leaves have for ages been in use to prepare from them an infusion for allaying local pain. An enormous quantity is annually collected and sold. The Peruvians mix the leaves with the forage of mules, to increase their power of enduring fatigue. Whether any of the many other species ·of Erythroxylon possesses similar properties, seems never yet to have been ascertained.

Eucalyptus Abergiana, F. v. Mueller.

Northern Queensland. A stately tree with unusually spreading branches of dense foliage. The quality of its timber has remained hitherto unknown, but the species will probably prove one of the most suitable among its congeners for tropical countries.

Eucalyptus amygdalina, La Billardière.*

South-Eastern Australia. Vernacularly known as Brown and White Peppermint-tree, Giant-Gumtree, and as one of the Swamp-Gumtrees. In sheltered springy forest-glens attaining exceptionally to a height of over 400 feet, there forming a smooth stem and broad leaves, producing also seedlings of a foliage different from the ordinary form of E. amygdalina, which occurs in more open country, and has small narrow leaves and a rough brownish bark. The former species or variety, which might be called Eucalyptus regnans, represents probably the loftiest tree on the globe. Mr. G. W. Robinson, surveyor, measured a tree at the foot of Mount Baw-Baw, which was 471 feet high. Another tree in the Cape Otway-ranges was found to be 415 feet high and 15 feet in diameter, where cut in felling, at a considerable height above the ground. Another tree measured 69 feet in circumference at the base of the stem; at 12 feet from the ground it had a diameter of 14 feet; at 78 feet a diameter of 9 feet; at 144 feet a diameter of 8 feet, and at 210 feet a diameter of 5 feet. Individual trees are known with a stem-circumference of 56 feet at 5 feet from the ground. The wood is fissile, well adapted for shingles, rails, for inner building-material and many other purposes, but it is not a strong wood. That of the smaller rough-barked variety has proved lasting for fence-posts. La Billardière's name applies ill to any of the forms of this species. Plants raised on rather barren ground near Melbourne have shown nearly the same amazing rapidity of growth as those of E. globulus; yet, like those of E. obliqua, they are not so easily satisfied with any soil. In the south of France this tree grew to a height of 50 feet in eight years. It has endured the frosts of the milder parts of England, with E. Gunnii and E. viminalis. In New Zealand it has survived the cold, where E. globulus succumbed. E. amygdalina, E. urnigera, E. coccifera, E. rostrata and E. corymbosa have proved more hardy than E. globulus, E. diversicolor, E. resinifera, E. longifolia and E. melliodora at Rome, according to the Rev. M. Gildas. E. coccifera being hardier than any other. Professor Ch. Naudin believes, that E. amygdalina will prove hardy along the western maritime districts of France as far north as Brittany; the ordinary variety proved also hardy in the mild clime of Arran (Rev. D. Landsborough); also along with E. globulus at Falmouth (G. H. Taylor), the typical rough-barked form enduring more frost than the silvan form E. regnans. The now well-known Eucalyptus-oil, the distillation of which was initiated by the writer, is furnished in greater or lesser proportion by all the different species. It was first brought extensively into commerce by Mr. Bosisto, who has the credit of having ascertained many of the properties of this oil for technic application. It is this species, which yields more volatile oil than any other hitherto tested, and which therefore is largely chosen for distillation ; thus it is also one of the best for subduing malarian effluvia in fever-regions, although it does not grow with quite the same ease and celerity as E. globulus. The respective hygienic value of various Eucalypts may to some extent be judged from the average percentage of oil in their foliage, as stated below, and as ascertained by Mr. Bosisto, at the author's instance, for the Exhibition of 1862:—

E. amygdalina	...	...	3·313 per cent. volatile oil.
E. oleosa ...	...	...	1·250 ,, ,,
E. Leucoxylon	...	...	1·060 ,, ,,
E. goniocalyx	...	...	0·914 ,, ,,
E. globulus ...	...	...	0·719 ,, ,,
E. obliqua ...	...	...	0·500 ,, ,,

The lesser quantity of oil of E. globulus is however compensated for by the vigor of its growth and the early copiousness of its foliage.

The proportion of oil varies also somewhat according to locality and season. E. rostrata, though one of the poorest in oil, is neverthless important for malaria-regions, as it will grow well on periodically inundated places and even in stagnant water not saline. According to Mr. Osborne's ·experiments initiated by myself, Eucalyptus-oils dissolve the following, among other substances, for select varnishes and other preparations: camphor, pine-resins, mastic, elemi, sandarac, kauri, dammar, asphalt, xanthorrhœa-resin, dragon's blood, benzoe, copal, amber, anime, shellac, caoutchouc, also wax, but not gutta-percha. These substances are arranged here in the order of their greatest solubility. The potash obtainable from the ashes of various Eucalypts varies from 5 to 27 per cent. One ton of the fresh foliage of E. globulus yields about 8¼ lbs. of pearl-ash ; a ton of the green wood, about 2¼ lbs.; of dry wood, about 4½ lbs. For resins, tar, acetic acid, tannin and other products and educts of many Eucalypts, see various documents and reports of the writer, issued from the Melbourne Botanic Garden.

Eucalyptus Baileyana, F. v. Mueller.

South-Queensland and northern part of New South Wales. A tree to about 100 feet high; bark remarkably tenacious (Bailey). The timber splits easily, yet is tough and durable, thus locally used for fence-posts and similar purposes (A. Williams). This species, unlike most of its congeners, can be grown to advantage on sandy soil. Branches more spreading and foliage more dense than of most other Eucalypts.

Eucalyptus botryoides, Smith.*

From East-Gippsland to South-Queensland. Vernacular name Bastard-Mahogany, and a variety called Bangalay, the latter generally found on coast-sands. One of the most stately among an extensive number of species, remarkable for its dark-green shady foliage. It delights in river-banks. Stems attain a height of 80 feet without a branch and a diameter of 8 feet. The timber usually sound to the centre, adapted for water-works, wagons, particularly for felloes, also knees of boats. Posts formed of it are very lasting, as no decay was observed in fourteen years; it is also well adapted for shingles. The Rev. Dr. Woolls, Mr. Kirton and Mr. Reader all testify to its general excellence.

Eucalyptus calophylla, R. Brown.

South-Western Australia, where it is vernacularly known as Red Gumtree. More umbrageous than most Eucalypts and of comparatively rapid growth. In its native forests it has quite the aspect of the eastern Bloodwood-trees. The wood is almost destitute of resin when grown on alluvial land, but not so when produced on stony ranges. It is preferred to that of E. marginata and E. cornuta for rafters, spokes and fence-rails, also used for handles and agricultural imple-ments; it is strong and light, but not durable underground. The bark is valuable for tanning, as an admixture to Acacia-bark; the seed-vessels of this and perhaps all other Eucalypts can be used for the same purpose. The stem of this tree may occasionally be observed to 10 feet in diameter; it is the only tree in West-Australia which yields copiously the fluid and indurating Eucalyptus-kino; this is soluble in cold water to the extent of 70 to 80 per cent. This species will only endure a slight frost; its. flowers are much frequented by the honey-bee.

Eucalyptus capitellata, Smith.

One of the Stringybark-trees. of South-East-Australia, extending into the dry mallee-country, attaining occasionally a height of 200 feet. The timber is principally used for fence-rails, shingles and rough building purposes. This species might with advantage be raised on wet sandy land.

Eucalyptus citriodora, Hooker.

Queensland. A handsome slender tree with a smooth white bark, supplying a useful timber. Succeeded remarkably well at Lucknow (Ridley), also at Zanzibar, where it grows at a tremendous rate (Sir J. Kirk), and thrives also in Bengal. According to notes of the late Mr. Thozet, a trunk 40 feet long and 20 inches in diameter broke after a flexion of 17 inches, under a pressure of 49 tons. This species combines with the ordinary qualities of many Eucalypts the advantage of yielding from its leaves a rather large supply of volatile oil (slightly more than one per cent.) of excellent lemon-like fragrance, in which respect it has, among about 120 species of Eucalypts, only one rival. Very closely allied to E. maculata, and perhaps only a variety. Particularly adapted for a tropical jungle-clime.

Eucalyptus cordata, La Billardière.

Southern Tasmania. Maximum height, 50 feet; flowering in a shrubby state already. The variety E. urnigera (J. Hooker) is particularly hardy, and may become of sanitary importance to colder countries in malarian regions, the foliage being much imbued with antiseptic oil. Greatest height of E. urnigera 150 feet; stem diameter to 6 feet (Abbott).

Eucalyptus cornuta, La Billardière.*

The Yate-tree of South-Western Australia. A large tree of rapid growth, preferring a somewhat humid soil. The wood is used for various artisans' work, and preferred there for the strongest shafts and frames of carts and other work requiring hardness, toughness and elasticity, and is considered equal to ordinary ash-wood. The tree appears to be well adapted for tropical countries, for Dr. Bonavia reports, that it attained a height of 8 to 10 feet in the first year of its growth at Lucknow, and that the plants did not suffer in the rainy season like many other Eucalypts. The dry wood sinks in water. E. occidentalis (Endlicher) is the flat-topped Yate, an allied and equally valuable species of South-Western Australia.

Eucalyptus corymbosa, Smith.

The principal Bloodwood-tree of New South Wales and Queensland. A tree attaining large dimensions; it has a rough furrowed bark and a dark-reddish wood, soft when fresh, but very hard when dry; very durable underground, and therefore extensively used for fence-posts, rails, railway-sleepers and rough building-purposes. The bark is rich in kino.

Eucalyptus corynocalyx, F. v. Mueller.

South-Australia, North-Western Victoria. The Sugar-Gumtree. A timber-tree, attaining a height of 120 feet, length of bole to 60 feet, circumference at 5 feet from the ground reaching 17 feet. The wood has come into use for fence-posts and railway-sleepers. Its durability is attested by the fact, that posts set in the ground fifteen years showed no sign of decay. The tree thrives well even on dry ironstone-ranges. The base of the trunk swells often out in regular tiers. The sweetish foliage attracts cattle and sheep, which browse on the lower branches, as well as on saplings and seedlings. Scarcely any other Eucalypt is similarly eaten (J. E. Brown). In ordinary culture the writer did not find this species of very quick growth; but Mr. Brown records, that under favorable circumstances it will grow one foot a month.

Eucalyptus crebra, F. v. Mueller.

The Narrow-leaved Ironbark-tree of New South Wales and Queensland. Wood reddish, hard, heavy, elastic and durable; much used in the construction of bridges and for railway-sleepers, also for wagons, piles, fence-posts. E. leptophleba and E. drepanophylla are closely allied species of similar value.. They all exude astringent gum-resin, resembling kino in appearance and property, in considerable quantity.

Eucalyptus diversicolor, F. v. Mueller.* (*E. colossea,* F. v. M.)

The Karri of South-Western Australia. A colossal tree, exceptionally reaching the height of 400 feet, with a proportionate girth of the stem. Mr. Muir measured stems nearly 300 feet long without a branch; widths of timber of as much as 12 feet can be obtained. Furnishes good timber for building, even for masts, likewise for planks; also valuable for shafts, spokes, felloes, fence-rails; it is elastic, but not so easily wrought as that of E. marginata. Mr. G. Simson does not deem the wood very durable, if exposed to moisture or if in contact with the ground. Its strength in transverse strain is equal to English oak. Wood exposed to the wash of the tide for twenty-six years continued quite sound. Fair progress of growth is shown by the trees planted even in dry exposed localities in Melbourne. The shady foliage and quick growth of the tree promise to render it one of our best for avenues. In its native localities it occupies fertile, rather humid valleys, and resembles there in habit the E. amygdalina *var.* regnans of South-Eastern Australia.

Eucalyptus Doratoxylon, F. v. Mueller.

The Spearwood-tree of South-Western Australia, where it occurs in sterile districts. The stem is slender and remarkably straight, and the wood of such firmness and elasticity, that the nomadic natives wander long distances, to obtain it as a material for their spears.

Eucalyptus eugenioides, Sieber. .

One of the Stringybark-trees of Victoria and New South Wales. The tree is abundant in some localities, and attains considerable dimensions. Its useful fissile wood is employed for fencing- and building-purposes. Systematically the species is closely allied to ·E. piperita.

Eucalyptus ficifolia, F. v. Mueller.*

South-Western Australia.¦ Although not a tree of large dimensions, this splendid species should be mentioned for the sake of its magnificent trusses of crimson flowers, irrespective of its claims as a shady, heat-resisting avenue-tree, not standing in need of watering. It bears a close resemblance to E. calophylla.

Eucalyptus globulus, La Billardière.*

The Blue Gumtree of Victoria and Tasmania, famed all over the world. The tree is, among evergreen trees, of unparalleled rapid growth, and attains exceptionally a height of 350 feet, furnishing a first-class wood. Ship-builders can get keels of this timber 120 feet long; besides this, they use it extensively for planking and many other parts of the ship. Experiments on the strength of various woods, instituted by Mr. Luehmann and the author, proved the wood of the Blue Gumtree in average of eleven tests to be about equal to the best English oak, American white oak and American ash. The best samples indeed carried as great a weight as hickory in transverse strain, the ordinary kind about as much as that of Eucalyptus rostrata, and more than that of E. macrorrhyncha, E. Gunnii, E. Sturtiana and E. goniocalyx, but did not come quite up to the strength of .E. melliodora, E. polyanthema, E. siderophloia and E. Leucoxylon. Bluegum-wood is also very extensively used by carpenters for all kinds of out-door work, joists and studs of wooden houses ; also for fence-rails, telegraph-poles, railway-sleepers (lasting nine years or more), for shafts and spokes of drays and a variety of other purposes. Mr. W. Tait, of Oporto, has recommended the wood for wine-casks, these requiring no soaking. The price of the timber in Melbourne is about 1s. 7d. per cubic foot. In South-Europe the E. globulus has withstood a temperature of 19° F., but' succumbed at 17° F.; it perished from frost at the Black Sea and in Turkestan, when young, according to Dr. Regel. The sirocco, however, does not destroy it. In Jamaica it attained 60 feet in seven years, on the hills ; in California it grew 60 feet in eleven years, in Florida 40 feet in four years, with a stem of 1 foot in diameter. In some parts of India its growth has been even more rapid ; at the Nilgiri-Hills it has been reared advantageously, where E. marginata, E. obliqua, E. robusta and E.

calophylla had failed. Its growth was there found to be four times as fast as that of teak, and the wood proved for many purposes as valuable. Trees attained a height of 30 feet in four years ; one tree, twelve years old, was 100 feet high, and 6 feet in girth at 3 feet from the ground ; to thrive well there it wants an elevation of not less than 4,000 feet. It has succeeded particularly well at elevations of from 2,500 to 7,000 feet in Central Mexico (Dr. Mariano Barccna). In Algeria and Portugal it has furnished railway-sleepers in eight years, and telegraph poles in ten years (Cruikshank). At Urana it grew 15 feet in two years, with irrigation (E. van Weenen). On the mountains of Guatemala it attained, in twelve years, a height of 120 feet and a stem-circumference of 9 feet (Boucard). According to the Rev. D. Landsborough, it proved hardy in the Isle of Arran. Mr. Ch. Traill notes it as thriving amazingly as far south as Stewart-Island. For window-culture in cold countries E. globulus was first recommended by Ucke ; for culture in hospital wards to counteract contagia, by Mosler and Goeze. Eucalyptus leaves generate ozone largely for the purification of air ; the volatile oil is very antiseptic. This tree, particularly when in an unhealthy state, is, at Melbourne, apt to be bored by the larva of a large moth (Endoxyla Eucalypti) and also by two beetles (Hapatesus hirtus and particularly Phoracantha tricuspis), as noticed by Mr. Ch. French. Seeds will keep for several years, admit of easy transmission abroad, and germinate quickly; but a tree of such celerity in growth and of such vast final dimensions wants necessarily soil open to great depth for full scope of its roots to attain unimpaired development. Mr. T. Waugh observed in South-Island, New Zealand, that plants, raised from locally ripened seeds, proved hardier than those raised from Australian ordinary seed.

Eucalyptus gomphocephala, De Candolle.*

The Tooart of South-Western Australia; attains a height of 120 feet, the clear trunk a length of 50 feet. The wood is tough, heavy and rigid, the texture close and the grain so twisted, as to make it difficult to cleave. It shrinks but little, does not split while undergoing the process of seasoning, and is altogether remarkably free from defects. It will bear exposure to all vicissitudes of weather for a long time, and is particularly valuable for large scantling, where great strength is needed; in ship-building it is used for beams, keelsons, stern-posts, engine-bearers and other work below the floatation; recommendable also for supports of bridges, framing of dock-gates and for wheelwrights' work; indeed it is one of the strongest woods known, whether tried transversely or otherwise (Laslett). This species, as well as E. odorata, E. fœcunda and E. decipiens, thrive best in limestone-soil.

Eucalyptus goniocalyx, F. v. Mueller.*

Generally known as Bastard-Boxtree and occasionally as Spotted Gumtree. From Cape Otway to the southern parts of New South Wales, rare near St. Vincent's Gulf (McEwin) and Flinders-Range (J. E. Brown). A large tree, which should be included among those for Eucalyptus-plantations. Its wood resembles in many respects that of E. globulus, and is comparatively speaking easily worked. For house-building, fence-rails and similar purposes it is extensively employed in those forest-districts, where it is abundant, and has proved a valuable timber. It is especially esteemed for wheelwrights' work (Falck). Our local experiments showed the strength greater than that of E. amygdalina and E. obliqua, but less than that of E. globulus. Melitose is formed occasionally on this tree and also on the following species.

Eucalyptus Gunnii, J. Hooker.*

Known as Swamp-Gumtree, the mountain-variety as Cider-tree. Victoria, Tasmania and New South Wales, ascending alpine elevations. In the low land along fertile valleys it attains a considerable size, and supplies a strong useful timber. It is this species, which survived severe frosts at Kew Gardens. Bees obtain unusually much honey from the flowers of this species. Cattle and sheep browse on the foliage. Timber found to be almost equal in strength to that of E. macrorrhyncha, E. rostrata and E. globulus. The other very hardy Eucalypts comprise E. pauciflora, E. alpina, E. urnigera, E. coccifera, and E. vernicosa, which all reach heights covered with snow for several months in the year. Succeeded well at Arran (Capt. Brown and Rev. D. Landsborough). Any cutting down of healthy Eucalypts in places of centres of populations, to substitute for them in a zone of evergreen vegetation northern deciduous trees extensively, is regrettable, as therewith the far superior hygienic value of the Eucalypts is lost.

Eucalyptus haemastoma, Smith.

One of the White Gumtrees of New South Wales and Southern Queensland, abundant in many localities. This species attains a very considerable size, and furnishes fencing and rough building material, also fuel of fair quality. Claims our attention particularly as fit for culture on sandy land, for which very few other Eucalypts are suited. A variety occurs with persistent stringy bark.

Eucalyptus hemiphloia, F. v. Mueller.*

South-Eastern Australia, particularly inland. A tree, reaching 90 feet in height and 4 feet in stem-diameter. Trunk generally not tall. Regarded as a timber-tree of great excellence. It is famous for the hardness and toughness of its timber, which is used for railway-sleepers, telegraph-poles, shafts, spokes, mauls, plough-beams and similar utensils. This passes at its places of growth as a "Box-tree" and a variety of it as "White Boxtree."

Eucalyptus Howittiana, F. v. Mueller.

Litoral North-Queensland. A tree, gaining a height of fully 100 feet, with remarkably umbrageous foliage; girth of stem towards the base to fully 12 feet. Wood comparable to that of the so-called "Boxtrees," but straighter in grain. This species would be particularly adapted for intra-tropical countries.

Eucalyptus largiflorens, F. v. Mueller.

South-Eastern Australia, principally in the inland-districts. One of the so-called Boxtrees, rising to a height of 120 feet as a maximum. Stem-diameter to 3 feet. Wood dark brown-red, excessively hard; fence-posts from this wood were found quite sound after 30 years.

Eucalyptus Leucoxylon, F. v. Mueller.*

The ordinary Ironbark-tree of Victoria and some parts of South-Australia and New South Wales. It attains a height of 100 feet, and supplies a most valuable timber, possessing great strength and hardness; it is much prized for its durability, is largely employed by wagon-builders for wheels and poles, by ship-builders for top-sides, tree-nails, the rudder (stock), belaying pins and other purposes; it is also used by turners for rough work. It proved to be the strongest of all the woods hitherto subjected to test by Mr. Luehmann and myself, bearing nearly twice the strain of American oak and ash, and excelling even hickory by about 18 per cent. It is much recommended for railway-sleepers and extensively used in underground mining work. It is likewise very extensively employed for the handles of axes and other implements by Victorian manufacturers. The price of the timber in the log is about 2s. 5d. per cubic foot in Melbourne. As it is, for some purposes superior to that of almost any other Eucalyptus, the regular culture of this tree over wide areas should be fostered, especially as it can be raised on stony ridges not readily available for ordinary husbandry. The wood is sometimes pale, in other localities rather dark. The tree is generally restricted to the Lower Silurian sandstone- and slate-formation with ironstone and quartz. Nevertheless, it accommodates itself to various geologic formations, thus even to limestone-ground. The bark is remarkably rich in kino-tannin, yielding as much as 22 per cent. in the fresh state, but much less after drying; the fresh leaves contain about 5 per cent. and the dried leaves 9 to 10 per cent. This kino-tannin is not equal in value to mimosa-tannic acid from Acacia-bark, but it is useful as a subsidiary admixture, when light-colored leather is not aimed at. As an astringent drug this kino is not without importance. The flowers are sought by bees, even more eagerly than those of most Eucalypts. E. Leucoxylon has, next to E. rostrata, thriven best about Lucknow (in India) among the species tried there for forest-culture. E. Sideroxylon is a synonym, referring particularly to the rough-barked variety.

Eucalyptus longifolia, Link.*

Extra-tropic Eastern Australia. A tree, known as the Woolly Butt, under favorable circumstances reaching 200 feet in height, the stem attaining a great girth. Mr. J. Reader asserts, that there is not extant a more useful timber; it stands well in any situation.

Eucalyptus loxophleba, Bentham.*

The York-Gumtree of extra-tropic West-Australia. Attains a height of about 100 feet, the stem a diameter of four feet. The wood is very tough, and preferably sought in West-Australia for naves and felloes. Even when dry it is heavier than water. This species passes into the earlier known E. foecunda (Endlicher).

Eucalyptus macrorrhyncha, F. v. Mueller.

The common Stringybark-tree of Victoria, not extending far into New South Wales. This tree attains a height of 120 feet, and is generally found growing on sterile ridges, not ascending higher mountains. The wood, which contains a good deal of kino, is used for joists, keels of boats, fence-rails and rough building purposes, also extensively for fuel. The fibrous dark-brown bark serves for roofs of huts and also for rough tying. The wood proved in our experiments here nearly as strong as that of E. globulus and E. rostrata, and considerably stronger than than that of E. obliqua. The fresh bark contains from 11 to 14 per cent. of kino-tannic acid (F. v. M. and Rummel).

Eucalyptus maculata, Hooker.

The Spotted Gumtree of New South Wales and Queensland. A tree, reaching 150 feet in height, the wood of which is employed in ship-building, wheelwrights' and coopers' work. The heart-wood is as strong as that of British oak (Rev. Dr. Woolls). Content with poor soil.

Eucalyptus marginata, Smith.*

The Jarrah or Mahogany-tree of South-Western Australia, famed for its indestructible wood, which is neither attacked by chelura, nor teredo, nor termites, and therefore much sought for jetties and other structures exposed to sea water, also for any underground-work, telegraph-poles, and largely exported for railway-sleepers. Vessels built of this timber have been enabled to do away with copper-plating. For jetties the piles are used round, and they do not split when rammed even into limestone or other hard foundations, provided the timber is of the best hard kind (Walker and Swan). The Government Clerk of Works at Perth observes, that he took up piles in 1877, which were driven for a whaling jetty in 1834, and that the timber was perfectly sound, although the place was swarming with teredo. At the jetty in Fremantle, piles thirty years old and others one year old could scarcely be distinguished. The durability of the timber seems largely attributable to Kino-red, allied to phlobaphen, of which it contains about 15 to 17 per cent. Of kino-tannin it contains 4 to 5 per cent. It is of a close grain and a slightly oily and resinous nature ; it works well, makes a fine finish, and is by local ship-builders considered superior to either sâl, teak or any other wood, except perhaps English oak or live-oak. In West-Australia it is much used for flooring, rafters, shingles; also for furniture, as it is easily worked, takes a good polish and then looks very beautiful. It is not too hard, and hence is more easily worked than E. redunca and E. loxophleba. The wood from the hills is darker, tougher and heavier than that from the plains. Well-seasoned timber weighs about 64 lbs.

per cubic foot; freshly cut, from 71 to 76 lbs. It is one of the least inflammable woods according to Captain Fawcett, and is locally regarded as one of the best woods for charcoal. Mr. H. E. Victor, C.E., of Perth, estimates the area covered at present by marketable Jarrah in South-Western Australia at nine million acres, and its yield at an average about 500 cubic feet of good timber per acre. The trees should be felled in autumn or towards the end of summer, in which case the timber will not warp. The tree grows chiefly on ironstone-ranges. At Melbourne it is not quick of growth, if compared to E. globulus or to E. obliqua, but it is likely to grow with celerity in mountain-regions. Massed in its native country it presents the features of the East-Australian stringybark-forests. Stems of this tree have been measured 80 feet to the first branch, and 32 feet in circumference at 5 feet from the ground. Instances are even on record of the stem having attained a girth of 60 feet at 6 feet from the ground, through the formation of buttresses.

Eucalyptus melanophloia, F. v. Mueller.

The Silver-leaved Ironbark-tree of New South Wales and Queensland. A middle-sized tree with a deeply furrowed bark and mealy white foliage. The timber is strong and durable, and used for telegraph-poles and railway-sleepers; it is however apt to rend, when exposed to the sun, unless well seasoned. Mr. Cosmo Newbery obtained from the bark 9 to 10 per cent. tannin.

Eucalyptus melliodora, A. Cunningham.*

The Yellow Boxtree of Victoria and some parts of New South Wales; of a spreading habit of growth, attaining a height of about 120 feet, with a comparatively stout stem. The wood resembles that of E. rostrata in texture, but is of a paler color, and not quite so durable; it is fully as strong, though second to that of E. Leucoxylon, E. siderophloia and E. polyanthema in this respect, but equalling that of E. globulus; it is esteemed for wheelwrights' and other artisans' work, in ship-building, and supplies excellent fuel; the young trees are used for telegraph-poles. Flowers much sought by bees.

Eucalyptus microcorys, F. v. Mueller.*

One of the Stringybark-trees of New South Wales and South-Queensland, mostly known as Tallow-wood by the colonists. It attains a great size; barrel up to 100 feet in length and to 7 feet in diameter. The wood is yellowish, free from kino-veins, easily worked by saw or plane; it is of a very greasy nature, so much so as to be quite slippery when fresh cut (Ch. Fawcett). This oily substance, very similar to viscin, of which it contains about 1 per cent., prevents the wood from splitting and twisting, though not from shrinking. The timber is also hard and durable underground, and is employed for railway-sleepers, wheelwrights' work, for knees and breasthooks in ship-building; the young trees serve for telegraph-poles. The foliage is remarkably rich in volatile oil. This species did particularly well at George-town in British Guiana, growing at a rate from 10 to 18 feet in a year while in a young state (Jenman).

Eucalyptus microtheca, F. v. Mueller.

Widely dispersed over the most arid extra-tropical as well as tropical inland-regions of Australia. Withstood unscorched a frequent heat of 156° F. in Central Australia. One of the best trees for desert-tracts; in favorable places 150 feet high. Wood brown, sometimes very dark, hard, heavy and elastic; it is prettily marked, hence used for cabinet-work, but more particularly for piles, bridges and railway-sleepers (Rev. Dr. Woolls).

Eucalyptus obliqua, L'Héritier.*

The ordinary Stringybark-tree of Tasmania, generally designated Messmate-tree in Victoria, attaining a height of 300 feet, with a stem more than 10 feet in diameter, growing mostly in mountainous country. The most gregarious of all Eucalypts from Spencer's Gulf to the southern parts of New South Wales, and in several varieties designated by splitters and other wood-workers by different names. Most extensively used for cheap fencing rails, palings, shingles, and any other rough wood-work, not to be sunk underground, nor requiring great strength or elasticity. The bulk of wood obtained from this tree in very poor soil is perhaps larger than that of any other kind, and thus this species can be included even in its native country, where it is naturally common and easily re-disseminated, among the trees for new forest-plantations in barren woodless tracts, with a view of obtaining a ready and early supply of cheap and easily fissile wood. The young trees are sometimes used for telegraph-poles. The fresh bark contains from 11 to 13½ per cent. kino-tannic acid.

Eucalyptus ochrophloia, F. v. Mueller.

Arid interior of sub-tropic Eastern Australia, on ground subject to occasional floods. A tree seldom over 30 feet high. Wood tough, serviceable for slabs, rails and rafters, but not lasting underground. A tree recommendable for hot and dry regions.

Eucalyptus odorata, Behr.

The Peppermint-tree of South-Australia. Reaching 70 feet in height, the stem 2½ feet in diameter. Timber hard, very durable; used for sleepers, posts and piles (J. E. Brown). The tree follows the limestone-formation, but occurs also in clayey loam; 1,000 lbs. of foliage gave 112 oz. of oil (Nitschke).

Eucalyptus oleosa, F. v. Mueller.

One of the smaller Eucalypts known as Mallee, extending from East- to West-Australia through the desert-regions. The essential oil, in which the foliage of this species is comparatively rich, dissolves india-rubber without heat, according to Mr. Bosisto. It is also one of the best solvents for amber and other fossil resins. The variety longirostris attains a height of 120 feet, with a stem of 70 feet without a branch, in West-Australia, where it is vernacularly known as Morrell. The wood is remarkably hard, splits freely, and is used for spars, rafters, fence-rails, wheelwrights' work and agricultural implements. It is of a red tinge, and sinks in water, even when dry. 1,000 lbs. of fresh foliage gave, in distillation, 62 oz. of oil (Nitschke). Of other Mallees E. gracilis gave 54 oz.; E. uncinata, 69 oz.; E. incrassata, 112 oz.

Eucalyptus paniculata, Smith.*

The Red Ironbark-tree of New South Wales. This species furnishes a hard durable wood, excellent for railway-sleepers. It is also much used for building and fencing, as it splits well and is lasting underground. All the trees of this series are deserving of cultivation, as their wood, though always excellent, is far from alike, and that of each species preferred for special purposes of the artisan.

Eucalyptus patens, Bentham.

The "Black-butt" of South-Western Australia. Attains a height of 120 feet and a stem-diameter of 6 feet. The timber is so tough as not to yield to ordinary splitting processes, therefore useful for various wheelwrights' work; it has proved also durable underground.

Eucalyptus pauciflora, Sieber. (*E. coriacea*, A. Cunningham.)

Vernacularly known as White-gumtree, Drooping-gumtree, or Swamp-gumtree. New South Wales, Victoria, Tasmania. A tree of handsome appearance, with a smooth white bark and generally drooping foliage; it attains considerable dimensions, grows best in moist ground, ascends to alpine elevations, and thus is one of the hardiest of all its congeners, enduring the winters of Arran (Rev. D. Landsborough); it shows a preference for basaltic soil. Horses, cattle and sheep browse readily on the foliage. It is locally a "stand-by" in bad pastoral seasons. Its timber is used for ordinary building and fencing purposes.

Eucalyptus phœnicea, F. v. Mueller.

Carpentaria and Arnhem's Land. Of the quality of the timber hardly anything is known, but the brilliancy of its scarlet flowers recommends this species for a place in any park-plantation of countries with a serene clime. For the same reason also E. miniata (Cunningham) from North-Australia, and E. ficifolia (F. v. M.) from South-Western Australia, should be brought extensively under cultivation.

Eucalyptus pilularis, Smith.

The Black-butt of South-Queensland, New South Wales and Gippsland. · One of the best timber-yielding trees about Sydney; of rather rapid growth (Rev. Dr. Woolls). It is much used for flooring-boards, also for railway-sleepers and telegraph-poles. Messrs. Camara and Kirton measured a tree in the Illawarra-district, which had a stem-girth of 45 feet and a height of about 300 feet.

Eucalyptus piperita, Smith.

New South Wales and Gippsland, often termed White Stringy-bark-tree. It grows to a considerable height, and its stem attains a diameter of four feet. The wood is fissile, and used for the same purposes as that of other Stringybark-trees. The foliage is rich in volatile oil. All Eucalypts with strong-scented foliage are useful as insecticides; the fresh leaves also purify the air of unsalubrious dwellings and hospitals.

Eucalyptus Planchoniana, F. v. Mueller.

South-Queensland and Northern part of New South Wales. A tree to about 100 feet in height, the stem reaching 3 feet in diameter. The foliage is dense. Timber sound, heavy, hard and durable, well adapted for sawing, but not easy to split (Bailey).

Eucalyptus platyphylla, F. v. Mueller.

Queensland. Regarded by the Rev. Julian Tenison-Woods as one of the best of shade-trees, and seen by him to produce leaves sometimes $1\frac{1}{2}$ feet long and 1 foot wide. This tree is available for open exposed localities, where any kinds of trees from deep forest-valleys would not thrive. It is closely allied to E. alba from Timor. The timber is curly and durable.

Eucalyptus polyanthema, Schauer.*

South-Eastern Australia. Generally known as Red Boxtree. A tree attaining a height of 150 feet; it furnishes an extremely hard and lasting timber, in great demand for mining purposes and railway-sleepers, also for wheelwrights' work. For fuel this wood is unsurpassed. It is extremely strong, excelling oak and ash, surpassed among Eucalypts in transverse strength, according to our experiments, only by E. Leucoxylon and E. siderophloia.

Eucalyptus populifolia, Hooker.

The "Bembil" or Shining-leaved Box-Eucalyptus. Dry inland-portions of Eastern Australia. Height not above 50 feet; but stem-diameter to 3 feet. Leaves deciduous for a short time in the year. Likes humid underground. Wood used for posts, handspikes, levers and other articles needing toughness; proved to be very durable (Bailey).

Eucalyptus punctata, De Candolle.*

The Leatherjacket or Hickory-Eucalypt of New South Wales. A beautiful tree, attaining a height of 100 feet or more, of rather quick growth, thriving even in poor soil. The wood is of a light-brown color, hard, tough and very durable; used for fence-posts, railway-sleepers, wheelwrights' work, also for ship-building (Woolls).

Eucalyptus Raveretiana, F. v. Mueller.*

Vernacularly known as Grey- or Iron-Gumtree. Queensland. A tree of the largest size, attaining a height of 300 feet and a stem-diameter of 10 feet; delights in the immediate vicinity of rivers or swamps.

It furnishes a very hard, durable, dark-colored wood, valuable for piles, railway-sleepers, and general building purposes (Thozet, O'Shanesy, Bowman). From cuts into the stem an acidulous almost colorless liquid exudes, available in considerable quantity, like that of E. Gunnii.

Eucalyptus redunca, Schauer.*

The White Gumtree of Western Australia, the Wandoo of the aborigines. Attains very large dimensions; stems have been found with a diameter of 17 feet. The bark is whitish, but not shining, imparting a white coloration when rubbed (Hon. J. Forrest). The tree is content with cold flats of comparatively poor soil, even where humidity stagnates during the wet season. It furnishes a pale, hard, tough, heavy and durable wood, highly prized for all kinds of wheel-wrights' work, and especially supplying the best felloes in West-Australia. The seasoned timber weighs about 70 lbs. per cubic foot.

Eucalyptus resinifera, Smith.*

The Red Mahogany-Eucalypt of South-Queensland and New South Wales. A superior timber-tree, of large size. Wood much prized for its strength and durability (Rev. Dr. Woolls). This Eucalyptus has proved one of the best adapted for a tropical clime; it grew 45 feet in ten years at Lucknow, but in the best soil it has attained 12 feet in two years (Dr. Bonavia). Proved in Italy nearly as hardy a E. amygdalina and E. viminalis, according to Prince Troubetzkoy, but is often confounded with E. siderophloia. A large-leaved variety extends far into the tropics.

Eucalyptus robusta, Smith.*

New South Wales and Southern Queensland, where it is known as Swamp-Mahogany. It attains a height of 100 feet and a stem-girth of 12 feet, bearing a really grand mass of foliage. Resists cyclones better than most of its congeners. The wood is remarkably durable, reckoned a fairly good timber for joists, also used for ship-building, wheelwrights' work and many implements, for instance such as mallets. The tree seems to thrive well in low, sour swampy ground near the sea-coast, where other Eucalypts look sickly but E. robusta the picture of health (W. Kirton).

Eucalyptus rostrata, Schlechtendal.*

The Red Gumtree of Southern Australia and many river-flats in the interior of the Australian continent, nearly always found on moist ground with a clayey subsoil. It will thrive in ground, periodically inundated for a considerable time, and even in slightly saline places. Attains exceptionally a height of 200 feet with a comparatively slight stem, but is mostly of a more spreading habit of growth than the majority of its tall congeners. Prof. Tate measured a tree on Mount Lofty which showed a stem-girth of 25 feet. Mr. R. G. Drysdale of the Riverina-district observed, that an exceptional temperature of

125° F. in the shade did not shrivel the foliage of this tree; it has also withstood the severest heat in Algeria better than E. globulus; and Dr. Bonavia found it to thrive well in the province of Oude in places, where E. globulus, E. obliqua and E. marginata perished under the extreme vicissitudes of the clime. It does not bear cold so well as E. amygdalina, succumbing when still young at a temperature below 23° F., as observed in Italy by Prince Troubetzkoy. In Mauritius and Réunion it resisted the hurricanes better than any other Eucalypt; in the latter island the Marquis de Chateauvieux observed it to grow 65 feet in six years, and it is always of more rapid growth than E. marginata, but less so than E. globulus. It grew also with remarkable rapidity in British Guiana (Jenman). It is recommended as an antiseptic tree for cemeteries in tropical countries. The timber is one of the most highly esteemed in all Australia among that of Eucalypts, being heavy, hard, strong and extremely durable, either above or under ground, or in water. For these reasons it is very much prized for fence-posts, piles and railway-sleepers. For the latter purpose it will last at least a dozen years, but if well selected much longer. It is also extensively employed by ship-builders for main-stem, stern-post, inner-post, dead-wood, floor-timbers, futtocks, transoms, knighthead, hawse-pieces, cant-, stern-, quarter- and fashion-timbers, bottom-planks, breast-hooks and riders, windlass and bow-rails. It should be steamed before it is worked for planking. Also largely used for felloes, buffers and posts and any parts of structures, which come in contact with the ground; not surpassed in endurance for woodbricks in street paving and for tramways. Next to the Jarrah from West Australia this is the best Eucalyptus-wood for resisting the attacks of the crustaceous chelura and limnoria, the teredo-mollusk and white ants, and it has the advantage of being considerably stronger, proving equal in this respect to American white oak. According to my experiments and those of Mr. Luehmann, it is surpassed in resistance to transverse strain by E. melliodora, E. polyanthema, and particularly E. sidero-phloia and E. Leucoxylon, though stronger than the wood of many other of its congeners. The kino of E. rostrata is far less soluble in cold water than that of E. calophylla, and is used as an important medicinal astringent. For other details of the uses of this and some Eucalyptus-trees, refer to the Reports of the Victorian Exhibitions of 1862 and 1867, also to the ten Decades of the Eucalyptographia. E. rostrata becomes already as spontaneous disseminated in Southern France, according to Prof. Ch. Naudin, whose important "Memoire sur les Eucalyptus 1883," should also be consulted regarding the characteristics, development, hardiness and uses of Eucalyptus.

Eucalyptus salmonophloia, F. v. Mueller.

The Salmon-barked Gumtree of South-Western Australia, attaining a height of 120 feet. The timber is good for fencing, while the foliage is available for profitable oil-distillation. The shining mixed whitish and purplish bark does not give off a white coloration like that of E. reduncas.

Eucalyptus saligna, Smith.

The Blue or Flooded Gumtree of New South Wales. A tall straight-stemmed species, attaining a stem-diameter of 7 feet. According to the Rev. Dr. Woolls the wood is of excellent quality, and largely used for building purposes. The tree is generally found on rich soil along river-banks.

Eucalyptus salubris, F. v. Mueller.

The Gimletwood or Fluted Gumtree of Western and Central Australia, living on poor dry soil and in the hottest desert clime. It is generally a slender-stemmed tree, sometimes to 100 feet high, and to 2 feet in stem-diameter, with scanty foliage. The bark is shining with a brownish tinge, and has broad longitudinal and often twisted impressions, or roundish blunt longitudinal ridges. The wood is hard and tough, but worked with comparative ease, heavier than water, even when dry. It serves locally for roof-supports, fence-posts and rails, poles and shafts. For xylography it seems better than Pear-tree-wood, and deserves attention for this purpose. The tree exudes kino.

Eucalyptus siderophloia, Benth'am.*

The Large-leaved or White Ironbark-tree of New South Wales and South-Queensland, attaining a height of 150 feet. According to the Rev. Dr. Woolls this furnishes one of the strongest and most durable timbers of New South Wales; with great advantage used for railway-sleepers and for many building purposes. It is likewise highly appreciated by wheelwrights, especially for spokes, also well adapted for tool-handles and various implements. Found by us to be even stronger than hickory, and only rivalled by E. Leucoxylon. It is harder than the wood of E. Leucoxylon, but for this reason worked with more difficulty. The Melbourne price of the timber is about 2s. 6d. per cubic foot in the log. The tree yields much kino. Mr. Newbery obtained from the bark 8 to 10 per cent. tannin. This species is often confounded with E. resinifera in culture.

Eucalyptus Sieberiana, F. v. Mueller. (*E. virgata,* Bentham, not Sieber.)

South-Eastern Australia. Vernacularly known as Mountain-ash in Gippsland and New South Wales, and as Ironbark-tree or Gum-top in Tasmania. A straight-stemmed tree, reaching 150 feet in height and 5 feet in stem-diameter. The wood is of excellent quality, strong and elastic, hence used for ship-building, implement-handles, cart-shafts, swingle-trees, also for fencing and for general building purposes. It splits freely, and is easy to work. It burns well, even when freshly cut. Systematically the species is very closely allied to E. hæmastoma, but much superior as a timber-tree.

Eucalyptus Staigeriana, F. v. Mueller.

North-Queensland. Durable. A rather small tree, allied to E. crebra. Wood reddish but twisted. Produces readily new shoots from the root (G. Wycliffe). Foliage delightfully fragrant, therefore available for the distillation of a superior cosmetic oil, reminding in odor of that Lippia; the perfume and flavor are so excellent as to render this oil eligible for table-condiments.

Eucalyptus Stuartiana, F. v. Mueller.

South-Eastern Australia. Known to the colonists as Apple-scented Gumtree. A medium-sized tree, with fibrous bark and drooping branches; foliage copious. Occurs on rather dry and sandy as well as on humid soil. The wood is mostly used for fencing and for fuel, but might also be turned to account for furniture, as it is of a handsome dark color, and takes a good polish (Boyle). According to our own observations here it is of nearly the same strength as that of E. rostrata and E. globulus, and somewhat stronger than that of E. amygdalina and particularly E. obliqua. This is one of the hardiest species, as first ascertained by Mr. T. Waugh.

Eucalyptus tereticornis, Smith.*

From Eastern Queensland, where it is termed Red Gumtree, to Gippsland, attaining a height of 160 feet. Closely allied to E. rostrata. The timber is esteemed for the naves and felloes of wheels. For telegraph-poles and railway-sleepers it is inferior to some of the Ironbark-trees, lasting a shorter time, and then not rarely decaying by dry rot. Quite under ground it remains sound much longer (Thozet), but much depends, as regards its durability, on the locality where it is obtained, and the manner of drying, a remark which applies also to many other Eucalypts.

Eucalyptus terminalis, F. v. Mueller.

The Bloodwood-tree of the northern parts of Australia, closely allied to E. corymbosa, attaining a considerable size. The wood is dark-red, hard and extremely tough, particularly fit for boards, as it does not crack. The tree resists the enormous desert-heat of Central Australia, where the shade-temperature ranges from 27° to 122° F., and where the annual rainfall in some years is only 2 inches and seldom more than 10 inches. Particularly adapted for dry tropical climes.

Eucalyptus tessellaris, F. v. Mueller.*

Central and Northern Australia and Queensland. The Moreton-bay-Ash. This tree reaches even on dry ridges a height of 150 feet, surpassing any other Central-Australian species in loftiness, and there resists the severest summer-heat (Rev .H. Kempe). Furnishes a brown, rather elastic wood, not very hard, easily worked, of great strength and durability, available for many kinds of artisans' work, and particularly sought for staves and flooring. The tree exudes much astringent kino (P. O'Shanesy). Several other species might yet be mentioned, particularly from tropical Australia, but we are not yet well enough acquainted with their technical value. All Eucalypts are eligible for the production of tar, pitch, acetic acid, paper-material, potash and various dye-substances.

Eucalyptus triantha, Link. (*E. acmenoides,* Schauer.)

New South Wales and Eastern Queensland. Known as White Mahogany. It attains a considerable height, with a stem reaching 4 feet in diameter, and is of rapid growth. The wood is used in the same way as that of E. obliqua, but is superior to it. It is heavy, strong, durable, of a light color, and has been found good for palings, flooring-boards, battens, rails and many other purposes of house-carpentry (Rev. Dr. Woolls).

Eucalyptus viminalis, La Billardière.

South-Eastern Australia. On poor soil only a moderate-sized tree, with a dark rough bark on the trunk, and generally known as Manna-gum tree; in rich soil of the mountain-forests it attains however gigantic dimensions, rising to a height of rather more than 300 feet, with a stem occasionally to 15 feet in diameter. It has there a cream-colored smooth bark, and is locally know as White Gumtree. The timber is light-colored, clear, and though not so strong and durable as that of many other kinds of Eucalyptus, is very frequently employed for shingles, fence-rails and ordinary building purposes. It is stronger than that of E. amygdalina and E. obliqua. The fresh bark contains about 5 per cent. kino-tannin. Professor Balfour observes, that a tree of this species has stood thirty years in the open air at Haddington (South-Scotland), attaining a height of 50 feet with a stem 8 feet in circumference at the base. Shelter against hard cold winds is in these cases imperative. This is the only species, which yields the crumb-like melitose-manna copiously. The wood of this, of E. globulus, E. melliodora and some others is occasionally bored by the larva of a large moth, Endoxyla Eucalypti, and also by two beetles, Phoracantha tricuspis and Hapatesus hirtus (C. French). For fuller information on Eucalypts consult my "Descriptive Atlas."

Euchlæna luxurians, Ascherson.* (*Reana luxurians,* Durieu.)

The Teosinte. Guatemala, up to considerable elevations. Annual. Highly recommendable as a fodder-grass for regions free of frost. A large number of stems, sometimes as many as 90, spring from the same root, attaining a height of 18 feet. The leaves grow to lengths of 3 feet and form a good forage. The young shoots, when boiled, constitute a fair culinary esculent. Dr. Schweinfurth harvested at Cairo from three seeds in one year about 12,000 grains; the plant requires about ten months to ripen seeds from the time of sowing. This grass, particularly in its young state, is remarkably saccharine. For scenic growth this stately plant is also recommendable. Vilmorin estimates one plant sufficient for feeding two head of cattle during twenty-four hours. Mons. Thozet, at Rockhampton, obtained plants 12 feet high and 12 feet wide in damp alluvial soil, each with 32 main-stalks bearing nearly 100 flower bunches. It is rather slower in growth than Maize, but lasting longer for green fodder, and not so hardy as Sorghum. Its growth can be continued by cutting the tufts as green fodder, thus tender feed is continued; also, it does not cause colic to horses and cattle. As a forage-plant it is without a rival in warm climes. It likes humid soil best, but also resists extreme dryness. It was first brought into notice by the Acclimatisation-Society of Paris, and introduced into Australia and Polynesia by the writer. In Fiji it is thriving most luxuriantly, forming, sown at 4 feet apart, impenetrable masses. The native parrots prey on the seeds, and horses are fond of this grass for forage (Hon. John Thurston). In cooler climes the Teosinte might well serve for ensilage, or as a big summer-grass. Each plant requires at least 16 square-feet of ground for its full development. Even in regions, where the seeds will not ripen, this huge grass should be annually re-sown on pastures overrun with noxious weeds, which would soon become suffocated. Euchlæna Mexicana might also be tested.

Euclea myrtina, Burchell.

South-Africa. Berry small, black, but edible. To us this plant would hardly be more than an ornamental bush.

Euclea Pseudebenus, E. Meyer.

Africa, down to extra-tropic regions. Yields the Orange-River Ebony.

Euclea undulata, Thunberg.

South-Africa. Berry small, red, edible. Other shrubby species from the same portion of the globe also yield esculent fruits, which under superior culture may vastly improve.

Eucryphia cordifolia, Cavanilles.

The Muermo or Ulmo of Chili. This magnificent evergreen tree attains a height of over 100 feet, producing a stem sometimes 6 feet in diameter. The flowers are much sought by bees. For oars and rudders the wood is preferred in Chili to any other (Dr. Philippi). We possess congeneric trees in Tasmania (E. Billardièri, J. Hooker) and in New South Wales (E. Moorei, F. v. M.).

Eugenia Australis, Wendland. (*E. myrtifolia*, Sims.)

East-Australia. A handsome bush with palatable fruit. Careful special culture would probably improve all Eugenia-fruits.

Eugenia cordifolia, Wight.

Ceylon, up to 3,000 feet elevation. Fruit edible, of 1 inch diameter.

Eugenia Hallii, Berg.

Quito. Fruit of large size, edible.

Eugenia Jambolana, Lamarck.

Southern Asia, Polynesia, East-Australia to extra-tropic latitudes. The fruit of this handsome tree is about cherry-size and edible; it may perhaps be improved by well directed skilful culture. Madame Van Nooten indeed in her splendid work on Java plants pictures fruits over 1½ inches long.

Eugenia maboides, Wight.

Ceylon, up to 7,000 feet elevation.   Fruit of the size of a small cherry (Dr. Thwaites). ·

Eugenia Malaccensis, Linné.

The large Rose-Apple. India. A tree, attaining a height of 40 feet, but fruiting already as a shrub. Although strictly a tropical · plant, it has been admitted into this list as likely adapted for warmer forest-regions in extra-tropic zones. The leaves are often a foot long. The large juicy fruits, of rosy odor, are wholesome and of agreeable taste. E. Jambos (L.), E. macrocarpa (Roxburgh), E. Javanica (Lamarck), E. aquea (Bnrmann), E. amplexicaulis (Roxburgh), also from Lower India, likewise produce edible fruit of good size, and may perhaps endure a cool clime.

Eugenia Nhanica, Cambessedes.   .   .

South-Brazil. The berries, which are of the size of plums, are there a table-fruit.

Eugenia pyriformis, Cambessedes.

Uvalho do Campo of South-Brazil.   Fruit of pear-size and edible.

Eugenia revoluta, Wight.

Ceylon, up to heights of 6,000 feet; berry 1 inch in diameter, edible.

Eugenia rotundifolia, Wight.

Ceylon, up to 8,000 feet; rejoicing therefore in a cool or even cold climate. The fruit of this and the allied E. calophylloides (E. calophyllifolia, Wight), which extends to Upper India, edible; so that of E. Arnottiana (Wight), which tree ascends to 7,000 feet.

Eugenia Smithii, ·Poiret.

From Gippsland to Queensland. A splendid large umbrageous tree, but not of quick growth, and requiring rich soil in river-valleys for its perfect development. The bark contains about 17 per cent. tannin. This fact may give a clue to the recognition of the same tan-principle in the barks of numerous other species of the large genus Eugenia.

Eugenia supra-axillaris, Spring.

The Tata of South-Brazil.   Fruit large, edible.

Eugenia uniflora, Linné. (*E. Michellii*, Lamarck.)

From Extra-tropical South-America extending to the Antilles. A tree of beautiful habit, with edible fruit of cherry-size. Dr. Lorentz mentions also as a sub-tropical Argentine fruit-species *E. Mato* (Grisebach). Hieronymus adduces similarly E. ligustrina (Willdenow) and E. edulis (Bentham), the fruit of the latter species being of apple-size.

Eugenia Zeyheri, Harvey.

South-Africa. A tree attaining about 20 feet in height. The berries are of cherry-size and edible. The relative value of the fruits of many Asiatic, African and American species of Eugenia remains to be ascertained; many of them doubtless furnish also good timber, and all more or less essential oil. All such, even tropical trees, should be tested in warm tracts of the temperate zone, inasmuch as many of them endure a cooler clime than is generally supposed.

Euonymus atropurpureus, Jacquin.

The "Waahoo" of Eastern North-America. A tall shrub, the bark of which is used in medicine, and from it the Euonymin, an oleo-resin as a cholagogue is prepared. To this species is closely allied E. occidentalis (Nuttall) from Oregon and California.

Euonymus Japonicus, Thunberg.

Japan. This evergreen shrub can be used advantageously for forming hedges; it is easily multiplied by layers and requires little care in cultivation.

Eupatorium purpureum, Linné.

Eastern North-America. " Queen of the Meadows." A perennial herb, easily naturalized; has come into medicinal use as a powerful diuretic; so also E. perfoliatum L., the " Thoroughwort" or " Boneset"; these with other species are also not unimportant as honey-plants; so of the same order in N. America species of Actinomeris and Prenanthes.

Eupatorium tinctorium, Grisebach.

Paraguay. A shrub of remarkably prolific and vigorous growth (E. H. Egerton). Competes almost with the indigo-plant for dye. It can be stripped of its leaves four times a year without injury to the plant.

Eupatorium triplinerve, Vahl. (*E. Ayapana*, Ventenat.)

Central America. A perennial somewhat shrubby herb, possibly hardy in the warmer parts of extra-tropical countries. It contains eupatorin and much essential oil peculiar to the plant. It stands locally in renown as a remedy against ophidian poisons, and evidently possesses important medicinal properties. A tanning extract is prepared for the English market from this herb, which contains about 20 per cent. tannic acid.

Euryale ferox, Salisbury.

From tropical Asia to Japan, ascending in Cashmir to 5,000 feet, extending northward in Amur to nearly 46° N. Though less magnificent than the grand Victoria Regia, this closely allied water-lily is much more hardy, and would live unprotected in ponds and lakes of a temperate climate. Though not strictly an industrial plant, it is not without utility, and undergoes some sort of cultivation in China for its edible roots and seeds. The leaves attain a width of 3 feet.

L

Eustrephus Brownii, F. v. Mueller.

East-Australia as far south as Gippsland. This climber produces sweet though only small tubers, which however are probably capable of enlargement through culture.

Euterpe andicola, Brongniart.

Bolivia. Ascends to 9,000 feet (Martius), an altitude higher than is reached there by any other palm unless E. Haenkeana and E. longivaginata (Drude). E. edulis (Martius) extends as far south as Minas Geraes in Brazil.

Euxolus viridis, Moquin. (*Amarantus viridis,* Linné.)

Temperate and tropical regions of Europe, Asia and Africa. Annual; quickly raised. Not without value as a spinage-plant. E. oleraceus is a cognate plant.

Excæcaria sebifera, J. Mueller. (*Stillingia sebifera,* Michaux.)

The Tallow-tree of China and Japan. The fatty coating of the seeds constitutes the vegetable tallow, which is separated by steaming. The wood is so hard and dense, as to be used for printing-blocks; the leaves furnish a black dye. The tree endures slight night-frosts, though its foliage suffers.

Exidia auricula Judae, Fries. (*Hirneola auricula Judae,* Fries.)

Widely spread over the globe. Of this edible fungus in 1871 alone the quantity exported from Tahiti to China represented a value of £7,600 (Simmonds). Doubtless this useful fungus is amenable to translocation and subsequent naturalization. We have it as indigenous in Australia also.

Exomis axyrioides, Fenzl.

South-Africa. A good salt-bush there for pastures (McOwan).

Fagopyrum cymosum, Meissner.

The perennial Buck-wheat or rather Beech-wheat of the Indian and Chinese highlands. Can be used with other species for spinage and grain; also a blue dye may be obtained from its leaves.

Fagopyrum emarginatum, Babington.

Chinese and Himalayan mountains, where it is cultivated for its seeds. Annual.

Fagopyrum esculentum, Moench.*

Central Asia, extending eastward to Manchuria, growing to an elevation of 14,000 feet in the Himalayas. The ordinary Buck-wheat, called Buch-Waizen in Germany, from the resemblance (in miniature) of the seeds to Beech-nuts ; hence also the generic name. This annual herb succeeds on the poorest land ; clayey soil yields more foliage but less grain. The crushed amylaceous seeds can be converted into a palatable and wholesome food by boiling or baking. Starch has also recently been prepared from the seeds as an article of trade. Fagopyrum can be raised with advantage as an agrarian plant for the first crop on sandy but not too dry heath-land, newly broken up, for green manure. It gives a good green-fodder, serves as admixture to hay, and is also important as a honey-plant. The period required for the cyclus of its vegetation is extremely short; thus it can even be reared on alpine elevations. In Norway it grows to lat. 67° 56′ (Schuebeler). The produce of this grain in the United States during 1879 was 13,140,000 bushels, valued at £1,636,000.

Fagopyrum Tataricum, Moench.*

Middle and Northern Asia. Yields for the higher mountain-regions a still safer crop than the foregoing; otherwise the remarks offered in reference to F. esculentum apply also to F. Tataricum; but the seeds of the latter are more thick-shelled, less amylaceous and less palatable.

Fagopyrum triangulare, Meissner.

In the Himalayan mountains, ascending naturally to regions 11,500 feet high. An annual. F. rotundatum (Babington) seems a variety of this species. It is cultivated for food like the rest.

Fagus betuloides, Mirbel.

Patagonia and Fuegia. An evergreen Beech, on the branches of which a peculiar edible fungus, Cyttaria Darwinii (Berkeley) occurs. It seems quite feasible, that the Cyttaria-fungs could be transferred from some beeches to others of different countries. Cyttaria Hookeri lives on Fagus antarctica (Forster). C. Darwinii forms for some months of the year a large share of the vegetable food of the Fuegians (Hieronymus).

Fagus Cunninghami, Hooker.

The Victorian and Tasmanian Beech. The Myrtle-wood of local trade. A magnificent evergreen-tree, attaining large dimensions, not rarely to 200 feet, but living only in cool, damp, rich forest-valleys. The wood is much used by carpenters and other artisans, and particularly liked for saddle-trees. It remains to be ascertained by actual tests in the forests, whether the allied tall evergreen New Zealand Beeches possess any advantage over this species for forest-culture; they are Fagus Menziesii, the Red Birch of the colonists; F. fusca and F. cliffortioides (J. Hooker), the Black Birches and F. Solandri (Hooker), the White Birch. A magnificent and peculiar beech, Fagus Moorei (F. v. Mueller), occurs in New South Wales on high mountains. Cyttaria Gunnii (Berkeley) is the Raspberry-fungus of the Tasmanian evergreen Beech; a Cyttaria occurs also on the New Zealand Beeches, as ascertained by the author of this work.

Fagus Dombeyi, Mirbel.

The Evergreen Beech of Chili, called there the Coigue or Coihue. Of grand dimensions. Canoes can be made out of its stem, large enough to carry 10 tons freight. The wood is still harder than that of the following species, with the qualities of which it otherwise agrees (Dr. Philippi). This species extends to the Chonos-group, and perhaps still further south, and thus may be of value even for Middle European forest-culture.   ·

Fagus ferruginea, Aiton.

North-American Beech. A large tree, with deciduous foliage, easily raised in woodlands. Wood variable according to localities. Well-seasoned wood, according to Simmonds, is remarkably hard and solid, hence employed for plane-stocks, shoe-lasts, tool-handles, · various implements and turneries.

Fagus obliqua, Mirbel.

The Roble of Chili, called Coyam by the original inhabitants. A tall tree with a straight stem, attaining 3 to 4 feet diameter. Wood heavy and durable, well adapted for posts, beams, girders, rafters and joists, but not for flooring. One of the few Chilian trees with deciduous foliage (Dr. Philippi). Its value as compared with that of the European Beech should be tested in forest-plantations. Cyttaria Berteroi (Berkeley) grows on the branches of this Beech.

Fagus procera, Poeppig.

Another deciduous Beech of Chili, where it passes by the name of Reulé or Rauli. Of still more colossal size than the Roble. Wood fissile, well adapted for staves; finer in grain than that of F. obliqua, and much used for furniture (Dr. Philippi).

Fagus silvatica, Linné.

The deciduous Beech of Britain, of most other parts of Europe and extra-tropical Asia. The trunk has been measured in height to 118 feet, the head to 350 feet in circumference. As far north as lat. 60° 23′ in Norway Professor Schuebeler found a tree over 70 feet high with a stem 12 feet in circumference; smaller trees grew even to lat. 67° 56′. · Apt to overpower any other kinds of trees in its native forests. The wood is hard, extensively used by joiners and ship-builders in their trade and by the manufacturers of various implements, especially for planes, shoe-lasts, keys and cogs of machinery, lathe-chucks, gun-stocks, staves, chairs, spoke-shaves, in piano-manufacture for bridges, likewise some portion of the work of organ-builders; enters also into the construction of harmoniums (beds of notes, pallets, rest-planks), further used for carved moulds and for wooden letters in large prints; · it is of rather difficult cleavage, great compactness and of considerable strength, and resists great pressure. Beech-tar contains a considerable proportion of paraffine; the ashes from any portion of this tree are rich in phosphate of lime. For trimming into copse-hedges many give preference to a purple-leaved variety for show. An allied Beech, Fagus Sieboldii (Endlicher), grows in Japan. In the warmer temperate zones all these could only be grown to advantage in springy mountain-forests.

Fatsia papyrifera, Bentham. (*Aralia papyrifera*, Hooker.)

Island of Formosa. The Rice-paper Plant, hardy in the lowlands of Victoria, and of scenic effect in garden-plantations. The pith furnishes the material for the so-called rice-paper, also for some sorts of solah-hats.

Ferula Assa foetida, L. (*Scorodosma foetidum*, Bunge.)

Persia, Afghanistan and Turkestan. This very tall perennial herb yields the ordinary medicinal assafetida. Ferula Narthex, Boissier (Narthex Assa foetida, Falconer), furnishes a very similar drug in Thibet. The cultivation of these plants in adequate climes seems not surrounded by any difficulties.

Ferula galbaniflua, Boissier.

Persia; on mountains 4,000 to 8,000 feet high. This tall perennial herb might be transferred to other sub-alpine regions, for obtaining locally from it the gum-resin galbanum.

Ferula longifolia, Fischer.

Southern Russia. The long aromatic roots furnish a pleasant vegetable (Dr. Rosenthal).

Ferula Sambul, J. Hooker. (*Euryangium Sambul*, Kauffmann.)

Turkestan, at elevations between 3,000 and 4,000 feet. A perennial herb, attaining a height of 9 feet or perhaps more. Yields the true Sambul-root, a powerful stimulant, with the odor of musk. It is also a scenic-decorative plant, and proved hardy in England.

Festuca Coiron, Steudel.

Chili.[1] A valuable perennial fodder-grass, according to the testimony of Dr. Philippi.

Festuca dives, F. v. Mueller.

Victoria, from West-Gippsland to Dandenong, towards the sources of rivers, ascending sub-alpine elevations. One of the most magnificent of all sylvan grasses, often 12 and sometimes when in rank growth on forest-brooks fully 17 feet long. Root perennial or perhaps of only two or three years' duration. This grass deserves to be brought to any forest-tracts in mild climes, as it prospers in shade; it assumes its grandest forms in deep soil along rivulets. The large panicle affords nutritious forage.

Festuca elatior, Linné.*

The Meadow-Fescue. Europe, North-Africa, Northern and Middle Asia. A perennial grass, attaining a height of several feet. There are several varieties of this species; the tallest follows rivers readily as far down as the tides reach; the ordinary form is well adapted for permanent pastures, has tender leaves, produces excellent, tasty, nutritious hay, and is early out in the season. Langethal places Meadow-Fescue above Timothy- and Foxtail-grass in value, though its copiousness is somewhat less. The seed is readily collected. The tall variety (arundinacea) will occupy wet land preferentially among the best of eligible fodder-grasses. It can be mixed advantageously with F. ovina, is superior to Rye-grass in production and improves with age. It succeeds also on humid and even swampy ground and in forest-land as well with sandy as a calcareous subsoil. Dr Curl observes, that this and some other Fescues grow vigorously in New Zealand, and yield pasturage there also in the cool season, when Rye-grass is nearly dormant. Chemical analysis, made in spring, gave the following results: Albumen 2·47, gluten 2·75, starch 0·50, gum 2·84, sugar 2·84 per cent. (F. v. Mueller and L. Rummel). F. arundinacea (Schreber), F. pratensis and F. loliacea (Hudson) are varieties of this species.

Festuca flava, F. v. Mueller. (*Poa flava*, Gronovius; *Tricuspis sesleriodes*, Torrey; *Uralepis cuprea*, Kunth.)

The tall Red-top Grass of the Eastern States of North-America. A perennial sand-grass with wide panicles. F. scabrella is one of the "Bunch-grasses" of Oregon and California.

Festuca gigantea, Villars.

Europe and Middle Asia. A good perennial forest-grass.

Festuca heterophylla, Lamarck.

Europe. This perennial grass is best fitted for cool forest-tracts. Recommended also among lawn-grasses.

Festuca Hookeriana, F. v. Mueller.*

Alps of Australia and Tasmania. A tall perennial grass, evidently nutritious, required to be tried for pasture-culture, and perhaps destined to become a new meadow-grass of colder countries elsewhere. Stands mowing and depasturing well; much liked by cattle, horses and sheep (Th. Walton).

Festuca litoralis, La Billardière.

Extra-tropical Australia and New Zealand. An important strong perennial grass for binding drift-sand on sea-shores.

Festuca Magellanica, Lamarck.

Extra-tropic South-America. Ascending the Andes to 12,000 feet, and contributing much to the fattening pasturage there (Hieronymus).

Festuca ovina, Linné.

Sheep-Fescue. Europe, Northern and Middle Asia, North-America; found also in South-America and the Alps of Australia and New Zealand. This species, like F. elatior, is obtainable with facility. F. duriuscula and F. rubra (Linné) are varieties. A perennial grass, thriving on widely different soils, even moory and sandy. It yields a good produce, maintains its virtue, resists drought, and is also well adapted for lawns and swards of parks. F. vaginata

(Willdenow) is a form particularly recommended by Wessely for sand-soil. Chemical analysis, made very late in spring, gave the following results: Albumen 1·86, gluten 8·16, starch 1·45, gum 2·14, sugar 5·05 per cent. (F. v. Mueller and L. Rummel).

Festuca purpurea, F. v. Mueller. (*Uralepis purpurea*, Nuttall ; *Tricuspis purpurea*, A. Gray.)

South-Eastern coast of North-America. A tufty sand-grass, but annual.

Festuca silvatica, Villars.

Middle and Southern Europe. A notable forest-grass. F. drymeia (Mertens and Koch); a grass with long creeping roots, is closely allied. Both deserve test-culture. Space does not admit of entering here into further details of· the respective values of many species of Festuca, which might advantageously be introduced from various parts of the globe for rural purposes.

Ficus Carica, Linné.

· ·The ordinary Fig-tree. Alph. de Candolle speaks of it as spontaneous from Syria to the Canary-Islands; Count Solms-Laubach confines the·nativity of the Fig-tree to the countries on the Persian Gulf. It attains an age of several hundred years. In warm temperate latitudes and climes a prolific tree. The most useful and at the same time the most hardy of half a· thousand recorded species of Ficus. The extreme facility, with which it can be propagated from cuttings, the resistance to heat, the comparatively early yield and easy culture recommend the Fig-tree, where it is an object to raise masses of tree-vegetation in widely treeless lands of the warmer zones for shade and fruit. Hence the extensive plantations of this tree made in formerly woodless parts of Egypt; hence the likelihood of choosing the Fig as one of the trees for extensive planting through favorable portions of desert-waste, where moreover the fruit could be dried with particular ease. Small cuttings went quite well chiefly by horse-post from Port Phillip to the central Australian Mission-stations, a distance as far as from Petersburg to the Black Sea, or from Bombay to Thibet, or from Capetown to Lake Ngami,· or from San Francisco to the Upper Missouri. Fig-trees can be grown even on sand-lands, at least as observed on the Australian south-coast. In ·Greece the average yield of figs per acre is about 1,600 lbs. (Simmonds). Caprification is unnecessary, even in ·some instances injurious and objectionable. Two main-varieties may be distinguished: that which produces two crops a year, and that which yields but one; The former includes the Gray or Purple Fig, which is the best, the White Fig and the Golden Fig, the latter being the finest in appearance, but not in quality. The main-variety, which bears only one crop a year, supplies the greatest quantity of figs for drying, among which the Marseillaise and Bellonne are considered the best. The Barnisote and the Aubique produce delicious large fruits, but they must be dried with fire-heat, and are usually consumed fresh. The ordinary drying is

effected in the sun. For remarks on this and other points concerning
the Fig, the valuable tract published by the Rev. Dr. Bleasdale
should be consulted. The first crop of figs grows on wood of the
preceding year; the last crop however on wood of the current year.
Varieties of particular excellence are known from Genoa, Savoy,
Malaga, Andalusia. For some further information, see among other
publications also that of the Hon. the Commissioner of Agriculture,
Washington, 1878.

Ficus columnaris, Moore and Mueller.

The Banyan-tree of Lord Howe's Island, therefore extra-tropical.
One of the most magnificent productions in the whole empire of
plants. Mr. Fitzgerald, a visitor to the island, remarks that the
pendulous aerial roots, when they touch the ground, gradually swell
into columns of the same dimensions as the older ones, which have
already become converted into stems, so that it is not evident, which
was the parent-trunk; there may be a hundred stems to the tree, on
which the huge dome of dark evergreen foliage rests, but these stems
are all alike, and thus it is impossible to say, whence the tree comes or
whither it goes. The aerial roots are rather rapidly formed, but the
wood never attains the thickness of F. macrophylla, which produces
only a single trunk. The allied F. rubiginosa of continental East-
Australia has great buttresses, but only now and then a pendulous
root, approaching in similarity the stems of Ficus columnaris. The
Lord Howe's Island Fig-tree is more like F. macrophylla than F.
rubiginosa, but F. columnaris is more rufous in foliage than either.
In humid, warm, sheltered tracts this grand vegetable living struc-
ture may be raised as an enormous bower for shade and for scenic
ornament. The nature of the sap, whether available for caoutchouc
or other industrial material, requires yet to be tested. A substance
almost identical with gutta-percha, but not like india-rubber, has been
obtained by exsiccation of the sap of F. columnaris (Fitzgerald). The
hardened sap of this species resembles in many respects that of F.
subracemosa and F. variegata, called Getah Lahoe, but differs apparently by its greater solubility in cold alcohol, and by the portion insoluble in alcohol being of a pulverulent instead of a viscid character.
The mode of exsiccation affects much the properties of the product.

Ficus Cunninghami, Miquel.

Queensland, in the eastern dense forest-regions to about 28° S.
Mr. J. O'Shanesy designates this as a tree of sometimes monstrous
growth, the large spreading branches sending down roots, which take
firm hold of the ground. One tree measured was 38 feet in circumference at 2 feet from the ground, the roots forming wall-like abut-
ments, some of which extended 20 feet from the tree. Several persons
could conceal themselves in the large crevices of the trunk, while the
main-branches stretched across a space of about 100 feet. A kind
of caoutchouc can be obtained from this tree. A still more gigantic
Fig-tree of Queensland is F. colossea, F. v. M., but it may not be

equally hardy, not advancing naturally to extra-tropical latitudes. This reminds us of the great Council-tree, F. altissima, from Java, where it grows in mountains on calcareous ground. F. eugenioides (F. v. M.) from North- and East-Australia, attains a height of 100 feet, and produces also columnar air-roots. It is comparatively hardy, reaching extra-tropic latitudes.

Ficus elastica, Roxburgh.*

Upper India, to the Chinese boundary, known as far as 28° 30' north-latitude. A large tree, yielding its milk-sap copiously for the kind of caoutchouc called Assam-Rubber. Roxburgh ascertained 70 years ago, that india-rubber could be dissolved in cajaput-oil (very similar to eucalyptus-oil), and that the sap yielded about one-third of its weight of caoutchouc. This tree is not of quick growth in the changeable and often dry clime of Melbourne; but there is every prospect, that it would advance rather rapidly in any mild humid forest-gullies, and that copious plantations of it there would call forth a new local industry. This tree has grown in Assam to 112 feet with 100 aerial roots in thirty-two years (Markham). In moist warm climes, according to observations in Assam by Mr. Gustave Mann, branches lopped off and planted will speedily establish themselves. Single branches attain a length of 50 feet; the root-crown will attain a diameter of 200 feet exceptionally (Haeckel). The import of caoutchouc into the United Kingdom in 1884 amounted to 198,000 cwt., representing a value of £2,266,000, of which F. elastica must have furnished a considerable proportion. Markham and Collins pronounce the caoutchouc of F. elastica not quite so valuable as that of the Heveas and Castilloas of South-America. Heat and atmospheric moisture greatly promote the growth of F. elastica. Like most other Fig-trees, it is easily raised from seed. A tree of F. elastica is tapped in Assam when twenty-five years old. After fifty years the yield is about 40 lbs. of caoutchouc every third year, and lasts till the tree is over 100 years old. The milky sap flowing from cuts in the stem yields nearly one-third of its weight of caoutchouc; the collected sap is poured into boiling water and stirred till it gets firm; or the sap is poured into large bins partly filled with water; the fluid caoutchouc-mass after a while floats on the surface, when it is taken out and boiled in iron-pans, after the addition of two parts of water, the whole being stirred continuously; after coagulation the caoutchouc is taken out and pressed, and if necessary boiled again, then dried, and finally washed with lime-water. The sap from cuts into the branches is allowed to dry on the trees (J. Collins). Dr. S. Kurz states, that F. laccifera (Roxburgh) from Silhet is also a caoutchouc-tree, and that both this and F. elastica yield most in a ferruginous clay-soil on a rocky substratum; further, that both can bear dryness, but like shade in youth. Several other species of tropical figs, American as well as Asiatic, are known to produce good caoutchouc, but it is questionable, whether any of them would prosper in extra-tropical latitudes; nevertheless for the conservatories of botanic gardens all such plants should be secured

with a view of promoting public instruction.   Te give some idea of
the vastly increasing extent, to which " Rubber " is now required, it
may be stated, that at Wetzell's factories in München and Hildesheim
alone during 1884 were produced 100,000 lbs. of surgical articles;
100,000 lbs. valves, buffers and washers; 150,000 lbs. hose and belting;
200,000 lbs. insertion-sheets and tucks-packings; 250,000 dozens of
fancy-colored balls, irrespective of other rubber-articles; this factory,
which exists since 1868, employing 600 workmen, and is operating
with machinery equal to 300 horses-power.   Ficus Vogelii (Miquel)
yields Rubber in West-Africa.

Ficus Indica, Linné.

The Banyan-tree of India, famed for its enormous expansion and
air-roots.   Although not strictly an utilitarian tree, it is admitted here
as one of the most shady trees, adapted for warm and moist regions.
At the age of 100 years one individual tree will shade and occupy
about one and a half acres, and rest on 150 stems or more, the main-
stems often with a circumference of 50 feet, the secondary stems
with a diameter of several feet.   At Melbourne the tree suffers
somewhat from the night-frosts.

Ficus infectoria, Willdenow.

India, ascending to 5,000 feet.   Probably hardy where frosts are
only slight, and then adapted for street-planting.   Brandis and
Stewart found its growth quicker than that of Siris or Albizzia procera.
F. religiosa (Linné) ascends to the same height, and is of quick growth
in moist climates.   It bears well the clime of Beloochistan.   It is one
of the trees, on which the lac insect largely exists.   The fruits of some
huge Himalayan species—for instance, F. virgata, F. glomerata
(Roxburgh) and F. Roxburghii (Wallich)—are edible.

Ficus macrophylla, Desfontaines.*

The Moreton-Bay Fig-tree, which is indigenous through a great part
of East-Australia.   Perhaps the grandest of Australian avenue-trees,
and among the very best to be planted, although in poor dry soil its
growth is slow.   In the latitude of Melbourne it is quite hardy in the
lowland.   The foliage may occasionally be injured by grasshoppers.
Easily raised from seed, the smallness of which admits of their very
easy transmission to remote places.   Average-growth in height at
Port Phillip, 30 feet in 20 years.

Ficus rubiginosa, Desfontaines.

New South Wales.   One of the most hardy of all Fig-trees, and
very eligible among evergreen shade-trees, particularly for promenades.
It is estimated, that the genus Ficus comprises about 600 species, many
occurring in cool mountain-regions of tropical countries.   The number
of those, which would endure a temperate clime, is probably not small.

Ficus Sycomorus, Linné.

The Sycomore-Figtree.   Egypt, Abyssinia, Nubia.   Copiously
planted along the roadsides of Egypt.   The evergreen shady crown,

with very spreading branches, extends to a width of 120 feet. Attains an enormous age. A tree at Cairo, which legends connect with Christ's time, still exists. Seven men with outstretched arms could hardly encircle the stem.

Fistulina hepatica, Fries.

Europe and Asia, East-Australia, This large esculent fungus grows generally on old oak-stems, but does accommodate itself to other habitations also. Its introduction elsewhere is worthy of trial.

Fitzroya Patagonica, J. Hooker.*

Chili, as far south as Chiloe. The Alerce of the Chilians. Grows on swampy, moory places. A stately tree, sometimes to 100 feet high; the diameter of the stem occasionally reaches the extraordinary extent of 15 feet. The wood is almost always red, easily split, light, does not warp, stands exposure to the air for half a century; in Valdivia and Chiloe almost all buildings are roofed with shingles of this tree (Dr. Philippi). The outer bark produces a strong fibre, used for calking ships. Like Libocedrus tetragona, this tree should be extensively planted in unutilized swampy moors of mountains.

Flacourtia Ramontchi, L'Héritier. (*F. sapida*, Roxburgh.)

India up to Beloochistan. This and F. cataphracta (Roxburgh) form thorny trees with somewhat plum-like fruits. They can be adopted for hedge-copses with other species.

Flemingia tuberosa, Dalzell.

Western India. The tubers of this herb are said to be edible. Another species, F. vestita (Bentham), is on record as cultivated in North-Western India, where it ascends to the temperate region up to 7,000 feet, for its small esculent tubers.

Flindersia Australis, R. Brown.

New South Wales and Queensland. With Araucaria Cunninghami and Ficus macrophylla, the tallest of all the jungle-trees of its localities, attaining 150 feet. Bark scaly; stem frequently with a diameter of 8 feet. Timber of extraordinary hardness (Ch. Moore). A noble tree for avenues. Rate of growth, according to Mr. Fawcett, about 25 feet in eight years.

Flindersia Oxleyana, F. v. Mueller.

The Yellow Wood of New South Wales and Queensland, called "Bogum Bogum" by the aborigines. Its wood is used locally for dye, also for staves as well as that of F. Australis, Tarrietia Argyrodendron, Stenocarpus salignus and Castanospermum Australe. Mr. C. Hartmann mentions, that F. Oxleyana attains a height of 150 feet, and supplies one of the finest hardwoods for choice cabinet-work. Other species occur, among which F. Bennettiana is the best for avenue-purposes.

Fœniculum officinale, Allioni.

The Fennel. Mediterranean regions, particularly on limestone-soil, extending to Central Asia, certainly wild in Turkestan (Dr. A. von Regel). A perennial or biennial herb, of which primary varieties occur, the so-called sweet variety having fruits almost twice as large as the other. The herb and fruits are in use as condiments and the latter also for medicine. The fruits are rich in essential oil, containing much anethol. Vilmorin found them to keep their vitality for about four years; he also remarks that the bleached leafstalks yield the Carosella-salad. A variety, F. dulce (Bauhin), yields its young shoots for boiling as a vegetable of sweetish taste and delicate aroma.

Fourcroya Cubensis, Haworth.

West-Indies and continental tropical America. A smaller species than the following, but equally utilized for fibre and impenetrable hedges. F. flavo-viridis (Hooker), from Mexico, is still smaller.

Fourcroya gigantea, Ventenat.

Central America. In species of Yucca, Agave, Dracæna, Cordyline, Phormium, Doryanthes and this as well as a few other Fourcroyas we have gigantic liliaceous and amaryllidaceous plants available industrially for fibre. Frost injures the leaves of this species. Development of flower-stalks extremely rapid, up to 30 feet high. Fibre often 3 feet long and of considerable tenacity. The fibre, produced in Mauritius by Messrs. Bourgignon and Fronchet, proved stronger than hemp and resisted decay in water. Mr. Boucard also testifies to the excellence of the fibre, which he describes as long, silky and solid, particularly adapted for luxurious hammocks and for cordage.

Fourcroya longæva, Karwinski and Zuccarini.

High mountains of Guatemala and Mexico, particularly at an elevation of about 10,000 feet. One of the most gigantic and magnificent of all liliaceous or amaryllideous plants, in volume only surpassed by Dracæna Draco, the Dragon-tree of the Canary-Islands. This is the principal high-stemmed species, the trunk attaining a height of 50 feet, and the huge panicle of flowers 40 feet more. It dies, like many allied plants, after flowering. The species is recorded here as a fibre-plant, but should also be cultivated for its ornamental grandeur.

Fragaria Californica, Chamisso and Schlechtendal.

California and Mexico. Closely allied to C. vesca.

Fragaria Chiloensis, Aiton.

Chili-Strawberry. In various of the colder parts both of North- and South-America. Almost incredible accounts have been published regarding the yield of the Chiloen Strawberry in the neighborhood of Brest, far exceeding the fecundity of any other strawberry.

Fragaria collina, Ehrhart.

Hill-Strawberry. In various parts of Europe. Cultivated in Norway to lat. 67° 56′ (Schuebeler); may be regarded as a mere variety of F. vesca. Fruit of a somewhat musky odor. ..

Fragaria grandiflora, Ehrhart. (*F. Ananas*, Miller.)

Ananas-Strawberry. Various colder parts of America. Closely allied to F. Chiloensis. Many of the large-fruited varieties have been derived from this species.

Fragaria Illinoensis, Prince.

North-America. Hovey's seedling and the Boston-kind from this plant. Is regarded by Professor Asa. Gray as a variety of F. Virginiana.

Fragaria pratensis, Duchesne. (*F. elatior*, Ehrhart.)

Cinnamon-Strawberry. Hautbois. In mountain-forests of Europe. F. moschata (Duchesne) is a variety of this species.

Fragaria vesca, Linné.

Wild Wood-Strawberry. Naturally very widely dispersed over the temperate and colder parts of the northern hemisphere, extending northward to Lapland and Iceland, southward to the mountains of Java, ascending the Himalayas to 13,000 feet (J. D. Hooker). From this typical form probably some of the other Strawberries arose. Middle forms and numerous varieties now in culture were produced by hybridization. These plants, though already abounding in our gardens, are mentioned here, because they should be naturalized in any ranges. Settlers, living near some brook or rivulet, might readily set out plants, which, with others similarly adapted, would gradually spread with the current. The minute seeds will retain their vitality for fully three years. A highland-variety, F. alpina (Persoon), furnishes fruit throughout the warm season, long after the other varieties cease bearing in gardens (Vilmorin).

Fragaria Virginiana, Miller.

Scarlet Strawberry. Eastern North-America, extending northward to 64° (Sir J. Richardson), therefore adapted for the coldest climates also, yet even fruiting well in Bermuda (Sir J. Lefroy).

Fraxinus Americana, Linné.*

The White Ash. Eastern North-America, extending from Florida to Canada. A large tree, which delights in humid forests. Trunks have been found 75 feet long without a limb, and 6 feet in diameter (Emerson). It is the best of all American Ashes, and of comparatively rapid growth. In Nebraska the stem attains to about 32 inches circumference at 2 feet from the base in 22 years (Furnas). Resisting extreme heat better than the common Ash. Timber largely exported; it assumes a red tint in age ; much valued for its toughness, lightness and elasticity, excellent for work subject to sudden shocks and strains, such as the frames of machines, carriage-wheels, agricultural implements, pick-handles, billiard-cues, fishing-rods, handles, chair-rails, shafts, staves, pulley-blocks, belaying-pins and oars; also for furniture and musical instruments. The young branches are utilized for mast-hoops. Baron von Mueller and Mr. J. G. Luehmann found the strength greater than that of our Blackwood-tree and of many Eucalypts, but not equal to that of E. Leucoxylon, E. siderophloia, E. polyanthema, the best E. globulus and hickory. Over-old wood not desirable. When once thoroughly seasoned, it does not shrink or swell, and is therefore preferred for flooring to any native timber in Virginia (Robb, Simmonds). The inner bark furnishes a yellow dye. The Red Ash (Fraxinus pubescens, Lamarck) and the Carolina-Ash (F. platycarpa, Michaux) are of smaller size, but F. pubescens may sometimes also become large.

Fraxinus Chinensis, Roxburgh.

It is this Ash, on which a peculiar wax is produced by Coccus Pela, perhaps also on some species of Ligustrum. About 40,000 lbs. are exported anually according to Bernardini. F. Mandschurica (Ruprecht) attains a height of 60 feet and a stem-diameter of 4 feet.

Fraxinus excelsior, C. Bauhin.*

The ordinary Ash of Europe and Western Asia, extending to the Himalayas and Thibet, there ascending to 9,000 feet. A tree of comparatively quick growth, known to attain an age of nearly 200 years. It is a very hardy tree, braving the winters of Norway to lat. 69° 40', though there only a shrub; but in lat. 61° 12' it attained still a height of 100 feet and a stem-diameter of 5 feet (Schuebeler). Rich soil on forest-rivulets or river-banks suits it best, although it also thrives on moist sand. Wood remarkably tough and elastic, used for agricultural and other implements, handles, ladders, drum-hoops, carriage-work, oars, axle-trees and many other purposes. Six peculiar kinds of Ash-trees occur in Japan, some also in the Indian highlands; all might be tried for industrial culture.

Fraxinus floribunda, Don.

Nepal-Ash. Himalaya, between 4,000 and 11,000 feet. It attains a height of 120 feet, girth of stem sometimes 15 feet. Serves not only as a timber-tree, but also as a fine avenue-tree. The wood much sought for oars, ploughs and various implements (Stewart and Brandis). For forest-plantations Ashes are best mixed with beeches and some other trees.

Fraxinus Oregana, Nuttall.

Californian and Oregon-Ash. A tree, reaching 80 feet in height, preferring low-lying alluvial lands. The wood of this fine species is nearly white, very tough and durable, often used for oars and handles of implements, also in carriage-building. Though allied to F. sambucifolia, it is very superior as a timber-tree. Ash-trees will grow readily in the shade of other trees.

Fraxinus ornus, Linné.

The Manna-Ash of the Mediterranean regions, extending to Austria and Switzerland. Height about 30 feet. Hardy still at Christiania. It yields the medicinal manna by incisions into the bark, which is done. only on one side of the stem each year. F. ornus is well adapted for a promenade-tree, and is earlier in foliage than F. excelsior, F. Americana and most other Ash-trees.

Fraxinus quadrangulata, Michaux.*

The Blue Ash of North-America, from Michigan to Tennessee. One of the tallest of the Ashes, becoming 70 feet high. Timber excellent, better than that of any other American species except the White Ash, hence .frequently in use for flooring and shingles. The inner bark furnishes a blue dye. The tree requires a rather mild clime and the most fertile soil.

Fraxinus sambucifolia, Lamarck.

Black or Water-Ash of Eastern North-America. Attains a height of 80 feet. Wood still more tough and elastic than that of *F. Americana*, but less durable when exposed; easily split into thin layers for basket-work; it is also comparatively rich in potash, like that of most of its congeners; for oars and implements it is inferior to that of the White Ash (Simmonds). F. nigra (Marsh) is the oldest name.

Fraxinus viridis, Michaux.

The Green Ash of Eastern North-America. Height reaching 70 feet. Wood excellent, nearly as valuable as that of the White Ash, but of less dimensions. The tree requires wet, shady woodlands. Especially recommended for street-planting by Dr. J. Warder. This species, like the preceding, is hardy as far north as Christiania in Norway (Schuebeler). Rate of circumferential stem-growth in Nebraska 30 inches in 22 years (Furnas).

Fuchsia racemosa, Lamarck.

Mountains of Hayti. One of the species with edible berries of very good taste. Another Fuchsia occurs in cold regions of Guatemala up to 10,000 feet high, with orange-colored flowers and with tasty wholesome berries, the latter an inch and a half long. F. excorticata (Linné fil.) of New Zealand extends there far south, and is perhaps as hardy as F. Magellanica (Lamarck); it rises to 30 feet, if not. higher, and gains a stem-diameter of 3 feet.

Garcinia Travancorica, Beddome.

Madras-Presidency, up to elevations of 4,500 feet. This seems to be the hardiest of the superior Gamboge-trees; hence there is some prospect of its prospering in forests of the warmer temperate zone.

Garuleum bipinnatum, Lessing.

South-Africa. A perennial herb of medicinal properties; praised like numerous other plants there and elsewhere as an alexipharmic, but all requiring close re-investigation in this respect.

Gaulthieria Myrsinites, Hooker.

Northern California, Oregon, British Columbia. The fruit of this procumbent shrub is said to be delicious. It would prove adapted for any of the Alps.

Gaulthieria Shallon, Pursh.

North-Western America. This handsome spreading bush would yield its pleasant edible berries in abundance, if planted on alpine mountains, where it would likely become naturalized. G. procumbens (L.) is the North-Eastern American Wintergreen used in medicine.

Gaylussacia frondosa, Torrey and Gray.

The Blue Tangleberry of Eastern North-America. A bush with deciduous foliage; fruit very sweet.

Gaylussacia resinosa, Torrey and Gray.

The Black Huckleberry of North-Eastern America. A dwarf shrub, with deciduous leaves. It likes swampy woodlands, and thus would find ample space in any forest-ranges. Berry of pleasant taste. Perhaps some of the South-American species may also produce edible fruits.

Geitonoplesium cymosum, Cunningham.

Through the whole East-Australian forests. It is mentioned here, to draw attention to the likelihood, that special culture may convert this into a culinary plant, as Mr. P. O'Shanesy found the young shoots to offer a fair substitute for Asparagus.

Gelsemium nitidum, Michaux.

Southern States of North-America, also in Mexico. "Yellow Jessamine." A twining shrubby plant of medicinal value, long since introduced into Australia by the writer, with numerous other plants of industrial or therapeutical importance. Active principle: gelsemin. The perfume of the flowers has also come into use as a cosmetic.

Genista monosperma, Lamarck.

Mediterranean regions. One of the best of Broom-brushes for arresting sand-drift. G. sphærocarpa, Lamarck, is of like use, and also comes from the Mediterranean Sea.

Genista tinctoria, Linné.

Europe, Northern and Western Asia. A perennial herb, of some medicinal use. From the flowers a yellow dye may be extracted, which with woad gives a good green, and comes well in for domestic dyeing, particularly of wool. A kind of Schüttgelb, different from the one prepared from Maclura tinctoria, is obtained from this Genista, known also as factitium-yellow, and perhaps not altogether to be superseded by picric acid or by Anilin-colors (G. Don; Rosenthal; Brockhaus).

Gentiana lutea, Linné.

Sub-alpine tracts of Middle and Southern Europe. A beautiful perennial herb, yielding the medicinal gentian-root. It could be easily raised in high mountains elsewhere. Chemical principles: gentian-bitter and gentianin. Medicinal gentian-root is also obtained from G. punctata, L., G. purpurea, L. and G. panonica, Scop., of the European Alps. Several native species are used as substitutes in North-America, particularly G. quinqueflora (Lamarck).

Geonoma vaga, Grisebach and Wendland.

West-Indies to Brazil. A dwarf decorative Palm, ascending mountain-regions to 3,000 feet.

Geum urbanum, Brunfels.

The "Avens" of Britain. Europe, North-Africa, extra-tropical and alpine Asia, South-Eastern Australia, North-America. A perennial herb; the powerful anti-dysenteric root, according to Muspratt, contains as much as 41 per cent. of tannic acid.

Gigantochloa Abyssinica, F. v. Mueller. (*Oxytenanthera Abyssinica,* Bentham.)

Tropical Africa. A tall species, ascending to considerable mountain-elevations.

Gigantochloa apus, Kurz. (*Bambusa apus,* Roemer and Schultes.)

Indian Archipelagus, at elevations under 5,000 feet. Height of stem reaching 60 feet. When young it is used for strings and ropes.

Gigantochloa aspera, Kurz.

Java. Found by Zollinger to attain a maximum-height of 170 feet.

Gigantochloa atter, Kurz.

Java, at elevations of from 2,000 to 4,000 feet. Height of stems reaching 70 feet. One of the species much grown for rural and industrial purposes.

Gigantochloa maxima, Kurz.

Java. Height sometimes 120 feet, the stems nearly a foot thick. One of the most extensively cultivated of all Asiatic Bamboos, ascending into mountain-regions.

Gigantochloa nigro-ciliata, Kurz. (*Oxytenanthera nigro-ciliata,* Munro.)

Continental and insular India. Stems to 130 feet long.

Gigantochloa robusta, Kurz.

Mountains of Java. Height to about 100 feet. Kurz noticed the early growth to be nearly 18 feet in a month, the principal branches only commencing when the shoot had reached a height of about 70 feet. Some Java-bamboos are known to measure 22 inches in girth at a height of about 120 feet.

M

Gigantochloa Thwaitesii, Kurz. (*Oxytenanthera Thwaitesii,* Munro.)

Ceylon, at cool elevations of from 4,000 to 6,000 feet. This pretty Bamboo reaches only 12 feet in height.

Gigantochloa verticillata, Munro. (*Bambusa verticillata,* Blume.)

.The Whorled Bamboo of India. It attains a height of fully 100 feet; in damp heat it grows at the astonishing rate of 40 feet in about three months, according to Bouché. The young shoots furnish an edible vegetable like G. apus and Bambusa Bitung.

Ginkgo biloba, Linné.* (*Salisburia adiantifolia,* Smith.)

Ginkgo-tree. China and Japan. A deciduous fan-leaved tree, to 100 feet high, with a straight stem to 12 feet in diameter. The wood is pale, soft, easy to work and takes a beautiful polish. The seeds are edible, and when pressed yield a good oil. The fruits, sold in China under the name of "Pa-Koo," are not unlike dried almonds, but the kernel fuller and rounder. Ginkgo-trees are estimated to attain an age of 3,000 years. Mr. Christy observes, that the foliage turns chrome-yellow in autumn, and that it is the grandest and most highly esteemed of all trees in Japan; it will grow in dry situations. In America it is hardy as far north as Montreal, in Europe to Christiania.

Gladiolus edulis, Burchell.

Interior of South-Africa. The bulb-like roots are edible, and taste like chestnuts when roasted.

Glaucium luteum, Scopoli.

Western and Southern Europe, Northern Africa and Western Asia. This fast-spreading biennial herb, now also naturalized on some of the Australian coasts, does good service in aiding to subdue drift-sand. The plant has also some medicinal value.

Gleditschia triacanthos, Linné.

The deciduous Honey-Locust tree of South-Eastern States of North-America. Height reaching 80 feet, trunk to 4 feet in diameter. Wood hard, coarse-grained, fissile, durable, sought principally for blocks and hubs. The tree is not without importance for street-planting. Rate of circumferential stem-growth in Nebraska, about 40 inches in 22 years at 2 feet from the ground (Furnas); growth in height at Port Phillip, about 35 feet in 20 years. Sown closely, this plant forms impenetrable, thorny, not readily combustible hedges. An allied species, the G. Sinensis, Lamarck (G. horrida, Willd.), occurs in East-Asia. The Water-Locust tree of North-America (G. monosperma, Walt.) will grow to a height of 80 feet in swamps. The flowers of Gleditschia exude much honey-nectar for bees.

Glycine hispida, Bentham. (*Soja hispida,* Moench.)

An annual herb of India, China and Japan. The beans afford one of the main ingredients of the condiment known as Soja; they are very oily, nutritious, and of pleasant taste when boiled. The plant endures slight frost (Wittmack). It is not subject to attacks of insects and fungs (Vilmorin). Oil is pressed from the seeds. Glycine Soja, Siebold and Zuccarini, is said to be a distinct plant, but probably serves the same purpose.

Glycyrrhiza echinata, Linné.

South-Europe and South-Western Asia. From the root of this herb a portion of the Italian liquorice is prepared. The Russian liquorice is also derived from this species. The root is thicker and less sweet than that of the following.

Glycyrrhiza glabra, Linné.

South-Europe, North-Africa, South-Western and Middle Asia. The extract of the root of this herb constitutes the ordinary liquorice. The plant grows most vigorously in adequate climes. Both this and the preceding are hardy in Norway to lat. 59° 55' (Schuebeler). Liquorice is of some utility in medicine and also used in porter-breweries. Chemical principle: glycyrrhizin.

Glyptostrobus heterophyllus, Endlicher. (*Taxodium heterophyllum,* Brongniart.)

China. An ornamental tree, allied to Taxodium distichum in some respects, and like that tree particularly fit for permanently wet ground. The Chinese plant it along the edges of canals and narrow creeks, the buttress of the tree standing actually in the moist mud (Dr. Hance).

Gmelina Leichhardtii, F. v. Mueller.

East-Australia. Grown now on a commercial scale for fancy timber-purposes in Queensland.

Gonioma Kamassi, E. Meyer.

South-Africa. This small tree furnishes the yellow Kamassi-wood, much sought for carpenters' tools, planes and other select articles of wood-work; also for wood-engraving, according to Dr. Pappe. Flowers deliciously fragrant.

Gordonia Lasianthus, Linné.

The Loblolly-Bay. South-Eastern North-America. A handsome tree, growing to a height of 60 feet; flowers snowy white. The wood is extremely light, of a rosy hue and fine silky texture, but unfit for exposure. The bark is extensively employed for tanning in the Southern States. Available for swampy coast-lands.

Gossypium arboreum, Linné.*

The Tree-Cotton. Upper Egypt, according to A. de Candolle, seemingly also Abyssinia, Sennaar and thence to Upper Guinea. A tall perennial species, but not forming a real tree, yielding cotton in the first season. Leaves long-lobed. Bracts with few teeth. Petals yellow, or in age pink or purple. Seeds brown, disconnected, after the removal of the cotton-fibre greenish-velvety. The cotton of long staple, but also a variety occurs with short staple. The New Orleans

M 2

cotton (G. sanguineum, Hasskarl) belongs to this species. The cotton-fibre is crisp, white, opaque and not easily separable. All Gossypiums can be regarded as honey-plants.

Gossypium Barbadense, Linné.*

Sea-Island Cotton. From Mexico to Peru and Brazil. Leaves long-lobed. Petals yellow. Seeds disconnected, black, after the removal of the cotton-fibre naked. The cotton of this species is very long, easily separable and of a silky lustre and always white. This species requires low-lying coast-tracts for attaining to perfection. Perennial, but yielding like the rest a crop already in the first season. Cultivated largely in the Southern States of North-America, also in South-Europe, Central and North-Africa, Queensland and various other countries. G. Kirkii (Masters), from Dar Salam, may be a wild state of G. Barbadense. The only other type of this genus in tropical Africa is G. anomalum according to Dr. Welwitch. The "Kidney-cotton" is a variety with more acuminate leaves. M. Delchevalerie has drawn attention to a new plant, tall in size and exceedingly prolific in bearing, raised in Egypt, called Bamia-cotton, which Sir Joseph Hooker regards as a variety of G. Barbadense. The Bamia Cotton-bush grows 8 to 10 feet high, ripens (at Galveston) fruit in four or five months, and produces 2,500 pounds of cotton and seed per acre. It is remarkable for its long simple branches, heavily fruited from top to bottom. Its cotton is pale yellow.

Gossypium herbaceum, Linné.*

Persia, Scinde, Cabul and some other parts of tropical and subtropical Asia. Much cultivated in the Mediterranean countries, also in the United States of North-America. Perennial. Leaves short-lobed. Petals yellow. Seeds disconnected, after removal of the cotton-fibre gray-velvety. Distinguished and illustrated by Parlatore as a species, regarded by Seemann as a variety of G. arboreum. Staple longer than in the latter kind, white-opaque, not easily separating. The wild type of this seems to be G. Stocksii (Masters). Even this species, though supposed to be herbaceous, will attain a height of 12 feet. The root of this and some other congeners is a powerful emmenagogue. A variety with tawny fibre furnishes the Nankin-cotton.

Gossypium hirsutum, Linné.*

Upland- or Short-staple Cotton. Tropical America, cultivated most extensively in the United States, Southern Europe and many other countries. Perennial. Seeds brownish-green, disconnected, after the removal of the cotton-fibre greenish-velvety. Staple white, almost of a silky lustre, not easily separable. A portion of the Queensland-cotton is obtained from this species. It neither requires the coast-tracts nor the highly attentive culture of G. Barbadense.

Gossypium religiosum, Linné.*   (*G. Peruvianum*, Cavanilles.)

Tropical South-America, southward to Chili. Kidney-Cotton, Peruvian or Brazilian Cotton. Leaves long-lobed. Petals yellow.

Seeds black, connected. The cotton is of a very long staple, white, somewhat silky, and easily separable from the seeds. A tawny variety occurs. This is the tallest'of all cotton-bushes, and it is probably this species, which occurs in the valleys of the Andes as a small tree, bearing its cotton, while frosts whiten the ground around.

Gossypium Taitense, Parlatore. (*G. religiosum*, Banks and Solander.)

In several islands of the Pacific Ocean. A shrub. Petals white. Seeds disconnected, glabrous after the removal of the fulvous cotton-fibre, which does not separate with readiness.

Gossypium tomentosum, Nuttall. (*G. Sandvicense*, Parlatore; *G. religiosum*, A. Gray.)

Hawaia. Perennial. Petals yellow. Seeds disconnected; after the removal of the tawny cotton-fibre fulvous-velvety, not easily parting with their cotton. The roots are a powerful remedial agent, which however should only be used in legitimate medical practice. The barks of Hamamelis Virginiana and Viburnum prunifolium are antidotes (Phares and Durham).

·For limitation of species and varieties Parlatore's " Specie dei Cotoni " (Florence, 1866) and Todaro's " Osservazioni su Cotone " may be consulted. Information on culture may be sought in Porter's "Tropical Agriculturist " and in Mallet's work on " Cotton " (London, 1862).

The following notes were written for the use and guidance of Victorian colonists:—

There are many parts of our colony, in which all these species of Gossypium could be cultivated, and where a fair or even prolific cotton-crop may be obtained. Good cotton for instance has been produced on the Goulburn-River, the Loddon, the Avoca and the Murray-River, particularly in places, where water could be applied. All cultivated kinds of cotton-plants are either naturally perennials or become such in favorable climes, although they may be treated strictly as annuals. Some of them will indeed in particular instances grow to the height of 20 feet. The geographical parallels, between which cotton-culture is usually placed, stretch in various girdles between 36° north latitude and 36° south latitude. According to General Capron, cotton is grown in Japan to 40° north latitude, but superior quality is not obtained north of 35°. ·

The cotton-culture in the Southern States of North-America utilized seven million acres before the civil war, cultivated by a million and a half of Negroes; India has now 14 million acres in cotton, as much as the United States in 1879, the yield in the latter being at an average nearly half a bale per acre, and the export thence in 1881 in value about 260 millions of dollars (J. R. Dodge); in 1883 the cotton-area of the United States was 16,777,993 acres; in 1882 the cotton-yield there 6,957,000 bales. The importations of cotton into the United Kingdom in 1884 amounted to 15½ million cwt., its value being £44,000,000; about two-thirds of this came from the United States. The primary advantages of this important culture are: a return in a few months, comparatively easy field-operations, simple and not laborious process of collecting the crop, and requirement or but little care in the use of the gin-machine in finally preparing the raw material for the market, the woolly covering of the seeds constituting the cotton of commerce. The oil obtained by pressure from the seeds is useful for various technic purposes, and the oil-cake can be utilized like most substances of similar kind as a very fattening stable-food. This oil can even be used quite well in domestic cookery (Colonel O. Nelson). Crushed cotton-seed cake without admixture is eaten by cattle and sheep with avidity. Of cotton seeds 212,000 tons were introduced into Great Britain in 1884, valued at £1,580,000, mostly from Egypt. Sea-Island cotton was raised to great perfection in the northern parts of Victoria fully twenty-five years ago from seeds extensively distributed by the writer; but the want of cheap labor has hitherto militated against the extensive cultivation of this crop, as well as that of tea and many other industrial plants. Cotton having been reared far away from the influence of the sea-air, it would be worthy of attempts, to naturalize various kinds of cotton in the oases of our deserts, irrespective of regular culture. Our native Gossypiums of the interior produce no fibre worth collecting. Cotton plants have a predilection for gently undulating or sloping ground, with light soil and a moderate supply of moisture. In the most favorable climes, such as that of Fiji, cotton produces flowers and fruit throughout the year, but the principal ripening falls in the dry season. From two hundred to three hundred plants or more can be placed on an acre. As many as seven hundred bolls have been gathered from a single plant at one time, twelve to twenty capsules yielding an ounce of mercantile cotton. Weeding is rendered less onerous by the vigorous growth of the plants. Cotton comes in well for rotation with other crops. Major Clarke has ascertained, that crossing cannot be effected between the oriental and occidental kinds of cotton. A high summer temperature is needed for a prolific cotton-harvest. Intense heat, under which even maize will suffer, does not injuriously affect cotton, provided the atmosphere is not dry in the extreme. The soil should not be wet, but of a kind that naturally absorbs and retains humidity, without over-saturation. In arid regions it is necessary, to irrigate the cotton-plant. Heavy rains at the ripening period are injurious, if not destructive, to the cotton-crop. Dry years produce the best returns, yet aqueous vapor in the air is necessary for the best yield. In colder localities the bolls or capsules continue to ripen after night-frosts prevent the formation of new ones. Porous soils, resting on limestones and metamorphic rocks, are eminently adapted for cotton-culture. The cane-brake-soil of the North-American cotton-regions absorbs ammonia to a prodigious extent.

Gourliaea docorticans, Grisebach.

The Chañar of Argentina. Bears sweet pleasant fruits, and yields a tough valuable wood (Dr. Lorentz).

Grevillea annulifera, F. v. Mueller.

West-Australia. A tall brush or small tree, with highly ornamental flowers. The seeds are comparatively large, of almond-taste, and the fruits produced copiously. The shrub will live in absolute desert-sands, where the other Australian proteaceous Nut-tree, Brabejum (Macadamia) ternifolium, could not exist. Well may we plead, that enlightened statesmanship should lastingly preserve at least on a few chosen spots also in South-Western Australia all the splendid Grevilleas and hundreds of other gay or remarkable plants, quite peculiar to that part of the world, where the endemism of vegetation is more singularly and strongly concentrated than anywhere else on the globe, unless in South-Africa and California; so that future generations also may yet be able, to contemplate at least the local remnants of a world of plants as charming as it is diversified and peculiar, before many of its constituents succumb by aggress of herds and flocks altogether.

Grevillea robusta, Cunningham.

A beautiful lawn-tree, indigenous to the sub-tropical part of East-Australia, rising to 150 feet, of rather rapid growth, and resisting drought in a remarkable degree; hence one of the most eligible trees, even for desert-culture. Cultivated trees at Melbourne yield now an ample supply of seeds. The wood is elastic and durable, valued particularly for staves of casks, also for furniture. The richly developed golden-yellow trusses of flowers attract honey-sucking birds and bees through several months of the year. The seeds are copiously produced and germinate readily. Rate of growth in Victoria, 20–30 feet in 20 years.

Grindelia squarrosa Dunal.

North-America in the middle-regions, but extending also far northward. A perennial balsamic herb, praised for medicinal virtues in its native lands. Several congeners occur from California and Mexico to Chili and Argentina. G. robusta (Nuttall) serves therapeutic purposes in California.

Guadua angustifolia, Kunth. (*Bambusa Guadua,* Humboldt and Bonpland.)

New Granada, Ecuador and probably other of the Central American States. This Bamboo attains a height of 40 feet, and might prove hardy in sheltered places of temperate low-lands. Holton remarks of this species, that it is, after the plantain, maize and cane, the most indispensable plant of New Granada, and that it might be called the lumber-tree, as it supplies nearly all the fencing and wood-work of most of the houses, and is besides manufactured into all kinds of utensils. The genus Guadua comprises the stoutest of all Bamboos.

Guadua latifolia, Kunth.* (*Bambusa latifolia,* Humboldt and Bonpland.)

One of the tall Bamboos of Central America, whence several other lofty Bamboos may be obtained, among them the almost climbing Chusqueas. This Guadua is stouter than any Indian Bamboo. In tropical America native Bamboos are planted for hedges.

Guevina Avellana, Molina. (*Quadria heterophylla,* Ruiz and Pavon.)

The evergreen Hazel-tree of Chili, extending to the Chonos-Archipelagus. One of the most beautiful trees in existence, attaining a height of 30 feet. The snowy white flower-spikes produced simultaneously with the ripening of the coral-red fruit. In the cooler southern regions the tree attains considerable dimensions. The wood is tough and elastic, and used particularly for boat-building (Dr. Philippi). The fruit of the allied South-African Brabejum stellatifolium (Linné) can only be utilized with caution and in a roasted state as an article of diet, because it is noxious or even absolutely poisonous in a raw state.

Guizotia oleifera, De Candolle.

India and probably also Abyssinia. Rantil-oil is pressed from the seeds of this annual herb, which yields its crop in three months. The oil is much used like Sesamum-oil, for culinary as well as for technic purposes.

Gunnera Chilensis, Lamarck.

Caraccas to Patagonia, chiefly on cliffs. A most impressive plant for scenic groups in gardens. Darwin measured leaves 8 feet broad and 24 feet in circumference. The acidulous leaf-stalks serve as a vegetable; the thick roots are used for tanning and dyeing. G. peltata (Philippi) is another large species, restricted to Juan Fernandez. G. macrophylla (Blume) is a native of Java and Sumatra, where it occurs on mountains up to 6,000 feet elevation.

Gymnocladus Canadensis, Lamarck.

The Chicot or Kentucky Coffee-tree. North-Eastern America. A timber- and avenue-tree, attaining a height of 80 feet; allied to Gleditschia, but, as the name implies, thornless. Delights in a rich soil and a sheltered position. Can be raised from cuttings of the roots. The wood is strong, tough, compact, fine-grained, and assumes a rosy color. The pods, preserved like those of Tamarinds, are said to be wholesome (Simmonds). Insects preying on the foliage of this tree are poisoned by it. It will bear the frosts of Norway to lat. 61° 17′ (Schuebeler).

Hagenia Abyssinica, Willdenow. (*Brayera anthelmintica,* Kunth.)

Abyssinia, at elevations from 3,000 to 8,000 feet. A tall tree, admitted in this list, because its flowers have come into medicinal use. It is moreover quite eligible for ornamental plantations.

Hancornia speciosa, Gomes.

Brazil, to far southern regions, on sandy plains. This small tree may prove hardy in extra-tropic regions free of frost. The good-sized berries are of very pleasant taste, and vernacularly known as Mangaihas. Most valuable is the very elastic rubber of this plant. Mr. Thomas Christie regards it superior to Para-Rubber, and worth at present 3s. per lb.

Hardwickia binata, Roxburgh.

India, up to elevations of nearly 4,000 feet. Maximum height of tree 120 feet. Wood from red-brown to nearly black, close-grained, exceedingly hard, heavy and durable; valued for under-ground work. The bark furnishes easily a valuable material for cordage. The tree can readily be pollarded for cattle-fodder (Brandis).

Harpullia Hillii, F. v. Mueller.

The Tulip-Wood of Queensland. One of the most important of the numerous kinds of trees indigenous there for select cabinet-work. H. pendula (Planchon) is equally valuable.

Hedeoma pulegioides, Persoon.

The Penny-royal of Eastern North-America. An annual herb of aromatic taste, employed in medicine. The volatile oil is also in use.

Hedera Helix, Linné.

The Ivy. Europe, Northern Africa, Western Asia as far as the Himalayas. Not to be omitted here, as it quickly forms evergreen walls over all kinds of fences and on sides of buildings without injuring any masonry; it is also a bee-plant for honey. Individual plants will live through several centuries. The variety with yellow-marked leaves is singularly ornamental. Resists the smoky air of cities (Loudon). Hederic acid is of medicinal value. A decoction of the leaves dyes hair black.

Hedysarum coronarium, Linné.*

The Soola-Clover. Southern Europe, Northern Africa. One of the best of perennial fodder-herbs, yielding a bulky return. It is also recommended as quite a handsome garden-plant.

Heleocharis sphacelata, R. Brown.

Australia, New Zealand and South-Sea Islands. This rush is well deserving, to be transferred to any swamps in warmer climes on account of its nutritious and palatable tubers.

Heleocharis tuberosa, Roemer and Schultes.

China, where it is called Matai or Petsi. This rush can be subjected to regular cultivation in ponds for the sake of its edible wholesome tubers. H. plantaginea (Brown) and H. fistulosa (Schultes) of tropical Asia and Madagascar are allied plants.

Helianthus annuus, Linné.*

The Sun-Flower. Peru. This tall, showy and large-flowered annual is not without industrial importance. As much as fifty bushels of seeds, or rather seed-like nutlets, have been obtained from an acre under very favorable circumstances, and as much as fifty gallons of oil can be pressed from such a crop. The latter can be used not only for machinery, but even as one of the best for the table; also for superior toilet-soaps and for painting; it belongs to the series of drying-oils. Otherwise the seeds afford an excellent fodder for fowl;

they are also used for cakes, and afford a substitute for coffee, according to Professor Keller. The leaves serve for fodder. The large flower-heads are important as yielding much honey   The stalks furnish a good textile fibre, and the blossoms yield a brilliant lasting yellow dye. About six pounds of seed are required for an acre. The plant likes calcareous soil. Important also for quickly raising vegetation around fever-morasses, the absorbing and exhaling power of this plant being very large (Dr. v. Hamm). The Sun-Flower, according to Lacoppidan, will exhale 1½ lb. of water during a hot day. Several North-American species may deserve rural culture. The return from a Sun-Flower field is attained within a few months. In Norway it can be grown to lat. 70° 4' (Schuebeler); yet it will, according to the Rev. H. Kempe, also endure the excessive summer heat of Central Australia better than any other cultivated herb yet tried there.

Helianthus tuberosus, Linné.*

Brazil.   Sun-Flower Artichoke, inappropriately passing under the name " Jerusalem-Artichoke," instead of " Girasol-Artichoke." The wild state, according to Professor Asa Gray, seems to be the North-American H. doronicoides (Lamarck). The tubers are saccharine and serve culinary purposes. As a fodder they increase the milk of cows to an extraordinary degree. The foliage serves well also as fodder. The plant is propagated from the smallest but undivided tubers, placed like potatoes, but at greater distances apart. The root is little susceptible to frost. The plant would be valuable for alpine regions. In Norway it can be grown successfully still at lat. 68° 24' (Schuebeler). The yield is as large as that of potatoes, with less labor, and continues from year to year in fairly treated land uninterruptedly and spontaneously. The stem is rich in textile fibre. The percentage of crystalline sugar is largest during the cold season, then 5–6 per cent. During the summer the starch-like inulin prevails. This plant can only be broughf to full perfection in a soil rich in potash.

Helichrysum lucidum, Henckel.   (*H. bracteatum,* Willdenow.)

Throughout the greater part of Australia. H. lucidum can be grown as a summer-plant to lat. 70° 4' in Norway (Schuebeler). The regular cultivation of this perennial herb would be remunerative, to supply its everlasting flowers for wreaths, just as those of H. orientale (Tournefort) from Candia are largely grown and sold in South-Europe, to provide wreaths for graves. Furthermore, the lovely Helipterum Manglesii (F. v. M.) from West-Australia could for the same purposes be reared on a large scale with several other Australian everlastings. Some South-African species of Helichrysum and Helipterum are also highly eligible for these purposes of decoration; as such may be mentioned Helichrysum fulgidum (Willdenow), H. sesamoides (Thunberg), H. vestitum (Lessing); Helipterum canescens, H. eximium and H. speciosissimum (De Candolle). Helichrysum apiculatum (De Candolle) affords herbage in the worst deserts of Australia.

Heliotropium Peruvianum, Linné.

Andes of South-America. A perennial somewhat shrubby plant. Among various species of Heliotrope this one can best be utilized for the extraction of the scented oil. Heliotropin obtainable from this and allied plants has been produced synthetically also.

Helleborus niger, Brunfels.

Forest-mountains of Middle and Southern Europe, particularly in sub-alpine regions. The Christmas-rose of British Gardens. A perennial handsome herb, remarkable for flowering even in cold countries during mid-winter. The roots are used in medicine; so those of H. viridis (Linné) from the same region, particularly where lime prevails in the soil.

Helvella esculenta, Persoon.

Europe. Dr. Goeppert notes among saleable Silesian mushrooms for table-use this species as well as H. gigas (Krombholz) and H. infula (Fries). Kohlrausch and Siegel found in H. esculenta when dried 26 per cent. of protein, against the following other results : in beef 39 per cent., in veal 44, wheat-bread 8, oatmeal 10, pulse 27 potatoes 5, various mushrooms often 33 per cent. Of course starch, sugar, inulin, pectin, gum and even fibre have to be further taken into consideration in these calculations on value of nutriments. The deleterious principle of H. esculenta needs to be removed by repeated treatment with boiling water, or by keeping the dried fungus for about a year before it is used for the kitchen. Dr. M. C. Cooke mentions as fair English substitutes of Morels Helvella lacunosa (Afzelius) and H. mitra (Linné). Bergner and Trog illustrate as edible among fungs of Switzerland H. crispa (Fries).

Hemarthria compressa, R. Brown.

Southern Asia, Southern Africa, extra-tropical Australia. This perennial grass, though somewhat harsh, is recommendable for moist pastures, and will retain a beautiful greenness throughout the year in dry climes ; highly esteemed by graziers in Gippsland (Victoria); it is not injured by moderate frost. H. uncinata is a mere variety, which grows down to high-water mark on estuaries of rivers ; also otherwise on somewhat saline ground. H. fasciculata (Kunth) occurs around the Mediterranean Sea. The genus is also represented by one species in the warmer litoral regions of America.

Heracleum Sibiricum, Linné.

Colder regions of Europe and Asia. A very tall biennial herb with leaves of enormous size. Recently recommended for sheep-fodder in alpine regions. This plant could also be turned to account for scenic effect in horticulture, as well as H. dulce (Fischer) of Kamtschatka.

Heterothalamus brunioides, Lessing.

Southern Brazil and Argentina. A dwarf shrub, furnishing the yellow Romerillo-dye from its flowers.

Hibiscus cannabinus, Linné. (*H. radiatus*, Cavanilles.)

Tropical Asia, Africa and Australia. An annual showy herb, yielding the Gambo-Hemp. Stems to 12 feet high, without ramification if closely sown. Rich soil on the Nile has yielded over 3,000 lbs. of clear fibre from one acre. The bearing strength is often found to be more than that of the Sunn-fibre. The leaves serve as sorrel-spinage. Several other Hibisci can be utilized in the same manner. Good fibre is also obtained from Sida rhombifolia (Linné).

Hibiscus esculentus, Linné.

Tropical Africa. A tall herb. The unripe mucilaginous seed-capsules are known as Ochro, Okra Bandakai or Gumbo, and used as a culinary vegetable. The summers of Victoria bring them to maturity. The Ochro can be preserved by being dried either in the sun or by artificial heat after previous slicing. The leaves of this and allied species can be used as pot-herbs. The seeds retain their vitality for about five years (Vilmorin). In hot moist countries also multiplied from cuttings, and there growing with amazing quickness. Dr. A. Gibson pronounces the Ochro quite a safe food, even when extensively consumed.

Hibiscus Ludwigii, Ecklon and Zeyher.

South-Africa. A tall, shrubby and highly ornamental species, desirable also as yielding a fibre of fair strength and toughness.

Hibiscus Sabdariffa, Linné.

Tropical Asia and Africa. A showy annual plant, occasionally of more than one year's duration, admitting of culture in the warmer temperate regions; it is however cut down by frost. It yields the Rosella-fibre. The acidulous calyces furnish a delicious sorrel and rosella-jellies, particularly relished in hot climes. H. punctatus (Dalzell and Gibson) is mentioned as an annual fibre-plant, occurring in Sindh and Mooltan; H. tetraphyllus (Roxburgh) is noted by Prof. Wiesner as an annual Indian fibre-plant also.

Hierochloa redolens, R. Brown.

South-Eastern Australia, there almost confined to the Alps; also in New Zealand, in the Antarctic Islands and the southern extremity of America. A tall, perennial, blady grass, with the odor of Anthoxanthum. It is worthy of dissemination on moist pasture-land in cool countries. H. borealis of the colder regions of the northern hemisphere accompanies H. redolens in the south, but is a smaller grass. These grasses are to some extent valuable for their fragrance as constituents of hay, the odorous principle, as in Anthoxanthum, Melilotus and Asperula, being cumarin. Hierochloas are particularly appropriate for cold, wet, moory grounds.

Hippocrepis comosa, Linné.

The Horse-shoe Vetch. Middle and Southern Europe, North-Africa. A perennial fodder-herb, not without importance. Likes stony ground, and delights like most leguminous herbs in limestone-soil. The foliage is succulent and nutritious. Professor Laugethal recommends it for a change after Sainfoin-pastures fail. It furnishes not quite as much but an earlier fodder.

Holcus lanatus, Linné.

Velvet-grass or Meadow-Softgrass, also known as Yorkshire-fog. Europe, North-Africa, Middle Asia. Indigenous in Norway to lat. 63° 34'. A well-known and easily disseminated perennial pasture-grass, of considerable fattening property. For rich soil better grasses can be chosen, but for moist, moory or sandy lands and also for forests it is one of the most eligible pasture-grasses, yielding an abundant and early crop; it is however rather disliked by cattle as well as horses. One of the best rural grasses in recently cleared forest-ground, not—like Cocksfoot-grass and particularly Rye-grass—apt to be attacked by caterpillars; also suited for suppressing bracken-ferns after they have been burnt down. Recommendable also for newly drained land. Does thrive according to the Rev. H. Kempe in the hottest and driest regions of Central Australia. The chemical analysis made in full spring gave the followg results:—Albumen, 3·20; gluten, 4·11; starch, 0·72; gum, 3·08; sugar, 4·56 per cent. (F. v. Mueller and L. Rummel).

Holcus mollis, Linné.

Creeping Softgrass. Of nearly the same geographic range and utility as the preceding species. Particularly adapted to sandy forest-land. Grown in Norway to lat. 63° 7' (Schuebeler).

Holoptelea integrifolia, Planchon. (*Ulmus integrifolia*, Roxburgh.)

The Elm of India, extending from the lowlands to sub-alpine regions. A large tree, with timber of good quality. Foliage deciduous.

Hordeum andicola, Grisebach.

Argentina. Pronounced by Prof. Hieronymus as an excellent pasture-grass of the Sierras.

Hordeum deficiens, Steudel.

The Red-Sea Barley. One of the two-rowed barleys cultivated in Arabia and Abyssinia. Allied to this is H. macrolepis, (A. Braun), a native of Abyssinia.

Hordeum distichon, C. Bauhin.*

Wild from Arabia to Central Asia (A. de Candolle). Cultivated as early as the stone-age (Heer). The ordinary Two-rowed Barley. To this species belong the ordinary English barley, the Chevalier, the Annat, the Dunlop, the Long-eared, the Black, the Large, the Italian and the Golden barley, along with other kinds. A variety with grains free from the bracts constitutes the Siberian- and the Haliday-barley, which however is less adapted for malt. Dry barley-flour, heated at the temperature of boiling water during several hours under the exclusion of atmospheric air, constitutes Hufeland's meal for invalids. Barley-culture can be carried on even in alpine regions. Marly and calcareous lands are particularly fit for rearing this cereal grass. It resists moderate spring-frosts. As much as 100 bushels of Cape-barley have been obtained from an acre of land in volcanic soil of Victoria as a first harvest.

Hordeum hexastichon, Linné.*

Orient. The regular Six-rowed Barley. In cultivation already during the stone-age (Heer). This includes among other varieties the Red, the Scotch, the Square- and the Bear-barley. Seeds less uniform in size than those of H. distichon. The so-called skinless variety is that, in which the grain separates from the bracts. Lange-thal observes, that it is most easily raised, requires less seed-grain than ordinary barley, has firmer stems, is less subject to the rust-disease and to bending down.

Hordeum secalinum, Schreber.*   (*H. nodosum*, Smith; *H. pratense*, Hudson.)

Europe, Northern and Middle Asia, North-America. Perennial. Famed as the best fattening grass of many of the somewhat brackish marsh-pastures on the North Sea. It never fruits, when kept down by cattle, and finally suppresses nearly all other grasses and weeds.

Hordeum vulgare, Linne.*

Orient. The Four-rowed Barley, though rather six-rowed with two prominent rows. Of less antiquity than H. distichon and H. hexastichon. Several varieties occur, among them: the Spring- and Winter-barley, Black barley, the Russian, the French, the Naked and the Wheat-barley. Pearl-barley is obtained from the winter-variety, which also surpasses Summer-barley in rigor of stems and rich and early yield, it being the earliest cereal in the season; the straw is copious and nutritious, and the grain is rich in gluten, hence far better adapted for flour than for malt. Summer-barley also passes under the name of Sand-barley. It is inferior in yield to H. distichon, but is content with a less fertile, even sandy soil, and comes to ripeness in a month's less time. In alpine regions it ripens with a summer of sixty or seventy days without frost. In Norway it can be grown to lat. 70° (Schuebeler). The Naked barley is superior to many other varie-ties for peeled barley, but inferior for brewing; the grain is also apt to drop (Langethal). Malt is important as an antiscorbutic remedy. Chemical principles of malt: asparagin, a protein substance, diastase, an acid and cholesterin fat. Maltine is a therapeutic extract.

Hordeum zeocriton, Linné.*

Central Asia. A Two-rowed Barley. To this species belong the Sprat, the Battledore, the Fulham- and the Putney-barley, the Rice-barley, the Turkish barley and the Dinkel. This species might be regarded as a variety of H. distichon. The grains do not drop spon-taneously, and this kind is securer than others against sparrows; requires however a superior soil and is harder in straw (Langethal).

Hovenia dulcis, Thunberg.

Himalaya, China, Japan. The pulpy fruit-stalks of this tree are edible. H. inæqualis (De Candolle) and H. acerba (Lindley) are mere varieties of this species.

Humulus lupulus, Linne.*

The Hop-plant. Temperate zone of Europe, Asia and North-America. Very hardy, being indigenous in Norway to lat. 64° 12' and cultivated to lat. 69° 40' (Schuebeler). This twining perennial unisexual plant is known to yield enormously on river-banks in rich soil or on fertile slopes, where irrigation can be effected. A pervious, specially alluvial soil, fertile through manure or otherwise, appliances for irrigation natural or artificial, and also shelter against storms are some of the conditions for success in hop-growth, and under such conditions the raising of hops will prove thus far profitable in countries and localities of very different mean-temperature. A dry summer season is favorable to the ripening and gathering of hops. On the Mitchell-River, in Gippsland, 1,500 lbs. have been obtained from an acre; on the King-River in Victoria even as much as 2,286 lbs. in one particular year. In Tasmania large crops have been realized for very many years. The plant might be readily naturalized on river-banks and in forest-valleys. The scaly fruit-catkins form the commercial hops, whose value largely depends on the minute glandular granules of lupuline. Hops impart their flavor to beer, prevent acetous fermentation, and precipitate albuminous substances from the malt principally by their tannic acid. Hop-pillows are recommended to overcome want of sleep. Many of the substitutes for hops are objectionable or deleterious. The refuse of hops of breweries possess double the value of stable-manure. Great Britain imported in 1884 nearly 13,000 tons of hops valued at £1,600,000. Active principles of hop-leaves and fruits: a peculiar volatile and bitter acid substance. The fibre of the stem can be made into cords and paper. The young shoots can be used for food, dressed like asparagus.

Hydnum coralloides, Scopoli.

Europe, Asia, North- and South-America. In Cashmere, where it inhabits hollow trunks of Pinus Webbiana, called the Koho-Khur. Common on dead wood in forests in the United States. Cooked, of excellent taste.

Hydnum imbricatum, Linné.

In pine-forests of Europe. A wholesome mushroom of delicious taste, which we should endeavour to naturalize in any pine-plantations. Other recommendable European species are, H. erinaceum (Persoon), H. album (Persoon), H. diversidens (Fries), H. auriscalpium (Linné), H. subsquamosum (Batsch), H. lævigatum (Swartz), H. violascens (Albertini), H. infundibulum (Swartz), H. fuligineo-album (Schmitz), H. graveolens (Brotero), H. Caput Medusæ (Nees), H. hystrix (Fries). These and some other edible fungi are given on the authority of Rosenthal's valuable work.   The Rev. M. J. Berkeley, Dr. Morren and Dr. Goeppert add Hydnum repandum (Linné) and H. suaveolens (Scopoli).

Hydrangea Thunbergi, Siebold.

Japan. The leaves of this shrub give a peculiar tea, called the "Tea of Heaven."

Hydrastis Canadensis, Linné.

The Yellow Puccoon or Golden Seal. Eastern North-America. A perennial herb, utilized in medicine. The root contains two alkaloids, berberin and hydrastin. The root-dye is of a brilliant yellow, admitting of its use with indigo for rich green colors.

Hymenæa Courbaril, Linné.

Tropical and Southern sub-tropical America. A tree of colossal size and remarkable longevity. Timber hard, extremely heavy, close-grained, used for select wheel-work, treenails, beams and planks, also in various machinery. Courbaril-wood exceeds the British oak four times in elasticity and nearly three times in resistance to fracture (Lapparents). A fragrant amber-like resin, known as West-Indian Copal, exudes from the stem. The Mexican trade-name of the resin is Coapinole. The beans of the pod are lodged in a mealy pulp of honey-like taste, which can be used for food. The chance of the adaptability of this remarkable tree to the warmer temperate zone needs to be ascertained. This is one of the Algaroba-trees.

Hymenanthera Banksii, F. v. Mueller.

South-Eastern Australia, New Zealand, Norfolk-Island. A tall spiny shrub, well adapted for close hedges, where rapid growth is not required. It stands clipping well. Flowers profusely fragrant, hence this plant is among those best to be chosen for maintaining successively a strong fragrance in gardens during the whole year in serene climes.

Hyoscyamus niger, Linné.

The Henbane. Europe, North-Africa, extra-tropical Asia. In Norway indigenous to lat. 63° 35′. An important medicinal herb of one or two years' duration. It contains a peculiar alkaloid,— hyoscyamin.

Hyphæne Argun, Martius.

Nubia, to 21° north-latitude. Possibly hardy anywhere on lowlands in the warmer temperate zone.

Hyphæne coriacea, Gaertner.

Equatorial Eastern Africa; the dichotomous Palm of the sea-coast-regions. It attains a height of 80 feet. Deserving of cultural trials in cooler latitudes.

Hyphæne crinita, Gaertner. (*H. Thebaica,* Martius.)

The Gingerbread-Palm or Doum-Palm. Abyssinia, Nubia, Arabia and Egypt, as far as 31° north latitude, and southward to the Zambesi, Nyassa and Sofala. In Arabia to 28° north latitude (Schweinfurth); up to the plateau of Abyssinia (Drude). It is much branched, and attains a height of about 30 feet. The mealy husk of the fruit is edible. Grows away from the sea.

Hyphæne ventricosa, Kirk.

Zambesi. Loftier than the other species. Stem turgid towards the middle. Fruit large. Perhaps not absolutely requiring a tropical clime.

Hypochœris apargioides, Hooker and Arnott.

Chili. A perennial herb. The root is used for culinary purposes like that of the Scorzonera Hispanica.

Hypochœris Scorzoneræ, F. v. Mueller. (*Achyrophorus Scorzoneræ,* D.C.)

Chili. Of the same use as H. apargioides. Allied species of probably similar utility exist in Western South-America.

Hyssopus officinalis, Linné.

South-Europe, South-Western Asia. A perennial herb, discarded and re-introduced in medicine. The essential oil of this herb is also used for some perfumeries.

Ilex Aquifolium, Linné.

The Holly. Europe, Western Asia. In some cold regions the only evergreen tree not coniferous. Known to have attained an age of more than 150 years, a height of 60 feet, and a stem-circumference of 8¼ feet. It yields a wood for ornamental turnery, mathematic and other instruments, remarkable for its almost whitish paleness. In Norway it is indigenous to lat. 63° 7', and in lat. 59° 45' it attained still a height of nearly 50 feet (Professor Schuebeler).

Ilex Cassine, Linné.

Southern States of North-America. A tea-bush, to which also remarkable medicinal properties are ascribed. Ilex opaca (Aiton) attains a height of over 50 feet in Alabama.

Ilex crenata, Thunberg.

Japan. The wood employed there for superior kinds of wood-cuts. This shrub proved hardy in Holland (C. Koch).

Ilex integra, Thunberg.

Japan. Bird-lime can be prepared from the bark of this and several other hollies; from this species at the rate of 10 per cent.

Ilex Paraguensis, St. Hilaire.

The Maté. Uruguay, Paraguay and Southern Brazil. This Holly-bush, which attains the size of a small tree, is inserted in this list rather as a stimulating medicinal plant than as a substitute for the

ordinary Tea-plant, although in its native country it is very extensively used as such. From the province of Parana alone more than
36 million pounds were exported in 1871, besides 9 million pounds
used for home-consumption; while in Rio Grande de Sul the local
provincial consumption is nearly four times as much, not counting
large quantities consumed by the aboriginal race. It is cheaper than
coffee or tea (about 5d. per pound), but an individual there uses
about 1 lb. per week. It has a pleasant aroma, can be taken with
milk and sugar, and is the favorite beverage in large portions of
South-America (Dr. Macedo Soares). The leaves destined for the
Maté are slightly roasted. I. Dahoon (Walter) and I. dipyrena
(Wallace) are used for the same purpose, and probably other hollies
may be found occasionally fair substitutes. I. theezans (Martius)
also yields in Southern Brazil a kind of Maté. Chemical principles:
coffein, quina-acid and a peculiar tannic acid, which latter can be
converted into viridin-acid.

Ilex verticillata, Gray. (*Prinos verticillatus,* Linné.)
    Eastern North-America. There the bark much used for medicinal
purposes, both externally and internally.

Illicium anisatum, Linné.
    China and Japan. The Star-Anis. An evergreen shrub or small
tree. The starry fruits used in medicine and as a condiment.
Their flavor is derived from a peculiar volatile oil with anethol.
This species and a few others also deserve culture as ornamental
bushes.

Illipe butyracea, F. v. Mueller. (*Bassia butyracea,* Roxburgh.)
    India, up to 4,500 feet. A tree, gaining a height of 50 feet. The
pulp of the fruit is edible. The seeds yield a soft fat.

Illipe latifolia, F. v. Mueller. (*Bassia latifolia,* Roxburgh.)
    The "Mahwa." Central India. A tree to 50 feet high, content
with dry, stony ground; enduring slight frost. The succulent corolla
affords a never-failing crop of nourishing food to the rural inhabitants. Each tree supplies 2 to 3 cwt., each hundredweight yields
on distillation about 3 gallons of spirit; essential oil is also obtained
from the corolla. The flowers are also used for feeding cattle; they
will keep for a long time. The seeds yield oil of thick consistence.
I. neriifolia is an allied species, which ascends to 4,000 feet.

Imperata arundinacea, Cyrillo.
    South-Europe, North-Africa, Southern and Eastern Asia, Aus-
tralia, Polynesia. The Lalong-grass of India. Almost a sugar-cane
in miniature. Valuable for binding sand, especially in wet localities.
Difficult to eradicate. Available also for thatching.

Indigofera Anil, Linné.
    Recorded as indigenous to the West-Indies, and extending naturally
through continental America from Carolina to Brazil. A shrub,

several feet high. Pods sicklé-shaped, short, compressed. One of the principal Indigo-plants under cultivation both in the eastern and western hemisphere. Only in thé warmer parts of the temperate zone can we hope to produce indigo with remunerative success. But many of the hardier species seem never yet to have been tested for pigment. Over 100 have already been recorded from extra-tropical Southern Africa alone. An Indigofera of Georgia, said to be wild, perhaps I. Anil, yields an excellent product. The pigment in all instances is obtained by maceration of the foliage, aeration of the liquid and inspissation of the sediment.

Indigofera argentea, Linné. (*I. cœrulea*, Roxburgh.)

Tropical and extra-tropical Northern Africa, Arabia and perhaps India. A shrub, several feet high, closely allied to I. Anil, and likewise a good Indigo-plant.

Indigofera tinctoria, Linné.*

Warmest parts of Asia, as far east as Japan; recorded also from tropical Africa and even Natal as wild and seemingly also indigenous to Northern Queensland. A shrubby plant, attaining a height of 6 feet. Pods straight, cylindical, many-seeded. Extensively cultivated in warm zones for indigo, and probably hardy in warm temperate regions. The plant is frequently sold fresh by the grower to the factories. The Indigo-plant requires a rich friable soil, neither too moist nor too dry. The seeds are sown in furrows about a foot apart, and in hot damp climes the plant can be cut in about two months, as soon as it begins to flower; in six or eight weeks it yields a second crop, and under favorable circumstances as many as four crops can be gathered in a year. The plants have to be renewed every year, as the old ones do not yield an abundant produce. Bright sunshine favors the development of the dye-principle, but frequent rains cause a more luxuriant growth (Hartwig). In 1884 Great Britain imported 104,000 cwt. of Indigo, valued at £2,484,000.

Inula Helenium, Linné.

The Elecampane. Middle and Southern Europe, Middle Asia eastward to Japan. A perennial herb. The bitter and somewhat aromatic root, for the sake of its stimulating and tonic properties used in medicine. It contains also the amylaceous inulin and the crystalline helenin. With many other large herbs adaptable for scenic effects.

Ipomœa Batatas, Poiret.* (*Batatas edulis*, Choisy.)

The Sweet Potato. Tropical South-America. First brought to Europe from Brazil. It has proved well adapted also for the southern part of Australia and for New-Zealand. The tuberous roots afford a palatable food, more nutritious than ordinary potatoes; they can also be well utilized for starch. Varieties with red, white and yellow roots occur. Each tuber weighs generally from 3 to 5 lbs., but may occasionally attain to 56 lbs. The yield is 200 to 300 bushels from an acre.

Ipomœa Batatilla, G. Don.

Cooler regions of Venezuela. The tubers serve as sweet potatoes. I. platanifolia (Roemer and Schultes), from Central America, and I. mammosa (Choisy), from Amboina, are similarly useful.

Ipomœa Calobra, Hill and Mueller.

Eastern Central Australia. Hardy in the South of France (Prof. Naudin). The stems cover the ground for a radius of several yards; the spots becoming picturesque by the showy large flowers for 8 months in the year; the tubers are formed at 4 or 5 yards' distance along running roots, weigh from 5 to 30 lbs., and are a fair esculent. The plant likes a ferruginous loam (Rev. Dr. Woolls).

Ipomœa costata, F. v. Mueller.

Central and North-Western Australia. Produces edible tubers.

Ipomœa graminea, R. Brown.

Tropical Australia. The root, called "Mallamak," is eaten by the natives either raw or cooked (Foelsche).

Ipomœa magapotamica, Choisy.

Southern Brazil and Argentina. The root attains several pounds weight, and serves as jalap. Propagation by pieces of the root or from cuttings of the underground-stem.

Ipomœa paniculata, R. Brown.

Almost a cosmopolitan plant on tropical coasts; so also indigenous to North-Australia and the warmer parts of East-Australia. The tubers of this species also are edible. If hardy, the plant would deserve cultivation in any mild extra-tropical countries.

Ipomœa pes caprae, Roth.

Tropical and sub-tropical sea-shores of both hemispheres. Can be used in coast-regions free of frost, to bind drift-sand. Preferentially chosen for this purpose by Colonel Worster in Madras.

Ipomœa purga, Wenderoth.

Mountains of Mexico. The true Jalap. This species yields the medicinal jalap-root. It has recently been cultivated with apparent success even at New York, and is therefore entitled to a trial in warm woodlands. Active principle: the resinous convolvulin. I. Orizabensis (Ledanois) also yields jalap, according to Hanbury.

Ipomœa simulans, Hanbury.

Mexico. From this species the Tampico-jalap, or rather the Sierra-Gorda-jalap, is derived. I. operculata (Martius) yields the Brazilian jalap.

Iris Florentina, Linné.

Countries around the Mediterranean Sea. The well-known "Orris-root" is obtained from this species. Of the same geographic range is Iris juncea (Poiret), the edible root of which is known by the name of·Zeloak among the Algerian natives (Simmonds). I. versicolor (Linné) of North-America is there drawn into medicinal use. .

Isatis indigotica, Fortune.

Northern China. Perennial, almost shrubby. Its use is similar to that of the following plant.

Isatis tinctoria, Linné.

" Dyer's Woad." From the Mediterranean regions through part of the Orient, apparently extending as far as Japan. In Norway it is hardy to lat. 67° 16' (Schuebeler). A tall herb of two years' duration. The blue dye is obtained from the fermented leaves. Woad succeeds best in rich limestone-ground. Contains luteolin. Many other species of Isatis, mostly Asiatic, may perhaps produce dye with equal advantage. Boissier enumerates twenty-eight kinds merely as Oriental.

Jacaranda mimosifolia, D. Don.

Brazil. This tree, with J. Braziliana and J. obtusifolia (Humboldt), furnishes a beautiful and fragrant kind of Palixander- or Palissandre-wood, and so do probably some other tropical American species. This wood is bluish-red, traversed by blackish veins. J. mimosifolia is hardy at Melbourne, soon recovering from the injuries of our slight nocturnal frosts, and thus may perhaps be reared with advantage in many of the warmer and moister regions of the temperate zone.

Jacksonia cupulifera, Meissner.

West-Australia. It might prove an advantage, to disseminate this small tree in arid desert-regions, as horses and cattle relish the foliage amazingly. Several other Jacksonias share the local renown, which this congener of theirs has acquired from its utility as a pasture-bush.

Jasminum grandiflorum, Linné.*

From India to Japan.. Flowers white. Extensively cultivated in South-Europe. It is planted in rows 3 feet apart. Leek, tuberoses and similar plants are used, to occupy the spare-ground for the first year; 1,000 plants in the second year after grafting produce 50 kilos (about 1 cwt.) of flowers in rich soil. Ten thousand lbs. can be produced on a hectare (nearly 2½ acres), which under very favorable circumstances will realize a profit of £230 per annum. Dr. Piesse records, that in very recent times at Grasse, Cannes and the adjoining villages about 100,000 lbs. of Jasmin-flowers were gathered annually for perfumery-purposes. The plants must be guarded against frost and exposure to wind (Dehérain). In France this jasmin is generally grafted on J. officinale. The bushes are richly manured and well watered. Ordinary cleft-grafting is practised, the stock being headed down to near the ground. A good workman and assistant will graft about 1,000 plants in a day. The delicate scent is withdrawn, either by fixed oil or fat through alcohol, if not required by itself, or it may be drawn over along with oil of orange-peel. The pecuniary yield obtainable from Jasmin-cultivation seems vastly overrated, even if inexpensive labor could be produced.

Jasminum odoratissimum, Linné.

Madeira. Shrubby like the rest. Flowers yellow. Used like the foregoing and following for perfumery. This may be prepared by spreading the flowers upon wool or cotton slightly saturated with olive oil or other fixed oil, and covering them with other layers so prepared. The flowers are renewed from time to time until the oil is thoroughly pervaded by the scent, when the latter is withdrawn by alcohol. Other modes of extracting the oil seem feasible.

Jasminum officinale, Linné.

From the Caucasus to China. Flowers white. This is the principal species cultivated in South-Europe for its scent. In Cannes and Nice about 180,000 lbs. of jasmin-flowers are produced annually for distillation (Regel). By Simmonnet's process the essence of jasmin is solidified as jasminin.

Jasminum Sambac, Aiton.

From India to Japan. It has the richest perfume of all. The bush attains a height of 20 feet, and is almost climbing. The flowers are white, and must be collected in the evening before expansion. The relative value of many other species of jasmin, nearly all from the warmest parts of Asia, seems in no instance to have been ascertained, so far as their oils or scents are concerned. The Australian species are also deliciously fragrant, amongst which J. lineare, Br., occurs in Victorian deserts; while also J. didymum, Forst., J. racemosum, F. v. M., J. simplicifolium, Forst., J. calcareum, F. v. M. and J. suavissimum, Lindl., reach extra-tropical latitudes.

Jubæa spectabilis, Humboldt.

The tall and stout Coquito-Palm of Chili, hardy still in Valdivia. Adapted for mild extra-tropical latitudes. A kind of treacle is obtained from the sap of this palm. A good tree will give 90 gallons of mellaginous sap (C. Darwin). The small kernels are edible. Stem reaching a height of 60 feet, turgid towards the middle; leaves sometimes 10 feet long. Has endured at Montpellier a winter cold ot + 10° F. (Osw. de Kerchove de Denterghem). Jubæa Torallyi ascends the Andes to 8,500 feet. First introduced into the colony of Victoria by the writer of this work.

Juglans cinerea, Linné.*

The Butternut-tree of Eastern North-America. About 50 feet high; stem-diameter to 4 feet. Growth of comparative celerity; admits of transplantation readily. Likes rocky places in rich forests, but is also content with poor soil. Wood lighter than that of the black walnut, durable and free from attacks of insects. It is particularly sought for furniture, panels of coaches, corn-shovels, wooden dishes and similar implements, as it is not heavy nor liable to split. Splendid for select posts and rails needing durability; it is soft and therefore easily worked. This tree with J. nigra endures even the severe frosts of St. Petersburg, where the Caryas can no longer be maintained (Regel). The kernel of the nuts is more oily than that of the ordinary walnut; taste similar to that of Brazil-nuts. The leaves, bark and husk are of medicinal importance, and so are those of other species. The sap is saccharine (C. Koch).

Juglans cordiformis, Maximowicz.

Japan. This species approaches in many respects J. Sieboldiana.

Juglans Mandschurica, Maximowicz.

Corea and Mandschuria. This Walnut is allied to J. cinerea of North-America. Wood splendid for cabinet-work. The nuts available as well for the table as for oil-factories.

Juglans nigra, Linné.*

Black Walnut-tree. Eastern North-America. Attains a height of 80 feet; trunk grows to 6 feet in diameter; fond of rich forest-land. Quicker of growth than the European walnut-tree, but the wood not so easily worked (Meehan). Maximum rate of circumferential stem-growth in Nebraska 4 feet at 2 feet from the ground in 16 years (Furnas). The tree will bear fruit after 10 years, giving, when ot large size, 10 to 15 or even 20 bushels in a season, realizing as much as 4 shillings per bushel. The tree is hardy still in Christiania, Norway. Wood most ornamental, purplish-brown, turning dark with age, strong, tough, not liable to warp nor to split; not attacked by insects. Supplies three-fourths of the material for hardwood-furniture in the United States (Sargent), and fetches there the highest price. Wood stored for many years is the best for gun-stocks, and used also for musical instruments. For the sake of its compactness, durability and its susceptibility to high polish, it is much sought for elegant furniture, stair-rails and other select purposes. Seeds more oily than the European walnut. The tree extends in a slightly altered variety to Bolivia and Argentina.

Juglans regia, Linné.*

The ordinary Walnut-tree of Europe, indigenous in Hungary (Heuffel) and Greece (Heldreich), extending from the Black Sea to Beloochistan and Burmah, and seemingly also occurring in North-China, preferentially in calcareous soil. It attains a height of fully 100 feet, and lives many centuries. Professor Schuebeler found it hardy in Norway to lat. 63° 35', bearing fruit occasionally; in lat. 60° 14' it attained still a height of nearly 50 feet and a stem-circumference of 13 feet. An aged walnut-tree at Mentmore had a circumference of 12½ feet at 4 feet from the ground, its branches spreading diametrically to about 100 feet (Masters). Wood light and tough, much sought for gun-stocks, the exterior of pianofortes and the choicest furniture. The shells of the nut yield a black pigment, the leaves serve also for dye-purposes and have come further into external medicinal use. Trees of select quality of wood have been sold for £600, the wood being the most valuable of Middle-Europe. In some departments of France a rather large quantity of oil is pressed from the nuts, which, besides serving as an article of diet, is used for the preparation of fine colors. To obtain first-class fruit, the trees are grafted in France (Michaux). An almost huskless variety occurs in the north of China. Can be grown in cold localities, as it lives up to 2,000 feet elevation in Middle-Europe. Nuts for distant transmission, to arrive in a fit state for germination, are best packed quite fresh in casks between layers of dry moss. The foliage yields a brown dye, and is administered occasionally also for medicinal effect.

Juglans rupestris, Engelmann.

From California to New Mexico, along the course of streams in rich moist soil. A handsome symmetrical tree of much utility, attaining a height of 60 feet and a stem-diameter of 3 feet (Dr. Gibbons). Hardy in Christiania still.

Juglans Sieboldiana, Maximowicz.

Throughout Japan, where it forms a large tree.

Juglans stenocarpa, Maximowicz.

From the Amoor-territory. Allied to J. Mandschurica.

Juniperus Bermudiana, Hermann.*

The Pencil-Cedar of Bermuda and Barbadoes. This species grows sometimes to 90 feet high, and furnishes a valuable red durable wood, used for boat-building, furniture and particularly pencils, also for hammer-shanks of pianofortes, on account of its pleasant odor and special fitness. It is almost the only native timber of Bermuda. It will thrive in the poorest soil, for instance coral-sand, and has a very great power to resist storms on account of the deeply penetrating roots, which may reach to 30 feet. Planks of 32 inches width have been obtained (Lieut.-General Sir J. H. Lefroy). Many of the plants in gardens called Thuya or Biotia Meldensis belong to this species.

Juniperus brevifolia, Antoine.

In the Azores, up to 4,800 feet; a nice tree with sometimes silvery foliage.

Juniperus Cedrus, Webb.

A tall tree of the higher mountains of the Canary-Islands.

Juniperus Chinensis, Linné.

In temperate regions of the Himalayas, up to an altitude of 15,000 feet, also in China and Japan. Hardy in Christiania (Schuebeler). This tree is known to rise exceptionally to about 100 feet, with a stem-girth of 13 feet; it is of comparatively rapid growth, furnishing

· a reddish, soft and fine-grained wood, suitable for pencils (Hoopes). Probably identical with it is the Himalayan Pencil-Cedar (Juniperus religiosa, Royle). The timber of some other tall Junipers needs tests.

Juniperus communis, Linné.

Colder parts of Europe, Asia, North-Africa and North-America, ascending the European Alps to 8,000 feet, the Indian mountains to 14,000 feet. In Norway it is indigenous to lat. 71° 10′, and under 60° 10′ it attains still a height of 40 feet (Professor Schuebeler). One of the three native Coniferæ of Britain. The berry-like fruits are of medicinal value, also used in the preparation of gin. Important for fuel in the coldest regions. Will grow on almost pure sand.

Juniperus drupacea, La Billardière.

Plum-Juniper. A very handsome long-leaved species, the "Habhel" of Syria. It attains a height of 30 feet, and produces a sweet edible fruit, highly esteemed throughout the Orient.

Juniperus excelsa, Bieberstein.

In Asia Minor, 2,000 to 6,000 feet above the sea-level. Extends to the Himalayas, where its range of elevation is from 5,000 to 14,000 feet. A stately tree, to 90 feet high. Trunk short, but of great girth, over 20 feet circumference being known according to Stewart and Brandis, who refer to this the J. Chinensis of Parlatore.

Juniperus flaccida, Schlechtendal.

In Mexico, at from 5,000 to 7,000 feet altitude. A tree reaching 30 feet in height, rich in sandarac-like resin.

Juniperus fœtidissima, Willdenow.

A tall and beautiful tree in Armenia and Tauria, also on the Balkan and in North-Greece, at from 5,000 to 6,500 feet.

Juniperus Mexicana, Schiede.

Mexico, at elevations from 7,000 to 11,000 feet. A straight tree, sometimes to 90 feet high; stem to three feet in diameter, exuding copiously a resin similar to sandarac.

Juniperus occidentalis, Hooker.

North-California and Oregon, ascending to 5,000 feet. A straight tree, to as much as 80 feet high, with a stem reaching often 3 feet in diameter. Wood pale, comparatively hard, close-grained, thrives well among rocks.

Juniperus Phœnicea, Linné.

South-Europe and Orient. A small tree, yielding an aromatic resin.

Juniperus procera, Hochstetter.

In Abyssinia and Arabia. A stately tree, furnishing a hard, useful timber.

Juniperus recurva, Hamilton.

On the Himalayas, from 7,500 to 15,000 feet. A tree attaining 80 feet in height according to J. Hoopes.

Juniperus sphærica, Lindley.

North-China. A handsome tree, sometimes to 40 feet high.

Juniperus Virginiana, Linné.

North-American Pencil-Cedar or Red Cedar, extending to 45° N.L. eastward and to 52° westward; likes limestone-subsoil. A drooping variety exists. Hardy still in Christiania. A handsome tree, rarely to 90 feet high, supplying a fragrant timber; it is dense, fine-grained, light and of pleasant odor; the inner part is of a beautiful red color; the outer is white; it is much used for pencils; one of the best of all woods for buckets, tubs and casks. Simmonds observes, that fence-posts of this wood last for ages. Of wonderful durability for railway cross-ties (Barney). The heartwood is almost imperishable (Vasey), nor is it bored by insects. The tree grows best near the sea, but is rather independent of soil and locality. Rate of growth in Nebraska according to Governor Furnas 26 inches stem-girth at 2 feet from the ground in 12 years. One cwt. of wood yields in distillation at an average 28 ounces of fragrant oil used for scented soaps (Piesse, Lubin.)

Juniperus Wallichiana, J. Hooker and Thomson.

From the Indus to Sikkim, at elevations from 9,000 to 15,000 feet. Attains a height of about 60 feet. Desirable for transfer to any Alps. Wood similar to that of J. excelsa (Stewart and Brandis).

Justicia Adhatoda, Linné.

India; enduring the climate of the lowlands of Victoria. This bush possesses anti-spasmodic and febrifugal properties. It can be utilized also as a hedge-plant.

Kentia Baueri, Seemann. (*Rhopalostylis Baueri,* H. Wendland and Drude.)

The Norfolk-Island Palm. Height to 40 feet.

Kentia Beccarii, F. v. Mueller. (*Nengella montana,* Beccari.)

On the mountains of New Guinea, up to 4,500 feet. This slender palm is only a few feet high and eligible for domestic decoration.

Kentia Belmoriana, Moore and Mueller. (*Howea Belmoriana,* Beccari.)

The Curly Palm of Lord Howe's Island; about 40 feet high. With its congeners evidently designed to grace our gardens, and to become also important for horticultural traffic abroad. K. Fosteriana is a close ally, restricted to the same island.

Kentia Canterburyana, Moore and Mueller. (*Hedyscepe Canterburyana,* H. Wendland and Drude.)

Umbrella-palm of Lord Howe's Island. Likewise a tall and hardy palm, growing at or below 2,000 feet altitude.

Kentia Moluccana, Beccari.

Ternate, at heights up to 3,500 feet. This noble and comparatively hardy palm attains a height of about 90 feet.

Kentia sapida, Blume. (*Rhopalostylis sapida,* H. Wendland and Drude.)

The Nika-palm of New Zealand and the Chatham-Islands. It rises to a height of about 40 feet, is one of the hardiest of all palms, and extends to the most southern latitude attained by any palm, being found down to 44° south. Proved hardy in Stewart's Island; Charl. Traill. The unexpanded flower-spikes can be converted into food as palm-cabbage.

Knightia excelsa, R. Brown.

The Rewa-Rewa of New Zealand. The wood of this tree is recommended as valuable for ornamental work and furniture (Campbell Walker).

Kochia eriantha, F. v. Mueller.

Proved an excellent fodder-herb for sheep on the hot and dry pastures of Central Australia, where the temperature in summer reaches 120° F. in the shade, and in the winter falls to 27° (Rev. H. Kempe). Several other Australian species of Kochia afford excellent pasture-fodder.

Kochia pubescens, Moquin.

South-Africa; there one of the best salt-bushes for pastures (McOwan).

Kochia villosa, Lindley.

In most of the depressed and saline regions of Australia, particularly inland, also on sand-lands. Renowned amongst occupiers of pasture-runs as the " Cotton-bush," strangely so called, on account of downy adventitious excrescences. This rather dwarf shrub resists the extremes of drought and heat of even the trying Central Australian clime. The roots sometimes penetrate into the ground to a depth of a dozen feet. With all other pasture animals also dromedaries like this and some other salt-bushes particularly for food; so also ostriches (Officer).

Kœleria cristata, Persoon.

Widely dispersed over the globe. A perennial grass of fair nutritive quality, sustaining itself on dry soil. The closely allied K. glauca can be sown with advantage on coast-sand.

Krameria triandra, Ruiz.

Chili, Peru and Bolivia, at elevations of from 3,000 to 8,000 feet. This pretty little shrub can be grown on sandy ridges in an equable clime. It produces the medicinal Ratanhia-root, well known also as a dentrifice, and used further for coloring wine. The root contains 38 to 43 per cent. tannin (Muspratt). Some other species have similarly astringent roots, particularly K. Ixine (Loefling), from Central

America and the West-Indies. Some could be chosen to aid in adorning and diversifying our gardens. Krameria is placed by Eichler among cæsalpinous Leguminosæ.

Lactuca sativa, C. Bauhin.

Southern Asia. The ordinary annual Lettuce, in use since remote antiquity. It is not without value for medicinal purposes, especially as a sedative. L. Scariola (Linné) seems to be the wild state of the garden-lettuce, and is a native of all the countries around the Mediterranean Sea. Mess. Dippe in Quedlinburg devote exclusively 130 acres to the culture of lettuce merely for seed. Mons. Vilmorin notes the seeds to retain their power of germination for about five years. L. altissima, Bieberstein, is 'a variety attaining a height of 9 feet. All yield lactucarium.

Lactuca virosa, Linné.

Middle and South-Europe, North-Africa, Middle-Asia. A biennial. The inspissated juice particularly of this lettuce forms the sedative lactucarium.

Lapageria rosea, Ruiz and Pavon.

The Copigué. Chili. Almost the only plant, which can exist in the area covered by the sulphurous smoke of the local smelting furnaces (Dr. R. O. Cunningham). A half-woody climber with large showy flowers. The berries, which attain the size of a hen's egg, are sweet and edible. The plant bears slight frosts.

Lardizabala biternata, Ruiz and Pavon.

Chili. A climber with stems of enormous length. Might be naturalized in forests for obtaining the tough fibre for cordage. In its native country the torrified stems are used instead of ropes, according to Dr. Philippi.

Laserpitium aquilegium, Murray.

Middle and Southern Europe. The stems of this perennial herb are edible. The fruits serve as a condiment.

Lasiocorys Capensis, Bentham.

South-Africa. Professor McOwan directs attention to the economy of this plant, it having a singular propensity of rendering rainwater retained in small gutters; the Lasiocorys compacts the detritus and impedes also soil washed onward, forming natural little catch-dams. The plant is bitter, hence not consumed by goats and sheep in plentiful times.

Lathyrus Cicera, Linné.

Countries at or near the Mediterranean Sea, also Canary-Islands. An annual, similar in its use to L. sativus, furnishing a tender palatable fodder on sandy soil. L. Clymenum, L., from the same regions, serves similar purposes.

Lathyrus macrorrhizus, Wimmer. (*Orobus tuberosus*, Linné.)

Europe, Western Asia. This herb would gradually establish pasturage in sterile forest-regions, and could with some allied species be disseminated in alpine elevations.

Lathyrus pratensis, Linné.

The Meadow-Pea. Europe, Northern and Middle Asia. Hardy in Norway to lat. 69° 40'. A good perennial pasture-herb. It can also be utilized for forest-pastures, like L. silvestris, L. The yield is considerable, and the herbage, though bitter, is relished by sheep. The plant spreads easily, particularly on fresh ground. L. tuberosus, L., can likewise be utilized as a fodder-herb; its tubers are edible, but very small; the plant is easily naturalized.

Lathyrus sativus, Linné.

The Jarosse. Middle and Southern Europe. An annual forage-herb. Can be grown in Norway to lat. 63° 26' (Schuebeler). Superior to vetches in quality of fodder and seed, but inferior in yield; according to Langethal's observations, content with a lighter soil, hence often chosen for first sowing on sand-lands. Lime in the soil increases the return. The seeds can only be used with great caution, as their frequent or continous use, like that of L. Cicera, induces paralysis, not only in horses, cattle and birds, but even in the human subject. The plant should therefore only be used for its herbage. The seeds will keep about five years. Probably other specimens of Lathyrus could advantageously be introduced.

Launæa pinnatifida, Cassini.

Coast of tropical Asia and East-Africa. A perennial herb, with creeping and rooting stems, arching from node to node (Hooker), by which means it keeps down drift-sand (Cleghorn, Bidie). In this respect the plant has rendered such good services on the Indian coasts, that its transfer to other shores in frostless zones seems desirable, particularly as it does not stray away from the coast to invade cultural lands.

Laurelia aromatica, Jussieu.

Southern Chili. A colossal tree, in Valdivia the principal one used for flooring. Wood never bored by insects, and well able to stand exposure to the open air, far superior to that of L. serrata (Bertero), the "Vouvan or Huahuoa," which tree predominates over L. aromatica in the far south of Chili (Dr. Philippi).

Laurus nobilis, Linné.

South-Europe and Asia Minor. The Warrior's Laurel of the ancients, generally called in Britain "Sweet Bay." Greatest height about 60 feet, but always displaying a tendency to emit suckers and rarely assuming a tree-like character (Loudon). The leaves are in much request for various condiments, and for their peculiar aroma these Bay-leaves cannot be replaced by any others, unless those of Lindera Benzoin and Machilus odoratissima.

Lavandula angustifolia, C. Bauhin. (*L. vera,* De Candolle.)

The principal Lavender-Plant. Countries around and near the Mediterranean Sea. Of somewhat shrubby growth; from it, by distillation, the best oil of lavender is prepared; the English being superior to others. It lives on dry soil, but is less hardy than the following, still it will grow in Norway to lat. 59° 55' (Schuebeler). A thousand plants will only yield about two quarts of oil (Piesse). The plants last about four years for distillation-purposes. The plantations should be renewed at intervals of three or four years. The soil should be calcareous (Vilmorin). Easily grown from cuttings also.

Lavandula latifolia, C. Bauhin. (*L. Spica,* De Candolle.)

South-Europe, North-Africa. This species is the richest yielder of oil. Hardy in Norway to lat. 67° 56'. The Lavenders are easily multiplied by slips. Seeds will keep for five years.

Lavandula Stœchas, Linné.

South-Europe, North-Africa. "Topped Lavender." This shrub can also be utilized for oil-distillation and other purposes, for which the two other Lavenders are used. The quality of the oil of these species seems to differ according to their locality of growth. Mr. James Dickinson, of Port Arlington, Victoria, informs us that this is the best plant known to him for staying sand. It grows much quicker than the Ulex; every seed which falls germinates, so that around each bush every stroke of the spade brings up lots of seedlings fit for transplantation. In mild regions it is five months in full flower annually, coming into bloom early. Bees are passionately fond of the nectar of the flowers. Mr. Dickinson calculates, that a ton of the finest-flavored honey can be obtained annually from an acre of this Lavender.

Lavatera arborea, Linné.

"Tree-Mallow" of the countries on the Mediterranean Sea. A tall biennial plant of rapid growth. The ribbon-like bast is produced in greater abundance and more rapidly than in most malvaceous plants, and is recommended for paper-material. Bears frost to 15° F. (Gorlie). The Tree-Mallow might easily be naturalized on sea-shores, where it would be useful as a quick shelter. Perhaps it might serve with allied plants for green manure. The bulky foliage has proved valuable for fodder, and so has that of Lavatera plebeja (Sims) of Australia.

Lawsonia alba, Lamarck.

North- and Middle-Africa, Arabia, Persia, India and North-Western Australia. The "Henne or Henna-bush." It may become of use as a dye-plant in regions free from frost. The orange pigment is obtained from the ground foliage. Mr. C. B. Clarke considers it one of the best hedge-plants in India, together with Dodonæa viscosa (Linné) and Odina Wodier (Roxburgh).

Leersia hexandra, Swartz.

Africa, South-Asia, warmer parts of America and Australia. Found by Mr. Bailey to be one of the most relished by cattle among aquatic grasses of East-Australia. In the Philippines regularly cultivated for fodder in the manner of rice. L. Gouini (Fournier) is a Mexican species.

Leersia oryzoides, Solander.

Middle and South-Europe, various parts of Asia, Africa and America. A perennial rather rough swamp-grass. Other Leersias from both hemispheres are deserving of introduction, if even only for the benefit of waterfowl.

Leonotis Leonurus, R. Brown.

South-Africa. The foliage of this highly ornamental bush deserves attention for therapeutic purposes, as, according to Professor Owen, the leaves, when used like tobacco, are highly stimulative.

Leontice thalictroides, Linné. (*Caulophyllum thalictroides*, Michaux.)

North-America. "Blue Cohosh," or "Pappoose-root." A perennial herb, the root of which is in medicinal use. The resin, extracted from the root, contributes the caulophyllin as an emmenagogue.

Lepidium latifolium, Linné.

Europe, North-Africa, Middle- and North-Asia. A perennial herb of peppery acridity, used for some select sauces.

Lepidium sativum, Linné.

The "Cress"-Orient. Annual. Irrespective of its culinary value, cress is of use as one of the remedies in cases of scurvy. Seeds will keep for several years. Active principle: a volatile oil and the bitter lepidin. The crisp-leaved variety preferentially reared.

Lepidosperma gladiatum, La Billardière.

The Sword-Sedge of the sea-coast of extra-tropical Australia. One of the most important plants for binding sea-sand, also yielding a paper-material as good as Sparta. Mr. Th. Christy has brought the "Cross" process for textile fabrics or paper-pulp recently into use, which consist in treating vegetable fibrous substances under strong pressure (15–30 lb. per square-inch) with water containing about 3 per cent. of Thiolyte.

Lepironia mucronata, Cl. Richard.

East-Australia, Malayan Archipelagus, East-Indies, South-China, Madagascar. This rush is cultivated (like Rice) in China for textile purposes, but in poor soils the manure impairs its strength. The plant renews itself by sprouts from its perennial root. It attains a height of 7 feet; the stems are beaten flat, to fit them to be woven or plaited for either bed-mats and bags or especially for mat-sails, the latter being the most extensively used for the junks in China; further, the plant is utilized for making the floor-matting, which is

exported in vast quantities to the United States, to be used in summer
for the sake of coolness, in preference to carpets (Dr. Hance). This
rush thus furnishes the raw material for a great manufacturing
industry. The dyeing of the mats yellow is effected with the flowers
of Sophora Japonica, under addition of alum; green with an acantha-
ceous plant, the Lam-yip (Blue Leaf), alum and sulphate of copper
(Dr. Hirst).

Leptospermum lævigatum, F. v. Mueller.* (*Fabricia lævigata*, Gaertner.)

The " Sandstay." Sea-shores and sand-deserts of extra-tropical
Australia, but not extending to Western Australia. This shrub or
small tree is the most effectual of all for arresting the progress of
drift-sand in a warm clime. It is most easily raised by simply
scattering the seeds on the sand in autumn and covering them loosely
with boughs, or better still by spreading lopped-off branches of the
shrub itself, bearing ripe seeds, on the sand.

Leptospermum, lanigerum, Smith.

South-Eastern Australia. This tall shrub or small tree can be
grown in wet semi-saline soil. It exercises antimalarian influences
on such places like Melaleuca ericifolia.

Lespedeza striata, Hooker and Arnott.*

China and Japan. Sometimes called "Japan-Clover." An annual herb, which in North-America has proved of great use. Professor Meehan states it to be identical with the "Hoop-Koop" plant,
and that it has taken possession of much waste land in the
Southern States. It grows there wonderfully on the hot dry soil,
and the cattle like it amazingly. Mr. Jackson observes, that it
spreads on spaces between forest-trees, covering the soil with a dense
permanent herbage. Dr. Carl Mohr says, that it stands drought well,
and thrives on sandy clay, but luxuriates on light calcareous soil.
It is impatient of frost (W. Elliott). The Department of Agriculture
of Washington (in 1878) regards it as rich in albuminous substances
as the best clovers.

Leucadendron argenteum, Brown.

The Silver-tree of South-Africa is included on this occasion, because it would add to the splendor of our woods, and thrive far better
there than in our gardens within the warm temperate zone. Moreover, with this tree many others, equally glorious, might be established in any mild forest-glens as a source of horticultural wealth,
were it only to obtain in future years a copious supply of seeds.
Mention may be made of the tall Magnolia-trees of North-America:
Magnolia grandiflora (Linné), 100 feet high; M. umbrella (Lamarck),
40 feet; M. acuminata (Linné), 80 feet; M. cordata (Michaux), 50
feet; M. Fraseri (Walter), 40 feet; M. macrophylla (Michaux), 40
feet; M. Yulan (Desfontaines), of China, 50 feet; M. Campbelli (J.
Hooker), of the Himalayas, 150 feet high, with flowers nearly a foot
across; M. sphærocarpa (Roxburgh), also of the Indian highlands,

40 feet; Stenocarpus .sinuosus (Endlicher), of East-Australia (the most brilliant of the Proteaceæ); the crimson and scarlet Ratas of New Zealand, Metrosideros florida (Smith); M. lucida (Menzies), M. robusta (Cunningham), 80 feet high; M. tomentosa (Cunn.), 40 feet; Fuchsia excorticata (L. fil), also from New Zealand, stem reaching 2 feet in diameter; and Rhododendron Falconeri (J. Hooker), from Upper India, 50 feet high, leaves 18 inches long.

Lewisia rediviva, Pursh.

North-Western America. The root of this herb is large and starchy, was formerly extensively used by the native inhabitants, and called by them "The Gift of the Great Spirit." The plant deserves trial culture.

Leyssera gnaphalioides, Linné.

South-Africa. A perennial herb of aromatic scent and taste. Much used there as a medicinal tea.

Liatris odoratissima, Willdenow.

Southern States of Northern America. A perennial herb, occurring in swampy places. The leaves are sometimes used, for the sake of their aromatic odor, to flavor tobacco and other substances (Saunders). L. spicata (Willdenow) is the "Button-Snakeroot," medicinal in the Eastern States.

Libocedrus Chilensis, Endlicher.

In cold valleys on the Southern Andes of Chili, at from 2,000 to 5,000 feet. A fine tree, sometimes 80 feet high, furnishing a hard resinous wood of a yellowish color. Libocedrus as a genus is hardly to be distinguished from Thuya, as pointed out more particularly by Bentham.

Libocedrus decurrens, Torrey.

White Cedar of California, growing on high mountains, in fine groves up to 5,000 feet, in what Hinchcliff calls the noblest zone of Coniferæ of the globe. Attains a height of fully 200 feet, with a stem to 25 feet in circumference. The wood is light, extremely durable and strong, used for exquisite cabinet-work, but also suitable for superior fence-rails and building purposes. According to Dr. Gibbons, the tree is well adapted for wind-breaks, and can be trained into tall hedges.

Libocedrus Doniana, Endlicher.

. Northern Island of New Zealand, up to 6,000 feet elevation. A forest-tree, reaching 100 feet in height, the stem 3 feet and more in diameter. The wood is hard and resinous, of a dark-reddish color, fine-grained, excellent for planks and spars.

Libocedrus tetragona, Endlicher.*

.On the Andes of Chili, at an elevation from 2,000 to 5,000 feet, growing as far south as Magellan's Straits, especially in moist moory localities. This species has a very straight stem, and rises to 120 feet. The wood, though soft and light, is resinous, and will resist underground-decay for a century and more, like that of Fitzroya Patagonica; for railway-sleepers this timber is locally preferred to any other (Dr. Philippi); it is also highly esteemed for various artisans' work; it is nearly white.

Ligustrum Japonicum, Thunberg.

The Japan-Privet. A shrub, evergreen or nearly so, promising to become a valuable hedge-plant. Hardy still in Christiania (Schuebeler). It grows readily from cuttings like the ordinary European Privet, Ligustrum vulgare (Linnè). Both will grow under trees, where scarcely anything else would live (Johnson).

Limonia acidissima, Linné.

India, up to 4,000 feet; hardy in England. This shrub or small tree has fruit of extreme acidity, but insignificant in size, which culture may enlarge.

Lindera Benzoin, Blume.

From Canada to the Gulf of Mexico, there called the Spice-Laurel. An aromatic bush, one of the hardiest of the order. The aroma of the foliage much like that of Bay-leaves.

Linum usitatissimum, Linné.*

The Flax-Plant. Orient. Perhaps indigenous also in South-Europe, and possibly derived fsom L. angustifolium (Hudson), which was cultivated in Switzerland already during the stone-age (Heer). A well-known annual, which yields linen-fibre and linseed-oil. Few plants find a wider congeniality of soil and climate, and few give a quicker return. Good and deep soil, particularly of forests, well-drained, is requiste for successful flax-culture. In Norway it is cultivated as far north as lat. 70° 3' (Prof. Schuebeler). The Flax belongs to the Potash-plants. Change of seed-grain is desirable. Thick sowing extends the length and flexibility of the fibre. To obtain the best fibre, the plant must be pulled, when the seeds commence to ripen. If the seeds are allowed in part to mature, then both fibre and seeds may be turned to account. If the seed is left to ripen completely, the fibre is generally discarded. The seed yields by pressure about 22 per cent. of oil. The residue can either be prepared as linseed-meal or be utilized as admixture to stable-fodder. The demand for both fibre and oil is enormous. Two principal varieties are under culture; a tall sort, with smaller flowers, closed capsules and dark seed; a dwarf sort, more branched (even if closely sown), with larger flowers and capsules, the seed-vessels opening spontaneously and with elasticity, while the seeds are of a pale color. None of the perennial species of Linum are so manageable in culture as the ordinary annual flax. Great Britain imported in 1884 of Flax 80,000 tons, worth over three million pounds sterling, and of Linseed 1,805,000 quarters, valued at £3,832,000.

Lippia citriodora, Kunth.

Peru, Chili, La Plata-States, Brazil. An evergreen shrub, yielding scented oil, used for condiments and for perfumery as "Verbena," the leaves fit for flavoring tea. L. Mexicana has come into therapeutic use, particularly as an expectorant.

Liquidambar Altingia, Blume.

At the Red Sea and on the mountains of India and New Guinea, ascending to about 3,000 feet. The tree attains a height of 200 feet. It yields the fragrant balsam known as Liquid Storax.

Liquidambar Formosana, Hance.

China. A silk-producing insect is reared on this tree (Hance).

Liquidambar orientalis, Miller. (*L. imberbe,* Aiton.)

Asia Minor. This tree also yields Liquid Storax, which is vanilla-scented, containing much styrol and styracin, and thus used for imparting scent to some sorts of tobacco and cigars, also for keeping moths from clothing. Its use in medicine is more limited than in perfumery.

Liquidambar stryaciflua, Linné.

The Sweet-Gum tree. In morasses and on the springs of forests of North-America, with a wide geographic range. Endures severe frost after the plant has attained considerable size. The crown of the tree attains vast dimensions; the stem to about 100 feet in height and to 10 feet in diameter. The wood is reddish-brown, very compact and heavy, fine-grained, durable, easily worked, little liable to warp, and admitting of a fine finish, with its pleasing tint, especially adapted for furniture. The terebinthine juice hardens, on exposure, to a resin of benzoin odor. The bark contains about 8 per cent. tannin.

Liriodendron tulipifera, Linné.

The Tulip-tree of North-America. One of the largest trees of the United States, and one of the grandest vegetable productions of the temperate zone. In deep fertile soil it sometimes attains a height of about 140 feet, with a straight clear stem reaching 9 feet in diameter. In Norway it is hardy to lat. 61° 17' (Schuebeler). The Tulip-wood is highly esteemed and very extensively used, wherever this tree abounds, uniting lightness with strength and durability. It is of a light-yellow color, fine grained, strong, compact, easily worked and takes a good polish. It is employed for house-building, inside as well as outside, for bridges, furniture, coach-building, implements, shingles, carriage-panels and a variety of other purposes. On account of its uniformity and freedom from knots and disinclination to warp or shrink, it is much used in Canada for railway-cars and carriage-building, chiefly for the panelling (Robb). The bark yields about 8 per cent. tannin. As this tree is difficult to transplant, it should be grown on the spot where it is to remain. Professor Meehan observes, that it is of quicker growth than the Horse-Chestnut tree and many

Maples. The flowers yield to bees much honey; indeed Mr. Langstroth speaks of the Tulip-tree as one of the greatest honey-producers in the world; as its large flowers expand in succession, new swarms will sometimes fill their hives from this source alone.

Lithospermum canescens, Lehmann.

North-American Alkanet. This, as the vernacular name indicates, offers a dye-root.

Lithospermum hirtum, Lehmann.

North-American Alkanna. A showy perennial herb; the root yields a red dye.

Lithospermum longiflorum, Sprengel.

North-America. A red pigment can also be extracted from the root of this species.

Livistona Australis, Martius.

East-Australia. The only Palm-tree in Victoria, occurring in East-Gippsland (in the latitude of Melbourne), and there attaining a height of about 80 feet. It endures the winters of South-France to 43° 32′ north lat. (Naudin). The young leaves can be plaited as a material for cabbage-tree hats. The seeds (of which about 200 are contained in one pound) retain their vitality far better than those of the Australian Ptychospermas. This palm can be transferred from its native haunts to very long distances for growth, by previously separating the main portion of the root from the soil, and leaving the plant for some months on the original spot, so as to remove it finally with new rootlets, retaining much soil. Some of the Indian Livistonas may be equally hardy; their stems often tower above the other forest-trees.

Livistona Chinensis, R. Brown,

South-China and Japan. A very decorative Fan-palm, and one of the hardiest of the whole order. In its native country, the hairy stem-covering of this palm is used for fixing lime-plaster to buildings (Christie).

Livistona Leichhardtii, F. v. Mueller.

North-Australia. Under this name might be combined L. inermis and L. humilis (R. Brown), neither name applying well to this finally tall palm with thorny leaf-stalks. The author of this work as well as Dr. Leichhardt saw it far inland in dry open not mountainous regions also; nevertheless it may need a moister clime than the following species.

Livistona Mariæ, F. v. Mueller.

Central and West-Australia, barely within the tropics. This noble Fan-palm attains 40 feet in height, and is likely to prove very hardy.

Lolium perenne. Linné.*

Europe, North-Africa, Western Asia. The well known perennial Rye-grass, mentioned here for the sake of completeness. In Norway it grows to lat. 65° 28' (Schuebeler). L. Italicum, Al. Br., the Italian rye-grass, seems to be only a variety. One of the most important of all pasture-grasses, also almost universally chosen for lawn-culture. It produces an abundance of seeds, which are readily collected and easily vegetate. It comes early to perfection. Nevertheless the produce and nutritive powers are considerably less than those of Dactylis glomerata, Alopecurus pratensis and Festuca elatior, but it pushes forward earlier than the last mentioned grass, while the ripening of seeds is less defective than in Alopecurus. The chemical analysis, made very early in spring, gave the following results:—Albumen, 3·36; gluten, 4·88; starch, 0·51; gum, 1·80; sugar, 1·80 per cent. (F. v. Mueller and L. Rummel). At the London Sewerage-Depôt, 60 tons of rye-grass were obtained from one acre (McIvor). Rye-grass, though naturally living but a few years, maintains its ground well, by the ease with which it disseminates itself spontaneously. Several sorts, which can scarcely be called varieties, are under cultivation. Rye-grass stands the dry heat of Australian summers well. It is likely to spread gradually over the whole of the Australian continent, and to play an important part in pasture, except the hottest desert-tracts. Sheep should not be continually kept on rye-grass pasture, as they may become subject to fits similar to those produced by L. temulentum, possibly due to the grass getting ergotized or otherwise diseased, as many observers assert. ·It is one of the best grasses to endure traffic on roads or paths, particularly on soil not altogether light, and is also one of the few among important grasses, which can be sown at any season in mild climes. The Italian rye-grass is preferably chosen as an early temporary shelter for tenderer but more lasting pasture-grasses, also furnishing a good collateral return the first season. Ordinary rye-grass can be cut several times in a season.

Lotus corniculatus, Linné.

" Bird's-foot Trefoil." Europe, North-Africa, Northern and Middle Asia, extra tropical Australia. Indigenous in Norway as far north as lat. 69° 58' (Schuebeler). A deep-rooting perennial herb, readily growing on pasture-land, sandy links and heathy places. This plant is well deserving cultivation on light inferior soil, on which it will yield a greater bulk of herbage than any of the other cultivated clovers; it is highly nutritious, and is eaten with avidity by cattle and sheep. From the great depth, to which its roots penetrate, it is not liable to be injured by drought. It well fills out vacant places between higher fodder-herbs on meadows; it is always somewhat saline and welcome among hay. L. tenuis (Kitaibel) is a valuable variety of the coasts. The nearly allied L. major (Scopoli) yields a still greater amount of herbage; it is particularly suited for bushy and moist localities, and it attains its greatest luxuriance on soils, which have some peat in their composition (Lawson). In Australia Lotus corniculatus shows a decided predilection for wet meadows.

Lotus tetragonolobus, Linné.

Countries on the Mediterranean Sea. Though annual, this herb is highly valued for sheep-pastures. The green pods serve even as a culinary vegetable. The allied L. siliquosus (Linné) is perennial, and occurs in a succulent form on sea-coasts. The seeds will retain their vitality for several years.

Loxopterygium Lorentzii, Grisebach.

La Plata-States. The bark, called Quebracho colorado, extensively used for tanning; latterly much exported to Europe. The length of time for the tanning process with this bark is only half that for oak-bark. The kino of this tree has come into therapeutic use.

Lupinus albus, Linné.

The White Lupine. ᾿Countries on the Mediterranean Sea, also in the Orient. An annual quick-growing. herb, valuable when young for fodder and also for green manure. In Norway it will grow to lat. 70° 22' north (Schuebeler). It is famed as the " Tramoso " in Portugal, to suppress sorrel and other obstinate weeds by its close and easy growth. The lentil-like seeds, after the bitter principle (lupinin) has been removed through boiling or soaking in salt-water,. are edible. It would lead too far, to enumerate here many others of the numerous species of lupines, of which unquestionably very many are eligible for agrarian purposes, while all are acceptable as hardy, elegant and easily grown garden-plants. One, L. perennis, L., extends in America to the Northern States of the Union and Canada; fourteen are recorded from South-Europe, seventeen from Brazil, and numerous species from other parts of America, where the limits of the genus are about Monte Video southward and about Nootka-Sound northward. The majority of the species are perennial. The Egyptian L. Termis (Forskael) and L. Græcus (Boissier) are closely allied to L. albus and of equal use; their flowers are bluish or blue.

Lupinus angustifolius. Linné.

Countries on the Mediterranean Sea. An annual blue-flowered species, preferable to L. luteus for grain-harvest. Hardy to lat. 70° in Norway. Some if not all lupines can be counted among honey-plants.

Lupinus arboreus, Sims.*

California. This has been used there for the reclamation of sand, on account of its long tap-roots, the latter having been traced to a depth of 25 feet, while the stems were only 3 feet high. The germination is easy and the growth rapid on the sand-downs. For aiding the young lupines during the first two months, to get hold of the sand, barley is sown with them, as the latter sprouts in a few days and holds the sand in the second week; the lupine subsequently covers the sand with a dense vegetation in less than a year.

Lupinus Douglasii, Agardh.

Oregon and California. Hardy in Norway to lat. 67° 56'. This somewhat woody species can be used for binding sand with L. Chamissonis, Escholtz (L. albifrons, Bentham) and many perennial lupines from other countries.

Lupinus luteus, Linné.*

The Scented Yellow Lupine. Countries in the vicinity of the Mediterranean Sea. Can be grown in Norway to lat. 70 °(Schuebeler). This annual species is predominantly in use as green manure through Middle Europe, to improve sandy soil; it is the best of all yet tested, and will do even on coast-drifts. It can also be employed like some other lupines as a fodder-herb, green as well as for hay; some lupines are also very valuable as pasture-herbs. Lupine-seeds are very fattening, when used as an addition to ordinary fodder, and are in this respect quite equal to oil-cake, while the foliage is said to be not inferior to that of clover and more bulky. Nevertheless some lupines have proved poisonous to sheep. About 90 lbs. of seeds are required for an acre. Langethal observes: "What the Sainfoin does for the poorest limestone or marly soil, that the Yellow Lupine carries out for sand-land." Lupines are not adapted for wet or moory ground, nor for limestone-formations, where most other leguminous fodder-plants do well. Mr. Joseph Augustin speaks of a yellow-flowering lupine which sometimes in the Azores attains a height of 12 feet in three months.

Lupinus varius, Linné.

The Blue Lupine. Also a Mediterranean annual, used like the above species; but a few others are under cultivation as Blue Lupines. Some of the American, particularly Californian species, are regarded even as superior to the Mediterranean kinds for agrarian purposes.

Lycium Afrum, Linné.

Africa and South-Western Asia. "The Caffir-Thorn." Can with many other species be utilized as a hedge-bush. It is ever-green, fiercely spiny, easily raised from seeds, readily transplanted, quick in growth, stands clipping well, seeds freely, is strong enough to resist cattle and close enough to keep off fowls. 1½ lbs. of seeds at a cost of 30 shillings suffices for a mile of hedging (Th. Lang).

Lycium barbarum, Linné.

Northern Africa and South-Western Asia. The most common kind grown in Europe for hedges. Is content with poor soil.

Lycium Europæum, Linné.

Countries around the Mediterranean Sea. Hardy in Norway to lat. 67° 56'. An excellent hedge-plant, particularly in sand-land, but emitting copious offshoots (C. Bouché). An allied congener is L. flaccidum (Moench).

Lycopodium dendroideum, Michaux.

North-America. This, with L. lucidulum (Michaux), has become there a great article of trade, being in request for bouquets and wreaths; both plants, after having been dyed of various colors, are used as ornaments in vases (Meehan). These club-mosses are mentioned here, to draw attention to similar species in other countries.

Lygeum Spartum, Linné.

Regions on the Mediterranean Sea. This perennial grass serves much like the ordinary Esparto-Grass, but is inferior to it.

Lyperia crocea, Ecklon.

South-Africa. The flowers of this shrub produce a fine orange dye, and are also in use for medicinal purposes.

Maba geminata, R. Brown.

One of the Ebony-trees in Queensland. Wood, according to M. Thozet, black towards the centre, bright red towards the bark, close-grained, hard, heavy, elastic and tough. It takes a high polish, and is recommended for veneers. Maba fasciculosa, F. v. M., has the outer wood white and pink. Several other species exist in Queensland, which may perhaps give good substitutes for Ebony-wood.

Macadamia ternifolia, F. v. Mueller.

The Nut-tree of sub-tropical Eastern Australia, attaining a height of 60 feet; hardy as far south as Melbourne; in forest-valleys probably of fair celerity of growth; endures slight frost. In favorable localities it bears fruit in seven years. The nuts have the taste of hazels.

Machilus odoratissima, Nees.

The "Soom-tree." From the Himalayas to Assam, Cochin-China, Burmah, Java and Sumatra, ascending to the cool elevation of 8,000 feet. A tree of considerable size. The Muga-Silkworm feeds on the foliage (Gamble). The leaves are pervaded by an orange-scent (Brandis).

Maclura aurantiaca, Nuttall.*

The Osage-Orange, or North-American Bow-Wood, or Yellow Wood. Texas, Arkansas, Louisiana. This thorny deciduous shrub or tree can be well trained into hedges. It is unisexual, and will in favorable localities on rich river-banks attain a height of 60 feet, with a stem 2 to 3 feet thick, thus becoming available as a timber-tree. It resists severe frosts. Rate of stem-growth in Nebraska about 1 inch a year circumferentially (Furnas). The saplings furnish stakes for vines, which are very lasting. The elastic wood serves well for bows, buggy-shafts, carriage-poles and similar articles. It is one of the most durable of all North-American woods, also valuable for all purposes, where toughness and durability are required (Dr. C. Mohr). The plant is not readily subject to blight or attacks of insects. It produces from the root a yellow dye. Mrs. Timbrell, at the suggestion of the author, has shown, that the foliage is as good a food for silkworms as that of the white mulberry, and the silk produced in no way inferior to ordinary silk. Leaves not too succulent are preferable. [Cf. also Riley, publications of the United States Department of Agriculture 1877]. M. tinctoria (D. Don) furnishes the Fustic-wood of Central and Southern America.

Maclura excelsa, Planchon.

West-Africa, on mountains up to 3,000 feet elevation. Height of tree often 150 feet. The wood is remarkably durable and tough, beautifully dark-brown and veined. Birds feed on the fruit.

Maclura Mora, Grisebach.

North-Argentina. A high tree. Wood greatly esteemed for its density and toughness; fruit edible (Dr. Lorentz).

Magnolia hypoleuca, Siebold.

Japan. A stately tree, with very large and whorled leaves. Trunk to a foot in diameter. Wood remarkably flexile; used for many kinds of utensils. Worthy of introduction as a magnificent garden-object (Christy).

Magnolia macrophylla, Michaux.

Eastern States of North-America. Although not cultivated for any special purposes of the arts or of technics, yet this tree is admitted into this list as one of the grandest of its kind, as well in foliage as flowers. It attains a height of about 60 feet; its leaves are from 1 to 3½ feet long, while its flowers attain a diameter of fully 1 foot. M. grandiflora, L., attains a height of 100 feet, and a stem-diameter of 3 feet on the Mississippi; it bears the winter of Philadelphia. M. acuminata, L., and M. Fraseri, Walter, are also large trees. The flowers of all exude much honey-nectar for bees.

Maharanga Emodi, A. de Candolle.

Nepal. The root produces, like that of Alkanna tinctoria, a red dye.

Malachra capitata, Linné.

Tropical America. A tall herb, annual, or of more than one year's duration. Its fibre is obtainable to lengths of 9 feet; it is of a silky lustre, and equal in technical value to Jute (O'Connor)..

Mallotus Philippinensis, J. Mueller. (*Rottlera tinctoria,* Roxburgh.)

South-Asia and East-Australia, in jungle-country, extending into New South Wales. A bush or tree, attaining, according to Mr. O'Shanesy, a height of about 60 feet. Though not of great importance, this plant should not be passed on this occasion, inasmuch as the powdery substance, investing the seed-capsules, constitutes the Kamala, which can be employed not only as an orange-dye, but also as an anthelmintic remedy. The Hindo silk-dyers produce the color by boiling the Kamala with carbonate of soda.

Mangifera Indica, Linné.

The Mango. South-Asia. An evergreen very shady tree, reaching 70 feet in height. Possibly it could be made to bear its delicious fruit in warm and humid forest-regions of sub-tropic zones. In the Himalayas its culture for fruit ascends to 3,500 feet just outside the tropics.

Manihot Aipi, Pohl.*

The Sweet Cassava. Tropical South-America, but traced as far south as the Parana-River. The root is reddish and harmless; it can therefore be used as a culinary esculent, without any preparation further than boiling, while its starch is also available for tapioca. It is a somewhat woody plant, several feet high, and too important to be left altogether unnoticed on this occasion, although we have no evidence, that it will be productive in a temperate clime. The Aipi has ligneous tough fibres, stretching along the axis of the tubers, while generally the roots of M. utilissima are free from this central woody substance.

Manihot Glazioui, J. Mueller.*

A native of Ceara, a coast-district of Brazil, in latitude 4°, possessing an arid climate for a considerable part of the year. This plant is evidently of a comparatively hardy character, and adapts itself readily to the exigencies of culture (D. Morris). It produces the Ceara-Rubber. Its cultivation is not difficult, and its growth remarkably rapid. It could only be grown in regions free of frost. Mr. Holtze, at Port Darwin, had the first grand success with this plant in Australia, seeds from Kew-Gardens having been placed by the author at his disposal; his plants attained a height of 12 feet in little more than a year. Perhaps the plant must be regarded as strictly tropical, and as then not admissible within the scope of this work. In Ceylon the plant has grown 20-30 feet in two years; the plants should be placed about 10 feet apart. It is best to wait with tapping till the trees are five years old (Keir Leitch).

Manihot utilissima, Pohl.*

The Bitter Cassava or Tapioca-Plant. Eastern Brazil. Closely allied to M. Aipi, producing varieties with roots of poisonous acridity and with roots perfectly harmless. The tubers attain a length of 3 feet; they can be converted into bread or cakes, the volatile poison of the milky sap being destroyed through pressing the grated root in first instance, the remaining acridity being expelled by heat. The starch, heated in a moist state, furnishes tapioca. Manihot is abundantly cultivated in many places, thus at Caraccas, where the singularly uniform temperature throughout the year is only 60° to 70° F. It is a very exhausting crop, and thus stands in need of rich soil and copious manuring. The propagation is effected by cuttings from the ligneous part of the stem. The soil, destined for Cassava, must not be wet. In warm countries the tubers are available in about eight months, though they still continue to grow afterwards. The growth of the plant upwards is checked by breaking off the tops. The Bitter Cassava is the more productive of the two. The yellowish tubers sometimes attain a weight of 30 lbs. They do not become soft by boiling, like Aipi.

Maoutia Puya, Weddell.

India, on mountains up to 4,000 feet. It is taller than Boehmeria nivea, and furnishes a similar fibre, which however is not so easily separated. This shrub belongs to a tribe of the Nettle-order not possessing burning acridity. None of the true nettles, such as the Girardinias, nor allied stinging plants, have been recommended in this index, although an exquisite fibre is derived from some, as the writer wishes to guard against the introduction of any burning species, which possibly might disseminate itself in a mischievous manner, and then probably could not again be suppressed.

Maranta aurundinacea, Linné.

The True Arrowroot-Plant, or more correctly "Aru-root," inasmuch as Aru-Aru is the Brazilian word for flour, according to Martius. West-Indies, Florida, Mexico to Brazil. The plant is introduced into this list not without hesitation, as it seems to require a tropical clime to attain perfection. It furnishes most of the West-Indian arrowroot-starch, although other species, such as M. nobilis, M. Allouya and M. ramosissima, are also cultivated for a similar starch contained in their tubers. Dr. Porcher observes, that it still flourishes as far north as Florida, producing even in the pine-lands from 200 to 300 bushels of tubers to the acre. General Sir John Lefroy found in Bermuda 100 lbs. of the root to yield 15 to 20 lbs. starch. M. indica (Tussac) is merely a variety.

Marlea Vitiensis, Bentham.

Fiji, New South Wales and Queensland. A middle-sized tree, generally with a gouty trunk; wood bright yellow with fine undulating rings, black towards the centre. Fruit edible (P. O'Shanesy). The generic name Stylidium by Loureiro has many years priority over that of Marlea by Roxburgh.

Marliera glomerata, Bentham. (*Rubachia glomerata*, Berg.)

The "Cambuca" of sub-tropical Brazil. The fruits attain the size of apricots, and are locally much used for food (Dr. Rosenthal).

Marliera tomentosa, Cambessedes.

Extra-tropical Brazil. The "Guaparanga." The sweet berries of this tall shrub are of the size of cherries.

Marrubium vulgare, L'Ecluse.

Middle and South-Europe, Northern Africa, Western Asia. The "Horehound." This tall perennial odorous herb, though in many countries quite a weed, is here also enumerated for completeness' sake. Its naturalization can nowhere be unwelcome, as it does not unduly spread, as it has important medicinal properties much dependent on the crystalline Marrubin, and as the flowers through much of the season afford to bees nectar for a pale excellent honey. The plant accommodates itself readily to any forlorn waste land.

Matricaria Chamomilla, Linné.

The annual "Chamomile." Europe, Northern and Middle Asia. A highly useful herb in medicine. In many parts of the European continent it is much more extensively employed than the ordinary perennial Chamomile. The infusion of the flowers has rather a pleasant taste without bitterness. The flowers serve as a tonic, and especially as a sudorific, and possess a peculiar volatile oil. In Norway this plant is grown as far north as lat. 70° 22' (Schuebeler).

Matricaria glabrata, De Candolle.

The South-African Chamomile. This annual herb is there in renown as an excellent substitute for the European Chamomile (Dr. Pappe).

Mauritia flexuosa, Linné.

From Guiana to Peru and Brazil. This noble Palm is known to ascend up to 4,000 feet along the Essequibo. As Palms, like Bamboos, prove to be among the hardier of tropical plants, experiments for naturalizing M. vinifera (Martius) might also be instituted. This attains a height of about 150 feet, has leaves sometimes 15 feet in length, and yields from the incised stem a copious sap, which forms a kind of wine by fermentation.

Maytenus Boaria. (Boaria Molinœ, De Candolle; Maytenus Chilensis, De Candolle.)

Chili. An evergreen tree, assuming considerable dimensions in the southern provinces. Wood extremely hard. Cattle and sheep browse with predilection on the foliage; hence the trees are cut down, when grasses become scarce through protracted snowfalls or in times of drought (Dr. Philippi).

Medicago arborea, Linné.

South-Europe, particularly Greece. This shrubby yellow Lucerne is of value for dairy-farmers, as it much promotes in cows the yield of milk. This genus includes several other species, valued as pasture-plants besides the present and those noted below.

Medicago lupulina, Linné.

The Black Medick. Europe, North-Africa and temperate Asia. An annual or biennial pasture-herb, easily grown, and not without nutritive importance. Langethal observes : " It effects for argillaceous soils, what the White Clover does for sandy moist soils. It will even succeed in moory ground, provided such contains some lime. It suits also particularly for sheep-pastures." It will thrive, where on account of poor soil lucerne and clover fail. In rich land its product is very copious. In Norway it will grow to lat. 63° 26'.

Medicago sativa, Morison*

The Lucerne, Purple Medick or Alfalfa. Orient and temperate Western Asia, now spread through Middle and Southern Europe and Middle Asia. The Romans brought it 470 years before the Christian era from Media, hence the generic name (A. de Candolle). A perennial fodder-herb of great importance, and largely utilized in most countries with a temperate clime; perhaps descended from the European and North-Asiatic Medicago falcata (Linné), the Yellow Medick, which also deserves naturalization, especially on light or sandy calcareous soil; but that plant is less productive than the true Lucerne, and does not resist occasional slight inundations so well, enduring however a rougher clime. Lucerne keeps green and fresh in the hottest season of the year, even in dry and comparatively barren ground and on coast-sands, but develops itself for field-culture with the greatest vigor on river-banks or when subjected to a judicious system of irrigation, particularly in soil rich in lime. Its deeply penetrating roots render the plant particularly fit for fixing embankments or hindering the washing away of soil subject to occasional inundations. The Peruvian variety (Alfalfa) resists drought and frost better than the original European Lucerne. Dr. Curl, of New Zealand, allows cattle to feed upon Alfalfa for two weeks, then takes them off and puts sheep on for two weeks, to eat the Alfalfa close to the ground; he then removes them and permits the Alfalfa to grow for a month, when he repeats the process. He allows five large cattle or twenty sheep to the acre. Lucerne is also an important honey-plant for bees. Much iron in the soil or stagnant-water is detrimental to lucerne-culture, while friable warm soil much promotes its growth. Langethal records instances of lucerne having yielded on the same field under favorable circumstances for fifteen years four or five cuts annually. The chemical analysis of the fresh herb, collected very early in spring, gave the following results : Starch 1·5, gum 2·1, unfermentable sugar 3, albumen 2·3, insoluble proteins 2·3, ash 2·3 per cent. (F. v. Mueller and L. Rummel). For sandy tracts a yellow variety.(M. media, Persoon) deserves preference. To show how enormously plants are affected in their mineral constituents by difference of soil, Lace has analyzed the ashes of lucerne (a) from granitic soil, (b) chalky soil with flints, (c) clayey with chalk, (d) very chalky, and found—

	a	b	c	d
	Per cent.	Per cent.	Per cent.	Per cent.
Silicic acid	·99	·41	·47	·58
Ferric oxide	·76	1·05	·29	·60
Magnesium carbonate	9·89	7·15	10·11	9·05
Calcium sulphate	4·50	3·04	7·51	6·80
,, phosphate	14·94	8·11	10·66	19·71
,, carbonate	13·42	48·15	49·68	30·19
Potassium carbonate	48·42	29·19	20·60	26·09
Potassium and sodium chlorides	6·67	2·90	·68	6·98

Medicago scutellata, J. Bauhin.*

Countries at and near the Mediterranean Sea, where this annual herb, as well as the allied M. orbicularis (Allioni) is regarded as a valuable fodder-plant (Caruel), without the disadvantage of their fruits adhering to fleeces like those of its prickly-fruited congeners. For this particular reason the author introduced these two plants into Australia, where in the dry hot inland-regions they have surpassed most other fodder-herbs in value and resistance to drought. They will also bear some frost.

Melaleuca ericifolia, Smith.*

South-Eastern Australia. This tall shrub or bushy tree is of importance for consolidating muddy shores; it will live in salty ground and water, almost like mangroves. I found it growing vigorously, where the water contained rather more than 2 per cent. chlorides, and the wet soil contained nearly 1½ per cent. chlorides (the contents of sea-water being from 3 to 4 per cent. chlorides, or about 2½ per cent. chloride of natrium). It yields also a comparatively large quantity of cajaput-oil. It admits of easy transplantation, even when full-grown. Myoporum insulare (R. Brown) and Leptospermum lanigerum (Aiton) can in like manner be used in tree-plantations for the sake of shelter on wet saline soil. The branches of M. ericifolia furnish the best native material in South-Eastern Australia for easily worked and lasting garlands. This species can be grown in swamps for hygienic purposes by subduing paludal malaria or fever-provoking effluvia.

Melaleuca genistifolia, Smith.

Northern and Eastern Australia. A tree, attaining 40 feet in height, flowering in a shrubby state already, fond of banks of water-courses. The copious flowers, according to Mr. C. French, attract bees to an extraordinary degree.

Melaleuca Leucadendron, Linné.

The Cajaput-tree of India, North- and East-Australia as far extra-tropical as 34° south latitude. This tree attains a height of about 80 feet, with a stem reaching 4 feet in diameter, on tidal ground; it can with great advantage be utilized on such areas and in salt-swamps for subduing malarian vapors, where no Eucalyptus will live. The lamellar bark protects it against conflagrations. The wood is fissile, hard and close-grained, regarded as almost imperishable underground, and resists the attacks of termites. It is well adapted for posts, wharf-piles, ship-building and various artisans' work. The allied Callistemons (C. salignus, D.C., 60 feet high, C. lanceolatus, D.C., 40 feet) produce a hard, heavy, close-grained wood, suitable for wheel-wrights' work and implements, proving very durable underground (W. Hill).

Melaleuca linarifolia, Smith.

Eastern Australia. A tree attaining finally a considerable height, deserving attention as eligible for wet saline land, on which it can be raised much more easily than Myoporum insulare. M. Thozet observes, that it occurs in places, where it is bathed by the tides; also that large saplings without roots can be transplanted. Thus it may be destined to aid, with several of its congeners and with Salicornias, Avicennias, Ægiceras, Batis, Suaedas and some other plants, to reclaim low muddy shore-lands from sea-floods. Foliage extremely rich in volatile antiseptic oil. M. squarrosa (Smith) of South-Eastern Australia, can be grown in fresh-water swamps, also to subdue miasmata. It attains exceptionally the height of 60 feet, with a stem two feet in diameter.

Melaleuca parviflora, Lindley.

Extra-tropical Australia. A tall bush or small tree. In mild climes one of the most important plants for fixing moving coast-sands.

Melaleuca styphelioides, Smith.

East-Australia. Height of tree reaching about 60 feet; stem-diameter 2½ feet. The timber is hard, close-grained, and stands well in damp situations. It is said, that the timber has never been known to decay (Queensland Exhibition, 1878). Tree adapted for swamps.

Melanorrhœa usitata, Wallich.

The Varnish-tree of Burmah, Munnipore and Tenasserim. Possibly· hardy in forest-valleys free of frost, as it ascends to 3,000 feet elevation. The hardened sap is used for a highly prized black varnish.

Melia Azadirachta, Linné. (*Azadirachta Indica,* Jussieu.)

From Persia to China, ascending the Himalayas to 5,000 feet. The "Neem-tree" attains a height of 50 feet, much planted on promenades. The powerful medicinal properties not unlike those of M. Azedarach. Furniture from its wood not attacked by insects. Leaves simply pinnate, the fresh ones issuing before the old ones drop.

Melia Azedarach, Linné.

Called "The Pride of India." South-Asia, North- and also East-Australia, and there to far extra-tropical latitudes, ascending the Himalayas to 6,000 feet. Height seldom over 40 feet; branches very spreading. As an avenue-tree not without importance, because it will successfully cope with dryness of clime and sterility of soil. It recommends itself also for retaining the foliage till very late in the season, and for producing an abundance of very fragrant spring-flowers, which may perhaps be worth distilling for essential oil. Grows from cuttings as well as seeds, the latter abundantly maturing. All parts of the plant are bitter; the bark with caution can be used as a purgative and anthelmintic; the leaves prove insecticidal (Hieronymus). The wood is considered of value for cabinet-work, also some kinds of musical instruments. A black-fruited Melia seems as yet little known. Casimir De Candolle distinguishes twelve species, their range being from India to Japan and Polynesia.

Melianthus major, Linné.

South-Africa. The leaves of this stately plant are very efficacious as antiseptics, also in cases of scald-head, ringworm and various other cutaneous diseases (Dr. Pappe). Its effect of promoting granulation is very remarkable (Dr. A. Brown). Flowers rich in honey, as indicated by the generic name. Will bear some frost.

Melica altissima, Linné.

Eastern Europe, Middle Asia. This perennial grass has recently come into use for pasture.

Melica ciliata, Linné.

Europe and Middle Asia. A perennial fodder-grass, particularly desirable for sheep. Best for dry gypsum- or lime-ground.

Melica nutans, Linné.

The Pearl-Grass. Europe, Northern and Middle Asia, enduring an alpine exposure and living also in the shade of forests. It will bear the clime of Norway to lat. 70° 28' (Schuebeler). It produces suckers, and affords good foliage in woody regions; so also does M. uniflora (Retzius). Several other species are on record from various parts of the globe, among which M. mutica (Walter), of North-America, seems to deserve special attention.

Melica sarmentosa, Nees.

Brazil and La Plata-States. A tall grass, climbing to a height of 12 feet in forests and on river-banks. Prof. Hieronymus speaks of M. aurantiaca (Desrousseaux), M. laxiflora (Cavanilles), M. macra (Nees), M. papilionacea (Linné), M. rigida and M. violacea (Cavanilles) as perennial Pampas-grasses, which though hard, are nutritious, and particularly sought by asses and mules.

Melicocca bijuga, Linné.

Central America, on mountains. So many sapindaceous trees of the Cupania-series have been shown by my own experiments to be hardy in a climate like that of Victoria, that this important member of the series could now also be admitted into this list. The pulp of the fruit is of a grape-taste; the seeds can be used like sweet chestnuts.

Melilotus alba, Desrousseaux.

The Cabul- or Bokhara-Clover. Europe, North-Africa, Middle Asia. Indigenous in Norway to lat. 60° 16' (Schuebeler). A biennial herb. On account of its fragrance it is of value as admixture to hay. It is also a good bee-plant, the white flowers lasting till late in the season. Odorous principle: cumarin.

Melilotus cœrulea, Lamarck.

South-Europe and North-Africa. Cultivated in Norway to lat. 70° 22'. An annual, very odorous fodder-herb. It forms an ingredient of the green Swiss cheese, which owes its flavor and color chiefly to this plant.

Melilotus officinalis, Desrousseaux.

Europe and Middle Asia. In Norway hardy to lat. 76° 17′. Biennial, or lasting through several years, if prevented from flowering. Contains also cumarin. An allied species is M. macrorrhiza (Persoon). Both serve purposes similar to those for which M. alba is employed. Grown on the coast it becomes less odorous. Honey from this and closely allied plants exquisite.

Melissa officinalis, Linné.

The Balm-Herb. Southern Europe and Western Asia. A perennial herb, valuable for its scent, which depends on a peculiar volatile oil. This herb is also important as a bee-plant. Readily propagated by division of the root.

Melocalamus compactiflorus, Bentham.

Martaban, ascending to 6,000 feet. This Bamboo attains a height of 25 feet, and is somewhat scandent.

Melocanna bambusoides, Trinius.

The Berry-bearing Bamboo, from Chittagong and other mountainous parts of India, as well as of the Archipelago. Height to 70 feet (Kurz). The fruit is very large, fleshy, size of a small pear, and contains a seed, which is said to be very pleasant eating (Masters). It is a thornless Bamboo, growing on dry slopes of hills. Height reaching 70 feet; circumference towards base 1 foot; growth beautifully erect.

Melocanna Travancorica. (*Beesha Travancorica*, Beddome.)

A Bamboo from Travancore, worthy of introduction.

Mentha arvensis, Linné.

Europe, extra-tropical Asia. The variety piperascens of this perennial herb constitutes the peculiar Japan Peppermint. From the distilled oil of this by refrigeration the crystalline menthol is obtained in unusually large proportion for medicinal purposes (E. Holmes, T. Christy).

Mentha laxiflora, Bentham.

Victoria and the most southern parts of New South Wales. This, the Australian "Forest-Mint," furnishes a peculiarly pleasant oil, not dissimilar to that of peppermint. A fair oil can also be distilled from M. Australis (R. Brown), the common "River-Mint" of South-Eastern Australia.

Mentha piperita, Linné.*

The "Peppermint." Middle-Europe. This well known perennial herb is important for its peculiar essential oil. This distilled oil is in considerable demand, and would be best obtained from plants cultivated in mountain-regions or naturalized along forest-rivulets. The annual production of oil of peppermint is estimated at 90,000 lbs., two-thirds of which are prepared in the State of New York (Masters).

Eminent authorities refer the peppermint as a variety to Mentha aquatica, L., the Water-Mint of Europe, North-America, West- and North-Asia, from which the true Crisp Mint (M. crispa, Linné) is again derived, as well as the Bergamot-Mint (M. citrata, Ehrhart).

Mentha Pulegium, Linné.

The true "Penny-royal." Europe, Western Asia, Northern Africa. A perennial scent-herb, yielding a peculiar ethereal oil. It likes moist soil. To be avoided on pastures, as not readidly repressed.

Mentha rotundifolia, Linné.

Western and Southern Europe, Northern Africa, Western Asia. Fond of wet places, which by the culture of this and other mints may be profitably utilized. In odor this mint approaches to Melissa. The French and Italian Crisp Mint is partly derived from this species. Closely allied to the following, and often regarded as a variety of M. viridis.

Mentha silvestris, Linné.

The "Horse-Mint." Europe, Northern Africa, temperate Asia. Perennial. One of the Crisp Mints is derived from this species. Hardy, like the three preceding species, to lat. 59° 55' in Norway, (Schuebeler).

Mentha viridis, Linné.

The "Spearmint." Middle and Southern Europe. Perennial. A particular sort of Crisp Mint (M. crispata, Schrader) belongs to this species. Readily propagated like other mints by division of the root.

Menyanthes trifoliata, Linné.

Inappropriately called the Bog-bean or Buck-bean. Europe, Northern and Middle Asia, North-America. In springy and spongy bogs. A perennial herb of great beauty, which could be naturalized with facility in any cold regions. Indigenous as far north as lat. 71° 10' in Norway (Prof. Schuebeler). The root is starchy. The whole plant is pervaded with a bitter principle, largely derived from menyanthin. The plant is used medicinally as a tonic and febrifuge.

Meriandra Abyssinica, F. v. Mueller. (*M. Benghalensis*, Bentham.)

Abyssinia, on high mountains. A shrub of penetrating odor; utilized much like sage.

Mesembrianthemum acinaciforme, Linné.

The "Hottentot-fig" of South-Africa. Under the same vernacular name is also comprised the distinct M. edule, L. Both should be transferred to any of the most inhospitable desert-regions, as they afford in the inner part of their fruit a really palatable and copious food. M. edule proved hardy in Southern England.

Mesembrianthemum aequilaterale, Haworth.

Australia and West-coast of America. This widely creeping species spreads readily over saline ground, whether clayey, sandy or rocky.

Mr. J. Clode observes, that sheep are very fond of this succulent plant, and require but little water when browsing on it; or in cool coast-districts they will do .without any. water even in summer, while thriving well on the foliage. Fruit with a sweetish edible pulp. This species with M. australe (Haworth) forms on the Australian coasts one of the most effectual first impediments to the influx of sea-sand; both should be encouraged in their growth at the very edge of the tide. Not less hardy than M. edule.

Mesembrianthemum capitatum, Haworth.

South-Africa. This perennial species, from the readiness and quickness of its growth, and from the abundance of its seeds and their easy dispersion, is one of the best for staying any rolling sea-sand (Dickinson). M. pugioniforme (Linné) and many other species serve the same purpose.

Mesembrianthemum crystallinum, Linné.

Countries at the Mediterranean Sea and South-Africa. Annual. Recently recommended as a spinage-plant. Can be grown on bare sand, which it helps to cover. Eaten by sheep. In Norway it will succeed northward to lat. 69° 18'.

Mesembrianthemum floribundum, Haworth.

South-Africa. This succulent perennial with many allied species from the same part of the globe is a far more important plant, than might be assumed, because "a good stretch of this is worth as much as a dam" (Professor McOwan). Succulent plants like these would live in sandy deserts, where storage of water may be impracticable.

Metrosideros tomentosa, Cunningham.

Northern Island of New Zealand. Could be grown for timber on rocky sea-shores. Height reaching about 80 feet; trunk stout, but comparatively short. The timber, according to Professor Kirk, deserves attention, as one of the most durable for the frame-work in ship-building, for jetties, docks, sills. Other species with dense wood, occurring in New Zealand, are M. lucida (Menzies) and M. robusta (Cunn.), both ornamental trees with crimson flowers.

Michelia excelsa, Blume.

In the Himalayas and other Indian mountains, up to 8,000 feet. It grows to a large size, supplying boards to three feet in width, and is one of the best timber-trees there. Foliage deciduous; flowers large, white; wood yellowish. M. lanuginosa (Wallich) ascends there also to temperate regions with M. Kisopa (Hamilton), M. Cathcartii (Hooker and Thomson), M. Champaca (Linné), M. punduana (H. & Th.) and M. Nilagirica (Zenker), all being tall trees.

Microseris Forsteri, J. Hooker.

The Native Scorzonera of extra-tropical Australia and New Zealand. A perennial herb deserving attention, as its root would probably enlarge and improve through culture. On alpine mountains the plant develops most luxuriantly during summer. The Australian aborigines use the root for food. The plant would prove hardy in Middle Europe.

Milium effusum, Linné.

English Millet-Grass. Europe, North- and Middle-Asia, North-America. Perennial, suited for damp forest-land particularly, the pastural capabilities of which it enhances. On river-banks it attains a height of 6 feet. It is relished by cattle. The seeds can be used like millet, the stems for the manufacture of superior straw-hats. It is a great favorite with pheasants and many other birds for the sake of its seeds, which ripen early in the season. Indigenous in Norway as far north as lat. 71° 7′ (Schuebeler). Nutritious hay to the extent of 3 tons on an acre obtainable from this grass (Coleman).

Mimosa rubicaulis, Lamarck.

India. A hedge-bush, almost inapproachable. It has proved hardy at Melbourne, enduring some frost.

Mimusops globosa, Gaertner.

Central America. Tree, reaching a height of about 120 feet, perhaps fit like many other Sapoteæ for frostless extra-tropic regions. The milky juice from cuts into the stem when exsiccated forms the Balata of commerce, a substance in its qualities allied both to Gutta-percha and India-rubber (Jenman).

Mimusops Sieberi, A. de Candolle.

West-Indies and Florida. Tree reaching 30 feet in height. Fruit of agreeable taste (Sargent).

Monarda didyma, Linné.

Eastern North-America. Hardy to lat. 59° 55′ in Norway. A perennial odorous herb, producing the medicinal Oswego- or Beebalm-Tea. M. fistulosa, L., and several others are also of very strong scent. Their volatile oil contains thymol.

Monarda punctata, Linné.

Eastern North-America, where it is called "Native Horse-mint." Bees extract an astonishing quantity of honey from this plant. M. citriodora (Cervantes) is an allied lemon-scented species, extending from the South-Western States to Mexico.

Monodora Angolensis, Welwitsch.

Tropical West-Africa, up to the comparatively cool elevation of 3,500 feet. A tree attaining 30 feet in height. The pleasantly aromatic seeds come into the market, like those of the following species; they measure about half an inch in diameter and are produced in numbers.

Monodora Myristica, Dunal.

West-Africa. A small tree. The seeds serve as nutmegs.

Morchella conica, Persoon.

Europe, Asia, Northern and Central America, northward to lat. 70° in Norway. With M. semilibera this Morel has been found in Victoria and New South Wales; its spread should be encouraged by artificial means, as it is a wholesome esculent. Kohlrausch and Siegel found 29 to 35 per cent. of protein in Morels when dried. European superior species, probably admitting of introduction, are: M. esculenta, M. Gigas, Pers., M. rimosipes, D.C., M. Bohemica, Krombh., M. deliciosa, Fries (which extends to Java) and M. patula, Pers., the Bell-Morell; but several others occur in other parts of the globe. Though these fungs show a predilection for pine-forests, they are not dependent upon them; thus the writer found M. conica (Persoon) in Eucalyptus-forests, and this late in the autumn. They can all be dried and preserved for culinary purposes.

Moringa pterygosperma, Gaertner.

The Horse-Radish Tree of India, abundant as far as the middle regions of the mountains. Height to about 20 feet, but thick-stemmed. The long pods are edible; the seeds are somewhat almond-like and rich in oil, which has no perceptible smell, and is esteemed by watchmakers particularly. Gum exudes from the stem. M. aptera (Gaertner) occurs from Abyssinia and Egypt to Arabia and Syria. M. Concanensis (Nimmo) is an allied species from the drier regions of North-Western India.

Morus alba, Linné.*

The White Mulberry-tree. Upper India and Western China. This tree in several varieties provides the food for the ordinary Chinese silk-insect (Bombyx Mori). Silk was produced in Italy 700 years ago, and this branch of industry has flourished there ever since. In China silk has been reeled for 4,500 years; this may demonstrate the permanency of an industry, which we wish to establish extensively anywhere under a similar sky. "One pound of silk is worth its weight in silver, and this pound may be produced (so far as the food of the Bombyx is concerned) from thirty pounds of mulberry-leaves or from a single tree, which may thus be brought to yield annually the material for 16 yards of Gros de Naples" (R. Thompson). The White Mulberry-tree is of extremely easy growth from cuttings, also readily raised from well-matured seeds. It is usually unisexual, and finally attains a very large size. It can still be grown in climes, where olives will no longer thrive. In Norway the tree bore seeds in lat. 59° 55' (Schuebeler). Spots for mulberry-culture must not be over moist, when the leaves are to be utilized for the Bombyx. In 1870, according to the *British Trade Journal,* the produce of cocoons amounted in Europe to £16,588,000; in Asia to £28,112,000; in Africa to £44,000; in the South-Sea Islands to £24,000; in America to £20,000—thus giving a general total of £44,788,000. In 1875 the yield of raw silk in the district of Rajshahye (British India) was estimated at £400,000, employing about 12,000 people, the plantations extending approximately over 150 square miles (Dr. S. Forbes Watson). In that district alone a quarter of a million people derive their support from the trade and other branches of the silk-industries. Great Britain imported in 1884 4,519,000 lbs. of raw silk, to a value of £3,341,000 of thrown silk; husks and waste to the value of £907,000, irrespective of manufactured silk-goods, the cost of which amounted to nearly eleven million pounds sterling. Superior varieties of mulberry can be grafted with ease on ordinary stock. M. Indica, L., M. macrophylla, Moretti, M. Morettiana, Jacq., M. Chinensis, Bertol., M. latifolia, Poir., M. Italica, Poir., M. Japonica, Nois., M, Byzantina, Sieb., M. nervosa, Del., M. pumila, Nois., M. tortuosa, Audib., as well as M. Constantinopolitana, Lamarck, with which, according to Prof. C. Koch, is identical M. multicaulis of Perrottet, are merely forms of M. alba, to which probably also M. Tatarica, L. and M. pabularia, Jacquin, belong. The variety known as M. Indica produces black fruits. The raising of Mulberry-trees has recently assumed enormous dimensions in California, where between seven and eight millions were planted since 1870. The process of rearing the silk-insect is simple, and involves no laborious exertions. The cocoons, after they have been properly steamed, dried and pressed, readily find purchasers in Europe, the price ranging according to quality from 3s. to 6s. per lb. The eggs of the silk-moth sell at a price from 16s. to £2 per ounce; in 1870, Japan had to provide two millions of ounces of silk-ova for Europe, where the worms had extensively fallen victims to disease. As an example of the profit to be realized, a Californian fact may be cited, according to which £700 were the clear gain from 3½ acres, the working expenses having been £93. The Commissioner of Agriculture of the United States has estimated that under ordinary circumstances an acre should support from 700 to 1,000 mulberry trees, producing, when four years old, 5,000 lbs. of leaves fit for food. On this quantity of leaves can be reared 140,000 worms, from which ova at a net-profit, ranging from £80 to £240 per acre, will be obtained by the work of one person. Mr. C. Brady, of Sydney, thinks the probable proceeds of silk-culture to be from £60 to £150 for the acre. The discrepancies in calculations of this kind are explained by differences in clime, soil, attention, treatment and also rate of labor.

A very palatable fruit is obtained from a variety cultivated in Beloochistan and Afghanistan. Morus Tatarica, L., resembles M. alba; its juicy fruit is insipid and small. The leaves are not generally used for silkworms. The white-fruited tree was found apparently wild by Dr. A. v. Regel, at Taschkent and Tutkaul; the stems there were 7 feet thick.

The results of Mr. Brady's experience on the varieties of Morus alba are as follows: In the normal form the fruits are white with a purplish tinge more or less deep; the bark is pale; the leaf is also of a pale hue, not very early, nor very tender, nor very abundant. It may be grown on moist ground, so long as such is drained, or it will live even on poor, loose, gravelly soil, bordering on running water. The Cevennes-variety is a free grower, affords a large quantity of leaves, though of rather thick consistence; all varieties of the Morus-Bombyx like these leaves, whether young or old; it is also called the Rose-leaved variety; the silk, which it yields, is substantial in quantity and also good in quality; it does best on rich dry slopes. The bushy Indian variety has a fine leaf of a beautiful green, which, though light in weight, is abundantly produced; it can be cut back to the stem three or four times a year; the leaves are flat, long and pointed, possess a fine aroma, and are relished by every variety of the ordinary silk-insect, though all do not thrive equally well on it; the silk derived from this variety is excellent, but not always so heavy in quantity as that produced from the rosy variety; it prefers rich, low-lying bottoms, is a greedy feeder, but may thus be made to cover an extraordinary breadth of alluvial or manured land in a marvellously short space of time. At Sydney Mr. Brady can provide leaves from this Indian variety all through the year by the removal of cuttings, which will strike their roots almost at any season; it also ripens seeds readily, and should be kept at bush-size; it requires naturally less space than the other kinds. A fourth variety comes from North-China; it has heart-shaped, flat, thickish leaves, which form very good food for the silkworm. Mr. Brady, as well as Mr. Martelli, recommend very particularly the variety passing under the name of Morus multicaulis for the worms in their earliest stages. The former recommends the Cape-variety also; the latter wishes likewise the variety called Morus Morettiana to be used on account of its succulent nutritious foliage, so well adapted for the insect, while yet very young, and also on account of producing the largest amount of food within the shortest time. The Manilla-variety, above mentioned as Morus multicaulis, comes into bearing several weeks earlier than most other sorts, and should therefore be at hand for early hatched worms. An excellent phytological exposition of the numerous varieties of the White Mulberry-tree is given in De Candolle's prodromus XVII. 238–245 (1873).

The Muscardine-disease is produced by Botrytis Bassiana, while the still more terrible Pebrine-disease is caused by a minute psorospermous organism. On the Pebrine Pasteur's researches since 1865 have shed much light. Countries like ours, happily free from these pests, can thus rear healthy silk-ova at a high premium for exportion.

The White Mulberry-tree with others, offering food to the silk-worms, such as the osage-orange, should be planted copiously everywhere for hedges or copses. A very soft textile fibre is obtained from the bark of the Chinese Mulberry-tree.

Morus celtidifolia, Humboldt.

From Peru to Mexico, ascending to 8,000 feet. The fruit of this Mulberry-tree is edible. M. insignis (Planchon), from New Granada, is a similar species; it reaches elevations of 11,000 feet; is therefore a plant fit also for the cold temperate zone, and deserves thus general cultural attention.

Morus nigra, Linné.*

The Black Mulberry-tree. South-Western Russia and Persia. Attains a height of about 60 feet. Highly valuable for its pleasant refreshing fruits. It is a tree of longevity, instances being on record of its having lived ·through several centuries; it is also very hardy, enduring the winter-cold of Norway to lat. 61° 15'; at Christiania it bore fruit (Schuebeler). Mr. John Hodgkins regards it as a superior tree for sandy coast-ridges. The leaves of this species also afford food for the ordinary silk-moth, and are almost exclusively used for this purpose in the Canary-Islands, although the produce therefrom is not always so good as that from M. alba. The tree occurs usually unisexual. M. atropurpurea (Roxburgh), from Cochin-China, is an allied tree. The cylindrical fruit-spike attains a length of 2 inches.

Morus rubra, Linné.*

The Red Mulberry-tree. Eastern North-America, North-Mexico. The largest of the genus, attaining a height of about 70 feet; it produces a strong and compact timber, of wonderful endurance underground, hence in demand for posts and railway-ties (General Harrison); also for knees of small vessels (Dr. C. Mohr) and a variety of other purposes. Fruit edible, sweet, large. The tree is still hardy in Christiania (Schuebeler). Rate of circumferential stem-growth in Nebraska 43 inches in 18 years (Furnas).

Mucuna Cochinchinensis, Bentham. (*Macranthus Cochinchinensis*, Loureiro.)

A climbing annual, which can be reared in the open air in England. Pods, cooked as a vegetable, taste like those of kidney-beans (Johnson).

Muehlenbergia diffusa, Willdenow.

Southern States of North-America. Perennial. Recorded among the good native fodder-grasses of Alabama by C. Mohr, thriving as well on dry hills as in low damp forest-ground. Prof. Killebrew mentions, that this grass in Tennessee carpets the soil in forests with a living green. M. glomerata (Trinius) is in the same region a pasture- and hay-grass, available in wet meadows (Dr. Vasey).

Muehlenbergia Mexicana, Trinius.

Southern parts of North-America. A perennial good fodder-grass, particularly fit for low humid ground, also in forests. Root creeping; stem much branched, bending down.

Murraya exotica, Koenig.

South-Asia, Polynesia, East- and North-Australia. This shrub or small tree is one of the best among the odoriferous plants in India (C. B. Clarke). M. Koenigii (Sprengel) ascends the Himalayas to 5,000 feet; its leaves are in frequent use as an ingredient of curries.

Musa Cavendishii, Lambert.* (*Musa regia*, Rumph; *Musa Chinensis*, Sweet; *Musa nana*, Loureiro.)

The Chinese Banana. A comparatively dwarf species, the stem attaining a height of only about 5 or 6 feet. Its robust and dwarf habit render it particularly fit for exposed localities, and this is one of the reasons, why it is so extensively cultivated in the South-Sea Islands. The yield of fruit is profuse (even as much as 200 to 300 fruits in a spike), and the flavor excellent. General Sir John Lefroy saw bunches of 80 lbs. weight produced in Bermuda, where the plant bears fruits all the year round. This, as well as M. sapientum and M. paradisiaca, still ripens its fruits in Madeira and Florida. The specific name, given by M. Loureiro, is entitled to preference. All Musas are grand honey-plants.

Musa corniculata, Rumph.*

Insular India. Fruits as large as a good-sized cucumber; skin thick; pulp reddish-white, firm, dry, sweet; an excellent fruit for cooking (Kurz). The Lubang-variety is of enormous size.

Musa Ensete, Gmelin.

Bruce's Banana. From Sofala to Abyssinia in mountain-regions. This magnificent plant attains a height of about 30 feet, the leaves occasionally reaching the length of fully 20 feet, with a width of 3 feet, being perhaps the largest in the whole empire of plants, exceeding those of Strelitzia and Ravenala, and surpassing even in quadrate-measurement those of the grand water-plant Victoria Regia, while also excelling in comparative circumference the largest compound frond of Angiopteris evecta or the divided leaf of Godwinia Gigas, though the leaves of some palms are still larger in circumference. The inner part of the stem and the young spike of the Ensete can be boiled, to serve as a table-esculent, but the fruit is pulpless. This plant produces no suckers, and requires several years to come into flower and seed, when it dies off like the Sago-Palm, the Caryota-Palm and others, which flower but once without reproduction from the root. It is probably the hardiest of all species, enduring slight frosts.

Musa Livingstoniana, Kirk.

Mountains of Sofala, Mozambique and the Niger-regions. Similar to M. Ensete; seeds much smaller. This superb plant requires no protection in favorable places in warm temperate climes, as it advances in its native country to elevations of 7,000 feet. This and a Musa of Angola, like M. Ensete, form no suckers.

Musa paradisiaca, Linné.*

The ordinary Plantain or Pïsang. India. Among the most prolific of plants, requiring the least care in climes adapted for its growth. Stem not spotted. Bracts purple inside. In this as well as M. Cavendishii and M. simiarum, new shoots are produced from the root, to replace annually the fruit-bearing stem. The fruit of this is often prepared by some cooking process. Very many varieties are distinguished, and they seem to have sprung from the wild state of M. sapientum. The writer did not wish to pass this and the allied plants unnoticed, as they will endure the clime in warmer localities of the temperate zone, where under careful attention they are likely to mature their fruit with regularity. They require rich and humid soil. Plantain-meal is prepared by simply reducing the dried pulp to powder; it is palatable, digestible and nourishing. M. sapientum,. L, the ordinary Banana or Sweet Plantain is a variety. It is one of the most important plants among those yielding nutritious delicious fruits. The stem is spotted; bracts green inside. The leaves and particularly the stalks and the stems of this and other species of Musa can be utilized for producing a fibre similar to Manilla-hemp, though not so strong. The fruit of this species is used chiefly unprepared; it is generally of a yellow color. Numerous varieties are distinguished. Under favorable circumstances as much as a hundredweight of fruit is obtained from a plant annually in tropical climes. At Caraccas, where the temperature is seldom much above or below 70° F., the plantain- and banana-plants are very productive, being loaded with fruits 12 to 15 inches long, on mountains about 5,000 feet high. In the dry Murray-regions of South-Eastern Australia the winter-temperature seems too low for the successful development of the plants except on sheltered spots; but bananas still ripen under the shelter of limestone-cliffs as far south as Swan-River in West-Australia. The plant matures its fruit also yet in the Canary-Islands. The banana requires infinitely less care within its geographic latitudes than the potato; contains along with much starch amply protein-compounds. The preparation of starch from bananas is lucrative, as the yield is copious. Many Indian populations live very extensively or almost exclusively on this fruit. In hot countries the tall Musas are sometimes reared as nurse-plants.

Musa simiarum, Rumph.* (*M. corniculata*, Loureiro; *M. acuminata*, Colla.)

From Malacca to the Sunda-Islands. About half a hundred marked varieties of this species, called mainly Pisangs in India, are under cultivation there, especially on the Archipelagus, while M. sapientum occurs wild more frequently on the mainland. Though the latter is principally cultivated on the Indian continent, yet it never equals in delicacy the cultivated forms of M. simiarum, the fruit of which sometimes attains a length of 2 feet (Kurz).

Musa textilis, Nees.

Philippine-Islands. This species furnishes the widely utilized Manilla-rope fibre; the plant was introduced first into Australia by the author, and may thrive in subtropic regions. It likes volcanic forest-land. Much of the fibre is exported to America for paper. About 1 lb. of fibre is annually obtained from each plant (Simmonds).

Musa troglodytarum, Linné. (*M. uranoscopos*, Rumph.)

India, and apparently indigenous also in the Fiji and other islands of the Pacific Ocean. The fruit-stalk of this species stands upright; the edible fruits are small, reddish or orange-colored; pulp gamboge-

yellow, mawkish-sweet (Kurz). The Chinese M. coccinea, Ait., a
dwarf ornamental species, has also the fruit-spike straight.

Myoporum lætum, Forster.

New Zealand, where it is called Ngaio by the aborigines, also in
the Chatham-Islands. As a shelter-tree it is equal to the Australian
M. insulare for the most exposed parts of the coast. It is excellent
for shade, and its wood takes a fine polish. It can be raised on the
beach from cuttings. Uprooted it will produce new roots, if covered
in near the sea. Sheep, cattle and horses browse on the foliage.

Myrica Californica, Chamisso and Schlechtendal.

Californian Sweet-gale Tree, attaining 40 feet in height and a
stem-diameter of 3 feet. The leaves, pervaded by a balsamic resinous-
oily principle, are of medicinal value (Dr. Kellogg). The root-bark
of several Myricas is also turned to therapeutic account.

Myrica cerifera, Linné.

The Wax-Myrtle or "Bay-Berry." Sandy sea-coast of Eastern
North-America. This shrub helps to bind the rolling sand; it has
fragrant leaves; the fruits are boiled, and the floating wax, which
can be converted into candles, is skimmed off. In Patagonia,
Argentina and Chili the scrophularinous Monttea aphylla, Bentham
(Oxycladus aphyllus, Miers), yields vegetable wax from its branches
(Lorentz).

Myrica cordifolia, Linné.

South-Africa. This bushy plant arrests the influx of the sea-sand;
it also yields wax from its fruits in remunerative quantity.

Myrica Faya, Aiton.

Madeira, Azores and Canary-Islands. A small tree. The drupa-
ceous fruits are used for preserves.

Myrica quercifolia, Linné.*

South-Africa. This, as well as M. cordifolia and M. serrata are
the principal wax-bushes there. Many other species from different
parts of the globe are available for trial-culture, but none have as yet
been discovered in Australia.

Myrica rubra, Siebold and Zuccarini.

China and Japan. The bark of this tree or shrub serves for a
brown dye; the fruit is edible.

Myrica sapida, Wallich.

Continental India, up to 7,000 feet, also in Borneo. A shady
evergreen tree. The fruit is one of the best of North-Western India,
and is there eaten by all classes (Edwin Atkinson).

Myrica serrata, Lamarck.

South-Africa. Shrub only about 3 feet high. Also wax-yielding.
The Myrica-wax is heavier, harder and more brittle than bees' wax,

but melts easier; it is obtained from the fruits throughout the cool season. The sowing of seeds is done after the first rain of the cool months has steadied the loose sand; it can also be multiplied from cuttings. The subterraneous trunk is creeping, and in age of considerable length (Dr. Pappe).

Myrrhis odorata, Scopoli.

The Sweet Chervil or Cicely. Mountains of Middle and Southern Europe and Asia Minor, particularly in forests. A perennial aromatic herb, used for salad and culinary condiments. It could be naturalized in forests, and would endure an alpine climate; a second species, M. occidentalis (Bentham) occurs in Oregon and Utah. Asa Gray keeps this with two additional Californian congeners in the genus Glycosma.

Myrtus acmenoides, F. v. Mueller.

Queensland. The fragrant leaves of this and of M. fragrantissima used locally for flavoring tea, according to Mr. P. O'Shanesy.

Myrtus communis, Linné.

Countries around the Mediterranean Sea. The Bridal Myrtle. This bush of ancient renown should not be passed; it is industrially in requisition for myrtle-wreaths.

Myrtus edulis, Bentham. (*Myrcianthes edulis,* Berg.)

Uruguay. A tree attaining a height of about 25 feet. Berries of 1½ inch diameter, of pleasant taste.

Myrtus incana, Berg.

La Plata-States. A dwarf shrub. The berries can be eaten raw, and are also made into a sweet rich jelly. M. sericea (Cambessedes) is an allied species from the same region.

Myrtus Luma, Molina.

South-Chili. A tree to fully 100 feet high in the virgin-forests. Wood very hard and heavy, much sought for press-screws, wheel-spokes and select implements (Dr. Philippi).

Myrtus Meli, Philippi.

South-Chili. Of the same use as the foregoing species, and in this manner most favorably contrasting with the numerous other myrtaceous trees of Chili.

Myrtus mucronata, Cambessedes.

La Plata-States. A low shrub. The leaves serve locally as a substitute for tea. The berries, which are of the size of muscatel grapes, are of pleasant taste, and therefore consumed by the native inhabitants (Hieronymus).

Myrtus nummularia, Poiret.

The Cranberry-Myrtle. From Chili to Fuegia, also in the Falkland-Islands. This trailing little plant might be transferred to the turfy moors of any alpine mountains. Sir J. Hooker describes the berries as fleshy, sweet and of agreeable flavor. Allied species occur in the cold zone of the Peruvian Andes.

Myrtus tomentosa, Aiton.

India and China. This showy shrub ascends to 8,000 feet. The berries are dark purple, of cherry-size, pulpy and of aromatic sweetness. Various other Myrtles with edible berries are known from different warm countries.

Myrtus Ugni, Molina.

The Chilian Guava. A hardy shrub, freely bearing its small but pleasantly aromatic berries.

Nageia (Podocarpus) amara, Blume.

Java, on high volcanic mountains. A large tree, sometimes to 200 feet high. Timber valuable.

Nageia (Podocarpus) andina, Poeppig. (*Prumnopithys elegans,* Philippi.)

The "Lleuque" of Chili. A stately tree, with clusters of edible cherry-like fruit. As might be expected from its native place, it will bear severe frost—0° F. (Gorlie). The wood is yellowish and fine-grained, and is chosen for elegant furniture-work.

Nageia (Podocarpus) bracteata, Blume.

Burmah, Borneo, Java, up to 3,000 feet. Generally reaching about 80 feet in height, with a straight trunk and horizontal branches. The close-grained wood is highly prized. The allied N. neriifolia from the Himalayas has proved hardy at Melbourne.

Nageia (Podocarpus) Chilina, Richard.

The "Manniu and Lahaul" of Chili and Peru, ascending to sub-alpine elevations. Height reaching 100 feet, with corresponding thickness of stem. Wood white, of excellent quality.

Nageia (Podocarpus) coriacea, Richard.

West-Indies. This tree attains a height of 50 feet, and advances to elevations of 8,000 feet. Other species of both hemispheres should be tested, beyond those here now mentioned.

Nageia (Podocarpus) cupressina, R. Brown.

Java and Philippine-Islands. Height of tree to 180 feet; furnishes a highly valuable timber.

Nageia (Podocarpus) dacrydioides, A. Richard.

In swampy ground of New Zealand; the "Kahikatea" of the Maoris, called White Pine by the colonists. Height to about 150 feet; diameter of stem often 4 feet. The white sweet fruit (fruit-stalklet) is eaten by the natives; the wood is pale, close-grained, heavy; it will not stand exposure to wet, but is one of the best for flooring-boards. The strength is equal to that of "Rimu," according to Professor Kirk; but it is more readily attacked by boring insects.

Nageia (Podocarpus) elata, R. Brown.

East-Australia to 35° S. A fine timber-tree, attaining a height of about 80 feet, with a stem 2 feet in diameter. The timber is soft, close-grained, free from knots, much used for joiners' work, also for spars. Market price in Brisbane £3 5s. to £3 10s. per 1,000 superficial feet (Queensland Exhibition, 1867).

Nageia (Podocarpus) elongata, L'Héritier.

South-Africa. With N. Thunbergi, Erythrina Caffra (Thunberg) and Ocotea bullata (Bentham), this is the tallest tree of Capeland and Caffraria, although it does not advance beyond 70 feet. The yellowish wood is highly valuable, deal-like, not resinous. The stems can be used for top-masts and yards of ships.

Nageia (Podocarpus) ferruginea, D. Don.

Northern parts of New Zealand. The "Black Pine" of the colonists; native name "Miro." Height reaching about 80 feet; it produces a dark-red resin of a bitter taste. The wood is of a reddish color, very hard; will stand exposure to sea-water. Fruit solitary.

Nageia (Podocarpus) Lamberti, Klotzsch.

Southern Brazil. A stately tree, yielding valuable timber.

Nageia (Podocarpus) macrophylla, D. Don.

The "Inou-maki" of Japan. A tree attaining about 50 feet in height. The nut-stalklets used for food there. The wood is white and compact, employed for carpenters' and joiners' work; the bark for thatching (Dupont).

Nageia (Podocarpus) nubigena, Lindley.

Southern Chili, generally a companion of N. Chilina, with which it agrees in its dimensions and the utility of its timber.

Nageia (Podocarpus) Purdieana, Hooker.

Jamaica, at 2,500 to 3,500 feet. This quick-growing tree attains a height of 100 feet.

Nageia (Podocarpus) spicata, Brown.

Black Pine or "Matai" of New Zealand. Fruit spicate. Tree sometimes to 80 feet high; wood pale or reddish, soft, close and durable; used advantageously for piles, machinery, stringers, braces, mill-wrights' work, house-blocks, railway-sleepers, also weatherboards and flooring-boards (Prof. Kirk).

Nageia (Podocarpus Thunbergi, Hooker.

South-Africa. Superior to N. pruinosa (E. Meyer) and even N. elongata in the quality of its wood; it is bright-yellow, fine-grained, and very handsome when polished (Dr. Pappe). Always a smaller tree than N. elongata according to Mr. Will. Tuck.

Nageia (Podocarpus) Totara, D. Don.*

New Zealand. A fine tree, to 120 feet high, with a stem reaching 20 feet in circumference; it is called Mahogany-pine by the colonists. The reddish, close-grained and durable wood is valuable both for building and for furniture; and is also extensively used for telegraph-posts; indeed it is considered the most valuable timber of New Zealand. Chosen for piles of bridges, wharves and jetties and in other naval architecture; the heart-wood resists decay and the attacks of the Teredo for a long time, according to Professor Kirk. It ranks below Kauri in strength, but equals it in durability. It is one of the most lasting woods for railway-sleepers. When used for piles, the bark should not be removed from the timber. Many other tall timber-trees of the genus Podocarpus or Nageia occur in various parts of Asia, Africa and America, doubtless all desirable; but the quality of their timber is not well known, though likely in many cases excellent. Nageia is by far the oldest published name of the genus.

Nardostachys Jatamansi, De Candolle.

Mountains of Bengal and Nepal, at elevations from 11,000 to 17,000 feet (J. Hooker). " The Spikenard." A perennial herb, famous in ancient times as a medicinal plant. The root contains an ethereal oil and bitter principle.

Nastus Borbonicus, Gmelin.

Réunion, where it forms a belt all round the mountains of the island, in a zone of 3–4,000 feet. This beautiful bamboo grows to a height of about 50 feet (General Munro). A second species, namely, N. capitatus (Kunth) occurs in Madagascar.

Nelumbo lutea, Caspary.* (*Nelumbium luteum*, Willdenow.)

The Water-Chinquepin. In Eastern North-America, northward to 44°; also in Jamaica. This magnificent perennial water-plant carries with it the type of Nelumbo nucifera, but seems more hardy, and thus better adapted for extra-tropical latitudes, the Pythagorean Bean not descending in Australia naturally beyond 23°, although this species also may perhaps be able to live in the warmer parts of the temperate zone. The tuberous roots of both species resemble somewhat the Sweet Potato and are starchy; the seeds are of particularly pleasant taste. The plants would be of great value as ornamental aquatics. The leaves of N. lutea are from 1 to 2 feet in diameter. The flower measures ½ to 1 foot across. The capsular fruit contains from twenty to forty nut-like seeds. The plant in congenial spots displaces nearly all other water-vegetation by the vigor of its growth.

Nelumbo nucifera, Gaertner.* (*Nelumbium speciosum*, Willdenow.)

The Pythagorean Bean or Sacred Lotus of the ancients. Egypt, on the Caspian and Aral Seas (46° N.); Persia; through India, where in Cashmere it occurs at an elevation of 5,000 feet; China, Japan; Amur (46° N); tropical Australia as far south as 23°. The occurrence of this grand plant at the Ima, at Pekin and at Astrachan proves sufficiently, that we can naturalize it in moderately cool climes, as has been done already by Marquis Ginoi at Doccia, near Florence.   The plant luxuriates even in New Jersey, where when once established Mr. Sturtevant saw it retaining its vitality while dormant through successive winters underneath ice 6 inches thick.   The Nelumbo requires deep water with a muddy bottom.   The large white or rosy flowers are very fragrant.   The seeds retain their vitality for several years.   According to the ancient Egyptian method, they are placed in balls of muddy clay and chaff, and then sunk into the water.   Not only the very palatable seeds serve for food, but also the stalks, which are used as a vegetable.   According to Moorcroft this plant contributes much to the sustenance of some thousands of people in Cashmere for eight months in the year.

Nepeta Cataria, Linné.

Europe, Western Asia.   The " Catmint."   This somewhat aromatic herb is valued in domestic medicine.   Apiarists praise it also, Quinby stating, that if there was any plant, which he should cultivate , especially, it would be this.

Nepeta Glechoma, Bentham.   (*Glechoma hederacea*, Linné.)

" The Ground-Ivy."   Europe, Western Asia.   This herb is still in great estimation as a pectoral medicine in many parts of Europe. It is also a honey-plant.   Perennial like most congeners.

Nepeta raphanorrhiza, Bentham.

Cashmere and Afghanistan.   The roots of this herb furnish a delicious vegetable (Dr. Aitchison); they have the taste of fresh almonds; large quantities are consumed by the inhabitants of the native countries of this plant.

Nephelium lappaceum, Linné.

South-India and Malay-Islands.   This rather tall tree furnishes the Rambutan- or Rampostan-fruit, similar to the Litchi- and Longan-fruit.   As one species of Nephelium is indigenous as far south as Gippsland (Victoria), and as all the species seem to require rather a moist mild forest-clime than great atmospheric heat, we may hope to bring this tree also to perfect bearing in favorable spots of a temperate clime.

Nephelium Litchi, Cambessedes.

Southern China, Cochin-China and the Philippine-Islands.   An evergreen middle-sized tree, producing the Litchi-fruit.   The pulpy arillus is of extremely pleasant taste, though not large.

Nephelium Longanum, Cambessedes.

India and Southern China.   Height of tree to about 40 feet.   The Longan-fruit is obtained from this tree; it is smaller than that of the Litchi-tree and less palatable.

Neurachne Mitchelliana, Nees.

The Mulga-Grass. In the arid interior of Eastern and South-Eastern Australia. With its companion, N. Munroi (F. v. M.) eligible as a perennial fodder-grass for naturalization in sandy or dry sterile land. It endures any extent of drought, but requires heavy rain to start anew (R. S. Moore).

Nicotiana glauca, Graham.

Argentina and Uruguay. This quickly-growing arborescent species can be raised on mere sand on the coast, as one of the best of plants to establish shelter and stay the shifting of the sand-waves.

Nicotiana multivalvis, Lindley.

The native Tobacco of the regions on the Columbia-River. An annual. This can be utilized for some inferior kinds of tobacco.

Nicotiana Persica, Lindley.

The Shiraz-Tobacco. Persia. Annual. This can be brought to perfection only in cool mountain-regions. The mode of culture is somewhat different from that of the ordinary tobacco. Moderate irrigation is favorable. The plants, when ripe, are cut off and stuck into the ground again until they become yellow; they are then heaped together for a few days in the drying-house; they are afterwards packed into thin strata and placed into bags for pressure and daily turning.

Nicotiana quadrivalvis, Pursh.

The native Tobacco of the Missouri. An annual.

Nicotiana repanda, Willdenow.

Cuba, Mexico, Texas. Annual. It is utilized for some of the Havanna-tobaccoes.

Nicotiana rustica, Linné.

Tropical America. Annual. Some exceptional sorts of East-Indian tobacco, of Manilla-tobacco and of Turkey-tobacco are derived from this particular species.

Nicotiana Tabacum, Linné.*

The ordinary Tobacco-plant of Central America. Annual. The tobacco-plant delights in rich forest-soil, particularly where limestone prevails, on account also of the potassium-compounds, which abound in soils of woodlands, and also because in forest-clearings that atmospheric humidity prevails, which is needful for the best development of the finest kinds of tobacco. Various districts, with various soils, produce very different sorts of tobacco, particularly as far as flavor is concerned; and again, various climatic conditions will greatly affect the tobacco-plant in this respect. We can therefore not hope, to produce for instance Manilla- or Havanna-tobacco in cooler latitudes; but we may expect to produce good sorts of our own also in Australia, more or less peculiar; or we may aspire to producing in our rich and frostless forest-valleys a tobacco similar to that of Kentucky, Maryland, Connecticut and Virginia. Frost is detrimental to the tobacco-plant; not only, particularly when young, must it be guarded against it, but frost will also injure the ripe crop. The scarcity of dew in some of the districts of Australia militates against the production of the best kinds, otherwise the yield as a rule is large, and the soil in many places well adapted for this culture. Leaves of large size are frequently obtained, but the final preparation of the leaf for the manufacturer must be effected by experienced skill. The cruder kinds are obtained with ease, and so are leaves for covering cigars. Virgin soil, with rich loam, is the best for tobacco-culture, and such soil should also contain a fair proportion of lime and potash, or should be enriched with a calcareous manure and ashes, or with well decomposed stable-manure. According to Simmonds the average yield in Greece is about 800 pounds of tobacco per acre. The seedlings, two months or less old, are transplanted. When the plants are coming into flower, the leading top-shoots are nipped off, and the young shoots must also be broken off. A few weeks afterwards the leaves will turn to a greenish yellow, which is a sign that the plants are fit to be cut, or that the ripe leaves can gradually be pulled. In the former case the stems are split; the drying is then effected in barns by suspension from sticks across beams. The drying process occupies four or five weeks, and may need to be assisted by artificial heat. Stripped of the stalks, the leaf-blades are tied into bundles, to undergo sweating or a kind of slight fermentation. It does not answer to continue tobacco-culture beyond two years on the same soil uninterruptedly. A prominent variety is Nicotina latissima, Miller, or *N.* macrophylla, Lehmann, yielding largely the Chinese, the Orinoco and the Maryland tobacco. Latakia-tobacco, according to Dyer, is prepared by submitting the leaves for several months to fumigation from fir-wood. Substances containing cumarin, particularly the Tonka-Bean (Dipterix odorata), are used to flavor tobacco and snuff. The dangerously powerful nicotin (a volatile acrid alkaline oily liquid) and nicotianin (a bitter aromatic lamellar substance) are both derived from tobacco in all its parts, and are therapeutic agents. The tobacco-plant has been grown as far north as lat. 70° 22′ in Norway (Schuebeler). The total quantity of tobacco, entered at the custom-houses for home-consumption in the United Kingdom during 1884 was over 52 million lbs., valued at £2,776,000.

Niemeyera prunifera, F. v. Mueller. (*Lucuma prunifera*, Bentham.)

The Australian Cainito. An evergreen tree, sparingly dispersed from the north of New South Wales through the coast-forests of Queensland. The fruit is of a plum-like appearance and edible. Culture is likely to improve its quality.

Nuphar multisepalum, Engelmann.

Western North-America. This Water-Lily produces nutritious seeds, which taste like Broom-Corn, and are used locally for food, but are more particularly valuable for waterfowl. Various species of

Nymphæa might be utilized in the same manner, irrespective of their value as decorative lake- or pond-plants. The author naturalized the British Water-Lily Nymphæa alba (Camerarius) in Victoria long ago. The very decorative N. gigantea (Hooker) extends naturally in Eastern Australia to 30° S.

Nyctanthes arbor tristis, Linné.

India, up to Assam. This arborescent shrub or small tree (to 30 feet) may be grown in almost any moist regions free from frost, for the exquisite fragrance of its flowers, from which essence of jasmin can be obtained.

Nyssa aquatica, Linné.

The Tupelo or Pepperidge. North-America. This large deciduous tree can be grown in pools and deep swamps, and is thus well adapted for aquatic scenery. The spongy roots serve as a substitute for cork and the floats of nets.

Nyssa multiflora, Wangenheim.

Eastern States of North-America, where it is called the Forest-Tupelo, or Black Gum-tree (Dr. Asa Gray); also called Sour Gum-tree. Attains a height of about 50 feet. Suited for forest-soil; has horizontal branches and a "light, flat spray, like the Beech." Can be propagated from cuttings. The wood is very hard, but light and almost unwedgeable; its serves for hubs of wheels, pumps, side-boards of carts, trays, bowls, dippers, mortars, wooden shoes, hatters' blocks and various turners' work. The foliage turns bright crimson in autumn. The fruits are pleasantly acidulous, like those of N. capitata (Walter) and of some other species, and often used for preserves.

Nyssa uniflora, Walter.

Eastern States of North-America. The Swamp-Tupelo. Wood soft, whitish; particularly adapted for trays, bowls and carving (C. Mohr), that of the roots very light and spongy, hence used for corks (Dr. Asa Gray). A shrub or small tree. The mucilaginous fruits are edible.

Ocimum Basilicum, Linné.

The ".Basil." Warmer parts of Asia and Africa. Will grow in Norway to lat. 63° 26' (Schuebeler). An annual herb, valuable for condiments and perfumery. Several varieties exist, differing considerably in their scent. A crystalline substance is also obtained from this and similar species. O. canum (Sims) is closely allied. Valuable, like many other aromatic Labiatæ, for bees. Seeds will keep for eight years.

Ocimum gratissimum, Linné.

Recorded from India, the South-Sea Islands and Brazil as indigenous. Somewhat shrubby. This is also a scent-plant like the following, and is one of the best of the genus. O. viride (Willdenow), from tropical Africa, seems a variety.

Ocimum sanctum, Linné.

Arabia, India, tropical Australia. A perennial herb. The odor of the variety occurring in North-Australia reminds of anise; the smell of the variety growing in East-Australia resembles that of cloves. O. tenuiflorum, L., seems to be another variety. Probably other species, cis- as well as trans-atlantic, can be used like Basil.

Ocimum suave, Willdenow.

East-Africa. A scrubby species.

Oenanthe Phellandrium, Lamarck.

Europe, Western and Northern Asia. A perennial swamp-plant, the fruitlets of which are of considerable medicinal value.

Olea Europæa, Linné.*

The Olive-tree. South-Western Asia; naturalized in the countries around the Mediterranean Sea. A tree not of great height, but of many centuries' duration and of unabating fecundity. In Corfu however it grows sometimes to a height of 60 feet, and forms beautiful forests. The well-known olive-oil is obtained from the fruit. Certain varieties of the fruit, preserved in vinegar or salt-liquid before perfectly ripe, are also much used for the table. For this purpose the fruit is generally macerated previously in water containing potash and lime. The gum-resin of the olive-tree serves as incense, it contains the crystalline olivil. The oil of the drupaceous fruit is a most important product of countries with a warm temperate climate. Its chemical constituents are: 30 per cent. crystalline palmitin; 70 per cent. olein, for which reason olive-oil belongs to those kinds, which are not drying. In pressing, the kernels must not be crushed, as then a disagreeable taste will be imparted to the oil. The wild variety of the olive-tree usually has short blunt leaves and thorny branches. Long-continued droughts, so detrimental to most plants, will affect the olive but slightly. It thrives best on a free, loamy, calcareous soil, even should it be strong and sandy, but it dislikes stiff clay. Proximity to the sea is favorable to it, and hill-sides are more eligible for its culture than plains. The ground must be deeply trenched. Manuring with well-decayed substances is requisite annually or every second and third year, according to circumstances. Irrigation will add to the productiveness of the plant. Captain Ellwood Cooper, of Santa Barbara, Southern California, obtained from orchards 10 years old sufficient fruit for 700 gallons of olive-oil to the acre, one-fourth of the produce paying for the expenses of preparing the soil, gathering the crop, pressing the oil and conveying it to market. Mons. Riordet distinguishes three main varieties, of which he recommends two: 1. The Cayon, a small-sized tree, which comes into bearing after three or four years, but bears fully only every second year; its oil is fine with some aroma. 2. The Pendulier, a larger tree, with long drooping branches, yielding an oil of first-rate quality. Mons. Reynaud, " Culture de l'Olivier," separates twelve varieties, as cultivated in

France, and recommends among them : 1. The Courniau or Courniale, also called Plant de Salon, bearing most prolifically a small fruit and producing an excellent oil. 2. Picholine, which by pruning its top-branches is led to spread over eight yards square or more; it is of weeping habit, yields a good oil in fair quantity, and resists the attacks of insects well. 3. The Mouraou or Mourette, a large tree also furnishing oil of a very fine quality. Olive-trees require judicious pruning immediately after the fruit is gathered, when the sap is comparatively at rest. They may be multiplied from seeds, cuttings, layers, suckers, truncheons and old stumps, the latter to be split. They can also be propagated from protuberances at the base of the stem, which can be sent long distances (Boothby). Tne germination of the seeds is promoted by soaking the nutlets in a solution of lime and wood-ash. The seedlings can be budded or grafted after a few years. Truncheons or estacas may be from one to many feet long and from one to many inches thick; they are placed in the ground horizontally. Some Olive-plantations at Grasse are worth from £200 to £250 per acre. For many details the tract on the "Culture of Olive and its Utilization," issued in Melbourne by the Rev. Dr. Bleasdale, should be consulted, as it rests largely on its author's observations during a long stay in Portugal; also the essay of Sir Samuel Davenport in Adelaide, and the treatise recently issued by Capt. Ellwood Cooper in San Francisco.

The following notes are derived from the important "Tratado del Cultivo del Olivo en Espana," by the Chev. Capt. Jose de Hidalgo-Tablada (second edition, Madrid, 1870). The olive-tree will resist considerable frost (5° F.) for a short time, provided that the thawing takes place under fogs or mild rain (or perhaps under a dense smoke). It requires about one-third more annual warmth than the vine for ripening its fruit. The Olive-zones of South-Europe and North-Africa are between 18° and 44° north latitude. An elevation of about 550 feet corresponds in Spain, as far as this culture is concerned, to one degree further north. Olives do not grow well on granitic soil. The fruit produced on limestone-formations is of the best quality. Gypsum promotes the growth of the tree. An equable temperature serves best; hence exposure to prevailing strong winds is to be avoided. The winter temperature should not fall below 19° F. The quantity of oil in the fruit varies from 10 to 20 per cent.; sometimes it even exceeds the latter proportion. In the Provence an average of 24 lbs. of olive-oil are consumed by each individual of the population annually; in Andalusia, about 30 lbs. G. Don mentions an aged tree near Gerecomis to have provided olives for 240 quarts of oil in one year. For obtaining the largest quantity of oil the fruit must be completely ripe. Hand-picked olives give the purest oil. Knocking the fruit from the branches with sticks injures the tree and lessens its productiveness the next year. Spain alone produces about 250,000,000 lbs of olive-oil a year. The imports of olive-oil into the United Kingdom in 1884 amounted to 17,000 tons, valued at £715,000; in 1883 the quantity was 31,000 tons, worth £1,194,000.

SPANISH VARIETIES.

A.—Varieties of early maturation, for colder localities:—

1. Var. *Pomiformis*, Clem.
   Manzanillo. (French: Ampoulleau.) Fruit above an inch in diameter, spherical, shining-black. Putamen broad and truncate.
2. Var. *Regalis*, Clem.
   Sevillano. (French: Pruneau de Catignac.) Fruit about an inch
   in diameter, ovate-spherical, blunt, bluish-black.
3. Var. *Bellotudo* or *Villotuda*.
   Fruit about an inch long, egg-shaped; pericarp outside dark-red,
   inside violet.
4. Var. *Redondillo*.
   Fruit ovate-spherical, nearly an inch long. Pericarp outside bluish-
   black, inside whitish. A rich yielder.
5. Var. *Ovalis*, Clem.
   Lechin, Picholin, Acquillo. (French: Saurine.) Fruit broad-oval,
   two-thirds of an inch long. A copious yielder.
6. Var. *Argentata*, Clem.
   Nevadillo blanco; Doncel; Zorzalena; Moradillo; Ojiblanco; Olivo
   lucio. Fruit broad-ovate, an inch long, very blunt, not oblique.
   Quality and quantity of oil excellent.
7. Var. *Varal blanco*.
   (French: Blanquette.) Fruit ovate-globular, three-fourths of an
   inch long, neither pointed nor oblique, outside blackish-red.
8. Var. *Empeltre*.
   Fruit ovate, an inch long, equable. Rich in oil of excellent quality;
   also one of the best sorts for pickles. Pericarp outside violet, inside
   whitish.
9. Var. *Racimal*.
   (French: Bouteillan, Boutiniene, Ribien, Rapugette.) Fruit violet-
   colored, globose-ovate, about an inch long; neither pointed nor
   oblique. Bears regularly also on less fertile soil, and is one of the
   earliest to ripen.
10. Var. *Varal negro*.
    Alameno. (French: Cayon, Nasies.) Fruit violet-black, spotted, glo-
    bose-ovate, nearly an inch long, somewhat pointed. Bears richly.
11. Var. *Colchonuaa*.
    Fruit spherical, outside red, inside whitish, an inch in diameter,
    slightly pointed. Produces a large quantity of good oil.
12. Var. *Ojillo de Liebre*.
    Ojo de Liebre. Fruit nearly spherical, outside violet-black, about
    one inch long, somewhat oblique. One of the less early varieties.
13. Var. *Carrasquena*.
    (French; Redouan de Cotignat.) Fruit black-red, almost spherical,
    about an inch long. Valuable both for oil and preserves, but
    liable to be attacked by various insects.

14. Var. *Hispalensis,* Clem.
Gordal; Ocal; Olivo real. Fruit black-grey, oblique, spherical, slightly oblique, measuring about au inch. Rather large and quick-growing tree. Fruit used in the green state for preserves, not used for table-oil.

15. Var. *Verdego.*
Verdial. (French: Verdal, Verdan.) Fruit black-violet, oblique, spheric, pointed, about one inch long. Furnishes good oil and resists the cold best of all.

B.—Varieties of late maturition, for warmer localities:—

16. Var. *Maxima,* Clem.
Madrileno; Olivo morcal. Fruit over an inch long, cordate-globose, strongly pointed. Less valuable for oil than for preserves.

17. Var. *Rostrata,* Clem.
Cornicabra. (French: Cournaud, Corniaud, Courgnale, Pl. de Salon, Pl. de la Fane; Cayon Rapunier, Grasse.) Strong and tall, less tender; Fruit blackish-red, over an inch long, oval, much pointed. Good for oil.

18. Var. *Ceratocarpa,* Clem.
Cornezuelo. (French: Odorant, Luquoise, Luques.) Fruit fully an inch long, oval, pointed. ·

19. Var. *Javaluno.*
Fruit black-grey, over an inch long, egg-shaped, somewhat oblique, gradually pointed. Rich in good oil; can also be chosen for preserves; much subject to attacks of insects.

20. Var. *Picudo.*
Fetudilla. Fruit fully an inch long, egg-shaped, blunt at the base, pointed at the apex, with black-grey pulp. Pericarp easily separable. Employed both for oil and preserves.

21. Var. *Nevadillo negro.*
Fruit egg-shaped, fully an inch long, with turned pointed apex. One of the richest of all varieties in yield. Endures considerable cold, and is not late in ripening.

All these Spanish varieties show rather long, lanceolate leaves of more or less width.

FRENCH VARIETIES.

(Some verging into the Spanish kinds.)

22. Var. *Angulosa,* Gouan.
Galliningue, Laurine. For preserves.

23. Var. *Rouget.*
Marvailletta. Produces a fine oil.

24. Var. *Atrorubens,* Gouan.
Salierne, Saverne. Fruit dusted white. Furnishes one of the best of oils.

25. Var. *Variegata*, Gouan.
Marbrée, Pigale, Pigau.   Purple fruit, with white spots.

26. Var. *Le Palma*.
Oil very sweet, but not largely produced.

27. Var. *Atrovirens*, Ros.
Pointue, Punchuda.   Fruit large, with good oil.

28. Var. *Rubicans*, Ros.
Rougette.   Putamen small.   Yield annual and large.

29. Var. *Alba*, Ros.
Olive blanche, Blancane, Vierge.   This, with many others omitted on this occasion, is an inferior variety.

30. Var. *Caillet rouge*.
Figanier.   Small tree.   Fruit large, red.   Oil good and produced in quantity.

31. Var. *Caillet blanc*.
Fruit almost white, produced annually and copiously, yielding a rather superior oil.

32. Var. *Raymet*.
Fruit large, reddish.   Oil copious and fine.   This variety prefers flat country.

· 33. Var. *Cotignac*.
Pardigniere.   Fruit middle-sized, blunt.   Oil obtained in quantity and of excellent quality.   This requires much pruning.

34. Var. *Bermillaon*.
Vermillon.   Yields also table-oil and resists cold well

This list was several years ago without permission copied into an official publication in another part of the globe, also without any allusion to Capt. Hidalgo-Tablada or the translator.

Many other apparently desirable varieties occur, among which the Italian Oliva d'Ogni Mese may be mentioned, which ripens fruit several times in the year, and furnishes a pleasant oil and also fruit for preserves.

Oncosperma fasciculatum, Thwaites.

Ceylon.   This Palm ascends there to 5,000 feet.   The very slender but prickly stem attains a height of 50 feet.   Desirable for scenic-culture.

Onobrychis sativa, Lamarck.*

The " Sainfoin, Esparsette or Cock's-head " Plant.   Southern and Middle Europe, South-Western and Middle Asia.   Hardy in Norway to lat. 63° 26' (Schuebeler).   A deep-rooting perennial fodder-herb, fond of marly soil, and living in dry localities.   It prepares dry calcareous soil for cereal culture.   Stagnant underground-humidity is fatal to this plant.   It prospers even, where Red Clover and Lucerne no longer succeed, and is richer in nutritive constituents than either, as shown already by Sir Humphrey Davy.   Sheep cannot be turned

out so well on young Sainfoin-fields as cattle. The hay is superior
even to that of Lucerne and Clover. The plant will hold out from
five to seven years (Langethal). It yields much honey for bees. O.
montana (De Candolle) is a dwarfer sub-alpine variety of limestone-
regions.

Onosma Emodi, Bentham. (*Maharanga Emodi*, A. de Candolle.)

Nepal. The root, like that of the Alkanna tinctoria, produces a
red dye.

Ophiopogon Japonicus, Ker. (*Flueggea Japonica*, Richard.)

The mucilaginous tubers can be used for food, a remark, which
applies to many other as yet disregarded allied plants.

Opuntia coccinellifera, Miller.

Mexico and West-Indies. The Cochineal-Cactus. On this and
O. Tuna, O. Hernandezii and perhaps a few others subsists the Coccus, which affords the costly cochineal-dye. Three gatherings can
be effected in the year. About 1,200 tons used to be imported annually into Britain alone, and a good deal to other countries, valued
at about £400 per ton. The precious carmin-pigment is prepared
from cochineal. Different Cochineal-Opuntias occur in Argentina
also. Some species of Opuntia will endure a temperature of 14° F.;
one even advances to 50° north latitude in Canada. Mr. Dickinson
observes, that many species are hardy at Port Phillip, growing even
in sand, overtopping by 10 feet the Leptospermum lævigatum, and
breaking it down by their great weight within a few yards of the
sea.

Opuntia Dillenii, De Candolle.

Central America. A Tuna-like Cactus, serving for uninflammable
hedges, and perhaps also for the rearing of the *Coccus Cacti*. It is
particularly eligible for barren land, but apt to stray beyond bounds in
hot countries.

Opuntia elatior, Miller.

Central America. A hedge-plant with formidable thorns.

Opuntia Ficus Indica, Miller.

Called inaptly, with other congeners, Indian Fig. Central America,
north as far as Florida. Serves for big hedges. Pulp of fruit
edible. Exudes a gum, somewhat like Tragacanth.

Opuntia Hernandezii, De Candolle.

Mexico. Also affords food for the *Coccus Cacti*.

Opuntia Missouriensis, De Candolle.

From Nebraska to New Mexico. Very hardy. Professor Meehan
found this Cactus covered with the Cochineal Coccus, and points to
the fact, that this insect will live through the intense cold, which
characterizes the rocky mountains of the Colorado-regions.

Opuntia Rafinesquii, Engelmann.

The Prickly Pear. North-America. The most northern of all species, extending to Lake Michigan. It resists severe frosts, as do also O. brachyantha, O. Comanchica, O. humilis (Mayer), O. Whipplei, O. oplocarpa, O. arborescens and Mammillaria Missouriensis (Loder, Meehan).

Opuntia spinosissima, Miller.

Mexico and West-Indies. Stem columnar, with pendent branches. Also a good hedge-plant. Harding recommends for hedges, besides these species, O. maxima (Miller) as the most repellent.

Opuntia Tuna, Miller.

West-Indies, Ecuador, New Granada, Mexico. Irrespective of its value as the principal cochineal-plant, this Cactus is also of use for hedges. It will attain a height of 20 feet. The pulp of the fruit is edible. With many other species hardy anywhere in Australia down to the south-coast. Of Cochineal Great Britain imported in 1884 14,100 cwt., value £80,000.

Opuntia vulgaris, Miller.

Central America, northward to Georgia, southward to Peru. Very hardy. Adapted for big hedges, and like the rest not inflammable, hence particularly valuable along railway-lines. The fruit almost smooth, eatable. A dye can also be prepared from its pulp and that of allied species. Numerous other species are industrially eligible for hedging purposes, but sometimes spreading beyond control.

Oreodoxa frigida, Humboldt.

Central America, ascending the Andes to 8,500 feet. This dwarf slender Palm may be chosen for domestic decoration.

Oreodoxa oleracea, Martius.

West-Indies, up to nearly 5,000 feet elevation. One of the most rapid growing of all Palms, rising to a height of 120 feet. In highly manured moist ground the Palm-cabbage, which in this species is of exquisite nut-flavor, can be obtained in two years (Imray, Jenman), should ever such a culture become desirable. Hardy in Florida (B. Smith).

Oreodoxa regia, Humboldt.

West-Indies. This noble Palm attains a height of 60 feet. It has proved hardy in Southern Brazil. The stem is thickened at the middle, and from it, as from that of O. oleracea, starch can be obtained.

Origanum Dictamnus, Linné.

Candia. Like the following, a scent-plant of somewhat shrubby growth.

Origanum Majorana, Linné.

North-Africa, Middle Asia, Arabia. A perennial herb, used for condiments, also for the distillation of its essential oil, much employed in French factories of scented soap (Dr. Piesse). In Norway it will grow to lat. 70° 22′ (Schuebeler).

Origanum Maru, Linné.

Palestine. Perennial and very odorous.

Origanum Onites, Linné.

Countries at and near the Mediterranean Sea. Somewhat shrubby and strongly scented.

Origanum vulgare, Linné.

The ordinary Marjoram. All Europe, North-Africa, Northern and Middle Asia. In Norway it is indigenous in lat. 66° 16′ (Schuebeler). A scented herb of perennial growth, containing a pleasant volatile oil. It prefers limestone-soil. Of importance also as a honey-plant. O. hirtum, Link, O. virens, Hoffmannsegg and O. normale, D. Don, are closely allied plants of similar use. Several other Marjorams, chiefly Mediterranean, are of value. Their seeds maintain vitality for a few years.

Ornithopus sativus, Brotero.

South-Europe and North-Africa. " The Seratella or Serradella." An annual herb, larger than the ordinary Bird's-foot clover, O. perpusillus, L. It is valuable as a fodder-plant on sterile particularly sandy soil. It requires no lime, but improves in growth on gypsum-land. A good honey-plant. It matures seeds near Christiania (Schuebeler). Has done particularly well in Hawaia.

Oryza latifolia, Desvaux.

Wild in Central America, but perhaps of Asiatic origin. This species is said to be perennial and to attain a height of 18 feet. It deserves trial-culture, and may prove a good fodder-grass on wet land in warm localities. O. perennis (Moench) seems closely allied. Bentham and J. Hooker are not inclined, to admit more than one species of Oryza. The present one is however maintained by Grisebach.

Oryza sativa, Linné.*

The Rice-Plant. South-Asia and North-Australia. Annual like most cereals. Many rivulets in ranges afford ample opportunities for irrigating rice-fields; but these can be formed with full advantage only in the warmer parts of extra-tropic countries, where rice will ripen as well as in Italy, China or the Southern States of the American Union. Among the numerous varieties of Indian rice may, be noted as prominent sorts: The Early Rice, which ripens in four months and·is not injured by saline inundations; the hardier Mountain-Rice, which can be raised on comparatively dry ground, and which actually perishes under lengthened inundation, but which is less productive; the Glutinous Rice, which succeeds as well in wet as in almost dry places, and produces black or reddish grains. In the rich plains of Lombardy, irrigated from the Alps, the average-crop is estimated at forty-eight bushels for the acre annually. According to General Capron the average-yield in Japan is fifty bushels per acre. The spirit, distilled from rice and molasses, is known as arrack. Rice-beer, known as " Sake," is extensively brewed in Japan, and is the principal fermented beverage used by the inhabitants. Rice-starch is now consumed in enormous quantities, particularly in Britain. Nearly 330,000 tons of rice to the value of £2,680,000 were imported into the United Kingdom during 1884. Rice-sugar, called "Ame" in Japan, constitutes there a kind of confectionery.

Oryzopsis cuspidata, Bentham.

South-Western parts of North-America. A perennial grass of easy dissemination. Tufts dense, hence one of the Bunch-grasses; thrives on soil too sandy and too dry for more valuable grasses (Dr. Vasey).

Oryzopsis panicoides, Bentham. (*Piptochætium panicoides*, E. Desvaux.)

Extra-tropical South-America. This with some congeners affords good pasturage in Chili and the La Plata-States (Hieronymus).

Osmanthus fragrans, Loureiro.

China and Japan. The flowers of this bush serve for oil-distillation like those of the Jasmine. The scent of one plant will perfume a whole conservatory (G. W. Johnson).

Osmitopsis asteriscoides, Cassini.

South-Africa. A camphor-scented shrub, much in use there for medicinal purposes (Dr. Pappe).

Ostrya carpinifolia, Scopoli.

" The Hop-Hornbeam." South-Europe and Orient. A deciduous tree, reaching 60 feet in height. Uses much like those of the following.

Ostrya Virginica, Willdenow.

"Lever-wood" Tree of Eastern States of North-America, also Mexico. Occasionally called Iron-wood; to 40 feet high, in rich woodlands. Wood singularly hard, close-grained and heavy, in use for levers, mill-cogs, wheels, mallets, wedges and other implements. Cattle browse on the foliage. The growth of the tree is very slow.

Osyris compressa, A. de Candolle.

South-Africa. One of the most valuable tans for finer leathers is provided there by the leaves and young twigs of this shrub or small tree. The bloom obtained from this tan is much like that imparted by Sumach.

Owenia venosa, F. v. Mueller.

Queensland; called locally Sour-Plum. A tree, approaching finally 40 feet in height, furnishing a wood of great strength. O. acidula, F. v. M., the "Rancouran," is a handsome tree, 50 feet high, with close-grained, nicely marked wood. Culture might improve the fruits.

Oxalis crassicaulis, Zuccarini.

Peru. This seems one of the best of those Wood-Sorrels, which yield a tuberous edible root. Amongst others, O. tuberosa (Molina) and O. succulenta (Barnsaud) from Chili, as well as O. carnosa (Molina) and O. conorrhiza (Jacquin) from Paraguay, might be tried for their tubers.

Oxalis crenata, Jacquin.

Peru and Bolivia; there the tubers largely consumed; they lose their acidity by being exposed to the sun, becoming sweet and containing a good deal of starch (Vilmorin).

Oxalis Deppei, Loddiges.

Mexico. The tubers of this Wood-Sorrel resemble small parsnips, and are not at all acid. The plant undergoes regular cultivation in some parts of its native country, and succeeds well in the south of England (Chambers). In Prof. Meehan's *Gardeners' Monthly,* August 1884, an Oxalis is mentioned as cultivated in California, which produced as much as 150 tubers in a season, their form being pear-like, from one original root.

Oxalis esculenta, Otto and Dietrich.

"Spurious Aracacha." Mexico, there with the preceding species and O. tetraphylla (Cavanilles), O. violacea (Linné) and several others producing tuberous, starchy, wholesome roots; the first-mentioned gives the largest yield. Propagated by subdivision of the root-stock. It requires a deep, rich, moist soil. In Norway it can be grown to lat. 70° (Schuebeler). As similarly useful, may be mentioned among many others, O. crenata (Jacquin) from Chili and O. enneaphylla (Cavanilles) from the Falkland-Islands and Magelhaen's Straits.

Oxytropis pilosa, De Candolle. (*Astragalus pilosus*, Linné.)

Europe, West-Asia. This perennial plant furnishes fair pasture-herbage; it is deep-rooted and content with almost absolute sand; the numerous other species—24 alone enumerated as Oriental by Boissier—should be tested. All these plants might be classed as Astragals. They as a rule are satisfied with poor soil.

Pachyma Cocos, Fries.

The Tuckahoe-Truffle or Indian Bread. North-America and East-Asia.

Pachyma Hœlen, Fries.

China. This large Truffle occurs particularly in the province of Souchong. Flavor most agreeable. Naturalization elsewhere to be tried.

Pachyrrhizus angulatus, Richard.

From Central America rendered spontaneous in many tropical countries. A climber, the horizontal starchy roots of which attain a length of 8 feet and a thickness of many inches. Dr. Peckolt records tubers of 70 lbs. weight. They keep, in dry ground, growing for five years, but such are then available only for starch, whereas annual tubers are the most palatable and yield 6 to 7 per cent. of starch. From the stems a tough fibre is obtained. The plant proved hardy at Sydney; it requires rich soil.

Paliurus ramosissimus, Poiret. (*P. Aubletia,* Schultes.)

China and Japan. A thorny tree, which could be utilized for hedging.

Paliurus Spina Christi, Miller. (*P. aculeatus,* Lambert.)

The Christ-Thorn. From the Mediterranean Sea to Nepal. A deciduous bush or finally tree, which can be trimmed into hedges.

Pandanus furcatus, Roxburgh.

This Screw-Pine occurs in India, up to heights of 4,000 feet, according to Dr. S. Kurz; hence it will be likely to bear a temperate clime, and give a stately plant for scenic group-planting. P. pedunculatus, R. Br., occurs in East-Australia as far south as 32°, and an allied tall species (P. Forsteri, Moore and Mueller) luxuriates in Howe's Island.

Panicum agrostoides, Muehlenberg.

North-America. One of the hardiest species, bearing the winter cold of New York. Can be utilized for muddy banks and undrained marshy meadows. Easily disseminated, forming large tall clumps, flowering as well from the joints as top; yields abundance of hay, but must be cut while young (Dr. Vasey).

Panicum altissimum, G. Meyer. (*P. elatius,* Kunth.)

From Mexico to Brazil. An almost woody species of arborescent habit, attaining a height of 30 feet. Panicles sometimes a foot and a half long. Evidently desirable for naturalization.

Panicum amarum, Elliot.

North-America. A perennial species, fit to be grown on drifting coast-sand.

Panicum atro-virens, Trinius. (*Isachne Australis,* R. Brown.)

South-Asia, East-Australia and New Zealand. A perennial grass, not large, but of tender nutritive blade, particularly fitted for moist valleys and woodlands.

Panicum barbinode, Trinius.

Brazil. Valuable as a fodder-grass.

Panicum brizanthum, Hochstetter.

From Abyssinia to Nepal. A large-grained perennial Millet-Grass.

Panicum coenicolum, F. v. Mueller.

Extra-tropic Australia. Valuable as an enduring grass for moist meadows.

Panicum compositum, Linné. (*Oplismenus compositus*, Beauvois.)

South-Asia, East-Australia, Polynesia, New Zealand. The growth of this soft-bladed and prolific grass should be encouraged in forest-ground.

Panicum Crus Galli, Linné.

The "Barnyard- or Cockshin-Grass." Occurring now in all warm countries, but probably of Oriental origin, as it seems not recorded in our ancient classic literature. Apparently spontaneous in North-Western Australia. A rich but annual grass of ready spontaneous dispersion, particularly along sandy river banks, also around stagnant water. P. colonum, L., and P. Crus Corvi, L., are varieties of it. Regarded by R. Brown as indigenous in Eastern and Northern Australia, where many other excellent fodder species occur, some perennial. It will succeed also on somewhat saline soil, particularly on brackish water-courses, likewise on moor-land. For rural rearing the short-awned variety should be chosen. On the lower Mississippi it has furnished as much as four or even five tons of hay from one acre. Cows and horses are very fond of this grass, whether fresh or dry (Professor Phares).

Panicum decompositum, R. Brown. (*P. lævinode*, Lindley.)

The Australian Millet. One of the most spacious of Australian nutritious grasses. The aborigines convert the small millet-like grains into cakes. It is the only grain stored by the nomads of Central Australia. This grass will thrive on poor soil with Eleusine cruciata (Lam.), coming after rains in one month to maturity in the torrid regions of Central Australia (Rev. H. Kempe). Hardly different from the North-American P. capillare, L., except in perennial roots. The allied P. trachyrrhachis (Bentham) from North- and East-Australia also constitutes a very good pasture-grass. Of similar value the exclusively Australian P. effusum, R. Br., and P. melananthum, F. v. M.

Panicum distichum, Lamarck. (*P. pilosum*, Swartz.)

Tropical Asia, Africa and America, Polynesia. This perennial grass is mentioned by Kurz among those yielding grain for human food in India.

Panicum divaricatissimum, R. Brown.

Australia, particularly in the warmer inland-regions. A good perennial grass, of easy growth on poor soil.

Panicum divaricatum, Linné.   (*P. bambusoides,* Hamilton.)

Central and Southern America. A grass of scandent habit, ascending high up in trees; desirable for naturalization in forests.

Panicum enneaneurum, Grisebach.

La Plata-States.   Prof. Hieronymus mentions this along with P. grumosum (Nees), P. laxum (Swartz), P. oblongatum (Grisebach) and P. rivulare (Trinius) as rendering the Pampas-pastures so nutritive.

Panicum flavidum, Retzius.

Southern Asia, tropical and Eastern sub-tropical Australia. A prolific seed-bearer, often prostrated by the weight of the seeds.

Panicum fluitans, Retzius.

Tropical Asia and Africa. This perennial grass, like P. spinescens (R. Brown) of East-Australia, ought to be naturalized along lakes, lagoons and rivers, particularly for the benefit of waterfowl.

Panicum foliosum, R. Brown.

India, East-Australia. Perennial. Mr. Bailey finds this to be one of the best grasses for river-banks.

Panicum frumentaceum, Roxburgh.

The "Shamalo- or Deccan-grass." Probably introduced from tropical Africa into South-Asia. A hardy summer-grass, having matured seeds even at Christiania (Schuebeler). It serves as a fodder-grass and produces also a kind of millet. In warm moist climes it ripens grains in 1½ months from the time of sowing. The grain much recommended by Mr. C. B. Taylor for culinary purposes.

Panicum Italicum, Linné.*   (*Setaria Italica,* Beauvois.)

This grass, notwithstanding its specific name, is of Indian origin, ascending the Himalayas to 6,500 feet, extending to China and Japan. It even ripens in cold climes, its seeds coming to perfection as far north as Christiania (Schuebeler). Reared in Switzerland since prehistoric ages; one of the five kinds of plants sown ceremoniously each year by the Emperor of China, according to an Imperial custom initiated 2,700 years before the Christian era (A. de Candolle). It is annual, attaining a height of 5 feet, and is particularly worthy of cultivation as a tender green fodder. It keeps weeds down, and is one of the most valuable of soiling plants; withstands drought well; yields early in the season a heavy crop of excellent hay, which dries easily (C. Mohr). The abundantly produced grain is not only one of the best for poultry, but that of some varieties can be utilized as millet. Considered by many a delicious grain for cakes and porridge.

The Brahmins hold it in higher esteem than any other grain (Dr. Ainslie); called in many places "Hungarian Millet." P. German-icum (Roth) is a form of this species. Allied is also the West-Indian Panicum (Setaria) magnum (Grisebach), which attains a height of 10 feet on margins of lagoons; and Panicum macrostachyum (Nees) of East-Australia, South-Asia and tropical America, the latter highly praised by Mr. R. L. Holmes in Fiji and by Prof. Hieronymus in Argentina.

Panicum Koenigii, Sprengel. (*P. Helopus,* Trin.)

Tropical and sub-tropical Africa, Asia and Australia. A good fodder-grass.

Panicum latissimum, Mikan.

Brazil. A highly ornamental grass. Leaves extremely broad, but hard; panicle very rich.

Panicum maximum, Jacquin.* (*P. Jumentorum,* Persoon.)

The Guinea-grass. Tropical Africa; elsewhere not indigenous. This perennial grass attains a height of 8 feet. It is highly nutritious and quite adapted for the warmer temperate zone, being hardy as far south as Buenos Ayres. In Jamaica it is the principal fodder-grass up to elevations of 5,000 feet, springing up over wide tracts of country almost to the exclusion of everything else. It forms large bunches, which when cut young supply a particularly sweet and tender hay; throws out numerous stolons; can be mown every six weeks; the roots can be protected in the ground against light frosts by a thin covering with soil. A favorite grass in tropical countries for stall-fodder. It is necessary, to guard against over-feeding with this grass solely. Succeeds even on poor clay-soil and on sea-sand. P. bulb-osum (Kunth) is a more hardy grass of the southern parts of North-America, regarded as a variety by General Munro, but remarkable for its thickly enlarged roots.

Panicum melanthum, F. v. Mueller.

Eastern Australia. A valuable perennial grass for pastoral purposes (Bailey).

Panicum miliaceum, Linné.* (*P. miliare,* Lam.)

The true "Millet." South-Europe, North-Africa, South-Asia, ascending the Himalayas to 11,000 feet, North-Australia. Cultivated in Southern Europe as early as the times of Hippocrates and Theo-phrastus; in Egypt prior to historic records, and in Switzerland during the stone-age. Annual, attaining a height of 4 feet. Several varieties occur, one with black grains. They all need a rich and friable soil, also humidity. Maturation very quick. It is one of the best of all grains for poultry, but furnishes also a palatable and nutritious table-food. It ripens even in Christiania (Schuebeler). In mild countries as much as 70 bushels of seeds have been harvested from an acre of land well cultivated with this grass (Ch. Flint).

R

Panicum molle, Swartz.* (*P. sarmentosum,* Roxburgh.)

Warmer parts of America, Africa and Asia. The Para-grass. A perennial, very fattening pasture-grass, of luxuriant growth, attaining a height of 6 feet (Grisebach). It is hardy at the Cape of Good Hope and other far extra-tropic regions.

Panicum myurus, Lamarck

Tropical Asia and America, North-Eastern Australia. A perennial aquatic grass, with broad-bladed foliage, fit for ditches and swamps. Regarded by Mr. Bailey as very palatable and nutritious to stock.

Panicum obtusum, Humboldt.*

The Mosquito- or Mezquite-grass of Mexico. Perennial, nutritious; has strong running stems, which at distances of two or three feet take root, and send up leafy bunches. Dr. Vasey thinks that this grass will show great endurance in droughty seasons.

Panicum parviflorum, R. Brown.

East-Australia. On dry hills a fine pasture-grass. P. bicolor and P. marginatum, R. Br., are likewise enumerated by Mr. Bailey among the nutritious grasses of East-Australia.

Panicum proliferum, Lamarck.*

Southern parts of North-America. Recorded by Steudel as perennial, by Chapman and others as annual. Evidently one of the hardier species, particularly eligible for wet brackish ground (Prof. A. Gray). Vegetates luxuriantly in the hottest part of the summer, the stems lengthening sometimes to 7 feet, soon bending and then rooting from the lower joints, throwing out numerous shoots from them, which grow rapidly, allowing of repeated cutting; stems thick, succulent, sweetish; panicles to 2 feet long. Through all stages of its growth this grass is much relished by horses and cattle (Dr. C. Mohr).

Panicum prolutum, F. v. Mueller.

South-Eastern Australia. Flourishes in the hottest weather; bears a large panicle of seed.

Panicum prostratum, Lamarck. (*P. setigerum,* Retzius.)

Egypt, South-Asia, North-Australia, perhaps also indigenous to tropical America. Perennial. Recommendable for pastures.

Panicum pygmæum, R. Brown.

East-Australia. Forms a soft, thick, carpet-like verdure in forest-shade (Bailey).

Panicum repens, Linné.

Near the Mediterranean Sea, also in South-Asia and North-Australia. Regarded by the Cingalese as a good fodder-grass. It is perennial and well suited for naturalization on moist soil, river-banks or swamps.

Panicum roseum, Steudel. (*Tricholæna rosea*, Nees.)

South-Africa. This perennial pretty grass promises to become with others of the section Tricholæna valuable for meadows in mild climes. It gets about 2 feet high. Mr. Danger counted nearly 300 stems on one plant in Gippsland.

Panicum sanguinale, Linné.

From Middle and South-Europe, Northern Africa, and Southern Asia, spread through all countries with a warm climate, but apparently also indigenous in North-and East-Australia. This is the "Crab-grass" of the Southern United States, where according to Mr. Hagenauer it is recognized as one of the most useful of all pasture-grasses; in Fiji it is also considered the best grass for pastures according to Mr. Holmes. It accommodates itself to swampy and shady places, readily disseminates itself on barren ground, and is likely to add to the value of desert pastures, although it is annual. Stock relish this grass. P. ciliare (Linné) and P. glabrum (Gaudin) are allied. Colonel Howard of Georgia says of the Crab-grass and Bermuda-grass, that they will live in spite of neglect, but when petted will make such grateful returns, as to astonish their benefactor.

Panicum semialatum, R. Brown.

Warmer regions of Asia, Africa and Australia. A superior tall pasture-grass, of easy dispersion in warm humid localities.

Panicum spectabile, Nees.*

The "Coapim" of Angola. From West-Africa transferred to many other tropical countries. A rather succulent, very fattening grass, famed not only in its native land, but also long since in Brazil. This grass, which was with the help of the great Kew establishment first obtained by the author for Australia and Polynesia, is according to Mr. R. L. Holmes "the wonder of all beholders in Fiji, strangling by its running roots almost everything in its course; at its original starting point forming a mass of the richest green foliage, over 6 feet high, gradually lowering to the outer border, where a network of shoots or runners is covering the ground; it roots at the joints, and sends up then a mass of the softest and most luscious fodder." In Fiji it runs over the ground at the rate of ten feet in three months. Readily propagated by pieces of the procumbent stem, which roots freely at each joint. Requires to be well fed down. It may be assumed, that at present about 300 well-defined species of Panicum are known, chiefly tropical and sub-tropical; very few extending naturally to Europe or the United States of North-America, Japan or the southern part of Australia. Though mostly from the hot zones, these grasses endure a cooler clime in many instances, and some of them would prove great acquisitions, particularly the perennial species. Numerous good kinds occur spontaneously in Queensland and North-Australia. Panicum is the genus richest in species among grasses.

Panicum striatum, Lamarck. (*P. gibbum*, Elliott.)

Southern States of North-America, West-Indies and Guiana. A perennial grass for swampy localities, valuable for pastoral purposes, according to C. Mohr, who mentions also P. anceps, L., and P. hians, Elliott, as good fodder-grasses.

Panicum tenuiflorum, R. Brown. (*Paspalum brevifolium*, Fluegge.)

South-Asia and East-Australia. It has a running stem and forms a good bottom as a pasture-grass (Bailey).

Panicum Texanum, Buckley.

Texas. The "Colorado Bottom-grass," also called "Green River-grass." One of the best of forage-grasses for horses, cattle and sheep; it is a sure crop, and produces 2 to 3 tons per acre (Vasey). The hay from this grass is of a superior kind; it can be cut twice in the season (Carrington). This grass is however annual, but very leafy and becomes decumbent and widely spreading: growth rapid, many stalks proceeding from the same root. Mr. P. Lea of Texas considers it the best native grass for hay there, and notes that all kinds of pasture animals like it preferentially, and that the hay from this proved also exceedingly nutritious, and that it subdues weeds readily. Mr. Ravenel of Carolina found it there to come up spontaneously, much better and larger than most other grasses. Height to 6 feet.

Panicum turgidum, Forskael.

Egypt, where this millet yields a bread-grain.

Panicum virgatum, Linné.

North-America. A tall perennial species, with a wide, nutritious panicle. Easily disseminated. Content with sandy soil, but likes some humidity. The foliage good for fodder when young. Frequent on the prairies, but it will grow even also on sandy sea-coasts. It passes in some places as Switch-grass.

Panicum viride, Linné. (*Setaria viridis*, Beauvois.)

Widely spread over many parts of the old world. Though annual, this grass is of value for the first vegetation on bare sand-land, over which, as well as over calcareous soil, it spreads with remarkable facility. The same may be said of Panicum glaucum and a few other related species.

Papaver somniferum, Linné.

The Opium-Poppy. Countries on the Mediterranean Sea. The capsules of this tall annual, so showy for its flowers, are used for medicinal purposes. From the minute, but exceedingly numerous seeds, oil of a harmless and most palatable kind can be pressed remuneratively; but the still more important use of this plant is for the preparation of opium. Both the black- and pale-seeded varieties can be used for the production of opium. The return of poppy-

culture, whether for opium or for oil, is obtained within a few months.
Mild and somewhat humid open forest-tracts proved most productive
for obtaining opium from this plant; but it can also be reared in colder
localities, good opium rich in morphia having even been obtained in
Middle Europe and the Northern United States, the summers there
being sufficiently long, to ripen the poppy with a well elaborated sap.
Indeed the plant matured its seeds as far north as lat. 69° 18′ in
Norway (Schuebeler). The morphia-contents in opium from Gippsland were on an average somewhat over 10 per cent. Opium was
prepared in the Melbourne Botanic Gardens for the Exhibition of
1866; but Mr. J. Bosisto and Mr. J. Hood have given first
commercial dimensions to this branch of rural industry in Australia.
The Smyrna-variety is particularly desirable for opium; it enables the
cultivator to get from 40 lbs. to 75 lbs. of opium from an acre,
generally worth 30s. to 35s. per pound. The ground for poppy-
culture must be naturally rich or otherwise be well manured; dressing
with ashes increases the fecundity of the plant. The seeds, about
9 lbs. to the acre, are generally sown broadcast mixed with sand. In
the most favorable places as many as three crops are obtained during
a season. The collecting of the opium, which consists merely of the
indurating sap of the seed-vessels, is commenced a few days after the
lapse of the petals. Superficial horizontal or diagonal incisions are
made into the capsules as they successively advance to maturity.
This operation is best performed in the afternoon and evenings, and
requires no laborious toil. The milky opium-sap, thus directed out-
wards, is scraped off next morning into a shallow cup, and allowed to
dry in a place away from sunlight; it may also be placed on poppy
leaves. From one to six successive incisions are made to exhaust the
sap, according to season, particular locality or the knife-like instru-
ment employed. In the Department of Somme (France) alone opium
to the value of £70,000 annually is produced and poppy seed to the
value of £170,000. Australian seasons as a rule are favorable for
collecting opium, and therefore this culture is rendered less precarious
here than in many other countries. Our opium has proved as good as
the best Smyrna-kind. The petals are dried for packing the opium.
The main-value of opium depends on its contents of morphia, for
which the genus Papaver, as far as heretofore known, remains the
sole source; but not less than fourteen alkaloids have been detected in
opium by the progressive strides of organic chemistry: codein,
metamorphin, morphia or morphin, narcein, narcotin, opianin,
papaverin, porphyroxin, xanthopin, meconidin, codamin, laudanin,
pseudo-morphin and thebain. It contains besides an indifferent
bitter principle, meconin and meconic acid (*vide* " Wittstein's Che-
mische Analyse von Pflanzentheilen," or my English edit., p. 163).
Various species of Papaver produce more or less opium and morphia. P. setigerum (De Candolle), supposed to be the wild state
of P. somniferum, was cultivated, evidently for the sake of the
seeds, by the lacustrine people of Switzerland prior to historic ages
(Heer).

Pappea Capensis, Ecklon and Zeyher.

South-Africa. The fruit of this tree is of the size of a cherry, savory and edible. The seeds furnish an oil similar to castor-oil in its effects (Prof. McOwan).

Pappophorum commune, F. v. Mueller.

Widely dispersed over the continent of Australia, occurring also in some parts of Asia and Africa. Perennial; regarded as a very fattening pasture-grass, and available for arid localities and almost rainless zones.

Parinarium Nonda. F. v. Mueller.

The " Nonda-tree " of North-Eastern Australia. Attains a height of 60 feet; its wood soft, close-grained, easily worked (W. Hill). May prove hardy in mild temperate climes, and may perhaps live in the dry and hot air of deserts, where it deserves trial-culture for the sake of its edible, mealy, plum-like fruit. A few other species with esculent drupes occur in different tropical countries.

Parkinsonia aculeata, Linné.

From California to Uruguay. A thorny shrub, clearly adapted for the warmer regions of the temperate zone, where it might be utilized with the following plant for evergreen hedges. The flowers are handsome. Six other species occur in various parts of America.

Parkinsonia Africana, Sonder.

South-Africa. A tall bush. '

Parrotia Jacquemontiana, Decaisne.

North-Western Himalayas, from about 3,000 to 8,500 feet elevation. This deciduous-leaved small tree merits attention. Its tough and pliable twigs are used for basket-work and preferable for the twig-bridges, the latter sometimes 300 feet long; hence this tree could be used for a variety of economic purposes (Stewart and Brandis). P. persica (C. A. Meyer) occurs on the Caspian Sea.

Parthenium integrifolium, Linné.

Eastern North-America. The flowering tops of this perennial bitter herb have come into use as a febrifuge (Houlton).

Paspalum ciliatum, Humboldt.

Tropical South-America. A perennial and lauded cereal grass.

Paspalum dilatatum, Poiret. (*P. ovatum,* Trinius.)

North- and South-America. Perennial; of excellent quality for fodder; keeps green during the hottest summer-time. Mr. Bacchus found it hardy up to a height of 2,000 feet in Victoria. It grew 4½ feet in little more than two months in New South Wales, after drought was followed by heavy rains. It is closely allied to the Mexican P. virgatum, L. Introduced into Australia by the writer with many other fodder-grasses.

Paspalum distichum, Linné.*

The " Silt-Grass." North- and South-America, except the colder regions; elsewhere probably introduced, though now widely naturalized also in the warmer coast-tracts of the eastern hemisphere. Possibly indigenous to Australia also. A créeping bank- or swamp-grass, forming extensive cushions. It keeps beautifully green throughout the year, affords a sufficiently tender blade for feed, and is exquisitely adapted to cover silt or bare slopes on banks of ponds or rivers, where it grows grandly; moderate submersion does not destroy it, but frost injures it; it thrives well also on salt-marshes. The chemical analysis made in spring gave the following results:— Albumen 2·20, gluten 7·71, starch 1·56, gum 1·64, sugar 5·00 (F. v. Mueller and L. Rümmel).

Paspalum læve, Michaux.

Southern States of North-America. Perennial. Prof. Phares states, that for successive years two tons of hay of this meadow-grass can be mown from an acre, it almost never seeding when regularly cut.

Paspalum lentiginosum, Presl.

Southern States of North-America and Mexico. Spreads by runners and forms a close turf (Dr. Vasey). P. Floridanum, Michaux, of the same region is a tall species of vigorous growth. P. purpurascens, Elliott, is leafy and succulent, and roots at the lower joints.

Paspalum notatum, Fluegge.*

From Virginia to Argentina. This is one of the best of fodder-grasses there, forming a dense, soft, carpet-like sward on meadows, and becoming particularly luxuriant and nutritious on somewhat saline soil (Lorentz). Closely cognate to P. distichum.

Paspalum platycaule, Poiret. (*P, compressum*, Presl.)

Warmer regions of North- and South-America. A nutritive pasture-grass, particularly eligible for sandy coast-lands (C. Mohr). The creeping stems become closely matted (Dr. Vasey).

Paspalum scrobiculatum, Linné.

Through the tropics of the eastern hemisphere widely dispersed, extending to South-Eastern Australia, New Zealand and Polynesia also. A valuable pasture-grass, which will grow on poor land, also on swampy ground. A superior variety is cultivated in India for a grain-crop. This grass furnishes a good ingredient for hay. Its stem sometimes attains a height of 8 feet. Rosenthal pronounces it pernicious, perhaps when long and exclusive use is made of this grass, or possibly when diseased through fungus-growth.

Paspalum stoloniferum, Bosc.

Central America. A fodder-grass of considerable value.

Paspalum undulatum, Poiret.

North- and South-America. Noticed by C. Mohr as valuable for fodder. A. Gray records it as annual.

Passiflora alata, Aiten.

Peru and Brazil. This Passion-flower and all the following (probably with some other species) furnish Granadilla-fruits. All the species here recorded are perennial, some woody and widely climbing.

Passiflora coccinea, Aublet.

From Guiana to the La Plata-States. The fruits are eaten raw or boiled (Hieronymus). This plant with probably many others of the genus contains a chemical principle of hypnotic value.

Passiflora coerulea, Linné.

South-Brazil and Uruguay. One of the hardiest of all Passion-flowers, and with many others well adapted for covering bowers, rockeries and similar structures. Many of the equatorial species come from mountainous regions, and may thus endure mild temperate climates.

Passiflora edulis, Sims.

Southern Brazil. Frost-shy, Fruit purple.

Passiflora filamentosa, Willdenow.

Southern Brazil.

Passiflora incarnata, Linné.

North-America, from Virginia and Kentucky southward. The fruits are called May-pops.

Passiflora laurifolia, Linné. (*P. tinifolia,* Jussieu.)

The Water-Lemon. From the West-Indies to Brazil.

Passiflora lingularis, Jussieu.

From Mexico to Bolivia. Professor Ernst of Caraccas says, that its fruit is one of the finest anywhere in existence.

Passiflora lutea, Linné.

North-America, from Pennsylvania and Illinois southward. With P. coerulea and P. incarnata among the hardiest of the genus. Berries small.

Passiflora macrocarpa, Masters.

Brazil and Peru. Mr. Walter Hill reports, having obtained fruits of 8 lbs. weight at the Brisbane Botanic Garden.

Passiflora maliformis, Linné.

From the West-Indies to Brazil.

Passiflora mucronata, Lamarck.

Brazil, extending far south. Fruit edible (Dr. Rosenthal).

Passiflora pedata, Linné.
From the West-Indies to Guiana. This is among the species, mentioned by Dr. Rosenthal as yielding edible fruits.

Passiflora quadrangularis, Linné.
Brazil. One of the most commonly cultivated Granadillas. The fruits attain a large size.

Passiflora serrata, Linné.
From the West-Indies to Brazil.

Passiflora suberosa, Linné. (*P. pallida*, Linné.)
From Florida to Brazil. A careful investigator, Dr. Maxw. Masters, has recently defined about 200 species of Passion-flowers.

Passiflora tiliifolia, Cavanilles.
Peru. Fruit edible, according to Dr. Rosenthal.

Paullinia sorbilis, Martius.
Brazil. A climbing shrub, possibly hardy in the warm temperate zones, where many tropical Cupaniæ and other sapindaceous trees endure the clime. The hard Guarana-paste of chocolate-color is prepared from the seeds by trituration in a heated mortar with admixture of a little water, kneading into a dough and then drying. This paste, very rich in coffein, serves for a pleasant beverage, and is also used medicinally.

Paulownia imperialis, Siebold.
Japan. A tree, hardier than Cercis Siliquastrum, of value for scenic effects. It will endure the climate of Norway to lat. 58° 58' (Professor Schuebeler). Rises in 10 years to 30–40 feet.

Peireskia aculeata, Miller.
The Barbadoes-Gooseberry. West-Indies. A tall shrub, adapted for hedges in localities free of frost. The cochineal-insect can be reared on this plant also. The berries are edible; the leaves available for salad. Several other species exist in tropical America, among which P. Bleo, Humb., is particularly handsome; but they may not all be sufficiently hardy for utilitarian purposes in an extra-tropical clime.

Peireskia portulacifolia, Haworth.
West-Indies. This attains the size of a fair tree.

Pelargonium odoratissimum, Aiton.
South-Africa. A perennial trailing herb, from the leaves of which a fragrant oil can be distilled. Pelargonium-oil is extensively produced in Algeria as a cheap substitute for attar of roses. There the rate of annual production of the "essence of geranium" being about 12,000 lbs. One ton weight of fresh leaves will yield about 2 lbs. volatile oil (Piesse). The same remark applies to the shrubby P. radula

and P. capitatum. The Kaffirs assert that these plants keep off snakes. Easily multiplied from cuttings. On this occasion may be pleaded also for the hundreds of distinct kinds of Pelargonium, many seemingly doomed to utter annihilation in South-Africa, unless indeed some statesman there by legislation will see these lovely plants protected at least on a few pristine spots within permanently reserved small areas, for the joy also of future generations; so also the hundreds of charming species of Ericas, only there to be found, claim immunity for ever against herds and flocks; and thus likewise might, under enlightened foresight, yet timely be saved and sheltered the numerous and gay Polygaleæ, podalyrious Leguminosæ, Phylicæ, Crassulæ, Mesembryanthema, Proteaceæ, Gnidia, and the incomparable Irideæ and Liliaceæ of the South-African Flora, so far at least, as not to be swept away altogether from the face of the globe !

Peltophorum Linnæi, Bentham. *(Cæsalpinia Brasiliensis,* Linné.)

A small tree, which provides the orange-colored Brasiletto-wood. This species likes dry calcareous soil (Grisebach). Endures the climate of Carolina.

Pennisetum latifolium, Sprengel.

Extra-tropical South-America. A tall perennial nutritious grass, forming large tufts, easily spreading from the roots or seeds. It is of quick growth.

Pennisetum villosum, R. Brown.

Abyssinia. A grass of decorative beauty, forming ample tufts; it is recommended by Dr. Curl for permanent pasture in New Zealand. With numerous other grasses it was introduced into Australia by the writer of this work. Proves hardy in Norway to lat. 67° 56' (Schuebeler).

Pennisetum thyphoideum, Richard.* *(Penicillaria spicata,* Willdenow; *Panicum cæruleum,* Miller.)

The Bajree or Pearl-Millet. Tropical Asia, Nubia and Egypt. An annual, requiring only about three months to ripen its millet-crop in warm countries. The stems are thick and reach a height of 6–10 feet; several being produced from one root, and each again forming lateral branches; the maximum-length of a spike is about a foot and a half; Colonel Sykes saw exceptionally 15 spikes on one plant and occasionally 2,000 seeds in one spike. Together with sorghum this is the principal cereal, except rice, grown in India by the native races. This grass requires a rich and loose soil, and on such it will yield upwards of a hundred-fold. It furnishes hay of good quality, though not very easily dried, and is also valuable as green fodder. In the United States cultivated as far north as Pennsylvania, and it matures seeds even as far north as Christiania in Norway (Schuebeler). Its fast growth prevents weeds from obtaining a footing. In very exceptional cases and under most favorable circumstances as regards soil and manure, the first cutting is in six or seven weeks, the stems up to seven feet high, giving at the rate of 30 tons green feed, or 6½ tons of hay per acre; in six or seven weeks more a second cutting is obtained, reaching 55 tons per acre of green feed, the grass being nine feet high; a third cut is got in the same season. Farm stock eat it greedily. One plant of pearl millet " is worth three of maize for fodder." Some of the many other species of Pennisetum are doubt-less of rural value. A plant allied to P. thyphoideum occurs in China, namely P. cereale (Trinius). This also affords millet or corn for cakes.

Pentzia virgata, Lessing.

South-Africa. A small cushion-like bush, recommended for establishment in deserts for sheep-fodder. It has the peculiarity, that whenever a branch touches the ground, it strikes roots and forms a new plant; this enables the species to cover ground rapidly (Sir Samuel Wilson). Valuable also for fixing drift-sand in water-rills, by readily bending over and rooting, thus forming natural little catch-dams to retain water (McOwan). Several other species occur in South-Africa.

Periandra dulcis, Martius.

Sub-tropical Brazil. The sweet root of this shrub yields liquorice.

Perilla arguta, Bentham.

Japan. An annual herb. An infusion of this plant is used for imparting a deep-red color to table-vegetables and other substances. In Japan the seeds are pressed for oil. P. ocimoides, L., of Upper India probably serves similar purposes. Some species of Perilla are suitable for ribbon-culture.

Persea gratissima, Gaertner.

The Avocado-Pear. From Mexico to Peru and Brazil in forest-tracts near the coast; but its real nativity, according to A. de Candolle, restricted to Mexico. Suggestively mentioned here as probably available for mild localities outside the tropics, inasmuch as it has become naturalized in Madeira, the Azores and Canary-Islands. A noble evergreen spreading tree. The pulp of the large pear-shaped fruit is of delicious taste and flavor. The fruit attains sometimes a weight of 2 lbs., and is generally sliced for salad. Its pulp contains about 8 per cent. of greenish oil. The seeds have come into medicinal use at the instance of Dr. Froehling, particularly through the efforts of Messrs. Parke and Davis, to whom we mainly owe the introduction of many other valuable new drugs into medicine.

Persea Teneriffæ, F. v. Mueller. (P. Indica, Sprengel.)

Madeira, Azores and Canary-Islands. This magnificent tree produces a beautiful, hard, mahogany-like wood, especially sought for superior furniture and turners' work. One of the most hardy trees of the large order of Laurinæ.

Peucedanum graveolens, Bentham.   (*Anethum graveolens*, Linné.)

The "Dill." South-Europe, North-Africa, Orient. Annual. The well-known aromatic fruitlets used as a condiment. In India known as Sowa. Distilled dill-oil is in use also for scented soaps.

Peucedanum officinale, Linné.

The Sulphur-Root. Middle and Southern Europe, Northern Africa, Middle Asia. Perennial. The root is used in veterinary medicine; it contains, like that of the following species, the crystalline peucedanin.

Peucedanum Ostruthium, Koch.   (*Imperatoria Ostruthium*, Linné.)

Mountains of Middle Europe. A perennial herb, which could be grown in alpine regions. The acid aromatic root is used in medicine, particularly in veterinary practice. It is required for the preparation of some kinds of Swiss cheese. P. Cervaria (Cusson) and P. Oreoselinum (Moench) are also occasionally drawn into medicinal use.

Peucedanum sativum, Bentham.   (*Pastinaca sativa*, Linné.)

The "Parsnep." Europe, Northern and Middle Asia. Biennial. The root palatable and nutritious. The wild root is somewhat acrid, and poisonous effects have occasionally resulted from its use. A variety is cultivated in the Channel Islands, with roots 3 or 4 feet long (Chambers).. A somewhat calcareous soil is favorable to the best development of this plant. It is very hardy, having been grown in Norway to lat. 70° 22'; it matured seeds as far north as lat. 67° 56' (Schuebeler). The culture is that of the carrot; for fodder the root surpasses that of the latter in augmenting milk (Langethal). A decoction of parsnep-roots ferments with sugar and yeast into a sparkling beverage, but requires casking for about a year (Baudinet).

Peucedanum Sekakul, F. v. Mueller.   (*Pastinaca Sekakul*, Russell; *Malabaila Sekakul*, Boissier.)

From Greece to Persia and Egypt. The root of this perennial herb was brought already under medical notice by Dr. Rauwolf; in a boiled state it affords a palatable esculent.

Peumus Boldus, Molina.

The Boldo of Chili. A small ornamental evergreen tree, with exceedingly hard wood, which is utilized for many kinds of implements. The bark furnishes dye-material. The fruits are of aromatic and sweet taste (Dr. Philippi).

Peziza macropus, Persoon.

Europe. Mentioned by Prof. Goeppert among the edible mushrooms, sold in Silesia along with P. repanda (Wahlenberg).

Phalaris aquatica, Linné.

Southern Europe and Northern Africa. Important as a perennial fodder-grass, fit for wet ground.

Phalaris arundinacea, Linné.

Temperate and colder regions of Europe, Asia and America; indigenous in Norway to lat. 70° 30'. Not without some importance as a reedy grass of bulky yield on wet meadows or in swampy places. A variety with white-striped leaves is a favorite as a ribbon-plant for borders.

Phalaris Canariensis, Linné.

The Canary-grass. An annual grass from the Canary-Islands, now widely dispersed as a spontaneous plant over the warmer zones of the globe. Thus it has also become naturalized in Australia. It will endure the climate of Norway to lat. 70° 22', bearing seed to lat. 63° 26' (Prof. Schuebeler). It is grown for its seeds, which form one of the best kinds of food for many sorts of small cage-birds. The flour is utilized in certain processes of cotton-manufacture, and liked even for some kinds of cakes. The soil for culture of the Canary-grass must be friable and not too poor. It is an exhaustive crop. As allied species of similar use, but mostly of less yield, may be enumerated : P. brachystachys (Link) from Italy, P. minor (Retzius) and P. trunctata (Gussone) from various countries on the Mediterranean Sea, the last-mentioned being perennial. Other species, including some from Asia, are deserving of trial. P. minor is recommended by Dr. Curl for permanent pastures, as it supplies a large quantity of fine, sweet, fattening foliage, relished by stock. It keeps green far into the winter in the climate of New Zealand. Chemical constituents here (in November): Albumen 1·59, gluten 6·14, starch 1·03, gum 6·64, sugar 2·86 per cent. (F. v. Mueller and L. Rummel); another analysis in the same month gave: Albumen 1·06, gluten 5·64, starch 0·98, gum 3·22, sugar 4·20 per cent.

Pharnaceum acidum, J. Hooker.

St. Helena. A dwarf perennial succulent plant, which might advantageously be naturalized on sea-shores, to yield an acid salad, perhaps superior to that of Portulaca oleracea.

Phaseolus aconitifolius, Jacquin.

India, up to 4,000 feet. A dwarf annual species. Dr. Forbes Watson admits it among the culinary beans of India. It will bear on arid soil. P. trilobus (Aiton), the Simbi-bean, is a still hardier variety, which becomes perennial.

Phaseolus adenanthus, G. Meyer. (*P. Truxillensis*, Humboldt; *P. rostratus*, Wallich.)

Almost cosmopolitan within the tropics, where, irrespective of navigation and other traffic, it becomes dispersed by migrating birds; truly spontaneous also in tropical Australia. A perennial herb with large flowers, resembling those of Vigna vexillata (Bentham). Cultivated for its seeds, which are rather small, but copiously produced. A variety with edible roots occurs.

Phaseolus coccineus, Kniphof.* (*P. multiflorus*, Willdenow.)

The Scarlet Runner. Tropical South-America. A twining showy perennial, as useful as the ordinary French-bean. Its seeds usually larger than those of the latter plant, purple with black dots, but sometimes also pure-blue and again quite white. The flowers occur sometimes white. The root contains a narcotic poison.

Phaseolus derasus, Schranck.

Brazil. There, next to maize, the most important and extensively used plant for human food (Dr. Peckolt). Sprengel refers this to P. inamænus (Linné), a variety of the following species.

Phaseolus lunatus, Linné.

The " Lima-bean." Perennial. Tropical South-America. A. de Candolle restricts the real nativity to that·part of the globe. Wittmack identifies beans from ancient graves in Peru as belonging to this species. The root is deleterious. Biennial according to Roxburgh. Much cultivated in the warm zone for its edible beans, which are purple or white. A yellow-flowered variety or closely allied species is know as the Madagascar-bean, and has proven hardy and productive in Victoria. P. perennis (Walter) from the United States of North-America is another allied plant. [1]

Phaseolus Max, Linné. (*P. Mungo*, Linné; *P. radicatus*, Linné.)

The "Green Gram." South-Asia and tropical Australia. An annual, very hairy plant, not much climbing. Frequently reared in India, when rice fails or where that crop cannot be produced. According to Sir Walter Elliot one of the most esteemed of Indian pulses. " It fetches the highest price and is more than any other in request among the richer classes, entering largely into delicate dishes and cake." Cultivated up to 6,000 feet (Forbes Watson). Col. Sykes counted sixty-two pods on one plant with from seven to fourteen seeds in each. The seeds are but small, and the herb is not available for fodder. This plant requires no irrigation, and ripens in two and a half to three months. In India, it yields the earliest pulse-crop in the season. The grain tastes well, and is esteemed wholesome. The harvest is about thirty-fold. Paillieux reminds us that the young sprouts serve as a delicate vegetable.

Phaseolus vulgaris, l'Obel.*

The ordinary Kidney-bean, or French-bean, or Haricot. Native country probably Western South-America, inasmuch as Professor Wittmack has recently identified beans from ancient graves at Lima as belonging to P. vulgaris. Though this common and important culinary annual is so well known, it has been deemed desirable to refer to it here, with a view of reminding our readers, that the kidney-bean is nearly twice as nutritious as wheat. The meal from beans might also find far-augmented use. As constituents of the beans should be mentioned a large proportion of starch (nearly half), then much legumin, also some phaseolin (which, like amygdalin, can be converted into an essential oil) and inosit-sugar. Lentils contain more legumin but less starch, while peas and beans are almost alike in respect to the proportion of these two nourishing substances. The kidney-bean can still be cultivated in cold latitudes and at sub-alpine elevations, if the uninterrupted summer warmth lasts for four months; otherwise it is more tender than the pea. The soil should be friable, somewhat limy and not sandy for field-culture. Phaseolus nanus, L. (the dwarf bean) and P. tumidus, Savi (the sugar-bean, sword-bean, or egg-bean) are varieties of P. vulgaris. Several other species of Phaseolus seem worthy of culinary culture. Haricot-Beans contain very decided deobstruent properties, which however are generally destroyed by too much boiling. To obviate this they should be soaked for 24 hours in cold water to which salt has been added, and then gently boiled for not more than 30 or 40 minutes in very little water (W. B. Booth). The seeds will retain their vitality fully three years.

Phleum pratense, Linne.*

The Timothy- or Catstail-grass. Europe, North-Africa, Northern or Middle Asia, ascends to 10,000 feet in Spain. One of the most valuable and most cultivated of all perennial fodder-grasses. Its production of early spring-foliage is superior to that of the Cock's-foot-grass. It should enter largely into any mixture of grasses for permanent pasturage. It will live also on moist and cold clay-ground. This grass, and perhaps yet more the allied Phleum alpinum, L., are deserving of an extensive transfer to moory mountain-regions. It is very hardy, having been found indigenous in Norway to lat. 70° (Professor Schuebeler). For hay it requires mowing in a young stage. The seed is copiously yielded and well retained, The greatest advantage from this grass arises, according to Langethal, when it is grown along with clovers. It thrives even better on sandy meadows than on calcareous soil; it will prosper on poorer ground than Alopocurus pratensis; the latter furnishes its full yield only in the fourth year, whereas the Phleum does so in the second. The Timothy-grass dries more quickly for hay and the seeds are gathered more easily, but it vegetates later, is of harder consistence, and yields less in the season after the first cut. Dr. Curl, of New Zealand, observes that, while many grasses and clovers, if eaten in their spring-growth, may cause diarrhœa in sheep, the Timothy-grass, when young, does not affect them injuriously.

Phœnix dactylifera, Linné.*

The Date-Palm. North-Africa, also inland; Arabia, Persia. This noble palm attains finally a height of about 80, exceptionally 120 feet. It is unisexual and of longevity. "Trees of from 100 to 200 years old continue to produce their annual crop of dates," though gradually at very advancing age at diminished rates. This palm seems to live through several centuries. Though sugar or palm-wine can be obtained from the sap, and hats, mats and similar articles can be manufactured from the leaves, we would utilize this palm beyond scenic garden-ornamentation only for its fruits. The date-palm would afford in time to come a real boon in the oases of desert-tracts, swept by burning winds, although it might be grown also in the valleys of mountains and in any part of lowlands free of severe frost. Several bunches of flowers are formed in a season, each producing often as many as 200 dates. In Egypt as many as 4 cwt. of dates have been harvested in one season from a single date-palm. Many varieties of dates exist, differing in shape, size and color of the fruit; those of Gomera are large and contain no seed. The unexpanded flower-bunches can be used for palm-cabbage and the fibre of the leaf-stalks for cordage. The town Elche, in Spain, is surrounded by a planted forest of about 80,000 date-palms, and the sale of leaves for decorative purposes produces a considerable income to the town, irrespective of the value of the date-fruits; and so it is at Alicante. As far north as the Gulf of Genoa also a date-forest exists. The ease with which this palm grows from seeds affords facilities in adapted climes to imitate these examples, and we certainly ought to follow them in all parts of Australia and in similar climes. The best dates are grown in oases, where fresh-water gushes from the ground in abundance and spreads over light soil of the desert subject to burning winds. The Zadie-variety produces the heaviest crop, averaging 300 lbs. to the tree; superior varieties can only be continued from offshoots of the root; these will commence to bear in five years and be in full bearing in ten years; one male tree is considered sufficient for half a hundred females. The pollen-dust is sparingly applied by artificial means. The pulpy part of the fruit contains about 58 per cent. of saccharine matter. It is estimated, that in Egypt alone four millions of date-palms exist, the produce of which is to a large extent consumed locally. The date-groves of Turkey produce annually 40,000 to 60,000 tons of dates in ordinarily good seasons (Le Duc). The date-palm will live in saltish soil, and the water for its irrigation may be slightly brackish (Surgeon-Major Colvill). Northern limit of date about 35° north latitude. Into Central Australia the date-palm was first introduced by the writer of this work. The variety "Datheres-sifia" ripens early in the season its fruits (Naudin). It is propagated from suckers. Its pungent rigidity protects this palm from encroachment of pasture animals; hence it can be disseminated without hedging.

Phœnix Hanceana, Drude.

South-China. This palm was buried for ten days under three feet of snow in the south of France without injury (Naudin).

Phœnix paludosa, Roxburgh.

India. A stout species, not very tall. Of value at least for decorative culture.

Phœnix pusilla, Gaertner.

India and South-China. A dwarf species, which bears the clime of the South of France without protection (Kerchove de Denterghem).

P. farinifera (Roxburgh) appears to be identical. It is adapted for sandy and otherwise dry and barren land, but prefers the vicinity of the sea. Berry shining black, with a sweet mealy pulp.

Phœnix reclinata, Jacquin.

South-Africa, in the eastern districts. A hardy species, but not tall, often reclining. It is adapted for ornamentation. The sweet coating of the fruit is edible (Backhouse).

Phœnix silvestris, Roxburgh.

India, almost on any soil or in any situation, down even to the edge of drift-sand on the coast. It has proved a very hardy species at Melbourne. Its greatest height is about 40 feet. Berries yellowish or reddish, larger than in P. pusilla. Where this palm abounds, much sugar is obtained from it by evaporation of the sap, which flows from incisions into the upper part of the trunk—a process not sacrificing the plant, as for 50 years the sap can thus be withdrawn. This palm-sugar consists almost entirely of cane-sugar. A kind of arrack is obtained by fermentation and distillation of this sap, and also from the young spikes. Each plant furnishes the juice for about 8 lbs. of date-sugar annually, but in some instances much more. About 50,000 tons of sugar a year are produced in Bengal alone from this and some other palms. The leaves are used for mats. It lives in drier regions than other Indian palms.

Phœnix spinosa, Thonning.

Tropical Africa, ascending mountain-tracts, thus perhaps hardy in milder extra-tropic regions. Sir John Kirk found, that the green bunches, if immersed in water for half a day, suddenly assume a scarlet hue, when the astringent pulp becomes edible and sweet.

Phormium tenax, J. R. and G. Forster.*

The Flax-Lily of New Zealand, where it grows as far south as 46° 30', occurring also in the Chatham-Islands and Norfolk-Island, though not on Lord Howe's Island. It is also found in the Auckland-Islands, nearly 51° south (Schur). It flowered in several places of England in exposed positions, and was not affected by severe frost (Masters). It perfected seeds even in the most northern of the Orkney-Islands (Traill) and will bear unhurt a temperature of 15° F.; the tops of the leaves become injured at 9° F. (Gorlie). It is desirable, that this valuable plant should be brought universally under culture, particularly on any inferior spare-ground or on the sea-beaches or any rocky declivities, where it may be left to itself unprotected, as no grazing animal will touch it. It is evident, that the natural growth will soon be inadequate to the demand for the plant. It is adapted for staying bush-fires, when planted in hedgerows. Merely torn into shreds, the leaves serve at once in gardens and vine-yards as cordage, and for this purpose, irrespective of its showy aspect, the Phormium has been distributed from the Botanic Garden of Melbourne during many years by the writer. From the divided

s

roots any plantation can gradually be increased, or this can be done more extensively still by sowing the seeds. In all likelihood the plant would thrive and become naturalized in Kerguelen's Land, the Falkland-Islands, the Faroe- and Shetland-Islands and many continental places of both hemispheres far into cold latitudes. Dr Traill records it having ripened seeds in the Orkney-Islands without protection. It has proved quite hardy in England. Among the varieties three are better characterized than the rest: the Tehore- the Swamp- and the Hill-variety. The first and the last mentioned produce a fibre fine and soft, yet strong, and the plant attains a height of only about 5 feet, whereas the Swamp-variety grows to double that height, producing a larger yield of a coarser fibre, which is chiefly used for rope- or paper-making. One of the most dwarf varieties is P. Colensoi (J. Hooker). As might be expected, the richer the soil the more vigorous the growth of the plant. Flooding now and then with fresh or brackish water is beneficial, but it will not live, if this is permanent. In swampy ground trenches should be dug, to divert the surplus of humidity. Fibre, free from gum-resin, properly dressed, withstands moisture as well as the best Manilla-rope. Carefully prepared, the fibre can be spun into various textile durable fabrics, either by itself or mixed with cotton, wool or flax. Elegant articles are woven from it by the Maoris. In October 1872 the sale of Phormium-fibre in London was 11,500 bales, ranging in price from £19 to £31. The tow can be converted into paper, distinguished for its strength and whiteness. The London price of Phormium-fibre for this purpose is from £10 to £20 per ton. A strong decoction of the root and leaf-bases used in surgery for dressing wounds with a view of producing ready and healthy granulation (F. A. Monkton).

For further details on the utilization of this plant, the elaborate report of the New Zealand Commission for Phormium should be consulted.

Photinia eriobotrya, J. Hooker.   (*P. Japonica*, Franchet and Savatier; *Eriobotrya Japonica*, Lindley.)

The "Loquat." China and Japan. This beautiful evergreen shrub or tree, remarkable for its refreshing fruit, is easily raised from seed; or superior varieties can, according to G. W. Johnson, be grafted not only on its own stock, but also on the Whitethorn, or better still on the Quince. It is also a grand bush for scenic ornamental effects. Hardy in England. Growth of celerity. In Southern Austria flowering during midwinter; the flowers are intensely fragrant, and do not suffer from a few degrees of frost; hence this plant is of particular horticultural importance, to contribute amply to garden-fragrance in winter-time. In South-Austria the Loquat is the earliest fruit of the season (Baron von Thuemen), bearing also copiously there. In Greece the fruits ripen already during May. Hemsley mentions this plant among the shrubs and trees, hardy in England, where however it does not mature its fruits. P. villosa, D.C., also yields edible native fruit to the Japanese.

Phyllanthus Cicca, J. Mueller. (*Cicca disticha*, Linné.)

Insular India. Hardy in Florida, 27° 30', where Mr. Reasmer finds it to be a desirable fruit-tree. The berries are small and acid, serving for jam. Mr. L. A. Bernays admits this plant among those recommended in his work on "Cultural industries for Queensland," 1883.

Phyllocladus rhomboidalis, Cl. Richard.

Celery-Pine of Tasmania. A stately tree, often to 60 feet high, with a stem 2 to 6 feet in diameter. The timber is particularly valuable for the masts and spars of ships. It will only grow to advantage in deep forest-valleys. Dwarfed in alpine elevations.

Phyllocladus trichomanoides, D. Don.

Celery-Pine of New Zealand; Maori-name, "Tanekaha." This tree attains a height of 70 feet, with a straight stem reaching 3 feet in diameter, and furnishes a pale close-grained timber, strong, heavy and remarkably durable, according to Professor Kirk, greatly valued for mine-props, struts, caps, sleepers, water-tanks, bridge-planks and piles, also spars; the Maoris employ the bark for dyeing red and black and yellow, according to admixtures. This species also ascends in a diminutive form alpine elevations.

Phyllostachys bambusoides, Siebold.

Himalayas, China and Japan. A dwarf Bamboo, but hardy; the yellowish canes available for excellent walking-sticks (Griffith).

Phyllostachys nigra, Munro.* (*Bambusa nigra*, Loddiges.)

China and Japan. Reaching 25 feet in height. The stems nearly solid and becoming black. Has withstood severe frost in the south of France and at Vienna. Known to have grown 16 feet in six weeks. Bamboo-chairs and walking-sticks often made of this species. A Japanese species of this bambusaceous genus proved hardy in Scotland. P. viridi-glaucescens and P. aurea are perfectly hardy in England (Munro); the latter withstood the severest winters of Edinburgh, with 0° F. (Gorlie).

Phymaspermum parvifolium, Bentham. (*Adenochana parvifolia*, De Candolle.)

South-Africa. A dwarf, somewhat shrubby plant, fit to be naturalized on mere sandy ground. Praised by Professor McOwan as equal in value to Pentzia virgata for sheep-pastures.

Physalis Alkekengi, Linné.

The Strawberry-Tomato or Winter-Cherry. Middle and South-Europe, North-Africa, Middle Asia, extending to Japan; said to have come originally from Persia. Ripening in Norway to lat. 63° 26' (Schuebeler). A perennial herb. The berry, which is red and of a not unpleasant taste, has some medicinal value. The leaves contain a bitter principle—physalin.

Physalis angulata, Linné.

In many tropical countries, extending as a native plant to the northern parts of the United States and to Japan. An annual herb. The berries yellowish, edible. P. minima, L. (*P. parviflora,* R. Br.), is closely allied, and extends also into tropical Australia.

Physalis Peruviana, Linné.

Temperate and tropical America, widely naturalized in many countries of the warmer zones. With double inaptness called the Cape-Gooseberry. A perennial herb; but for producing its fruit well it requires early renovation. . The acidulous berries can be used as well for table-fruit as for preserves. Doubtless several other kinds of Physalis can be utilized in the same manner. In colder countries the P. Peruviana becomes annual. Seeds will keep for eight years (Vilmorin).

Physalis pubescens, Linné.

Warmer regions of North- and South-America. Though annual, worth cultivation on account of its acidulous fruits, called the Gooseberry-Tomato or inaptly Barbadoes-Gooseberry, under which name also the very similar P. Barbadensis (Jacquin) is comprised.

Pilocarpus pinnatifolius, Lemaire.

The principal Jaborandi-plant of tropical and sub-tropical Brazil. The leaves and bark of this shrub, which contain essential oil and a peculiar alkaloid, are famed as an agreeable, powerful and quickly acting sudorific. Recommended as a specific in diphtheria and supposed to be also reliable in hydrophobia. This bush is likely to endure the clime of milder temperate forest-regions (Continho, Baillon, Hardy, Guebler). Like P. simplex, also an active sialogogue. Pilocarpin contracts the pupil, and stimulates powerfully the salivary glands.

Pimpinella Anisum, Linné.

The Anise-plant. Greece, Egypt, Persia. An annual. The seed-like fruits enter into various medicines and condiments, and are required for the distillation of oil, rich in anethol. The herbage left after obtaining the seeds serves for fodder. The plant will bear seeds in Norway up to lat. 68° 40′ (Schuebeler). The seeds will retain their power of germination for three years (Vilmorin).

Pimpinella saxifraga, Linné.

Europe, Northern and Middle Asia. A perennial herb; its root used in medicine; a peculiar volatile oil can be distilled from the root. P. magna, L., is a closely allied species, and P. nigra, W., is a variety. The root of the last is particularly powerful.

Pimpinella Sisarum, Bentham. (*Sium Sisarum,* Linné.)

Middle and Eastern Asia. A perennial herb. The bunches of small tubers afford an excellent culinary vegetable. The taste is sweet and somewhat celery-like. . The roots endure frost.

Pinus Abies, Du Roi.* (*Pinus Picea*, Linné.)

Silver-Fir, Tanne. Middle and South-Europe, extending to the Caucasian mountains, ascending the Pyrenees to 6,000 feet. It will endure the climate of Norway to lat. 67° 56′ (Schuebeler). A fine tree, already the charm of the ancients, attaining about 200 feet in height and 20 feet in circumference of stem, reaching an age of 300 years. It furnishes a most valuable timber for building as well as furniture, and in respect to lightness, toughness and elasticity it is even more esteemed than the Norway-Spruce, but it is not so good for fuel or charcoal; it is pale, light, not very resinous, and is mostly employed for the finer works of joiners and cabinet-makers, for sounding boards of musical instruments, largely for toys, also for lucifer-matches, for coopers' and turners' work, and for masts and spars. It also yields a fine white resin and the Strasburg-turpentine, similar to the Venetian. Besides the above normal form the following two main-varieties occur:—P. Abies *var.* Cephalonica, Parlatore (*P. Cephalónica*, Éndlicher), Greece, 3,000 to 5,000 feet above the sea. A tree 60 feet high, with a stem-circumference of 10 feet. The wood is very hard and durable, and much esteemed for building. General Napier mentions, that in pulling down some houses at Argostoli, which had been built 150 to 300 years, all the wood-work of this fir was found as hard as oak and perfectly sound. The very resinous wood probably of a variety of this or an allied species was used by Stradivari and his sons for making the famous Italian violins in the last century.—P. Abies *var.* Nordmanniana, Parlatore (*P. Nordmanniana*, Steven), Crimea and Circassia, to 6,000 above the sea. Can be grown in Norway to lat. 61° 15′. This is one of the most imposing firs, attaining a height of about 100 feet, with a perfectly straight stem. It furnishes a valuable building-timber. The Silver-Fir is desirable for mountain-forests. It will grow on sand, but only half as fast as P. Pinaster. In Britain the upward growth is about 50 feet in 30 years.

Pinus alba, Aiton.

White Spruce. From Canada to Carolina, up to the highest mountains. It resembles *P. picea*, but is smaller, at most 50 feet high. Exudes a superior resin. It bears the shears well, when trained for hedges, which are strong, enduring and compact (J. Hicks). The bark richer in tannin than that of the Hemlock-Spruce. The timber well adapted for deal-boards, spars and many other purposes, but on the whole inferior to that of the Black Spruce. The tree grows in damp situations or swampy ground. Eligible for alpine regions. Hardy in Norway to lat. 67° 56′. P. Engelmanni (Parry) is closely akin; it occurs in British Columbia; stem to 3 feet in diameter; wood excellent and durable (Dr. G. Dawson).

Pinus albicaulis, Engelmann.

British Columbia and California, ascending to 9,000 feet. Akin to P. flexilis. Extremely hardy, resisting the most boisterous weather, but never exceeding 40 feet in height (Sir Jos. Hooker). Wood light, soft, rather brittle though close-grained (Prof. Sargent). Fruit-strobiles nearly globular, purplish, with short and thick scales. Bark whitish, scaly.

Pinus Alcockiana, Parlatore.

Japan, at an elevation of 6,000 to 7,000 feet. A fine spruce, often to 120 feet high, with very small blue-green leaves; the wood is used for light household-furniture. P. tsuga and P. polita ascend there to the same height (Rein).

Pinus amabilis, Douglas.

California Silver-Fir. North-California, Oregon, British Columbia, at elevations of from 4,000 to 7,000 or even 10,000 feet. A handsome fir, to 200 feet high, circumference of stem to 24 feet; the stem is branchless up to 100 feet. The tree passes under the name or the " Queen of the Forests " (Lemmon). The wood is elastic, strong and hard, fit for masts and spars; it has a peculiar red color; spikes, nails and bolts hold firm and never corrode in it (Dufur). Very closely allied to P. nobilis and also to P. grandis. Hemsley records as distinct from this P. lasiocarpa (Hooker), which gains a height of fully 250 feet and has branchlets with yellowish bark.

Pinus aristata, Engelmann.

California, at elevations of 8–10,000 feet in the Sierras. A pine, attaining about 75 feet in height, the stem three feet in diameter; leaves extremely short (Gibbons). Fit for any alpine country.

Pinus Arizonica, Engelmann.

Arizona, California. This pine differs from P. ponderosa in glaucous branchlets, thinner leaves constantly in fives and of different structures, and in thicker and shorter fruit-cones, with greater prominence on the scales (Engelmann, Sargent, Perry).

Pinus Australis, Michaux.*

Southern-Pine, also called Hard Pine, Georgia, Yellow Pitch-Pine, Long-leaved Yellow or Broom-Pine. Southern States of North-America. The tree attains a height of about 100 feet, and requires soil open to a great depth, and follows the " stratified drifts," consisting of gravel, sand and clay (Prof. Mohr). It furnishes a superior timber for furniture and building, also for naval architecture, railway-ties and flooring, particularly eligible also for very tall flagstaffs; thus yields the principal yellow pine-wood of the lumber-trade. The wood is compact, straight-grained, very durable, of delicate shades of yellow and brown (C. Mohr), and has only a slight layer of sapwood. The tree is not so quick of growth as many other pines. According to Dr. Little the tree produces 30,000 feet of first-class timber per acre. It is this species, which forms chiefly the extensive pine-barrens of the United States, and yields largely the American turpentine, as well as resin, pitch and tar. Great Britain in 1884 imported 23,000

tons oil of turpentine, value £560,000, and 73,500 tons resin, value
£376,000; by far the greatest portion of these two articles came from
the United States, where P. Australis would yield a large share. In
1883 the value of oil of turpentine from the United States to Britain
was £533,000, and of resin £386,000. A solution of this oil in
alcohol is known as camphin. The turpentine is obtained by
removing in spring and summer stripes of bark by chipping and collecting the effluence into appropriate boxes particularly applied. The
first yield is the best; in the fourth year the tree becomes exhausted
(Prof. C. Mohr). The average annual yield during this time is 20
lbs. Porcher observes, that the tree shoots up devoid of branches for
sometimes as much as 60 feet, and he calls it "one of the greatest
gifts of God to man." The tree prevails, according to C. Mohr,
where the silicous constituents of the drift-soil mingle with the out-
crops of tertiary strata, and he observes, that forests of this pine
cause grateful showers with wonderful regularity through all seasons.
The emanations from pines, particularly the very resinous species,
are antimalarian and antiseptic, as proved by residences near pine-
forests, and by the use of hospital buildings constructed by pine-
wood.

Pinus Ayacahuite, Ehrenberg. (*P. Loudoniana,* Gordon.)

In Mexico, at an elevation of 8,000 to 12,000 feet. An excellent
pine, to 150 feet high, with a stem-diameter of three to four feet.
It has the habit of P. excelsa, and is equal to it in its own line
of beauty (Beecher) and in hardiness, yielding a much esteemed
white or reddish timber. Its cones are among the very largest,
measuring as much as 15½ inches in length (Sir J. Hooker). Furnishes a fragrant balsamic turpentine and resin, the latter used as
incense (O. Finck).

Pinus Balfouriana, Jeffrey. (*P. aristata,* Engelmann.)

The "Fox-tail or Hickory-Pine." California to Colorado, up to
12,000 feet elevation. Height reaching 100 feet; trunk-diameter
reaching 5 feet. Wood close-grained, tough, very strong (Sargent).

Pinus balsamea, Linné.

"Balsam-Fir, Balm of Gilead-Fir." Canada, Nova Scotia, south
to New England, Pennsylvania and Wisconsin. An elegant tree, to
40 feet high, which with Pinus Fraseri yields Canada-balsam (Balsam of Firs), the well-known oleo-resin. The timber is light, pale,
soft and useful for furniture and implements. The wood is also of
very particular value for superior violins (Dr. R. Tannasch). Rate of
circumferential stem-growth in Nebraska 26 inches at 2 feet from the
ground in 12 years (Governor Furnas). The tree does not attain a
very great age. Sends a pleasant odor through the forest, regarded as
salubrious, to especially phthisic patients, a remark which applies to
many other pines. It thrives best in cold swampy places. Eligible
for alpine regions; in Norway it is hardy to lat. 63° 26'.

Pinus bracteata, D. Don.

Southern California, up to 6,000 feet. A very handsome fir, attaining about 150 feet in height, forming a slender, perfectly straight stem, not more than two feet in diameter. A somewhat older name is P. venusta (Douglas). The resin is used for incense. The young shoots, according to Hemsley, injured by spring-frosts in Britain.

Pinus Brunoniana, Wallich. (*P. dumosa*, D. Don.)

Himalaya, descending to 8,000 and ascending to 10,500 feet. This fir attains a height of about 120 feet, and the stem a circumference of 28 feet (Sir J. D. Hooker). Particularly eligible for alpine tracts. The timber is pale and soft, and does not stand exposure well.

Pinus Canadensis, Linné.

"Hemlock-Spruce." In Canada and over a great part of the United States, on high mountains, as well as on undulating land. A very ornamental fir, to about 100 feet high, with a pale cross-grained wood, remarkably durable when used for submerged water-works; also employed for railway-ties. According to Vasey it is one of the most graceful of spruces, with a light and spreading spray. Schacht saw aged stems on which 440 wood-rings could be counted. Can be kept trimmed for hedges. Next to P. Strobus it is the highest pine of the Eastern States of North-America. The tree is extremely valuable on account of its bark, which is much used as a tanning material, containing 9 to 14 per cent. tannin; this is much liked as an admixture to oak-bark for particular leathers of great toughness, wearing strength and resistance to water. The extract of the bark for tanning fetches in the London market from £16 to £18 a ton, and is imported to the extent of 6,000 tons a year; the bark is stripped off during the summer-months. The young shoots are used in making spruce-beer. P. Caroliniensis is the Hemlock-Spruce of Carolina. The bark is in medicinal use also.

Pinus Canariensis, C. Smith.*

Canary-Pine. Canary-Islands, forming large forests at an elevation of 5,000 to 6,000 feet. A tree, reaching the height of 80 feet, with a resinous, durable, very heavy wood, not readily attacked by insects. It thrives well in Victoria, and shows celerity of growth. Will endure an occasional shade-temperature of 118° F. (W. J. Winter). Growth in height at Port Phillip 45–50 feet in 20 years.

Pinus Cedrus, Linné.*

Cedar of Lebanon. Together with the Atlas-variety on the mountains of Lebanon and Taurus, also in North-Africa and Cyprus. The tree grows to a height of about 100 feet, with a heavy trunk sometimes 46 feet in circumference (Booth) and attains a very great age. Goeppert and Russegger allot to Lebanon-Cedars an age reaching to the commencement of the Christian era. The wood is of a light-reddish color, soft, almost inodorous, easy to work, and much esteemed for its durability.

Pinus Cedrus, *rar.* Deodara.*

Deodar-Cedar. On the north-western Himalaya-Mountains, also in Afghanistan, 3,000 to 12,000 feet above the sea-level. A majestic tree, reaching a hight of more than 300 feet, and sometimes over 40 feet in circumference of stem. The wood is of a light-yellow color, very close-grained and resinous, strongly and agreeably scented, light, extremely durable, well resisting the vicissitudes of a changeable clime, and furnishes one of the best building-timbers known. Pillars of Kashmir-mosques made of this wood are found sound after 400 years, and bridges of still greater antiquity are in existence. White ants hardly ever attack the heartwood. Boats built of this wood have lasted about forty years. It is also extensively used for canal-edges and for railways. The tree should not be felled too young. It yields a good deal of resin and turpentine. A humid clime very much accelerates the growth of this pine, which would come best and · quickest to its development in forest-ranges. Deodars will endure, when not too young, an exceptional temperature of 118° F. in the shade (W. J. Winter). Rate of growth at Port Phillip, 40 to 50 feet height in 20 years.

Pinus Cembra, Linné.

On the European Alps, also in Siberia and Tartary, extending to Kamtschatka, the Kuriles and arctic America. Less hardy than P. Laricio, although from high Alps; still it grows to a height of 60 feet at Christiania (Schuebeler). The "Zirbel-Pine" attains a height of about 120 feet, the stem gets fully 4 feet in diameter, but the growth is slow. The wood is of a yellow color, soft and resinous, of an extremely fine texture, and is extensively used for carving and cabinet-work. The seeds are edible, and when pressed yield a great quantity of oil, as much as 47 per cent., according to Schuppe. A particular turpentine is also obtained from this pine, called Carpathian-balsam.

Pinus cembroides, Zuccarini. (*P. Llareana,* Schiede and Deppe.)

Mexican Swamp-Pine. A small tree to 30 feet high, growing at elevations from 8,000 to 10,000 feet. The timber is not of much use, but the seeds are edible and wholesome, and have a very agreeable taste.

Pinus Cilicica, Antoine and Kotschy.

Cilician Silver-Fir. Asia Minor. 4,000 to 6,500 feet above sea-level. A handsome tree of pyramidal growth, to 160 feet high. Quite hardy in climes like that of Vienna. The wood is very soft, and used extensively for the roofs of houses, as it does not warp.

Pinus concolor, Engelmann.

The great White Silver-Fir. North-Western America, at elevations of 8,000 to 9,000 feet. A fir reaching 150 feet in height; trunk to 4 feet in diameter. The wood is tough, eligible for building-purposes and other substantial work (Vasey). It does not warp, shrinks hardly at all, makes choice ceilings, and needs less paint than most other timber (Kellogg).

Pinus contorta, Douglas. (*P. Bolandri*, Parlatore.)

On high damp ranges in California, Oregon and British North-Western America; also abundant on the mountains of Colorado; very eligible for clothing rocky hill-sides (Meehan). In California this pine forms dense thickets along the coast, and is in this respect as valuable as P. Laricio, P. Pinaster and P. Halepensis in Europe, as a shelter-tree in stormy localities. Dr. Gibbons remarks of this pine, which vernacularly is called Tamarak or Hack-me-tack, that its size has generally been underrated. At the foot of the Sierra and on mountains 8,000 feet high he saw it in great numbers, forming one of the most stately of forest-pines, not rarely attaining a height of 150 feet and 4 feet in stem-diameter. The timber is pale, straight-grained and very light; there considered the best and most durable material for dams and for general building purposes. It furnishes sea-ports with piles and masts; yields also railway-ties. Its value is beyond calculation. Dr. G. Dawson notes, that the cambium-layer is so saccharine as to afford food to the autochthones. This species includes P. Murrayana, Balfour.

Pinus Coulteri, D. Don.

California, on the eastern slope of the coast-range, at elevations from 3,000 to 4,000 feet. A pine of quick growth, attaining a height of about 100 feet, with a trunk about 4 feet in diameter; with P. Ayacahuite, P. Montezumæ, P. Lambertiana, P. Sabiniana and P. excelsa it has the largest cones of all pines, comparable in size and form to sugar-loaves. The nuts are nutritious (Vasey).

Pinus Cubensis, Grisebach. (*P. Elliottii*, Engelmann.)

Swamp-Pine, Slash- or Bastard-Pine. Higher mountains of Cuba, also in the Southern States of Eastern North-America. Allied to P. Tæda. Likes moist, sandy, flat lands. Height of tree to 120 feet, of clear stem to 70 feet; growth comparatively quick, overpowering P. Australis. Yields some turpentine and resin (Prof. C. Mohr).

Pinus densiflora, Siebold and Zuccarini.

The "Akamatsou-Pine" of Japan, where it forms along with P. Thunbergi extensive forests at 1,000 to 2,000 feet above sea-level. It is hardy at Christiania. Attains an age of several centuries (Rein). The timber is excellent for building; it is less resinous than that of P. Thunbergi (Dupont).

Pinus Douglasii, Sabine.[*]

Oregon-Pine or Fir, called also the Yellow Pine or Fir of Puget-Sound, where it yields the principal timber for export, and is therefore of great commercial value in the lumber-trade. It extends from Vancouver's Island and the Columbia-River through California to Northern Mexico, from the coast up to the higher mountains of 9,000 feet. The maximum height known is nearly 400 feet; the greatest diameter of the stem 14 feet. Can be grown very closely, when the stems will attain, according to Drs. Kellogg and Newberry, a height

of over 200 feet without a branch. A densely wooded forest will
contain about 36 full-grown trees to an acre. The timber is fine and
clear-grained, heavy, strong, soft, and hence easily worked, yet firm
and solid, splendid for masts and spars, ships' planks and piles; also
valuable for flooring, being for that purpose regarded as the best of
California (Bolander). It will bear a tension of 3 to 1 as compared
with the Sequoias. It is the strongest wood on the North-Pacific
coast, both in resisting horizontal strain and perpendicular pressure.
Sub-alpine localities should be extensively planted with this famous
tree. It requires deep and rich soil, but likes shelter; its growth is as
rapid as that of the larch; it passes in various localities as Black and
Red Spruce. Both in clayey and light soil it attains 50 feet in about
eighteen years; it requires however a moist forest-clime for rapid
growth.

Pinus edulis, Engelmann.

New Mexico. A pine, not tall, but very resinous. Wood easily
split. One of the best for fuel (Meehan). It yields the " Pino "-
nuts, which are produced in immense quantities and are of very
pleasant flavor (Sargent). Closely cognate to P. monophylla.

Pinus excelsa, Hamilton.*

The Lofty- or Bootan-Pine. Himalaya, forming large forests, at
from 5,000 to 12,500 feet elevation; also in Macedonia and Monte-
negro. A fine tree, at length about 150 feet high, furnishing a valuable, close-grained, soft and easily workable wood, ranking among
Himalayan pine-woods for durability next to Deodar-timber (Stewart
and Brandis); the wood is also highly recommended for patterns in
foundries, further for levelling-staves and cot-planks (Watson). This
pine also furnishes a good quantity of turpentine. Under cultivation
it shrinks before a fierce summer-sun (Beecher); but will bear the
winter of Christiania (Schuebeler). Cones often 15 inches long
(Sir J. Hooker). This tree produces seeds early and copiously;
disseminates itself easily even on steep bare declivities (Brandis).

Pinus firma, Antoine.

Northern Japan, at 2,000 to 4,000 feet above the sea-level in
humid valleys. A lofty tree of the habit of the Silver-Fir. The
timber is pale, soft and fine-grained, employed particularly by
coopers and upholsterers.

Pinus flexilis, James.

The White Pine of the Rocky Mountains, also known as the Bull-
Pine. From New Mexico to British Columbia, ascending to 13,000
feet. Prefers the limestone-formation. A valuable fir for cold
regions. It attains a height of 150 feet, according to Dr. Gibbons,
but Mr. J. Hoopes states, that it is of slow growth. Wood pale, soft
and compact, of fine texture, according to Prof. Sargent intermedi-
ate between that of P. Strobus and P. Lambertiana. Dr. G. Dawson
noted, that the seeds afford food to the autochthones.

Pinus Fortunei, Parlatore.   (*Abies Jezoensis*, Lindley.)

China, in the neighbourhood of Foo-Chow-Foo. A splendid fir, to 70 feet high, somewhat similar in habit to P. Cedrus.

Pinus Fraseri, Pursh.

Double Balsam-Fir. On high mountains of Carolina and Pennsylvania. This tree, which grows to a height of about 20 feet only, yields with P. balsamea the well-known Canada-balsam. The tree is hardy at Christiania still.

Pinus Gerardiana, Wallich.

Nepal Nut-Pine. In the north-eastern parts of the Himalayas at an elevation of 10,000 to 12,000 feet, extending to Afghanistan. With P. Deodara, P. excelsa, P. Webbiana, P. Smithiana and Juniperus excelsa reaching the highest regions of pine-forests in Southern Asia. The tree attains a height of 60 feet, with a comparatively short stem, exceptionally 10 feet in girth, and produces very sweet edible seeds, also turpentine. Hoopes refers to it as remarkable for the copiousness of its resin. In reference to the nut-seeds the proverb prevails at Kunawar, " One tree a man's life in winter."

Pinus glabra, Walter.

From Carolina to the Mississippi. Allied to P. mitis. It attains, according to Prof. C. Mohr, a height of about 80 feet. Dr. Porcher compares the wood to that of P. Strobus.

Pinus grandis, Douglas.

From California to British Columbia. . Great Silver-Fir, also known as the Yellow-Fir. A splendid quick-growing fir, to 200 feet high and upwards, growing best in moist valleys of high ranges. The stem occasionally attains a diameter of 7 feet at 130 feet from the ground, and of 6 feet at 200; concentric wood-growth of stem as much as 1 inch in a year; height reached by the tree in Wales 75 feet in 33 years (A. D. Webster). Trees occur of 15 feet stem-diameter and 320 feet high. The wood is pale and soft, too light and brittle, according to Dr. Vasey, for general purposes; while Prof. Brewer asserts, that it is employed for boards, boxes, cooperage, and even much sought for ship-building, but it seems fit only for inside work; it is of pleasant scent. Rate of upward growth in favorable places about 2½ feet in a year. Hardy in England (Hemsley). P. subalpina (Englemann) is closely cognate.

Pinus Griffithii, Parlatore.   (*Larix Griffithii*, J. Hooker and Thomson.)

The Himalayan Larch. Descends to 8,000 feet and ascends to 12,000 feet. Timber pale, soft, without distinct heartwood, one of the most durable of all pine-timbers (Stewart and Brandis). P. Ledebourii (Endlicher) is the Siberian Larch.

Pinus Halepensis, Miller.   (*P. maritima*, Lambert.)

Aleppo-Pine. South-Europe and North-Africa, South-Western Asia. This well-known pine attains a height of 80 feet, with a stem to 5 feet in diameter. The timber of young trees is pale, of older trees dark-colored; it is principally esteemed for ship-building, but also used for furniture. The tree yields a peculiar kind of turpentine, as well as a valuable tar. Although ascending mountains in South-Europe to the height of 4,000 feet, it thrives best in sandy coast-lands, where in ten years it will measure 25 feet, and finally will become a larger tree than on firmer lands. M. Boitel has published a special work on the importance of this pine for converting poor sand-land into productive areas, referring also to P. silvestris and P. Laricio for the same purpose. According to Mr. W. Irvine Winter it will resist an occasional heat of 118° F. in the shade. We find the Aleppo-Fir one of the best of evergreen avenue-trees in Victoria, as first proved by the writer. It is content with the poorest and driest localities, and also here comparatively rapid in growth.

Pinus Hartwegii, Lindley.

Mexico, 9,000 to 14,000 feet above sea-level. A pine, reaching 150 feet in height, with a very durable wood of a reddish color; it yields a large quantity of resin.

Pinus Hookeriana, McNab. (*Abies Hookeriana*, Murray.)

California, at 5,000 to 6,000 feet elevation. A fir, allied to P. Pattoniana, but distinct (Dr. McNab). Height of tree to about 300 feet, stem perfectly straight. Wood hard, of a reddish color, with handsome veins. Not a resinous tree. Hardy in Middle Europe.

Pinus Hudsonica, Poiret. (*P. Banksiana*, Lambert.)

Grey Pine. Colder parts of North-America, both eastern and western up to 64° north latitude. Height of tree as much as 40 feet; in the cold north only a shrub. The wood is light, tough, resinous and easily worked.

Pinus inops, Solander.

Eastern North-America. The Jersey-Pine. A tree content with barren soil, attaining a height of 40 feet, available for fixing drift-sand on coasts. Easily disseminated. Remarkably rich in resin, hence to be classed with pines most desirable for sanitary plantations. Wood reddish-yellow. P. Virginiana (Miller) is by far the eldest name.

Pinus Jeffreyi, Murray.

California. A pine, to 150 feet in height. Hardy at Christiania, Norway. The glaucous branchlets of aromatic fragrance with thinner and greyish leaves, the greater size of the fruit-cones with thin and recurved spines to the scales, the larger nutlets and more numerous cotyledons separate this pine from P. ponderosa (Engelmann, Sargent, Perry).

Pinus Jezoensis, Antoine.

Amur and Japan. This spruce is closely allied to the N. W. American P. Sitchensis. Picea Ajanensis (Fischer) is identical. For synonyms see Dr. Masters' essay in the Journal of the Linnean Society 1881.

Pinus Kæmpferi, Lambert.

Chinese Larch; also called Golden Pine. North-Eastern China. This is the handsomest of all the larches; it forms a transit to the cedars. Resists severe frost. It is of quick growth and attains a height of 150 feet. The leaves, which are of a vivid green during spring and summer, turn to a golden-yellow in autumn. The wood is very hard and durable.

Pinus Kasya, Royle.

Kasya and also Burmah, from 2,000 to 7,000 feet. Closely related to P. longifolia. Attains a height of 200 feet. Wood very resinous, somewhat fibrous, rather close-grained, pale-brown with darker waves (Kurz).

Pinus Koraiensis, Siebold and Zuccarini.

Kamtschatka, China and Japan. A handsome pine, often to 40 feet high, producing edible seeds.

Pinus Lambertiana, Douglas.*

Shake- Giant- or Sugar-Pine. British Columbia and California, mostly at great altitudes. A lofty tree, of rapid growth, upwards of 300 feet high, with a straight stem attaining 60 feet in circumference. It holds, in most places, preëminence in beauty and size over accompanying pines, and reaches an age of 600 years (Dr. Vasey). It thrives best in sandy soil, and produces a soft, pale, straight-grained wood, which for inside-work is esteemed above any other pine-wood in California, and obtained in large quantities; it is especially used for shingles, flooring and for finishing purposes by joiners and carpenters. The tree yields an abundance of remarkably clear and pure resin, of sweet taste, eaten even by the natives. The cones may be 19 inches long; the seeds are edible. This pine would come to perfection best in the humid regions of higher mountains. P. reflexa (Engelmann) is an allied large species with smaller fruit, occurring in Arizona.

Pinus Laricio, Poiret.* (*P. maritima*, Miller.)

Corsican Pine. South-Europe, ascending to about 6,000 feet. It attains a height of 150 feet. A splendid shelter-tree in the coldest regions. It will succeed on stiff clay as well as on sandy soil, even on sea-sand. The wood is pale, towards the centre dark, very resinous, coarse-grained, elastic and durable, and much esteemed for building, especially for water-works; valuable also for its permanency underground. There are three main-varieties of this pine, namely, P. L. Poiretiana in Italy, P. L. Austriaca in Austria, P. L. Pallasiana on the borders of the Black Sea. The tree grows best in calcareous soil, but also in poor sandy soil, where however the timber is not so large nor so good. It yields all the products of P. silvestris, but in greater quantities, being perhaps the most resinous of all pines. Assumed to attain an age of 500 years (Langethal). The Austriaca variety attained a stem-girth of nearly 2 feet in 10 years when cultivated in Nebraska (Governor Furnas). This species is regarded by some as even preferable for timber-rearing to P. silvestris and P. Larix. Rate of upward growth 1½–2 feet in a year.

Pinus Larix, Linné.

Common Larch. On the European Alps, up to 7,000 feet. A tree of quick growth in cool localities; adapted to poor soil, its foliage as in all larches deciduous. It attains often a height of 100 feet, sometimes rising even to 160 feet, and produces a valuable timber of great durability, which is used for land- and water-buildings, and much prized for ship- and boat-building; for staves of wine-casks almost indestructible, not allowing the evaporation of the spirituous contents (Simmonds), also much employed for pumps. The Briancon-Manna exudes from the stem. Larch-trees, cut in Bohemia, have shown over 500 annual rings in their wood (Langethal). Larch-timber lasts three times longer than that of the Norway-Spruce, and although so buoyant and elastic it is tougher and more compact; it is proof against water, not readily igniting, and heavier and harder than any deal (Stauffer). The Venetian houses, constructed of larch-wood, showed for almost indefinite periods no symptoms of decay. This wood is also selected for the most lasting panels of paintings. The bark is used for tanning and dyeing. The tree is also of great import-ance for its yield of Venetian turpentine, which is obtained by boring holes into the stem in spring; these fill during the summer, supply-ing from half to three-quarters of a pint of turpentine. In Piedmont, where they tap the tree in different places, and let the liquid continu-ally run, it is said, that from seven to eight pints may be obtained in a year; but the wood suffers through this operation. The larch is grown in Norway to lat. 66° 5'; in 63° 26' a tree still attained a height of over 70 feet (Professor Schuebeler). P. L. *var.* Rossica, the Russian Larch, grows principally on the Altai-Mountains, from 2,500 to 5,500 feet above sea-level. The species would be important for upland, particularly alpine country, even for peatbogs.

Pinus leiophylla, Schiede and Deppe.

At elevations of from 7,000 to 11,000 feet on the mountains of Mexico. A pine to as much as 90 feet high. A very resinous species, according to Mr. Hugo Finck. The wood is excessively hard.

Pinus leptolepis, Endlicher.

The Karamatsou or Japan-Larch. In Japan, between 35° and 48° north latitude, up to an elevation of 9,000 feet. Never a very tall tree. The timber, when mature, reddish-brown and soft; it is highly valued by the Japanese.

Pinus longifolia, Roxburgh.*

Emodi-Pine or Cheer-Pine. On the Himalayan mountains, from 2,000 to 9,000 feet. A handsome tree, with a branchless stem for 50 feet, the whole tree attaining a maximum-height of somewhat over 100 feet, the girth of the stem 12 feet. Does not like much shade. Growth in height at Port Phillip about 40 feet in 20 years. The wood is resinous, and the red variety useful for building; it yields a quantity of tar and turpentine. The branches are used for torches by the rural population of its native country (Dr. Brandis). The tree stands exposure and heat well. According to W. J. Winter it endures an occasional shade-temperature of 118° F.

Pinus Massoniana, Lambert.

China. A good-sized pine, with widely spreading ramifications. The wood, when well-seasoned, is much employed as material for tea-boxes. Prof. C. Koch regards P. sinensis (Lambert) a distinct species.

Pinus Merkusii, Junghuhn.

Burmah, Borneo and Sumatra, chiefly at elevations of from 3,000 to 4,000 feet. A tall pine. The only species of Pinus, which extends south of the equator, closely related to P. Massoniana. Wood exceedingly resinous (Brandis); stems valuable for masts and spars, according to Mr. Gamble. Weight of wood about 50 lbs. per cubic foot. The resin of this pine resembles Damar (Wiessner).

Pinus Mertensiana, Bongard.

Western Hemlock-Spruce. North-Western America. The wood is pale, tough and very soft, but is often used for building. This fir gains a height of about 200 feet, with a stem 4 to 6 feet in diameter; the bark is in great repute for tanning; the roots yield strong fibres, even for seines and nets; the tender sprigs are the characteristic ingredient required for making spruce-beer locally; the tree yields also much resin (Dr. Kellogg). Though naturally so tall a tree, it can young be trained and trimmed to the best of garden-hedges of a lovely green (Prof. Bolander).

Pinus mitis, Michaux.*

Yellow Pine of Eastern North-America, extending to Missouri and Texas, called also Short-leaved Pine, in contrast to P. Australis. In dry sandy and more particularly somewhat clayey soil attaining a height of about 90 feet; rapid in growth; eligible for rocky ridges. Wood yellowish, compact, hard, durable, fine-grained, moderately resinous, valuable for flooring, cabinet-work and ship-building. According to Dr. Vasey it commands a higher price even than that of P. Strobus. P. glabra (Walter) is closely allied to P. mitis, and fit for growth on low hummocks. Seeds smaller than those of the North-East American pines, hence easier of transit in quantity (Meehan).

Pinus monophylla, Torrey.

Stone- or Nut-Pine of California, on the Sierra Nevada and Cascade-Mountains, up to 6,500 feet. It thrives best on dry limestone-soil. The large seeds are edible, of almond-like taste, and consumed in quantity by the natives. Height of tree generally about 35 feet, but occasionally as much as 80 feet; stem not of great thickness. This species is not of quick growth. Wood pale, soft, very resinous, much used for charcoal.

Pinus montana, Du Roi. (*P. pumilio,* Hænke.)

On the Alps, Pyrenees and Carpathians, also in Greece, up to the highest points of woody vegetation, covering large tracts, and thriving on the poorest soil. In Norway it will live to lat. 70° 4′ (Schuebeler). This pine grows to about 25 feet height, but in favorable localities to 50; it yields much oil of turpentine. The wood is used largely for carving. Only available to advantage for highlands. The oil, distilled from the foliage of this and many other pines is a safe anthelmintic (Dr. H. Pinkney).

Pinus Montezumæ, Lambert. (*P. Devoniana,* Lindley; *P. Grenvilleæ,* Gordon.)

Mexico. A handsome pine, to 80 feet high; wood pale, soft and resinous. Cone attaining a length of 15 inches (Dr. Masters).

Pinus monticola, Douglas.

From British Columbia to California, at an elevation of 7,000 feet. This pine thrives best in poor soil of granite-formation, and attains a height of about 200 feet, with a stem often 7 feet thick. The wood is pale, close-grained, similar to that of P. Strobus. Dr. Gibbons observes, that this species is less than half the size of P. Lambertiana, but in all other respects resembles it. Woodmen are very pronounced in their statement, that there are two kinds of sugar-pine, both growing in close proximity to each other. Mr. J. Hoopes states, that the wood is similar to White Pine, but tougher. Dr. G. Dawson says, that the aborigines use also the seeds of this pine for food.

Pinus muricata, D. Don.

Bishop's Pine. California. Found up to 7,500 feet. This hardy pine grows ordinarily to about 40 feet, but reaches 120 feet under favorable circumstances. It might be utilized for wind-breaks (Dr. Gibbons). Hardy in Middle Europe.

Pinus nigra, Aiton. (*Abies rubra,* Michaux.)

Black Spruce. Eastern America, occurring extensively between 44° and 53° north latitude. In Norway it will grow to lat. 63° 45′ (Schuebeler). This tree, which is termed Double Spruce by the Canadians, likes humid sheltered localities, attains a height of about 70 feet, and furnishes a light elastic timber of pale color, excellent for yards of ships; largely sawn into boards and quarterings; has also come extensively into use for paper-pulp. The spruce-lumber of eastern markets in the United States is chiefly furnished by this species (Sargent). The young shoots are used for making spruce-beer, and the small roots serve as cords. The tree prefers poor and rocky soil, but a humid cool clime, and is best available for mountainous localities inaccessible to culture. Mr. Cecil Clay estimates,

T

that 20,000 cubic feet of timber can be obtained from this tree on one acre of ground.

Pinus nobilis, Douglas.

Noble White Fir; but also known as Red Fir. Oregon and its vicinity, where it forms extensive forests at 6,000 to 8,000 feet. A majestic tree, attaining a height of 320 feet (Brewer, Gardner), with regular horizontal branches. Timber splendid. P. magnifica, Murray, is a variety. Hardy in Middle Europe.

Pinus Nuttallii, Parlatore.

The Oregon-Larch, at elevations of from 3,000 to 6,000 feet. According to Dr. Gibbons, one of the most graceful trees. Stem straight, frequently 200 feet to the first limb. Timber readily fissile, very strong and durable (Dufur), tough, light and elastic; it can also be employed in water-work (Dr. Kellogg). Tree only available for cool mountain-regions to serve commercial final purposes.

Pinus obovata, Antoine. (*P. Schrenkiana,* Antoine.)

North-Eastern Europe and Northern Asia. Somewhat like the Norway-Spruce. Wood soft and pale, locally used for furniture and household-implements, also for packing-boxes of great durability (Regel).

Pinus orientalis, Linné.

Sapindus-Spruce. Asia Minor, ascending to 6,600 feet, thus becoming alpine. Hardy in Christiania. The tree rises to about 80 feet, and somewhat resembles the Norway-Spruce. The wood is exceedingly tough and durable.

Pinus Parryana, Engelmann.

California. One of the pines with edible nuts.

Pinus parviflora, Siebold and Zuccarini.

The "Imekomatsou." Kuriles and Japan. A middle-sized pine of longevity; ascends to alpine heights. Much used as an avenue-tree. Wood valuable for furniture and boat-building. It is harder than that of P. Thunbergi and P. densiflora (Dupont).

Pinus Pattoniana, McNab.

California, restricted to elevations above 5,000 feet and advancing thence to the glacier-region in a gradually dwarfed state. This fir rises to a maximum height of 150 feet, the stem enlarging sometimes at the base to a diameter of 13 feet (Jeffrey).

Pinus patula, Schiede and Deppe.

Mexico, at elevations of from 6,000 to 12,000 feet. A graceful pine, becoming 80 feet high.

Pinus pendula, Solander. (*P. microcarpa,* Lambert.)

Small-coned American Larch, Black Larch or Tamarack. From Labrador and Canada to Virginia. Delighting in swampy ground.

A pine of pyramidal growth, to 100 feet high. The timber is pale, heavy, resinous, and as highly valued as that of the common larch; it is close-grained, well adapted for underground work; it combines lightness, strength and durability; much sought by ship-builders, as for knees, bends and ship-garlands it cannot be surpassed (Robb); much in use also for railway-ties. Rate of circumferential stem-growth in Nebraska two feet in ten years (Furnas). P. laricina (Du Roi) is by far the oldest name for this larch, as pointed out by Prof. C. Koch.

Pinus picea, Du Roi.* (*P. Abies*, Linné.)

Norway-Spruce, Fichte. Middle and Northern Europe and Northern Asia, rising from the plains to an elevation of 4,500 feet, and forming extensive forests. It exceeds even the birch in endurance of cold. Indigenous in Norway to lat. 69° 30' (Schuebeler). Adapted to most kinds of soil. The tree attains a height of 150 feet or even more, and furnishes an excellent timber, commonly known under the name of White Deal, for building and furniture, for masts, spars, ladders and oars. Stems of 6 feet diameter are on record with more than 200 wood-rings. It also produces the Burgundy-pitch in quantity, while the bark is used for tanning. Though enduring dry summers, this spruce would have to be restricted for timber-purposes to damp mountains. A variety with pendent branches occurs. Hemsley mentions other forms of this spruce, and indeed many varieties of other species of Pinus. Britain alone imported in recent time pinewood to the value of nine millions sterling annually, of which P. picea must have furnished a considerable portion.

Pinus Pinaster, Solander.* (*P. maritima*, Poiret and De Candolle.)

Cluster-Pine. From the shores to the mountains of the countries on the Mediterranean Sea. The tree rises to about 60 feet in height. The wood is soft and resinous; it yields largely the French turpentine. Among the best of pines for consolidating sandy coasts, and for converting rolling sands into pastoral and agricultural land. For ease of rearing and rapidity of growth one of the most important of all pines. Average-growth at Port Phillip 40 feet in 20 years. On the testimony of Mr. J. Hoopes, it does not thrive well on calcareous soil. W. J. Winter observed, that P. Pinaster and the allied P. Pinca can withstand an occasional shade-temperature of 118° F. A tree 60 to 70 years old, heavily tapped, yields 12 to 16 lbs. of turpentine, equal to 4 lbs. of resin, the rest being oil of turpentine (Simmonds). The tree comes into full flow of turpentine at about 25 years, and the tapping process, if only a slight one, is endured by this pine for an enormous length of time. Thus the annual production of resin from a good tree fluctuates between 5 and 8 lbs. The quantity of resin gathered in France during 1874 was about sixty million pounds (Crouzetter-Desnoyers). The felling of up-grown pines, planted with wise foresight for antimalarian and other hygienic purposes at places of centres of population, can but be regarded as most

T 2

reprehensible, when the simple reason of such destruction consists in replacing the pines by other perhaps more fashionable but less sanitary trees.

Pinus Pinceana, Gordon.

Mexico, up to 9,000 feet above the sea-level. A very remarkable pine, frequently to 60 feet high, having drooping branches like the Weeping Willow. Most desirable for cemeteries.

Pinus Pindrow, Royle.

Himalayan mountains, 7,000 to 12,000 feet above the sea-level. A fine, straight-stemmed fir, becoming 190 feet high; cones purple. Considered by Stewart and Brandis a variety of P. Webbiana.

Pinus Pinea, Linné.*

Stone-Pine. Countries bordering on the Mediterranean Sea, extending to the Canary-Islands. Height of tree 80 feet; top rather flat. The wood is whitish, light, but full of resin, and much used for furniture, naval architecture and general building purposes. The seeds are edible, but of a resinous though not disagreeable taste; they should be left in the cones until they are about to be used, as otherwise they speedily become rancid; they only ripen in their third year. This pine grows as easily and almost as quickly as the Cluster-Pine. The bark contains much tan-principle.

Pinus Pinsapo, Boissier.

Spanish Fir. Spain and North-Africa, at from 3,000 to 6,000 feet elevation. A tree to 70 feet high, with branches from the ground. The timber is similar to that of the Silver-Fir and resinous.

Pinus polita, Antoine.

Japan and Kurile-Islands. A tall superb spruce, forming large forests on the mountain-ranges (A. Murray). Resists severe frost. Allied to P. Smithiana.

Pinus ponderosa, Douglas.* (*P. Benthamiana*, Hartweg.)

Yellow Pitch-Pine or Trucker-Pine. North-Western America. Height of tree often to 225 feet, with a stem reaching 24 feet in circumference. Growth comparatively quick. The wood is yellowish, hard, strong, durable and heavy; for general purposes it is preferred to that of any other pine, and also largely used in mining operations. There are fine groves of this tree up to 5,000 feet elevation in California, but the variety P. Engelmanni (Parry) ascends to 12,000 feet. The bark contains a considerable quantity of tanning substance. Wood pale and soft, neither knotty nor resinous, much esteemed for cabinet-work (Hoopes); it is of great strength, and used for floors, joists and much other work in carpentry. Gibbons relates, that the wood, with the bark adherent, exposed to the weather, will decay within a year, but that when stripped and covered with soil it is readily preserved. Dr. Kellogg, who aptly calls this tree herculean, saw logs, which had been in the ground twelve years, quite sound. This pine has proved well adapted even for rather dry localities in Victoria, but is there slow of growth.

Pinus Pseudo-Strobus, Lindley.

Mexico, up to 10,000 feet. This pine is superior in appearance to any other Mexican pine; height to about 80 feet.

Pinus pungens, Michaux.

South-Eastern States of North-America. Although seldom over 50 feet high, this pine has the recommendation of being of remarkably quick growth, especially in early life.

Pinus Pyrenaica, Lapeyrouse. (*P. Brutia*, Tenore.)

In the countries at the Mediterranean Sea, ascending to 5,000 feet. A pine, of quick growth, to 80 feet in height; the wood is pale and dry, almost free from resin, and of considerable value. The tree commences to bear fruits in about a dozen years already.

Pinus radiata, D. Don.* (*P. insignis*, Douglas.)

California. A splendid dark-green pine, fully to 100 feet high, with a straight stem, occasionally 8 feet in diameter. It is the quickest growing of all pines, a seedling one year old being strong enough for final transplantation; it has been noticed to grow fully 5 feet annually in light soil near Melbourne. Mr. J. Dickinson found it to attain a height of 70 feet with a stem-girth of 5 feet in 13 years at Port Phillip. According to Mr. W. J. Winter it will endure unhurt exceptional exposure to 118° F. in the shade. In the United Kingdom it suffers greatly from the attacks of the Pine-Beetle, Hylurgus piniperda (Lawson). The wood is tough, and is sought for boat-building and various utensils. This tree can be utilized for obtaining tar and pitch. It bears exposure to the sea at the very edge of the coast. Produces fruit-cones only at somewhat advanced age. Mr. J. Kruse, on the author's suggestion, subjected the foliage to distillation, obtaining oil in 0·01 quantity, of 0·845 specific gravity, of 293° F. boiling-point, and a pleasant odor reminding of Geneva-gin.

Pinus reflexa, Engelmann.

California. Allied to P. flexilis, belonging to the Strobus section, but with large inappendiculated nutlets.

Pinus religiosa, Humboldt.

Oyamel-Fir. Mexico, from 4,000 to 11,500 feet above the sea-level, thus reaching the limits of arboreous vegetation. A magnificent tree with somewhat silvery leaves, growing to a height of 150 feet, stem reaching 6 feet in diameter. The wood is particularly well fitted for shingles and laths. This species endures the winters of Middle Europe.

Pinus resinosa, Solander.

Red Pine. North-America, principally Canada and Nova Scotia, but extending to Pennsylvania. It attains a height of 150 feet, the stem a diameter of 2 feet. It is of rapid growth, and on account of the red-barked stem very ornamental (Sargent); delights in sandy soil; the wood is hard, fine-grained, heavy and durable, not very resinous, and is used for ship-building and structures of various kinds.

Pinus rigida, Miller.*

American Pitch-Pine. From New England to Virginia. It grows to a height of 80 feet; the timber from gravelly or rocky soil heavy and resinous, from damp alluvial soil light and soft; used for building; but the tree is principally important for its yield of turpentine, resin, pitch and tar. It is suitable for sea-shores; it will also grow in the driest localities, as well as in swamps, nor is it readily susceptible to injury from fire. Professor Meehan mentions this as the most rapid grower among North-East American pines. With P. Tæda among the most oleous and resinous pines, to be disseminated million-fold in such extensive malarial regions as cannot be readily or profitably drained, to subdue miasmata by the copious evolution of the double oxyde of hydrogen and ozone. The first trees in Australia were reared by the writer of this work.

Pinus rubra, Lambert.

The Red Spruce. North-Eastern America. Allied to P. alba and P. nigra. Wood reddish-brown.

Pinus Sabiniana, Douglas.*

Californian Nut-Pine or White Pine. From California to the Rocky Mountains. Height to 150 feet; stem frequently 5 feet in diameter. The wood is pale and soft; according to Dr. Gibbons it is hard and durable when seasoned, with close and twisted grain, and contains much resin; for fuel, when well distributed heat is requisite, far surpassing all other Californian woods in value; yields an abundant supply of excellent turpentine, and thence again by distillation a superior oil (Dr. Kellogg). The clustered heavy cones attain a length of one foot. The seeds are edible; they are produced in great profusion, and constituted formerly a large portion of the winter-food of the native tribes. Proves even in dry localities of Victoria to be of quick growth.

Pinus selenolepis, Parlatore.

Japan, up to elevations of 7,000 feet. This fir rises to a height of about 150 feet. It is known also as Veitch's Fir.

Pinus serotina, Michaux.

Pond-Pine. Southern States of Eastern North-America, in morassy soil, principally near the sea-coast. It gets to be 50 feet high. The wood is soft. Of importance as antimalarian for fever-swamps. Regarded by Prof. Meehan as an extreme form of P. rigida.

Pinus Sibirica, Turczaninow. (*P. Pichta*, Fischer.)

Siberian Pitch-Fir. Russia, westward to the Volga, eastward to Kamtschatka, ascending the Altai-mountains to 5,000 feet. This pine reaches a height of about 50 feet.

Pinus silvestris, C. Bauhin.

Scotch Fir or Pine, Foehre, Kiefer. Europe, Northern and Western Asia, reaching to 70° north latitude, ascending the Alps to 6,000 feet, extending south-eastward to the Black Sea, thriving best in sandy soil. Of all trees the one, which needs the least of mineral aliment from the soil; hence adapted for pure sand, where it forms twice as much humus within the same time as Robinia pseudacacia or poplars, while its wood is much more valuable. More easily transplanted than any other European species (Wessely). A very valuable tree, becoming fully 100 feet high, usually growing to an age of about 120 years, but sometimes getting much older; thus a venerable tree at Schandau, blown down by a storm, showed 463 annual rings. It is important for masts and spars. The Red Baltic, Norway- or Riga-deals are obtained from this pine, as well as a large portion of the European pine-tar and pitch. Great Britain, in 1884, imported 173,000 barrels of tar, valued at £130,000, to a great extent furnished by this tree. A kind of vanillin is prepared from the cambium-sap of this pine. Its cones have come into use for tanning in France. Proves well adapted even for the drier parts of Victoria. Maximum rate of growth in Nebraska according to Governor Furnas (in Prof. Meehan's *Gardeners' Monthly*) 3 feet stem-circumference at 2 feet from the ground in 10 years. The leaves of pines can be well converted into material for pillows and mattresses, with the great recommendation of healthfulness for such a purpose. All fir-forests are antimiasmatic and salubrious for hectic patients, in consequence of the di-oxyde of hydrogen evolved from their terebinthine emanations. The annual importation of tar and resin from Coniferæ into Britain approached, recently, one million sterling in value, of which P. silvestris must have contributed a large share.

Pinus Sitkensis, Bongard. (*P. Menziesii*, Douglas.)

North-Western America. The Blue Spruce of California, also called Tideland-Spruce, ascending to elevations of 9,000 feet, of rapid growth in congenial soil. A very handsome tree, which furnishes soft, light, pale and fine-grained timber, used largely for piles (Dr. Gibbons). It thrives best in moist ground. According to Professor Brewer, instances are on record of trees having attained a height of over 300 feet, and a stem of 7 feet in diameter at 100 feet from the base. From an exceptionally large tree 100,000 shingles were obtained, besides 58 cords of wood.

Pinus Smithiana, Lambert. (*P. Khutrow*, Royle.)

Himalaya-mountains, at elevations from 6,000 to 11,000 feet, extending to Afghanistan and to China; this spruce, known vernacularly also as Kutro- or Morinda-Spruce, attains a height of 150 feet, and the stem a girth of 21 feet. The wood is pale, even and straight-grained, but only durable under shelter, but for inside-work greatly in use. Hardy in Middle Europe.

Pinus Strobus, Linné.*

Weymouth-Pine or American White Pine. North-Eastern America, growing on any soil, but particularly adapted for deep, rich ground in mountain-valleys; known to reach a height of 270 feet, with a stem as much as 8 feet in diameter. It is the principal pine of the lumber-trade of the Eastern States. One of the finest among ornamental conifers. The wood is soft, whitish or yellowish, light, free from knots, almost without resin, easy to work, very durable, and much esteemed for masts, bridges, frames of buildings, windows, ceilings, flooring, oars, cabinet-work and organ-pipes. The tree yields American turpentine and galipot. Mr. Cecil Clay cut exceptionally 40,000 feet of its timber on an acre of ground in the Virginian mountains. The sap-wood is remarkably thin. The tree endures the climate of Norway to lat. 61° 15' (Professor Schuebeler). Maximum rate of circumferential stem-growth in Nebraska 2½ feet in 12 years (Governor Furnas). The softest and least resinous of pine-woods can advantageously be converted into paper-material as an admixture to other substances; in Europe the wood of P. picea and P. Abies is preferentially used for this purpose.

Pinus Tæda, Linné.

Frankincense- or Loblolly-Pine. Florida, Carolina and Virginia, westward to Texas, in moist loamy-sandy soil, attaining a height of about 120 feet. The timber is liked for pumps, but liable to warp and decay in buildings on exposure (Sargent). Stems sought for masts (Prof. Mohr). The tree yields turpentine in good quantity, though of inferior quality, and exudes much resin; it likes regions near the coast; hence can be well utilized for raising fir-forests on shore-lands, especially as this pine takes readily possession of cleared forest-ground, and by quick growth overpowers other young trees (Prof. C. Mohr).

Pinus tenuifolia, Bentham.

Mexico, at an elevation of about 5,000 feet, forming dense forests. Height of this pine to nearly 100 feet; stem to 5 feet in diameter.

Pinus Teocote, Chamisso and Schlechtendal.

Okote- or Torch-Pine. Mexico, from 5,000 to 11,000 feet above the sea-level. Tree often to 150 feet high; stem to 4 feet in diameter. It yields the Brea-turpentine from which locally resin and oil of turpentine are obtained; the wood is remarkably durable.

Pinus Thunbergii, Parlatore.

Japan. A tall pine with wide ramifications. Closely cognate to P. Massoniana. The most common of all trees in Japan, called there the "Matsu" or "Kouromatsou." It attains a stem-diameter of 6 feet, a height of 100 feet, and reaches an age of several centuries. It prefers sandy soil. Splendid for avenues (Rein). It supplies a resinous, tough and durable wood, used for buildings and furniture, but suitable only for indoor-work (Veitch). The roots, when burned with the oil of Brassica Orientalis, furnish the Chinese lampblack.

Pinus Torreyana, Parry.

California. An average-cone of this pine will contain about 130 seeds, weighing 3 ounces ; they are edible (Meehan).

Pinus Tsuga, Antoine.

Northern provinces of Japan, 6,000 to 9,000 feet above the sea. This very hardy spruce-fir grows to a height of only 25 feet. Its timber is highly esteemed for superior furniture, especially by turners. It is of a yellowish-brown color. P. Araragi (Siebold) is the oldest name for this species as pointed out by Prof. C. Koch.

Pinus Webbiana, Wallich.

King-Pine, Dye-Pine. Himalaya-mountains, at an elevation of from 7,000 to 13,000 feet, extending to Afghanistan. A splendid fir, reaching a height of 150 feet, the stem a circumference of 30 feet. Will bear a good deal of shade (Dr. Brandis). The wood is pale, soft, coarse-grained and very resinous, on the testimony of Mr. Webb equalling in texture and odor the Bermuda-Cedar. The natives extract a splendid violet dye from the cones. The oldest name for this species is P. spectabilis (D. Don).

Pinus Williamsonii, Newberry.

California and Oregon, up to 12,000 feet. Height of tree reaching 150 feet. Timber very valuable (Vasey). Many other pines, eastern as well as western, not alluded to on this occasion, are worthy of especial utilitarian inquiries.

Piptadenia Cebil, Grisebach. (*Acacia Cebil*, Grisebach.)

La Plata-States. A tree attaining 60 feet in height, there furnishing a tan-bark of fair strength (Hieronymus).

Piptadenia rigida, Bentham.

Sub-tropical and extra-tropical South-America. This acacia-like tree furnishes the angico-gum, similar to gum-arabic. The wood, according to Saldana da Gama, serves for naval constructions.

Pipturus propinquus, Weddell.

Insular India, South-Sea Islands and warmer parts of East-Australia. This bush is higher and rather more hardy than Boehmeria nivea, but in fibre it is similar to that plant. P. velutinus, Wedd., is closely allied. The few other species serve probably as well for fibre.

Pircunia dioica, Moquin.

Southern Brazil and La Plata-States.   The Ombu.   A deciduous
tree, for shady avenues grown in South-Europe, as well as in many
tropical countries; shown by the writer of this work to be hardy in
the lowlands of Victoria.   It attains a height of about 60 feet and is
comparatively quick of growth.

Piscidia erythrina, Linné.

West-Indies.   Florida "Jamaica-Dogwood."   A tree, reaching a
height of about 30 feet.   The bark has come into medicinal use,
particularly as an hypnotic.

Pisonia aculeata, Linné.

Tropical and sub-tropical countries of both hemispheres, extending
as a native plant into New South Wales.   This rambling prickly bush
can be chosen for hedge-copses.

Pistacia Lentiscus, Linné.

The Mastic-Tree.   Mediterranean regions.   A tall evergreen bush,
exuding the mastic-resin, mostly through incisions into its bark.   In
Morocco the plant is extensively used for hedges also.   The deciduous
P. Atlantica, Desf., yields likewise mastic.

Pistacia Terebinthus, Linné.

Countries around the Mediterranean Sea.   A tall bush or small
tree with deciduous foliage.   The fragrant Cyprian or Chio-turpentine
exudes from the stem of this species.

Pistacia vera, Linné.

Syria and Persia.   A deciduous tree, sometimes to 30 feet high,
yielding the Pistacio-nuts of commerce, remarkable for their green
almond-like kernels.   The galls from this tree are of technic value.

Pisum sativum, Linné.*

The Common Pea.   South-Western Asia.   Matures seeds as far
north as 70° 22' in Norway (Schuebeler).   Cultivated even by the
ancient Greeks and Trojans (Virchow, Wittmack).   This annual of
daily use could hardly be left unnoticed on this occasion.   Suffice it
to say, that the herbage as a nutritious fodder deserves more attention
than it receives.   The green fruit contains inosit-sugar and cholestrin
fat.   For field-culture a sandy-calcareous loam should be chosen for
this plant, to ensure rich and safe harvests.   Peas retained their
vitality afer four years' exposure to the extreme frosts of Polaris-
Bay.   A second species, P. Aucheri (Jaubert and Spach), which is
perennial, occurs in alpine elevations on the Taurus.

Pittosporum tenuifolium, Banks and Solander.

New Zealand.   This with P eugenioides has proved very suitable
for tall garden-hedges, for which these and several other species
were first brought into notice by the writer.   Unhurt by a cold of 9° F.
(Gorlie).

Pittosporum undulatum, Ventenat.

South-Eastern Australia. This tree with P. bicolor (Hooker) produces a wood well adapted for turners' purposes and also to some extent a substitute for boxwood. The flowers furnish a highly fragrant volatile oil on distillation. Under very favorable circumstances attaining a height of 80 feet (De la Motte).

Planera aquatica, Gmelin.

South-Eastern States of North-America. An elm-like tree, which can be chosen for plantations in wet localities. The wood is hard and strong.

Plantago lanceolata, Linné.

Europe, Western Asia, Northern Africa. The Rib-herb or Plaintain-herb. This perennial weed disseminates itself readily, and is recommended by some ruralists, though neither by Langethal nor Morton, as valuable on very poor pasture land; the allied P. media (L.) is of similar use and so perhaps P. major (Camerarius), all of equal geographic range; the seeds are much liked by cage-birds.

Plantago Psyllium, Linné.

Countries around the Mediterranean Sea, extending to Austria and Persia. An annual herb. The seeds render water very mucilaginous, and come thus into requisition for the preparation of silk-ware, for imparting gloss to colored paper, and for cotton-printing, irrespective of some medicinal utility (Wiesner). The same may be said of P. arenaria (Waldstein and Kitaibel). Both species could be easily naturalized on sandy coast-land.

Platanus occidentalis, Catisbye.*

The true Plane-Tree of Eastern North-America; also known as Buttonwood. More eligible as an avenue-tree than as a timber-tree. Height reaching about 100 feet; diameter of stem at times to 14 feet. Wood dull-red, light, not readily attacked by insects: used in the manufacture of pianofortes and harps; cuts into very good screws, also presses, dairy-utensils, windlasses, wheels and blocks. The young wood is silky-whitish and often handsomely mottled (Robb). The tree likes alluvial river-banks, and has been successfully planted in morassy places, to cope with miasmatic effluvia.

Platanus orientalis, Linné.*

The Plane-Tree of South-Europe and Middle Asia. Hardy in Norway to lat. 58° 8' (Schuebeler). One of the grandest trees for lining roads and for street-planting, deciduous like the other planes, rather quick of growth, and not requiring much water. Attains a height of 90 feet and a stem-circumference of occasionally 70 feet, reaching an age of over 800 years. It resists the smoke in large towns, such as London, better than any other tree, growing vigorously even under such disadvantage. The wood is well adapted for furniture and other kinds of cabinet-work. Propagation from seeds or cuttings. Growth in height at Port Phillip 30–40 feet in 20 years. An evergreen plane was mentioned already by Plinius as occurring in Candia (Sir J. Hooker) and has lately been rediscovered.

Platanus racemosa, Nuttall.

The Californian Plane-Tree. A good promenade-tree, which according to Professor Bolander grows more rapidly and more compact than P. occidentalis. Wood harder and therefore more durable, also less liable to warp. According to Dr. Gibbons the tree attains a height of about 100 feet and a stem-diameter of 8 feet; the wood is brittle; in use however by turners.

Plectocomia Himalaiana, Griffith.

Sikkim, up to 7,000 feet, extending to 27° north latitude. This Rattan-Palm requires moist forest-land. Its canes are not durable; but the plant is an object worthy of scenic horticulture, and would prove the hardiest among its congeners. P. elongata (Blume) ascends, according to Drude, to 4,500 feet.

Plectocomia macrostachya, Kurz.

Tenasserim, at about 3,000 feet elevation, therefore most likely hardy in temperate lowlands.

Plectronia ventosa, Linné.

South-Africa. A hedge-bush, like P. ciliata (Sonder) and P. spinosa (Klotzsch).

Poa Abyssinica, Jacquin.

The Teff of Abyssinia. An annual grass. The grain there extensively used for bread of an agreeable acidulous taste.

Poa airoides, Koeler. (*Catabrosa aquatica*, Beauvois.)

The Water Whorl-grass. Europe, North-Africa, Northern and Middle Asia, North-America. A creeping grass, suitable for pastures subject to inundation.

Poa alpina, Linné.

Alpine and Arctic Europe, Asia and North-America. Deserves to be transferred to other higher mountains as a nutritious perennial pasture-grass. P. Sudetica (Haenke) and P. hybrida (Gaudin) are mentioned also as excellent alpine grasses.

Poa aquatica, Linné. (*Glyceria aquatica*, Smith.)

Europe, Northern and Middle Asia, North-America. This conspicuous Water-grass attains a height of about 6 feet. It is perennial, and deserves naturalization in our swamps. It produces a large bulk of foliage, and may be disseminated for fodder-purposes. On the testimony of Dr. Curl this is one of the best feeding grasses in New Zealand.

Poa Bergii, Hieronymus.

La Plata-States. Supplies excellent fodder there, with some species of the section Eragrostis.

Poa Billardieri, Steudel.

Extra-tropical Australia. A perennial rigid grass, of some value for saline meadows.

Poa Brownii, Kunth.* (*Eragrostis Brownii*, Nees.)

Tropical and Eastern extra-tropical Australia, It is here mentioned as a valuable perennial species, keeping beautifully green in the driest Australian summer, even on poor soil; indeed the Missionary Pastor Kempe pronounces it to be the best of all grasses on the Central-Australian pastures. The section Eragrostis of the genus Poa contains numerous species in the hotter parts of the globe. Of these many would doubtless be hardy far beyond the tropics, and prove of value on pastural land.

Poa cæspitosa, G. Forster.

Extra-tropical Australia and New Zealand, ascending alpine elevations. A tufty grass, available throughout the year for pasture-feed, when young or when offering flowering or seeding-stalks, or when presenting tender varieties; the rougher varieties utilized by the aborigines for nets and cordage. Resisting drought. Well worthy of being naturalized in other parts of the globe.

Poa Canadensis, Beauvois.

The Rattlesnake-grass of South-Eastern America. A valuable swamp-grass.

Poa Chinensis, Koenig.

Southern and Eastern Asia, East-Australia. Recommended by Mr. F. M. Bailey as a valuable pasture-grass, perhaps on account of its tender panicles. Poa bulbosa, L., of Europe and Western Asia, and P. compressa, L., of the same regions, will grow in pure sand.

Poa cynosuroides, Retzius.

North-Eastern Africa, Southern Asia. A harsh perennial grass, not serviceable for fodder, but mentioned by Royle as a fibre-plant of North-Western India, where it is valued as a material for ropes.

Poa digitata, R. Brown.

South-Eastern and Central Australia. Valuable for fixing wet river-banks and slopes. It forms large stools. Cattle and horses relish the young shoots.

Poa distans, Linné.

Europe, North-Africa, Middle and Northern Asia, North-America. Perennial. It is one of the limited number of tender grasses, suited for moist saline soil, and thus affords pasturage on coast-marshes.

Poa fertilis, Host. (*P. serotina*, Ehrhart.)

Europe, Northern Asia, North-America. Perennial. Important for wet meadows, even with sandy subsoil. Its foliage is tender, tasty and nourishing. In mixtures of grasses it keeps up the growth late into the autumn; it will prosper also on sandy and saline soil.

Poa fluitans, Scopoli. (*Glyceria fluitans*, R. Brown.)

The Manna-grass. Europe, North-Africa, Middle and Northern Asia, North-America, East-Australia. Perennial. Excellent for stagnant water and slow-flowing streams. The foliage is tender. The seeds are sweet and palatable, and in many countries are used for porridge. This grass is indigenous in Norway northward to lat. 69° 9' (Schuebeler).

Poa foliosa, J. Hooker.

Auckland- and Campbell-Island, as well as Southern New Zealand, reaching almost to the glacier-region. Perennial, forming large mounds. Prof. Kirk calls it a noble species, producing an immense yield of foliage. Mr. Buchanan also speaks of the fattening food afforded to horses and cattle by this grass.

Poa Forsteri, Steudel. (*Dactylis cæspitosa*, Forster.)

The Tussock-grass. Fuegia, Falkland-Islands, South-Patagonia. Introduced by Sir Joseph Hooker into the Hebrides, and by Mr. Traill into the Orkney-Islands. Delights, according to Mr. Ingram, in deep, boggy and mossy land, even when exposed to sea-spray. Cultivated plants might be dressed with some salt. Thrives in cold countries near the sea in pure sand, at the edge of peat-bogs. It would probably prosper on alpine moors. It is perennial, and reaches a height of nine feet. It is very nutritious and much sought by herds. The base of the stem is nutty and edible. An allied species is P. Cookii (J. Hooker) from Kerguelen's Island.

Poa maritima, Hudson.

Europe, North-Africa, Northern Asia, North-America. Roots long and creeping. This grass can also be depastured and grown on brackish meadows.

Poa nemoralis, Linné.

Europe, Northern and Middle Asia, North-America. This perennial grass can be raised on shady forest-land, as the name implies; but it accommodates itself also to open places, and will grow even among dry rocks. It endures alpine winters. According to Lawson no better grass exists for displacing weeds on pleasure-lawns; the same may be said of Poa compressa, L.

Poa nervata, Willdenow.

Southern States of North-America, called in Alabama the Manna-grass. Perennial. Valuable for pastures in low forest-land (C. Mohr).

Poa palustris, Linné. (*Poa scrotina,* Ehrhart.)

Europe, Northern Asia, North-America. A perennial grass, allied to P. nemoralis, excellent for moist meadows and river-banks. P. fertilis (Host) is a mere variety of this species.

Poa pectinacea, Michaux. (*Eragrostis pectinacea,* Gray.)

Middle and Southern States of North-America. This perennial grass spreads rapidly over dry ground and even coast-sands. C. Mohr regards it as valuable for pastures, and mentions as such also Eragrostis nitida (Chapman) and E. tenuis (Gray).

Poa pratensis, Linné.*

The ordinary English Meadow-grass. A perennial species, with creeping roots, fit for any even very dry meadows, thriving early, and able to live also in alpine localities. In Greenland it is indigenous to lat. 80-81° N. L. (Nathurst). Better adapted for pasture than hay. It is suitable for moor-land, when such is laid dry; although it flowers only once during the season, it offers a nutritious fodder, even on comparatively poor soil; it resists drought, forms an excellent sward, and can be used with advantage for intermixing with other pasture-grasses. In the United States it is known as the Kentucky Blue Grass or Pennsylvania Green Grass, and is considered one of the best for lawns by Professor Meehan, as it will crowd out all weeds in time.

Poa trivialis, Linné.

Europe, North-Africa, Middle and Northern Asia. Also a good perennial grass for mixture on pasture-land. One of the best grasses for sowing on ground recently laid dry. Recommendable also as a lawn-grass. Sinclair regarded the produce of this Poa as superior to many other kinds, and noticed the marked partiality, which horses, oxen and sheep evince towards it. To thrive well, it wants rather moist and rich soil and sheltered places. It is a later grass than P. pratensis, well adapted for hay, and gives good after-growth (Langethal).

These few species of Poa have been singled out as recommendable, because they are well tested. Future experiments beyond Europe will add others to lists of recommendations like this.

Podachænium alatum, Bentham. (*Ferdinanda eminens,* Lagasca.)

Central America, up to a height of about 8,000 feet. A tall shrub; on account of the grandeur of its foliage in requisition for scenic effects.

Podophyllum peltatum, Linné.

Eastern North-America, where it is known as the Mandrake. Hardy in Christiania. A perennial forest-herb of importance for medicinal purposes. The root contains the bitter alkaloid berberin. Podophyllum Emodi (Wallich), occurring in the Indian mountains at heights of from 6,000 to 14,000 feet, can probably be used like the American species.

The berries of both are edible, though the root and leaves are poisonous. A third species, P. pleianthum, has been described by Dr. Hance from Formosa.

Pogostemon Patchouli, Pelletier.

Mountains of India. A perennial herb, famed for its powerful scent, arising from a volatile oil. P. parviflorus and P. Heyneanus (Bentham) belong to this species. One cwt. of the herb yields about 28 ounces oil (Piesse); but the essence is chiefly obtained by enfleurage.

Polianthes tuberosa, Linné.

Mexico. The Tuberose. Valuable for perfume. Available late in the season; thus one of the plants most required to maintain garden-fragrance in serene climes throughout the year. The gathering of flowers of "Tubereuses" at Grasse, Cannes and adjacent villages alone comes annually to about 20,000 lbs. (Piesse).

Polygala crotalaroides, Hamilton.

Temperate Himalaya and Khasia. Praised as an ophidian alexipharmic. To several other species both of the eastern and western hemisphere similar properties are ascribed, but we are almost entirely without any reliable medical testimony on these and many other supposed vegetable antidotes against snake-poison. Doubtless this small perennial herb possesses therapeutic virtues like many of its congeners.

Polygala Senega, Linné.

The Senega Snake-Root. Eastern North-America. A perennial herb. The root is of medicinal value.

Polygaster Sampadarius, Fries.

South-Eastern Asia. One of the most palatable of all truffles.

Polygonum tinctorium, Loureiro.

China and Japan. An annual herb, deserving attention and local trials, as yielding a kind of indigo; one of the most important dye-plants of Japan. It can be cultivated in cold climes, being hardy still at Christiania. Its growth would be vigorous. Various Polygonums contain tannin, P. amphibium (Linné) as much as 11½ per cent. (Masters).

Polyporus giganteus, Fries.

Europe. Dr. Goeppert records this and also the following species as allowed to be sold for food in Silesia: *P. frondosus*, Fr., *P. ovinus*, Fr., *P. tuberaster*, Fr., *P. citrinus*, Pers. Dr. Atkinson mentions as edible among the fungs of Cashmere P. squamosus (Fries). Bergner and Frog illustrate P. confluens (Fries) among the esculent fungs of Switzerland; near relatives of all these occur in Australia also.

Pophyra vulgaris, Agardh.

Temperate and cold oceans. This largely cosmopolitan seaweed is mentioned here, because in Japan it undergoes regular cultivation.

For this purpose branches of Quercus serrata are placed in shallow bays, where Porphyra occurs, during spring, and the crop is obtained from October to March, the seaweed being cônsumed in its young state. It grows best, where fresh water enters the sea. Porphyra contains about 26 per cent. of nitrogenous substances (with more than 4 per cent. of nitrogen) and about 5 per cent. of phosphate of potash. In Japan, according to the catalogue of the International Exhibitions of Sydney and Melbourne, the following Algæ are also consumed for food; Gloiopeltis intricata, G. capillaris, Laminaria saccharifera, two species of Phylloderma, Phyllitis debilis, Kallhymenia dentata, Capea elongata, Alaria pinnatifolia, Gracilaria lichenoides, G. confervoides, Enteromorpha compressa, species of Cystoseira and Halochloa, Codium tomentosum, Mesogloia decipiens and Gelidium corneum.

Populus alba, Dodoens.

The Abele or White Poplar, indigenous to South-Eastern Europe, North-Africa and Northern and Middle Asia, extending to North-China, growing on the Himalayas up to 10,000 feet, ceasing at 4,000 feet. In Norway it is hardy to lat. 67° 56' (Prof. Schuebeler). Height reaching 90 feet. Emits suckers. It has proved an excellent avenue-tree, even in comparatively waterless situations, and the partial whiteness of its foliage gives a pleasing effect in any plantation. A Silver-Poplar at Slowitz attained a basal stem-diameter of 20 feet, indicating according to Pannewitz an age of probably 400 years. The wood is pale, with a reddish tinge, brown near the centre, soft and light. It can be used for flooring; it is particularly sought for trays, bowls, bellows and shoe-soles; also, according to Porcher, for wooden structures under water. "Sparterie" for plaiting is obtained from the wood-shavings. The wood of this and some other poplars is easily converted into paper-pulp, which is cheaply bleached. Lines of poplars along forest-streams prevent or impede the progress of wood-conflagrations. The roots of poplars spread widely. P. canescens (Smith), the Grey Poplar, is either a variety of the Abele or its hybrid with the Aspen, and yields a better timber for carpenters and millwrights.

Populus angulata, Aiton.

Eastern North-America. The "Water-Poplar" or Carolina-Poplar. Acquires a height of about 70 feet; branches very spreading; hence this species well adapted as a promenade-tree.

Populus balsamifera, Linné.

The Tacamahac or Balsam-Poplar of the colder, but not the coldest parts of North-America; also in Siberia and on the Himalayan Mountains, where it ranges from 8,000 to 14,000 feet. "The balmiest of all trees" called by Dr. Kellogg. It will endure the winters of Norway to lat. 69° 40' (Schuebeler). It attains a height of 80 feet. The tree may be lopped for cattle-fodder (Stewart and Brandis). Professor Meehan says that it will grow near the ocean's brink. Its variety is P. candicans (Aiton). Acknowledged as a distinct species by Wesmael.

U

Populus ciliata, Wallich.

Himalaya, from 4,000 to 10,000 feet. Height as much as 70 feet, with a straight trunk, which attains 10 feet in girth.

Populus Euphratica, Olivier.

From Algeria dispersed to the Himalayas and Songaria, up to 13,500 feet. Height to 50 feet. Wood harder than that of most poplars, the inner wood turning blackish in old trees. It is used for planking and boat-building (Stewart and Brandis), also for beams, rafters, boxes, panelling, turnery. Cattle will browse on the leaves. This is the Willow of the 137th Psalm (C. Koch).

Populus Fremontii, S. Watson.

California and adjoining States on river-banks. Tree, attaining about 150 feet in height and 4½ feet in stem-diameter; leaves large. Much lauded for shading road-sides and promenades, for which however the staminate trees should only be selected. Wood less white than that of P. tremuloides, excellent for dry goods, fruit- butter- and salt-boxes, trays, bowls and other articles; outer bark a fair substitute for cork. The foliage brightens splendidly in autumn. Wood convertible into paper-pulp (Dr. Kellogg).

Populus grandidentata, Michaux.

The Soft Aspen. Eastern North-America. To 80 feet high. Wood whitish, soft, very light; can be ground into pulp for paper. The oldest name seems P. deltoides, Marsh.

Populus heterophylla, Linné.

The Downy Poplar of North-America, passing also by the name of Cottonwood. Height often 60 feet. The wood is very pale, soft and fissile. All poplars, like willows, are very important to eliminate miasma by absorbing humidity to an enormous extent from stagnant swampy localities; they are likewise good scavengers of back-yards.

Populus monilifera, Aiton.* (*P. Canadensis,* Moench.)

The Cottonwood-tree of North-America, extending to New Mexico. Height to 150 feet; stem to 8 feet in diameter. Moench's name is the oldest for this species (C. Koch). One of the best poplars for the production of timber, which is soft, light, easy to work, suited for carving and turnery; it is durable if kept dry, and does not readily take fire. The wooden polishing-wheels of glass-grinders are made of horizontal sections of the whole stem, about one inch thick, as from its softness the wood readily imbibes the polishing material. It is also useful for rails and boards, and supplies a fair fuel. Judge Whitning says, that it has no rival in quickness of growth among deciduous trees. Governor Furnas found the stem-girth in Nebraska reaching to 93 inches in eleven years at 2 feet above ground. Recommended by Wessely, together with P. alba and P. nigra, for fixing drift-sand, on which these poplars never become suffocated. It is advisable, to obtain cuttings from male trees only, for planting

along streets or near dwellings, as the minute downy seeds of the
female trees are copiously wafted through the air, and may have irri-
tant effects on the respiratory organs of the frequenting people. P.
angustifolia (James) is regarded by Wesmael as a mere variety of this
species.

Populus nigra, C. Bauhin.

The European Black Poplar, extending spontaneously to China;
in the Himalayas up to 12,500 feet. The spreading variety is one
of the best of trees for lining roads. This species includes P. dilatata
(Aiton), or as a contracted variety P. fastigiata (Desfontaines),
the Lombardy-Poplar. Greatest height 150 feet. Growth rapid, like
that of all other poplars, or even more so. At Bensberg a Black
Poplar formed in 80 years a stem 19 feet in circumference; at Wip-
pach a hollow stem showed a breadth of 48 feet. In warm zones the
growth is still more rapid than in Middle Europe, as is the case with
the majority of trees. Wood soft, light and of loose texture, used
for joiners, coopers and turners' work; also for matches; furnishing
furthermore superior charcoal for gunpowder. Bark employed in
tanning, producing a fragrant leather; it is however not rich in tannic acid. The tree requires damp soil. It retains its foliage longer
than most poplars through the season.

Populus tremula, C. Bauhin.

The Aspen. Europe, North-Africa, Northern Asia to Japan.
Height reaching to about 100 feet, stem-circumference to 12 feet; age
130 years or more. Emits suckers; content with sandy soil, if not
too dry. The aspen is very hardy; in lat. 70° in Norway a tree still
attained a height of 60 feet (Schuebeler). The aspen-wood is whitish
and tender, and in use by coopers and joiners. Like the wood of
other poplars much sought for paper-mills as an admixture to the
pulp. In Japan it is used for engraving rough works and posters.
In Sweden largely employed for matches. A variety of this tree
with pendent branches occurs.

Populus tremuloides, Michaux.

The North-American Aspen. Ascends to alpine elevations of
about 10,000 feet; easily disseminated. Height to as much as 50
feet. The wood is whitish, soft, readily worked, and can be converted
into paper-pulp; also of this a weeping variety occurs; the tree extends westward to California. All poplars might be planted in gullies,
like willows, to intercept forest-fires; also generally on river-banks.
They are also valuable honey-yielders (Prof. Cook). All can easily
be propagated from cuttings, and are of quick growth.

Populus trichocarpa, Torrey and Gray.

From British Columbia to California. One of the "Cottonwood"-
trees." The stem attains a diameter of 5 feet, and is used by the
autochthones for canoes (Dr. G. Dawson).

Portulacaria Afra, Jacquin.

South-Africa. A shrub, rising to 12 feet, called "Spekboom." Affords locally the principal food for elephants; excellent also for sheep-pasture, according to Professor McOwan; hence this succulent shrub may deserve naturalization on stony ridges and in sandy desert-land, not readily otherwise utilized.

Pouzolzia tuberosa, Wight.

India. The turnip-shaped root of this herb is edible. The plant may prove hardy in extra-tropic frostless regions, and its root may improve in culture.

Prangos pabularia, Lindley.

Plateaux of Mongolia and Thibet. A perennial fodder-herb, much relished by sheep, eligible for cold and arid localities and deserving naturalization on alpine pasture-grounds. Other perennial species exist near the Mediterranean Sea, on the Atlas, the Caucasus and the Indian highlands. P. pabularia is regarded by some as the Silphium of Arrianus.

Prestoa pubigera, J. Hooker. (*Hyospathe pubigera*, Grisebach.)

Trinidad. At an elevation of about 3,000 feet (Krueger). The stem of this palm attains only about 12 feet in height. Valuable among the dwarf palms, now so much sought for table- and window-decoration.

Pringlea antiscorbutica, W. Anderson and R. Brown.*

The Cabbage or Horse-radish of Kerguelen's Island. · The perennial long roots taste somewhat like horse-radish. The leaves in never-ceasing growth are crowded cabbage-like into heads, beneath which the annual flower-stalks arise. The plant ascends mountains in its desolate native island to the height of 1,400 feet, but luxuriates most on the sea-border. To arctic and other antarctic countries it would be a boon. Probably it would live on our Alps. Whalers might bring us the roots and seeds of this remarkable plant, which seems never to have entered into culture yet. The plant was used as cabbage by the celebrated Captain Cook and all subsequent navigators, touching at yonder remote spot, and it proved to possess powerful properties against scurvy. Sir Joseph Hooker observes, that Pringlea can sectionally be referred to Cochlearia. The whole plant is rich in a pungent volatile oil. Through culture important new culinary varieties, may probably be raised from this plant. This vegetable in its natural growth tastes like mustard and cress ; but when boiled it proved a wholesome and agreeable substitute for the ordinary cabbage.

Priva lævis, Jussieu.

Chili, Argentina. A perennial herb, the small tubers of which can be used for food (Philippi).

Prosopis alba, Grisebach.

La Plata-States. A tree, rising finally to about 40 feet, with a stem-diameter to 3 feet. The fruit, known as Algaroba blanca, is considered wholesome and nutritious. The tree yields also tan-bark. P. nigra (Hieronymus) serves in Argentina similar purposes also.

Prosopis dulcis, Kunth.

From California and Texas to the southern parts of the La Plata-States. Vernacularly known as the Cashaw- Mesquite- or Algaroba-tree. A thorny shrub, growing finally to a tree of 30 feet height, with a stem 2½ feet in diameter; adapted for live-fences. The wood is durable and of extraordinary strength and excessive hardness, fit for select furniture particularly, assuming when polished the appearance of mahogany. This is one of the species yielding the sweetish Algaroba-pods for cattle-fodder, and utilized even in some instances for human food. The pods of the various kinds of Prosopis are adapted only for such animals as chew the cud, and thus get rid of distending gases (R. Russell). Argentina Algaroba-pods contain, according to Sievert, 25 to 28 per cent. grape-sugar, 11 to 17 per cent. starch, 7 to 11 per cent. protein, of organic acids, pectin and other non-nitrogenous nutritive substances 14 to 24 per cent. They are also comparatively rich in potash, lime and phosphoric acid. A sparkling drink called Aloja is made of the fruits. This and some allied species yield the Algarobylla-bark for tanning; the leaves contain, according to Sievert, 21 per cent. tannin. The pods also of several species are rich in tannic acid. Mere varieties, according to Bentham, are: P. horrida, P. juliflora, P. siliquastrum, P. glandulosa. Particularly the latter variety exudes a gum not unlike gum-arabic, and this is obtained at times so copiously, that children could earn two to three dollars a day in Texas while gathering it, latterly about 40,000 lbs. being bought by druggists there. A short communication on the American Algaroba-trees was presented to the Parliament of Victoria by the writer in 1871. Pods of some Prosopis, used as fodder, have caused the death of horses in Jamaica by overfeeding.

Prosopis pubescens, Bentham.

The Tornillo or Screw-bean. Texas, California, Mexico. The pods ripen at all seasons and contain much saccharine nutritive substance (J. S. Gamble). Likely available for hedges with other species of other countries. Seeds can be converted into food (Sargent). Not resisting climatic vicissitudes so well as P. dulcis.

Prosopis spicigera, Linné.

India, extending to Persia. A thorny tree, also producing edible pods and enduring some frost. It attains a height of 60 feet, but is of slow growth (Brandis). Serves for hedge-lines. It can be chosen for desert-land (Kurz).

Prosopis Stephaniana, Kunth.

Syria and Persia. A shrubby species for hedge-growth.

Prostanthera lasiantha, Labillardière.

South-Eastern Australia and Tasmania. Confined to the banks of forest-streams. The only one among more than 2,500 Labiatæ which becomes a good-sized tree, reaching a height of fully 60 feet. Wood useful for many technologic purposes. The leaves of this and its many congeners afford on distillation aromatic oils.

Protea mellifera, Thunberg.

South-Africa. This tall bush is deserving a place among the plants of this work, not only in view of its gaudy ornamental aspect, but also on account of the richdom of honey-nectar in its large inflorescence.

Prunus Americana, Marshall. (*P. nigra*, Aiton.)

Canada, Eastern United States of America. A thorny tree, furnishing the Yellow and Red Plum of North-America. Hardy in Norway northward to lat. 65° (Schuebeler). The fruit is roundish and rather small, but of pleasant taste. All kinds of Prunus are important to the apiary.

Prunus Amygdalus, J. Hooker.* (*Amygdalus communis*, Linné.)

The Almond-tree. Countries around the Mediterranean Sea and South-Western Asia ; really indigenous on the Anti-Lebanon, in Kurdestan, Turkestan and perhaps on the Caucasus (Stewart). Both the sweet and bitter almond are derived from this species. The cost of gathering the crop in South-Europe is about 20 per cent. of its market-value. Their uses and the value of the highly palatable oil, obtained by pressure from them, are well known. This oil can well be chosen as a means of providing a pleasant substitute for milk during sea-voyages, by mixing with it, when required, half its weight of powdered gum-arabic, and adding then successively, while quickly agitating in a stone-mortar, about double the quantity of water; thus a palatable and wholesome sort of cream for tea or coffee is obtained at any moment. There exist hard- and soft-shelled varieties of both the sweet and bitter almond. Almonds can even be grown on seashores. The tree bears the climate of Christiania in Norway (Professor Schuebeler). The crystalline amygdalin can best be prepared from bitter almonds, through removing the oil by pressure, then subjecting them to distillation with alcohol, and finally precipitating with ether. The volatile bitter almond-oil—a very dangerous liquid—is obtained by aqueous distillation. Dissolved in alcohol it forms the essence of almonds. This can also be prepared from peach kernels. The almond-tree is one of the aptest, to be chosen as a standard of comparison with other kinds of trees (as well as other plants) for records of synchronous flowering time.

Prunus Armeniaca, Linné. (*Armeniaca vulgaris*, Lamarck.)

The Apricot-tree. China, as already indicated by Roxburgh, not indigenous in Armenia. Cultivated up to 10,000 feet in the Himalayas. Professor C. Koch points to the alliance of this tree to

P. Sibirica (Linné), and he considers P. dasycarpa (Ehrhart) to be a hybrid between the apricot- and plum-tree. A variety of apricot occurs with a sweet kernel. Cold-pressed apricot-seeds yield an oil much like that of almonds. Muspratt found as much as 24 per cent. tannin in the bark. The Chinese P. Mume, Sieb. and Zucc., is a peculiar apricot-tree.

Prunus Caroliniana, Aiton.

South-Eastern States of North-America. Porcher regards it as one of the most beautiful and manageable evergreens of the States. It can be cut into any shape, and is much employed for quick and dense hedges. It can be grown on coast-land.

Prunus cerasifera, Ehrhart. (*P. Myrobalanus*, Desfontaines.).

The Cherry-Plum tree. Countries at and near the Caspian Sea. The fruits known also as Mirabelle-Plums, whence long ago the objectionable designation Myrobalane-Plum arose. Among all kindred species it is this one, which flowers earliest, indeed before the development of its leaves, hence its claims for decorative horticulture. On this and some other cultivated species see also Koch's Dendrologie, 1869.

Prunus Cerasus, Linné.

The Cherry-tree. Orient, especially in the countries near the Caspian Sea. The name applies strictly only to the species, distinguished by never assuming large dimensions, by emitting suckers. by smoothness of leaves and austerity and acidity of fruit. P. avium (Linné), the sweet-fruited Cherry-tree, seems naturally to extend as far as Middle Europe, and attains a high age, when the stem may acquire a diameter of 4 feet, produces no suckers and has downy more wrinkled leaves, irrespective of some few other discrepancies. It afforded its fruit already to the ancient inhabitants of Switzerland in pre-historic times (Heer, Mortillet), and the tree was cultivated by the early Greeks also, according to historic records (A. de Candolle). It is hardy in Norway to lat. 66° 30' (Schuebeler). In the Himalayas it is cultivated up to 12,000 feet. The tree enjoys everywhere a remarkable immunity from insect-attacks.

Prunus Chisasa, Michaux. (*P. angustifolia*, Marsh.)

North-America, west of the Mississippi. On the prairies it is only 3 to 4 feet high. Fruit spherical, red, rather small, with a tender usually agreeable pulp. Other species with edible fruit occur in North-America, such as P. pumila and P. Pennsylvanica (Linné), but their fruits are too small, to render these plants of importance for orchard-culture, though they also may become enlarged by rural treatment. Marsh's name is the oldest.

Prunus demissa, Walpers.

California. The Wild Plum of Utah. Worthy of improving cultivation. It fruits abundantly, often when only 2 or 3 feet high. It is of near affinity to P. Virginiana.

Prunus domestica, Linné.

Plum-tree, Damson-tree, Prune-tree. From the Black Sea to Western China. In the countries at the Mediterranean Sea numerous varieties were cultivated even at the commencement of the. Christian era. In Norway this species endures the winter to lat. 64° (Professor Schuebeler). The wood is sought for musical instruments and select turnery.

Prunus ilicifolia, Nuttall.

California. In deep rich soil, valuable for evergreen hedges of intricate growth. Fruit about ½ inch diameter, red or black, of a pleasant sub-acid flavor, but somewhat astringent (Gibbons).

Prunus insititia, Linné.

The Bullace. Middle and Southern Europe, North-Africa, Western Asia to the Himalayan mountains. Professor Heer has proved, that the lacustrine Swiss of the stone-age were already acquainted with the Bullace as well as the Sloe. This species yields some of the Damascene-Plums. P. cerasifera seems descended from P. insititia, and this again may be the original wild plant. of P. domestica (Loudon, J. Hooker).

Prunus Lauro-Cerasus, Linné.

The Cherry-Laurel. Persia and adjoining countries. A tall shrub or small tree, with evergreen remarkably shining foliage; the latter, as not quickly shrivelling, valuable for garlands and for other decorative purposes. From the leaves the medicinal laurel-water is distilled.

Prunus Lusitanica, Linné.

The Portugal Cherry-Laurel. A small tree, seldom over 30 feet high, not of strictly industrial value, but mentioned here as one of the very hardiest among evergreen trees not coniferous.

Prunus Mahaleb, Linné.

South-Europe and South-Western Asia. It deserves some attention on account of its scented seeds and also odorous wood, the latter used in turnery for pipes and other articles. The flowers are in use for perfumes. The tree is hardy in Norway to lat. 63° 26′. The kernels are used for making marasquino-liqueur (Prof. Wittstein).

Prunus maritima, Wangenheim.

The Beach-Plum of Eastern North-America. A shrubby species, of service not only for covering coast-sands, but also for its fruit, which is crimson or purple, globular, measuring from ½ to 1 inch. Information on these and other varieties and on orchard-fruits in general may be sought in Hogg's "Fruit-Manual."

Prunus Padus, Linné.

The Birds' Cherry-tree. Europe, Northern and Western Asia, extending to the Himalayas and the mountains of Northern Africa. A small tree. Foliage deciduous; the leaves distilled for medicinal purposes, the bark also utilized therapeutically.

Prunus Persica, J. Hooker. (*Amygdalus Persica,* Linné.)

China, not really indigenous to Persia, as ascertained by Alph. de Candolle. The Peach-tree, as delightful through its early flowering as through the ready yield of its luscious fruit. Not quite so hardy as the Almond-tree in cooler climes, its near ally, though enduring the clime of England. In the southern of the United States peaches are not rarely turned to account for alcoholic fermentation and distillation (Rhind). The Nectarine, which is characterized by smooth fruits, is a variety merely. The bark used as an anthelmintic. The necessity of reducing the genus Amygdalus to that of Prunus was indicated in 1812 already by Stokes (Bot. Mat. Med. III. 101) and in 1813 by F. G. Hayne (Arznei-Gewaechse IV. 38).

Prunus Pseudo-Cerasus, Lindley. (*P. Puddum,* Roxburgh.)

The "Sakura" of Japan, extending to Upper India. A large shady tree, the stem attaining two feet in diameter, charming to view when bearing its profusion of flowers. The fruit is of the size of small cherries and of pleasant and refreshing taste, though never quite sweet (Wallich). This is this tree, which supplies mainly the wood so extensively required for xylography in Japan (Dupont).

Prunus serotina, Ehrhart.

The Black Cherry-tree of Eastern North-America. Fruit slightly bitter, but with a pleasant vinous flavor; wood compact, light, easily worked, not liable to warp (Sargent), very valuable for cabinet- and sash-makers (A. Gray). In Virginia and Alabama the tree attains a height of about 100 feet, with a stem 4 feet in diameter; it prefers rich porous soil in the upper parts of valleys. Wood pale-red, dense, fine-grained; when polished as beautiful as mahogany-wood (Robb and Simmonds). Will live on the poorest soil, and even within the salt-spray of the coast. Readily raised from seeds and transplanted; not succumbing under rough usage (Sargent).

Prunus spinosa, Linné.

The Sloe or Blackthorn. Wild in many parts of Europe. Indigenous in Norway to lat. 60° 8'; but it will endure the winter even to lat 67° 56' (Schuebeler). Hardly at all liable to be attacked by insects. With its flowers it is one of the earliest plants to announce the spring. Its tendency, to throw out suckers, renders the bush less adapted for hedges of gardens than of fields, but these suckers furnish material for walking-sticks. The small globular fruits can be made into preserves. Perhaps the fruit of some of the species from Eastern Asia, California and tropical America may be improved by horticultural skill. The sloe and others might with advantage be naturalized on forest-streams.

Prunus tomentosa, Thunberg.

Northern China. A very hardy species with cherry-like edible fruits.

Prunus Virginiana, Linné.

The Choke Cherry-tree of the Eastern United States. In a mild clime and fertile soil this tree attains a height of about 100 feet and a stem-circumference of 16 feet. Endures the winters of Norway to lat. 67° 56' (Schuebeler). The wood is compact, fine-grained, and not liable to warp when perfectly seasoned, of a dull light-red tint, deepening with age. The fruit finally loses its acerbity. The bark used in medicine.

Psamma arenaria, Roemer and Schultes.* (*P. littoralis*, Beauvois; *Calamagrostis arenaria*, Roth.)

The Morram, Marrem or British Bent-grass. Sand-coasts of Europe, North-Africa and North-America. One of the most important of reedy grasses with long descending roots, to bind moving drift-sands on the sea-shore, for the consolidation of which this tall grass and Elymus arenarius are chiefly employed in Europe. It delights in the worst of drift-sands, and for its full development gradual accumulation of fresh sands around it becomes necessary (Wessely): hence it never gets suffocated. The plant will by gradual upgrowth finally form stems and roots, sanded in to a depth of fully 100 feet. Psamma Baltica (R. & S.) from the Baltic- and North-Sea, serves the same purpose. Both can also be used in the manner of Sparta for paper-material, for tying and for mats. Like Elymus arenarius, they are not touched by grazing animals. P. arenaria collects the sand-heaps at the tops of ridges, while the Elymus fastens their sides.

Psidium acidum, Martius.

Higher regions on the Amazon-River. A tree, at length 30 feet high; its guava-fruit pale yellow and of apple-size.

Psidium Araca, Raddi.

From the West-Indies and Guiana to Peru and Southern Brazil, where it is found in dry high-lying places. This is one of the edible guavas, already recorded by Piso and Marcgrav. The greenish-yellow berry is of exquisite taste.

Psidium arboreum, Vellozo.

Brazil, province of Rio de Janeiro. The guava-fruit of this plant measures about one inch, and is of excellent flavor.

Psidium Cattleyanum, Sabine.*

The Purple Guava. Brazil and Uruguay. One of the hardiest of the guava-bushes, attaining finally a height of 20 feet. The purple berries are seldom above an inch long, but, as well known, of delicious flavor and taste, resembling thus far strawberries. P. buxifolium (Nuttall) of Florida, seems nearly related to this species.

Psidium chrysophyllum, F. v. Mueller. (*Abbevillea chrysophylla,* Berg.)

The Guabiroba do Mato of South-Brazil. This tree attains a height of about 30 feet. The fruit is generally not larger than a cherry. Perhaps other species of the section Abbevillea would be hardy and worthy of cultivation.

Psidium cinereum, Martius.

Brazil, provinces Minas Geraes and Sao Paulo. Also yielding an edible fruit.

Psidium cordatum, Sims.

The Spice-Guava. West-Indies. This attains the height of a tree. Its fruit is edible. Probably hardy in sub-tropic regions.

Psidium cuneatum, Cambessedes.

Brazil, province Minas Geraes. Fruit greenish, of the size of a Mirabelle-plum.

Psidium grandifolium, Martius.

Brazil, provinces Rio Grand do Sul, Parana, Sao Paulo, Minas Geraes, where the climate is similar to Southern Queensland. A shrub of rather dwarf growth. The berries edible, size of a walnut.

Psidium Guayava. Linné.* (*P. pomiferum,* Linné; *P. pyriferum,* Linné.)

The large Yellow Guava. From the West-Indies and Mexico to South-Brazil. This handsome evergreen and useful bush should engage universal attention anywhere in warm lowlands, for the sake of its aromatic wholesome berries, which will attain the size of a hen's egg, and can be converted into a delicious jelly. The pulp is generally cream-colored or reddish, but varies in the many varieties, which have arisen in culture, some of them bearing all the year round. Propagation is easy from suckers, cuttings or seeds. Many other berry-bearing Myrtaceæ of the genera Psidium, Myrtus, Myrcia, Marliera, Calyptranthes and Eugenia furnish edible fruits in Brazil and other tropical countries; but we are not aware of their degrees of hardiness. Berg enumerates as esculent more than half a hundred from Brazil alone, of which the species of Campomanesia may safely be transferred to Psidium.

Psidium incanescens, Martius.

Brazil, from Minas Geraes to Rio Grand do Sul. This guava-bush attains a height of 8 feet. Berry edible.

Psidium lineatifolium, Persoon.

Mountains of Brazil. Berry about 1 inch in diameter.

Psidium malifolium, F. v. Mueller. (*Campomanesia malifolia,* Berg.)

Uruguay. Berry about 1 inch in diameter.

Psidium polycarpon, Al. Anderson.*

From Guiana to Brazil, also in Trinidad. A comparatively small shrub, bearing prolifically and almost continuously its yellow berries, which are of the size of a large cherry and of exquisite taste.

Psidium rufum, Martius.

Brazil, in the province of Minas Geraes, on sub-alpine heights. This guava-bush gains finally a height of 10 feet, and is probably the hardiest of all the species producing palatable fruit.

Psophocarpus tetragonolobus, De Candolle.

Tropical Africa, perhaps to Madagascar. A climber with annual stem; pods to one foot long, used as peas. P. palustris (Desvaux) is closely allied, and has shorter pods. Likely to ripen fruits also outside the tropics.

Psoralea esculenta, Pursh.

North-America. This herb is mentioned here, as its tuberous roots, known as the Prairie-Turnip, may be capable of great improvement by cultivation, and of thus becoming a valuable esculent.

Psychotria Eckloniana, F. v. Mueller. (*Grumilia cymosa*, E. Meyer.)

South-Africa. Dr. Pappe describes the wood of this tree as of a beautiful citron-yellow.

Pterocarpus Indicus, Roxburgh.

The Lingo of China and India. A tree of considerable dimensions, famed for its flame-red wood. It furnishes also a kind of dragon-blood resin.

Pterocarpus marsupium, Roxburgh.

India, ascending in Ceylon and the Circars to fully 3,000 feet altitude; hence this tree would doubtless grow without protection in those tracts of the temperate zone, which are free from frost. The tree is large when in its final development; its foliage is deciduous. It exudes the best medicinal kino, which contains about 75 per cent. of tannic acid. P. santalinus (Linné fil.) which provides the Saunders or Red Sandal-Wood, is also indigenous to the mountains of India and important for dye-purposes.

Pterocarya fraxinifolia, Kunth.

From Central Asiatic Russia to Persia. A kind of Walnut-tree, which, with P. stenoptera (Cas. de Candolle) on Dr. Hance's recommendation, should be adopted as trees for both ornament and timber, and so perhaps also the Japanese species, P. rhoifolia (Siebold and Zuccarini).

Ptychosperma Alexandræ, F. v. Mueller.

The Alexandra-Palm. Queensland, as well in tropical as extra-tropical latitudes. The tallest of Australian palms, and one of the noblest forms in the whole empire of vegetation. Aged it exceeds 100 feet in height, and is likely destined to grace many shady moist groves yet outside the tropics, so long as they are free from frost, as this palm seems less tender than most others. The demand for seeds has already been enormous; for long voyages they are best packed into the sawdust of resinous kinds of wood.

Ptychosperma Arfakiana, Beccari.

New Guinea, reaching elevations of 5,000 feet in comparatively temperate regions. Height as much as 30 feet.

Ptychosperma Cunninghami, Hermann Wendland.

East-Australia, as far south as Illawarra; thus one of the most southern of all palms. This also is a very high species, destined to take a prominent position in decorative plantations even far beyond the tropics. Several congeners occur in Fiji and other islands of the Pacific Ocean, and others again might be obtained from India, but they are probably not so hardy as those just mentioned. Though strictly speaking of no direct industrial value, these palms are important for horticultural trade, and are objects eminently fitted for experiments in acclimation.

Ptychosperma disticha, Miquel. (*Areca disticha,* Griffith.)

Assam, up to 4,000 feet.

Ptychosperma elegans, Blume. (*P. Seaforthia,* Miquel; *Seaforthia elegans,* R. Brown.)

Litoral forests of tropical Australia. Also a magnificent Feather-palm. Its leaflets are erose. It may prove hardy in mild extra-tropic regions.

Ptychosperma Musschenbroekiana, Beccari.

Ternate, Insular India, up to 3,000 feet. Height of this palm reaching 90 feet. Almost sure to be hardy in sheltered localities of the warmer temperate zone.

Pueraria Thunbergiana, Bentham.

Japan. There starch is prepared from the tubers of this climber. The fibre of the bark is woven locally into cloth (Dyer).

Pueraria tuberosa, De Candolle.

Southern Asia, up to 4,000 feet. A tall woody twiner. Its large tubers are edible, and might improve by culture.

Pugionium cornutum, Gaertner.

From the Caspian Sea to China. This herb is grown by the Mongols as a vegetable (Hance).

Punica Granatum, Linné.

The Pomegranate. North-Africa and South-Western Asia, in the Himalayas up to 6,000 feet. Well-known for its showy habit, rich-colored flowers, peculiar fruit and medicinal astringency, but much overlooked regarding its value as a hedge-plant. The bark contains 32 per cent. tannin (Muspratt), and is also used for dyeing the yellow Morocco-leather (Oliver). The peel of the fruit serves likewise for dye. For therapeutic purposes particularly the root-bark is administered. Concerning pelletierin and other alkaloids from the root-bark, ample information is given in Husemann's and Hilger's Pflanzenstoffe (1884).

Pycnanthemum incanum, Michaux.

North-America. A perennial herb, in odor resembling both Penny-royal and Spearmint. It likes to grow on rocky woodland, and on such it might be easily naturalized.

Pycnanthemum montanum, Michaux.

The Mountain-Mint of North-America. A perennial herb of pleasant, aromatic, mint-like taste. These two particular species have been chosen from several North-American kinds to demonstrate, that we may add by their introduction to the variety of our odorous garden-herbs. They may also be subjected with advantage to distillation.

Pyrularia edulis, Meissner.

Nepal, Khasia, Sikkim. A large umbrageous tree. The drupa-ceous fruit is used by the inhabitants for food. A few other species occur in Upper India, one on the high mountains of Ceylon, and one in North-America. The latter, P. pubera (Michaux), can be utilized for the oil of its nuts.

Pyrus aucuparia, Gaertner.

Europe, Northern and Middle Asia. The Rowan or Mountain-Ash. Height seldom over 30 feet. Wood particularly valuable for machinery and pottery-work, also crates.

Pyrus coronaria, Linné.

The Crab-Apple of North-America. This showy species is mentioned here as worthy of trial-culture, since it is likely that it would serve well as stock for grafting. Best grown in glades. Wood nearly as tough for screw-work as that of the pear-tree (Robb).

Pyrus communis, Linné.

The Pear-tree. Middle and Southern Europe, Western Asia. Well known even at the time of Homer; and many varieties were cultivated in Italy at the commencement of the Christian era; pears were available also to the lacustrine people of Switzerland, Lombardy and Savoy, but seemingly not so extensively as the apple. Prof. C. Koch regards the Chinese Pyrus Achras (Gaertner), which is the oldest name for P. Chinensis of Desfontaines and Lindley, as the wild plant, from which all our cultivated varieties of pears have originated. The pear-tree is cultivated up to 10,000 feet in the Himalayas; like the apple-tree, it sets no fruit in tropical regions, but on the other hand it will bear a good deal of frost, being grown in Norway to lat. 63° 52'. The tree attains an age of over three hundred years, fully bearing. At Yarmouth, a tree over 100 years old has borne as many as 26,800 pears annually; the circum-ference of its crown is 126 feet (Masters). Pear-wood is used by wood-engravers, turners and instrument-makers. A bitter gly-cosid, namely phlorrhizin, is attainable from the bark of apple- and pear-trees, particularly from that of the root; while a volatile alkaloid, namely trimethylamin, can be prepared from the flowers. Pyrus auricularis, Knoop (P. Polveria, L.), the Bollwiller-Pear, is a hybrid between P. communis and P. Aria, Ehrhart. Curious fruits have been produced latterly in North-America by the hybridization of the apple with the pear. The generic writing of Pirus is inadmissible, as even Plinius used both Pirus and Pyrus in his writings, and as the latter wording was already adopted by Malpighi and fixed for the genus by Linné. The flowers of all the leading European fruit-trees afford nectar for honey to bees.

Pyrus Cydonia, Linné. *(Cydonia vulgaris, Persoon.)*

The Quince. Countries at the Caspian Sea. Reared in South-Europe from antiquity ; in the Himalayas its culture reaches to 5,500 feet elevation. The Portuguese variety bears extremely large fruit. The preserved quince is one of the most agreeable of fruits. The seeds impart copiously to water a tasteless mucilage. Quinces are not readily attacked by sparrows.

Pyrus Germanica, J. Hooker. *(Mespilus Germanica, Linné.)*

The Medlar. Southern Europe, Western Asia. Of this species a variety exists with large fruits of particularly pleasant taste. The ordinary medlar-fruits become edible after some storage. A large-fruited variety of excellent taste is cultivated in South-Europe. P. Maulei (Masters) is a closely cognate plant, with golden-yellow edible fruit, particularly fit for preserves.

Pyrus Japonica, Thunberg.

Japan. One of the prettiest of small hedge-bushes, and one of the earliest flowering. Under favorable circumstances it will produce its quince-like fruit. It is one of the early species, so valuable to the apiarist.

Pyrus Malus, Linné.

The Apple-tree. Europe, Western Asia, ascending the Himalayas to 11,000 feet. Shown to have been in culture already in Switzerland and Northern Italy prior to historic records, though Professor C. Koch regards neither the wild and variable crab-trees nor the pear, as original denizens of Middle and Northern Europe, but simply as strayed from cultivation and degenerated. Koch traces some sorts of cultivated apples to P. pumila (Miller) of South-Western Asia; as other original forms he notes the P. dasyphylla (Borkhausen), P. silvestris and P. prunifolia (Willdenow) of Middle and Western Asia. This tree is one of longevity; Mr. H. C. Hovey gives records of an apple-tree in Connecticut, which at the age of 175 years measured about 14 feet in circumference at 3 feet from the ground, the diameter of the top of the tree being over 100 feet. In Prof. Meehan's *Gardeners' Monthly* is a record of the fecundity of an apple-tree in New England, given by Mr. W. S. Platt, of Cheshire; its eight branches spread over six rods, and five of the branches bore in one

year over 100 bushels of apples, the bearing taking place alternately
with the other three branches. The value of the annual import of
American apples into the United Kingdom has risen to two millions
sterling. In Europe apple-trees and other fruit-trees are occasionally
bored by the Scolytus destructor. Succulent apples contain about 70
per cent. of juice, a remark which may serve in calculating the yield
of cider. Apple-trees will endure the winters of Norway to lat. 65° 28'
(Schuebeler). The best dried apples and similar fruits are obtained
by submitting them, according to the new American method, to a blast
of cold air. The United States sent to England in the season 1880-1
about 1,350,000 barrels of apples, irrespective of the large quantity
sent by Canada.

Pyrus nivalis, Jacquin.

The Snow-Pear. Middle and Southern Europe. This would be
adapted for orchards in higher mountain-regions. The fruit becomes
soft and edible through exposure to snow. P. amygdaliformis
(Villars) or P. Kotschyana (Boissier) are probably the wild state of
this tree. Pear-cider is often made of the fruit of this species.

Pyrus rivularis, Douglas.

The Crabapple-tree of North-Western America. Fruit prized by
the aborigines for food (G. Dawson); likely amenable to cultural
improvements. Dr. C. Koch draws attention to the probable identity
of P. Toringo (Siebold) from Japan.

Pyrus salicifolia, Linné.

Greece. Turkey, Persia, South-Western Russia. Hardy at Christiania. Though its fruit, which softens slowly, is edible, this tree is
mainly utilized as a superior stock for grafting.

Quercus Ægilops, Linné.*

South-Europe, also Syria. A nearly evergreen tree of the size of
the British oak. The cups, known as Valonia, used for tanning and
dyeing; the unripe acorns, called Camata or Camatena, for the same
purpose. Valonia is largely exported from Smyrna to London (33,802
tons in 1876). Greece used to produce annually 10,000 tons, worth
as much as £18 per ton. The supply is inadequate to present
demand. 34,450 tons of Valonia, worth about £526,000, were imported into the United Kingdom in 1884. Valonia (Wallones)
produces a rich bloom on leather, which latter also becomes less
permeable to water (Muspratt). The ripe acorns are eaten raw or
boiled. This oak is also recommended as a fine avenue-tree. It
bears considerable frost. The wood is capital for furniture. Dr.
Kotschy separates Q. Ægilops into several species, of which A.
Græca, Q. oophora and Q. Vallonea yield the mercantile article.

Quercus agrifolia, Née.

California and Mexico. One of the most magnificent among evergreen oaks, with dense, wide-spreading foliage. The thick bark
available for tanning. According to Dr. Gibbons this tree attains a

height of about 100 feet, a stem-diameter of 8 feet, and a crown of 125 feet breadth. Wood-cutters distinguish two varieties, one with red and one with pale wood. It grows naturally near the sea, and luxuriates in the deep soil of valleys, but also on the tops of mountains. The value of its timber is not fully appreciated. Although brittle when green and perishable if exposed to the weather, it becomes almost as hard and strong as live-oak, if properly seasoned, and is especially adapted for ships' knees.

Quercus alba, Linné.*

The White or Quebec-Oak. From Canada to Florida, west to Texas. A most valuable timber-tree, becoming fully 100 feet high; diameter of stem to 7 feet, trunk sometimes 65 feet long to first branch. Rate of stem-growth in Nebraska according to Governor Furnas 29 inches circumferentially in 22 years. Attains a great age; succeeds best in rich woodlands; and is of quicker growth than the English oak. The timber is pliable, most durable, one of the very best of all woods for casks, also of first-class value for cabinet-work, for machinery, spokes, naves, beams, plough-handles, agricultural implements, carriages, flooring, basket-material (Sargent) and railway-ties (Robb); it is also largely employed in ship-building; the young saplings serve for hoops and whip-handles. The bark contains about 8 per cent. tannin, and is used also in medicine.

Quercus annulata, Smith.

Upper India. A large evergreen oak, which provides a very good, timber. It does not ascend quite so high as Q. incana. Q. spicata (Smith), another very large Indian oak, ascends only 5,000 feet; it is known also from Borneo, Java and Sumatra.

Quercus aquatica, Walter.

North-America. Height of tree often 60 feet; it furnishes a superior bark for tanning. This oak should be chosen for planting in wet ground or for bordering streams. Although the wood is not of much value, yet the tree is a great favorite as a shade-tree, being of rapid growth and fine outline. Prof. C. Koch identified this with the true Q. nigra of Linné.

Quercus bicolor, Willdenow.

Southern White-Oak. South-Eastern States of North-America. Closely allied to Q. Prinus, but vernacularly distinguished as Basket-Oak; it thrives best in deep, damp forest-soil, and is regarded as the most important hardwood-tree in the Gulf-region; height reaching 120 feet, stem-length to 70 feet. The growth comparatively slow; wood similar in applicability to that of the white oak; it is split readily into thin strips of great strength and flexibility for rough baskets (Dr. C. Mohr).

Quercus Castanea, Née.

The Mexican Chestnut-Oak. Evergreen. It furnishes edible acorns.

x

Quercus Cerris, Linné.

Turkey or Moss-cupped Oak. Southern Europe, South-Western Asia. Hardy still at Christiania. Of the height of the English oak; in suitable localities of quick growth. The foliage deciduous or also evergreen or nearly so. The wood available for wheel-wrights, cabinet-makers, turners, coopers, also for builders generally. It is still firmer and harder than that of the British oak; the sap-wood larger, the heartwood of a more saturated brown, and the large rays more numerous, giving it a most varied and beautiful wainscot-grain (Dr. Brandis, Prof. C. Koch).

Quercus Chinensis, Bunge.

Northern China. One of the hardiest among the evergreen oaks.

Quercus chrysolepis, Liebmann.

California. According to Dr. Vasey this evergreen oak rarely exceeds 50 feet in height, but supplies the hardest oak-wood on the Pacific coast. Dr. Gibbons observes, that it holds a primary rank among Californian forest-trees, but is of sparse occurrence; in suitable soil on the sides of mountains it is of giant-growth, spreading out in magnificent proportions. In toughness and density of wood it repre-sents the live-oak of Florida, being thus highly useful to imple-ment-makers, wheelwrights and machinists; the ivory-like appearance of the wood befits it particularly for inlaying (Dr. Kellogg).

Quercus coccifera, Linné.

The deciduous Kermes-Oak of South-Europe, North-Africa and South-Western Asia. So called from the red dye, furnished by the Coccus ilicis from this oak. It also supplies tanners' bark con-taining about 8 per cent. tannin (Muspratt). The huge and ancient Abraham's Oak belongs to this species. The tree likes rich wood-lands.

Quercus coccinea, Wangenheim.

The Black Oak of North-America. Height to about 100 feet; stem-diameter to 5 feet. Foliage deciduous. The tree thrives best in rich woodlands and moist soil. The timber is almost as durable as that of the white oak, and in use for flooring and other carpenters' work. Rate of growth about the same as that of the red oak. The yellow dye, known as quercitron, comes from this tree; it is much more powerful than that of woad (Bancroft). With alumina the tinge of the bark is bright yellow, with oxyde of tin it is orange, with oxyde of iron it is drab (Porcher). Q. velutina (Lamarck) or Q. tinc-toria (Bartram) has been called a variety of this. According to Sargent, it produces timber of close grain and great durability, utilized for carriage-building, cooperage and various constructions; the bitter inner bark yields a yellow dye. The bark of the variety called scarlet oak is practically far inferior in value to that of the black oak (Meehan). Bark contains about 8 per cent. of tannic

acid. Dr. Eugelmann found the black oaks twice as rapid in growth
as the white oaks of the United States. Bartram's oak (Q. heterophylla) is, according to him, a hybrid between the willow-oak and
scarlet oak. Hybrid oaks produce acorns capable of germination.

Quercus cornea, Loureiro.

China. An evergreen tree, at length 40 feet high. Acorns used
for food.

Quercus corrugata, Hooker.

Mexico. Attains a height of about 80 feet. The acorns are as
large as those of Q. Skinneri.

Quercus cuspidata, Thunberg.

Japan. A magnificent evergreen oak, grand in its proportions,
bears acorns in bunches or strings, of very sweet taste when baked
like chestnuts, but only of small size (F. C. Christy). These acorns,
boiled or roasted, are regularly sold in Japan for food (Rein).

Quercus densiflora, Hooker and Arnott.

Californian Chestnut-Oak. A large evergreen tree of beautiful
outline, dense foliage and compact growth. Very hardy, having
withstood the severest winters at Edinburgh with a temperature of
0° F. (Gorlie). Bark very valuable for tanning; wood however
subject to rapid decay (Prof. Bolander).

Quercus dentata, Thunberg.*

Manchuria, Northern China, Japan. This is one of the species, on
which the Oak-silkworm (the Yama Mayon) lives. Franchet and
Savatier enumerate 22 distinct species of oaks as indigenous to
Japan.

Quercus Douglasii, Hooker and Arnott.

The Blue Oak, California. Stem reaching 7 feet in circumference
(Brewer). Resembles the white oak in the quality of its timber,
but this particularly used in wheelwrights' work.

Quercus dilatata, Lindley.

From the Himalayas to Afghanistan, at elevations from 4,500 to
10,000 feet. Evergreen. Height becoming 100 feet; crown very
shady; branches lopped for sheep-fodder. The hard, heavy, elastic
and durable wood much used for building purposes and implements
(Major Madden), easily worked, and but little apt to warp and rend
(Dr. Brandis).

Quercus falcata, Michaux.

South-Eastern States of North-America. Known as Spanish Oak.
A tree, attaining a height of 80 feet, with a stem 5 feet in diameter.
Foliage diciduous. It lives in dry sandy ground, and can also be
utilized for sea-coasts. Produces an excellent tanners' bark, and also
galls for superior ink. The wood is finer grained and more durable

than that of Q. rubra, and used for staves, railway-carriages and in ship-building (C. Mohr). Prof. C. Koch points out, that Q. cuneata (Wangenheim) is the oldest name for this species.

Quercus Garryana, ·Douglas.

North-Western America, along the coast between the 38th and 50th degrees. A tree, to 100 feet high or more, with a stem often 6 feet in diameter. This, with Q. Douglasii and Q. lobata, passes as California White Oak. The timber is remarkably pale for an oak, hard and fine-grained, of great strength and durability, well suited for almost every kind of construction, for which the white or the European oak is employed. The acorns, being sweet and agreeable, form an excellent mash for hogs.

Quercus glabra, Thunberg.

Japan. Evergreen. The acorns are consumed for food by the Japanese.

Quercus glauca, Thunberg.

The Kashi of Japan. A truly magnificent evergreen tree, to 80 feet high. The hard and close-grained wood is chosen there for select tools, particularly planes and utensils (Christy).

Quercus Ilex, Linné.

The Holly-Oak of South-Europe ; extending also to Algeria and to the Himalayas, which it ascends up to about 10,000 feet. Height of tree rather less than that of the English oak, but occasionally it is very lofty. Wood in use for ship-building and wheelwrights' work, bark for tanning. From varieties of this tree are obtained the sweet and nourishing Ballota- and Chestnut-acorns, as much as 20 bushels occasionally from one tree in a season.

Quercus incana, Roxburgh.

Himalayas, at elevations between 3,000 and 8,000 feet. A beautiful evergreen tree of great dimensions. Young branchlets in spring, as noted by Dr. Brandis, from whitish to lilac-colored. Mr. Simmonds reminds us that a silkworm (Antheræa Roylei), producing large cocoons, lives on this oak. In its native localities Q. lanuginosa (D. Don) is associated with it. Q. lamellosa (Smith) of the same region attains a height of about 120 feet, with a straight trunk to 60 feet with a girth of 15 feet (Brandis).

Quercus infectoria, Olivier.

Countries around the Mediterranean Sea, extending to Persia. A tree deciduous in its foliage. The galls of commerce are chiefly obtained from this species. A variety or closely allied species Q. Lusitanica (Webb) or Q. Mirbeckii (Durieu) reaches a height of 120 feet, with a stem-girth of 20 feet. Some states of this are almost evergreen, and then particularly eligible as promenade-trees.

Quercus lancifolia, Roxburgh (not Chamisso nor Bentham).

A tall evergreen timber-tree of the Himalayas. Wood valued for its durability; its medullary rays exceedingly fine (Brandis).

Quercus lobata, Née.

California. The Sacramento White Oak. A tree finally about 150 feet high, with a stem six feet in diameter, with wide-spreading branches, which often bend to the ground. Hardy in Middle Europe (C. Koch). The wood is brittle when green, but hard and tough when seasoned; its value has been much underrated (Gibbons). The acorns of this oak used to form a large proportion of the winter-food of the aboriginal inhabitants of North-California.

Quercus lyrata, Walter.

The Overcup-Oak of the South-Eastern States of North-America, extending from South-Illinois to Florida and Louisiana. A tree of majestic size, with a stem to four feet in diameter. Lately recommended as valuable for timber-cultivation, especially in wet ground.

Quercus macrocarpa, Michaux.

The Burr-Oak of Eastern North-America. Tree to about 70 feet high; stem-diameter sometimes 8 feet. Hardy at Christiania. The timber regarded by some almost as good as that of the white oak. The bark contains about 8 per cent. tannin. Circumferential stem-measurement after 22 years' growth 3½ feet in Nebraska (Furnas).

Quercus macrolepsis, Kotschy.*

Greece. This evergreen oak also yields Valonia, being closely allied to Q. ægilops. A. de Candolle unites it with Q. Græca of Kotschy.

Quercus magnolifolia. Née.

Mexico, in cooler mountain-regions. From Née's note it would appear, that he saw on this oak the numerous caterpillars, which construct ovate cocoons eight inches long, consisting of a kind of grey silk, which was there locally manufactured into stockings and handkerchiefs.

Quercus Mongolica, Fischer.*

Manchuria and Northern China. It is on this tree and on Q. serrata and Q. dentata, that the silk-insect peculiar to oak-trees mainly, if not solely, is reared, as shown by Dr. Hance.

Quercus Muehlenbergii, Engelmann.

Middle and Eastern States of North-America. A middle-sized tree; its wood compact, strong, durable for posts and railway-ties (Sargent).

Quercus palustris, Du Roi.

The Pin-Oak or Marsh-Oak of South-Eastern North-America. Hardy at Christiania. Height at length 80 feet; of quick growth.

The wood is fine-grained, strong and tough; it is ornamental for furniture on account of the strong development of medullary rays.

Quercus Phellos, Linné.

The Willow-Oak of the South-Eastern States of North-America. In low damp forest-land attaining a stem-girth of 12 feet. The wood is hard, compact and very elastic, suitable for railway-carriages and many other structures (Dr. C. Mohr). The acorns available for food. A variety or closely allied species is the Shingle-Oak, Q. imbricaria, Michaux. The comparative value of the very numerous Cis- and Trans-Atlantic oaks, but little as yet understood in the eastern world either for avenue-purposes or timber-plantations, should be tested with practical care. Even recently oaks have been discovered on the south-eastern mountains of New Guinea at not very high elevations.

Quercus Prinus, Linné.

The Swamp-Oak or Chestnut-Oak. South-Eastern States of North-America. A tree, becoming 90 feet high; aged stem as much as 15 feet in girth (Meehan). The tree is hardy in Norway to lat. 59° 55'. Foliage deciduous. Wood strong and elastic, but more porous and of a coarser grain than that of the white oak; according to Porcher it is easy to split and not hard, used for building purposes, also cooperage. A red dye is produced from the bark; the latter is one of the most important among oak-bark for tanning, furnishing a very solid and durable leather.

Quercus Robur, Linné.*

The British Oak. Extending through the greatest part of Europe, also to Western Asia, attaining a great age and an enormous size. It endures the frosts of Norway as far north as 65° 54'; while in lat. 59° 40' a tree measured was 125 feet high and 25 feet in circumference of stem (Schuebeler). Over 700 sound annual rings have been counted, and it has even been contended, that oaks have lived through 1,500 years. At Ditton's Park, owned by the Duke of Buccleugh, is an ancient oak, assumed to be 600 years old, with a stem-circumference of 30 feet at some distance (a few feet) from the ground (Dr. Masters and Th. Moore). Oaks have been known to gain a stem 12 feet in diameter at the base, 10 feet in the middle and 5 feet at the main branches. Two varieties are distinguished; 1. Q. sessiliflora (Salisbury), the Durmast-Oak, with a darker, heavier timber, more elastic, less fissile, easier to bend under steam. This tree is also the quicker of the two in growth, and lives in poorer soil. Its bark is richer in medicinal dyeing and tanning principles. Extract of oak-bark for tanners' use fetches about £18 per ton in the London market; the best oak-bark yields 16 to 20 per cent. tannin. 2. Q. pedunculata (Ehrhart). This variety supplies most of the oak-timber in Britain for ship-building, and is the best for cabinet-makers' and joiners' work. In Britain it is sometimes attacked by Scolytus multistriatus. Mr. W. Winter noticed, that the British oak withstood an occasional shade-temperature of 118 degrees F. in Riverina,

New South Wales. The long continued adherence of dead leaves in the cool and most verdant season renders this oak not so well adapted for pleasure-grounds in the warmer parts of the temperate zone as many others, particularly evergreen oaks. The English oak is however of quicker growth than many other species. At Port Phillip it attains to a height of 40–50 feet in 20 years. The galls, produced by Cynips calicis, are sought for particular tanning, and called in Germany Knoppern. The best oak-bark for tanning is obtained from trees 12–36 years old (Prof. Wiesner).

Quercus rubra, Linné.

The Red-Oak of Eastern North-America. Height reaching about 100 feet; diameter of stem 4 feet. A tree, content with poor soil. The wood, though coarse, is of rigidity, and has not the fault of warping ; it is of fair value for staves (Simmonds), and even building purposes, but variable in quality according to soil and clime (Sargent). The bark is rich in tannin. Autumnal tint of foliage beautifully red. The acorns, which are produced in great abundance, are relished by hogs. The tree is hardy still at Christiania. Circumferential stem-measurement at 2 feet from the ground after 22 years about 38 inches (Furnas).

Quercus semecarpifolia. Smith.

In the Himalayas and adjoining ranges up to about 10,000 feet. The largest of the oaks of India, upwards of 100 feet high, with a stem often 18 feet in girth. Leafless annually for a short time, not quick of growth. It furnishes a hard and heavy timber of fair quality.

Quercus serrata, Thunberg.*

One of the twenty-three known Japanese Oaks; extending to China and Nepal. Hardy in Middle Europe. A good avenue-tree, though deciduous. It yields the best food for the Oak-silkworm (Bombyx Yamamai). It is recommended to pack acorns intended for far distances in wooden cases between dry moss or sand, to secure retention of vitality; moreover they must be quite fresh, when packed.

Quercus sideroxyla, Humboldt.

Mountains of Mexico, up to about 8,000 feet elevation. An oak of great size; timber compact, almost imperishable in water. Q. lanceolata, Q. chrysophylla, Q. reticulata, Q. laurina, Q. obtusata, Q. crassipes, Q. glaucescens, Q. Xalapensis, Humb. and Q. acutifolia, Née, are among the many other highly important timber-oaks of the cooler regions of Mexico. No printed record seems extant, concerning the technology of the numerous Mexican oaks, though doubtless their respective values are well known to local artisans. According to the Abbé and Surgeon Liturgie, one of the Mexican oaks, near San Juan, nourishes a Bombyx the cocoons of which are spun by the natives into silk (Tschichatchef).

Quercus Skinneri, Bentham.*

Mexico. On limestone-soil, in the temperate region at 7,000–8,000 feet elevation. "Cozahual." Acquiring a height of 150 feet; thickness of stem to 12 feet. Wood yellowish, remarkably durable and elastic, not excelled in value by that of any other oak. Bark rich in tannin (Hugo Finck). Foliage deciduous. The acorns of this oak measure nearly 6 inches in circumference, and are available for feeding various domestic animals.

Quercus stellata, Wangenheim. (*Q. obtusiloba*, Michaux.)

The Post-Oak of North-Eastern America. Content with poor and even sandy soil, but not a large tree. Can be reared on sea-shores. On account of its very durable and dense wood it is much in requisition there for posts, and is particularly prized for ship-building, also sought for railroad-ties.

Quercus Suber, Linné.*

The Cork-Oak of South-Europe and North-Africa. It is evergreen and attains an age of fully two hundred years. Hardy in the lowlands of England. After about twenty years it can be stripped of its bark every six or seven years; but the best cork is obtained from trees over forty years old. Height of the tree finally about 40 feet. Acorns of sweetish taste. Mr. W. Robinson found that young cork-oaks, obtained from the writer, made a growth of 4 feet yearly in the humid Western Port-district of Victoria. The bark of Q. pseudo-suber (Santi) is inferior for cork, but the closely-allied Q. occidentalis (Gay), which is hardier than Q. Suber, produces an excellent cork-bark.

Quercus Sundaica, Blume.

One of the oaks from the mountains of Java, where several other valuable timber-oaks exist. The existence of oaks on the north-western mountains of New Guinea has been demonstrated by Dr. Beccari; hence, in all probability, additional valuable evergreen species will be obtainable thence for our arboreta and forests.

Quercus Tozæ, Bosc.

South-Europe. One of the handsomest oaks, and one of the quickest in growth. Will live in sandy soil and emits suckers. It furnishes superior tanners' bark.

Quercus virens, Linné.*

The Live-Oak of North-America, extending northward only to Virginia, occurring also in Mexico. One of the hardiest of the evergreen species. Likes a coast-climate and a soil rich in mould. Becomes 60 feet high, with a stem sometimes to 9 feet in diameter. Supplies a most valuable timber for ship-building; it is heavy, compact, fine-grained; it is moreover the strongest and most durable yielded by any American oaks. Like Q. stellata, it lives also on sea-shores, helping to bind the sand, but it is then not of tall stature. Q. Virginiana (Miller) is the oldest name for this oak, as pointed out by C. Koch. Of many of the 300 oaks, occurring in the western and eastern portions of the northern hemisphere, the properties remain unrecorded and perhaps unexamined; but it would be important to introduce as many kinds as possible for local test-growth.

Quercus Wislizenii, A. de Candolle.

Mexico, at an elevation of about 7,000 feet, also reaching California. A magnificent tree with dense foliage, the stem attaining finally a circumference of 18 feet (Prof. Sargent).

Quercus Xalœpensis, Humboldt and Bonpland.

Mexico, ascending to 5,000 feet, preferring for localities poor soil, but of ferruginous clay, where little else will grow. Height to 80 feet; it is a quick grower; its timber will endure only under roof; the tree gives a heavy crop of acorns (Hugo Finck).

Quillaja saponaria, Molina.

Chili. A colossal tree, fit not only for loamy but also sandy and peaty soil. The bark is rich in saponin, and therefore valuable for dressing wool and silk, also for various cleansing processes.

Rafnia amplexicaulis, Thunberg.

South-Africa. The root of this bush is sweet like liquorice, and is administered in medicine. Rafnia perfoliata (E. Meyer), also from South-Africa, furnishes likewise a medicinal root.

Raphanus sativus, Linné.

The Radish. Temperate Asia, southward to the Himalayas, up to 16,000 feet, eastward to Japan. In Norway it can be grown northward to lat. 70° 22′ (Prof. Schuebeler). R. caudatus, L., the radish with long edible pods, is regarded by Dr. Th. Anderson as a mere variety, and he thinks, that all are sprung from the ordinary R. Raphanistrum, L., of Europe. All radishes succeed best in a calcareous soil, or their culture must be aided by manure rich in lime. The root of the black radish is comparatively rich in starch. The seeds, according to Vilmorin, will keep about five years.

Remirea maritima, Aublet.

Intra-tropical coast-regions around the globe. A perennial creeping sedge for binding sand.

Reseda Luteola, Linné.

The Weld. Middle and Southern Europe, Middle Asia, North-Africa. An herb of one or two years' duration. Likes calcareous soil. A yellow dye (luteolin) pervades the whole plant. The plant must be cut before the fruit commences to develop, otherwise the pigment will much diminish.

Reseda odorata, Linné.

The true Mignonette. North-Africa and Syria. A favorite garden-herb of one or very few years duration. The delicate scent can best be concentrated and removed by enfleurage. To be counted also among the honey-plants. Mess. Dippe in Quedlinburg devote regularly about 50 acres to rearing of mignonettes for seeds.

Rhagodia Billardieri, R. Brown.

Extra-tropical Australia. An important bush for binding moving sand on sea-shores. Resists the severest gales as well as the spray of the sea.

Rhagodia nutans, R. Brown.

Southern, Eastern and Central Australia. This, as well as the allied R. hastata, is a good fodder-herb for saltbush-runs. Some other species, mostly shrubby, are equally valuable.

Rhamnus Alaternus, Linné.

Countries around the Mediterranean Sea. A hedge-shrub, becoming arborescent, thus gaining a height of 20 feet. It strikes readily from cuttings. G. Don admits it as a splendid honey-plant.

Rhamnus alnifolius, L'Héritier. (*R. Purshianus,* D.C.)

From Oregon to California and British Columbia. Allied to R. Carolinianus (Walter). Reaches a height of about 20 feet. Leaves deciduous. This species furnishes as "Cascara Sagrada" its bark famed for cathartic properties; the fruits are also powerfully aperient.

Rhamnus catharticus, C. Bauhin.

The Buckthorn. Middle and Southern Europe, North-Africa, Middle Asia. It can be utilized as a hedge-plant. The berries are of medicinal value, as indicated by the specific name. The foliage and bark can be employed for the preparation of a yellow and green dye; the juice of the fruit mixed with alum constitutes the "sap-green" of painters. The plant is hardy in Norway to lat. 60° 48'. R. Dahuricus (Pallas) is a closely cognate species.

Rhamnus chlorophorus, Lindley.

China. From the bark a superior green pigment is prepared. R. utilis, from the same country, serves for the like purpose. This kind of dye is particularly used for silk, and is known as Lokao.

Rhamnus Frangula, Linné.

Europe, North-Africa, Northern and Western Asia. Endures the climate of Norway to lat. 64° 30' (Schuebeler). A tall shrub, with deciduous leaves. The flowers are particularly grateful to bees (G. Don). The wood one of the very best woods for gunpowder. Recommended by Sir Joseph Hooker to be grown on the coppice-system for this purpose. The bark is valuable as a cathartic; yields also a yellow dye. R. purpureus (Edgeworth) is an allied Himalayan species.

Rhamnus Græcus, Reuter.

Greece. From this shrub, and to no less extent from the allied R. prunifolius (Sibthorp) are derived the green dye-berries collected in Greece, according to Dr. Heldreich. These shrubs grow on stony mountains up to 2,500 feet.

Rhamnus infectorius, Linné.

On the Mediterranean Sea and in the countries near to it. Hardy still at Christiania. The berry-like fruits of this shrub are known in commerce as Graines d'Avignon and Graines de Perse, and produce a valuable green dye. Other species seem to supply a similar dye-material; for instance, R. saxatilis, L., R. amygdalinus, Desf., R. oleoides, L., R. tinctorious, W. & K., all from the Mediterranean regions and near them.

Rhapidophyllum Hystrix, Wendland and Drude. (*Chamærops Hystrix,* Fraser.)

The Blue Palmetto of Florida and Carolina. A hardy dwarf Fan-palm.

Rhapis flabelliformis, Linné fil.

China and Japan. This exceedingly slender palm attains a height of only a few feet. The stems can be used for various small implements. It is one of the best plants for table-decorations. It bears the climate of the South of France to 43° 32' N. lat. (Naudin).

Rhaponticum acaule, De Candolle.

On the Mediterranean Sea. A perennial herb. The root is edible.

Rheum australe, D. Don.* (*R. Emodi,* Wallich; *R. Webbianum,* Royle.)

Himalayan regions up to 16,000 feet. From this species at least a portion of the medicinal Rhubarb is obtained, its quality depending much on the climatic region and the geological formation, in which the plant grows. Should we wish to cultivate any species here for superior medicinal roots, localities in our higher and drier alpine tracts should clearly be chosen for the purpose. Hayne regards the presence of much yellowish pigment in the seed-shell as indicating a good medicinal rhubarb-plant. As much as 5 lbs. of the dried drug are obtainable from a single plant several years old. An important orange-red crystalline substance, emodin, allied to chrysophanic acid, occurs in genuine rhubarb. Medicinal rhubarb-root is now also grown in England.

Rheum officinale, Baillon.*

Western China and Eastern Thibet on the high table-land. Height of stem sometimes to 10 feet; circumference of foliage reaching 30 feet; blade of leaf 2 feet long and broad (Balfour). It furnishes most of the true Turkey-Rhubarb, not merely from the root but also from the woody stem. Suited for mountainous regions. Recommended also as a scenic plant by Regel. Hardy at Christiania.

Rheum palmatum, Linné.*

From insular to alpine North-Eastern Asia. Attains a height of
9 feet. A variety from the Tangut-country of Mongolia or North-
Thibet, found by Col. Przewvalski, yields an excellent medicinal root,
known as the Kiakhta- or Khansu-Rhubarb (Maximowicz)—indeed
the best Russian Rhubarb. The plant is valuable also for decorative
effect. For medicinal culture alpine valleys with soil rich in lime
are needed (Sir Rob. Christison). For indications of the literature
on medicinal rhubarbs see among lexicographic works particularly
B. D. Jackson's "Vegetable Technology," London Index-Society,
1882.

Rheum Rhaponticum, Linné.

From the Volga to Central Asia. This species, together with R.
Tartaricum, L. fil., R. undulatum, L. and a few others, all Asiatic
(one extending to Japan), provide their acidulous leaf-stalks and
unexpanded flower-mass for culinary purposes. Rhubarb leaves can
also be used in the manner of spinage. Propagation generally by
division of root. The soil for rhubarb-plants, intended to yield
kitchen-vegetable, must be deep and rich.

Rhizopogon magnatum, Corda.

Europe. One of the edible truffles sold in the markets of Middle
Europe, with R. rubescens, Tulasne.

Rhododendron maximum, Linné.

North-Eastern America. Attains a height of about 20 feet. Irrespective of its being a fine acquisition for any garden-copses, this
bush seems of industrial importance, because Mr. C. Forster asserts,
that the wood of this and the allied Kalmia latifolia, L., is surpassed
only by the best boxwood. This may give a clue to other substitutes for that scarce commodity, needed so extensively by the wood-
engraver.

Rhus aromatica, Aiton.

North-America, from the Atlantic to the Pacific Ocean, northward
to Canada. A straggling bush. The aromatic foliage important for
medicinal purposes.

Rhus caustica, Hooker and Arnott. (*Lithræa venenosa,* Miers.)

Chili, where it is called the Litre. A small or middle-sized tree,
the very hard wood of which is used for wheel-teeth, axletrees and
select furniture. The plant seemed neither caustic nor otherwise
poisonous (Dr. Phillippi).

Rhus copallina, Linné.

Eastern North-America, extending to Canada. A comparatively
dwarf species. This can be used for tanning. A resin for varnishes
is also obtained from this shrub.

Rhus coriaria, Dodoens.

The Tanner's Sumach. Countries around the Mediterranean Sea extending to temperate Western Asia. The foliage of this shrub or small tree, simply dried and reduced to powder, forms the sumach of commerce. It is remarkably rich in tannic acid, yielding as much as 30 per cent., and is extensively used for the production of a superior Corduan or Maroquin-leather and pale-colored leathers and dress-goods. Sumach allows the leather to carry more grease (Ballincnt). Price in Melbourne £24 to £36 per ton. It thrives best in loose calcareous soils, and cannot endure stagnant water. The strongest sumach is produced on dry ground. The cultivation presents no difficulty. A gathering can be obtained from suckers in the first year. The duration of sumach-fields under manure extends to fifteen years. Sumach can also be used for ink and various, particularly black, dyes. Under favorable circumstances as much as a ton of sumach is obtained from an acre. Sumach from Melbourne-plants was shown already at the Exhibition of 1863.

Rhus cotinoides, Nuttall.

Arkansas and Alabama. A tree, rising to 40 feet. The inner bark and the wood valuable for yielding a yellow dye (C. Mohr).

Rhus cotinus, Linné.* (*Cotinus coggyria*, Scopoli.)

The Scotino. In the countries on the Mediterranean Sea, extending to Hungaria and to the Himalayas. The wood of this bush furnishes a yellow pigment. The Scotino, so valuable as a material for yellow and black dye, and as a superior tanning substance, consists merely of the ground foliage of this plant. It contains up to 24 per cent. tannin. The plant endures the Norwegian winters northward to lat. 67° 56' (Prof. Schuebeler).

Rhus glabra, Linné.

North-America, extending to 54° north latitude; in Norway hardy to lat. 58° 8'. This sumach-shrub will grow on rocky and sterile soil. It produces a kind of gall, and can also be used as a substitute for the ordinary sumach. This species can be easily multiplied from suckers. It will live on poor soil, and is rich in the quality and long lasting yield of honey from its flowers (Quinby). American sumachs contain generally from 15 to 20 per cent., or occasionally up to 26 per cent. tannin. [On value of American Sumachs see Special Report No. 26, U. S. Department of Agriculture, 1880.] Employed also for therapeutic purposes.

Rhus lucida, Linné.

South-Africa. This shrub proved in Victoria of peculiar adaptability for forming hedges; it is evergreen, close growing, and stands clipping well. About half a hundred South-African species are known, of which probably some could be utilized like ordinary sumach; but hitherto we have remained unacquainted with the nature and degree of any of their tanning and coloring principles.

Rhus rhodanthema, F. v. Mueller.

East-Australia, on river-banks. A tree finally to 70 feet high; stem often 2 feet in diameter. Wood dark-yellow, soft, fine-grained, beautifully marked, much esteemed for cabinet-work. Worth £5 to £6 per 1,000 feet in Brisbane (W. Hill).

Rhus semialata, Murray.

China and Japan, extending to the Himalayas. Attains a height of 40 feet. This species produces a kind of nutgalls. It is apt to spread beyond ready control in rich soil. The stem will finally reach the thickness of a foot or more; the wood is tough and durable but stringy, prettily marked with dark edging.

Rhus succedanea, Linné.

The Japan Wax-tree, extending to China and the Himalayas, there up to 8,000 feet. The produce of this tree has found its way into the English market. The crushed berries are steamed and pressed, furnishing about 15 per cent. of wax, which consists mainly of palmatin and palmitic acid. Rhus silvestris (Siebold & Zuccarini) and R. vernicifera yield there a similar wax.

Rhus typhina, Linné.

The Staghorn-Sumach. Eastern North-America, extending to Canada. Hardy in Norway to lat. 61° 17'. This species will become a tree of about 30 feet height. Its wood is of orange tinge. Through incisions into the bark a kind of copal is obtained. The leaves may be used like ordinary sumach. This bush can be reared on inferior land. The leaves of American sumachs must be collected early in the season, if a clear white leather like that from Sicilian sumach is to be obtained. This can be ascertained by the color of the precipitate effected with gelatine. Some of the American and also other sumachs are important to apiarists.

Rhus vernicifera, De Candolle.

Extends from Nepal to Japan. It forms a tree of fair size and yields the Japan-varnish. In India it ascends to 7,000 feet; but Stewart and Brandis are doubtful, whether the Japan species (R. Vernix, L.) is really identical with the Indian. The fruit yields vegetable wax. R. Wallichii (J. Hooker) of the Himalayas is a cognate species.

Ribes aureum, Pursh.

From Arkansas, Missouri, Oregon to Canada. Endures the cold of Norway to lat. 70° (Schuebeler). This favorite bush of garden-shrubberies would probably along forest-streams produce its pleasant berries, which turn from yellow to brown or black. Professor Meehan mentions a variety or allied species from Utah, with berries larger than those of the black currant; they are quite a good table-fruit, and of all shades from orange to black, and this variety remains constant from seeds. Allied to this is R. tenuiflorum (Lindley) of California and the adjoining States, with fruits of the size of red currants, of agreeable flavor and either dark-purple or yellow color. R. aureum, R. palmatum and some other strong American species have come into use as stocks, on which to graft the European gooseberry (C. Pohl).

Ribes Cynosbati, Linné.

The Prickly-fruited Gooseberry-bush of Canada and the Eastern States of the American Union. The berries are large. There is a variety not so objectionably burlike-prickly. R. Cynosbati has been hybridized with R. Grossularia, and the sequence has been a good result (Saunders).

Ribes divaricatum, Douglas.

California and Oregon. One of the gooseberry-bushes of those countries. Can be grown in Norway to lat. 69° 40′. Berries smooth, black, about one-third of an inch in diameter, pleasant to the taste. Culture might improve this and many of the other species. R. Nuttalli (R. villosum, Nuttall, not of Gay nor of Wallich) is an allied plant, also from California.

Ribes floridum, L'Heritier.

The Black Currant-bush of North-Eastern America. The berries resemble in odor and taste those of R. nigrum. Allied to this is R. Hudsonianum (Richardson) from the colder parts of North-America.

Ribes Griffithi, J. Hooker and T. Thomson.

Himalaya, at heights from 10,000 to 13,000 feet. Allied to R. rubrum, bearing similar but larger berries of somewhat austere taste. R. laciniatum (H. & T.) is likewise a Himalayan species with red berries, and so is R. glaciale (Wallich). Furthermore, R. villosum, Wall. (R. leptostachyum, Decaisne), comes from the Indian highlands and seems worthy of practical notice.

Ribes Grossularia, Linné.*

The ordinary Gooseberry-bush. Europe, North-Africa, extra-tropical Asia, extending to the Chinese boundary (Regel), on the Himalayan mountains up to a height of 12,000 feet; in Norway enduring the cold to lat. 62° 44′. This plant, familiar to everyone, is mentioned here merely to indicate the desirability of naturalizing it in any sub-alpine regions, where it is not indigenous already.

Ribes hirtellum, Michaux.

North-America, particularly in the New England-States, extending to Canada. It likes moist ground. Yields the commonest smooth gooseberry there.

Ribes nigrum, Linné.

The Black Currant-bush. Europe, Middle and Northern Asia, North-America, ascending the Himalayan and Thibetan mountains to a height of about 12,000 feet; also particularly fit to be dispersed through forests in elevated situations. Hardy in Norway to lat. 69° 30'.

Ribes niveum, Lindley.

One of the Oregon Gooseberry-bushes. Berries small, black, of a somewhat acid taste and rich vinous flavor. Hardy to lat. 67° 56' in Norway.

Ribes orientale, Desfontaines.

From Syria to Afghanistan, up to an elevation of about 11,000 feet. The leaves emit a pleasant perfume (C. Koch). The berries act as a powerful purgative (Dr. Aitchison).

Ribes rotundifolium, Michaux.

Eastern North-America, as far as Canada. Hardy at Christiania. Yields part of the smooth gooseberries of the United States. The fruit is small, but of delicious taste. Unlike the ordinary gooseberry, not subject to mildew. Careful cultivation has gradually advanced the size of the fruit (Meehan).

Ribes rubrum, Linné.

The ordinary Red Currant-bush. Europe, North-America, Northern and Middle Asia, in the Himalayan mountains, ceasing where R. Griffithi commences to appear. One of the best fruit-plants for jellies and preserves, that can be chosen for colder mountain-altitudes. It endures the climate of Norway to lat. 70° 30' (Prof. Schuebeler). The root-bark contains phlorrhizin. Perhaps other species than those recorded here, among them some from the Andes, may yet deserve introduction, irrespective of showiness, for their fruits.

Richardia Africana, Kunth. (*R. Aethiopica,* Rosenthal.)

The "Calla" of gardens. From the Nile to the Cape of Good Hope. Important for scenic effects, particularly on the margins of waters. Easily moved at all seasons. The fresh root contains about 2 per cent. of starch.

Richardsonia scabra, Kunth.

From Mexico to Brazil. As an herb for pastures and hay-crop appreciated in localities with sandy soil (C. Mohr). It has spread over the Southern States of North-America.

Ricinus communis, Linné.*

The Castor-Oil Plant. Spontaneous in the tropical and sub-tropical zones of Asia and Africa, but hardly in South-Europe, originating according to A. de Candolle in North-Eastern Africa. A shrubby, very decorative plant, attaining the size of a small tree. At Christiania it grew to 12 feet in height and bore fruit, and it is reared as a summer-plant even to lat. 68° 7' (Prof. Schuebeler). It was well known to Egyptians four thousand years ago, and is also mentioned in the writings of Herodotus, Hippocrates, Dioscorides, Theophrastos, Plinius and other ancient physicians, philosophers and naturalists. The easy and rapid growth, the copious seeding, and the early return of produce render this important plant of high value in the warm temperate zone, more particularly as it will thrive on almost any soil, and can thus be raised even on arid places, without being scorched by hot winds. Recently recommended for staying bush-fires and for keeping off noxious insects and blights from plantations. It may thus become an important plant also for culture in desert-tracts, and is evidently destined to be in countries with cheap labor one of the most eligible plants to furnish oil for technical uses, particularly for lubricating machinery, irrespective of the value of its oil for medicinal purposes. The scalded leaves, applied externally, have long been known as particularly active on the mammary glands as a powerful galactagogue; the foliage is also in use as an emmenagogue; the root-bark has purgative properties. The seeds contain about 50 per cent. oil. To obtain the best medicinal oil, hydraulic pressure should be employed, and the seeds not be subjected to heat; the seed-coat should also be removed prior to the extracting process being proceeded with. A screw-press suffices however to obtain the oil for ordinary supplies. By decantation and some process of filtration it is purified. For obtaining oil to be used for lubrication of machinery or other technological purposes, the seeds may be pressed and prepared by various methods under application of heat and access of water. For lubrication it is one of the most extensively used of all oils. Castor-oil is usually bleached simply by exposure to solar light, but this procedure lessens to some extent the laxative properties of the oil. It dissolves completely in waterless alcohol and in ether, and will become dissolved also in spirit of high strength, to the extent of three-fifths of the weight of the latter. Solutions of this kind may become valuable for various technical purposes, and afford some test for the pureness of the oil. If pressed under heat it will deposit margaritin. Heated in a retort about one-third of the oil will distil over, and a substance resembling india-rubber remains, which saponizes with alkalies. Other educts are at the same time obtained, which will probably become of industrial value. These facts are briefly mentioned here merely to explain, that the value of this easily produced oil is far more varied than is generally supposed; and this remark applies with equal force to many other chemical compounds from vegetable sources, briefly alluded to in this present enumerative treatise. The seeds contain also a peculiar alkaloid—ricinin. The solid chemical compound of castor-oil is the crystalline isocetic acid (a glycerid). The oil contains also a non-crystalline acid, peculiar to it (ricinoleic acid). For the production of a particular kind of silk the Ricinus-plant is also important, inasmuch as the hardy Bombyx Arrindi requires the leaves of this bush for food. Even a few of the seeds, if swallowed, will produce poisonous effects. The root-bark has also been drawn into use as a purgative (Bernays).

Y

Robinia Pseudacacia, Linné.

The North-American Locust-Acacia, ranging from Alleghany to Arkansas. Height reaching 90 feet. Hardy to lat. 63° 26′ in Norway. The hard and durable wood is in use for a variety of purposes, and particularly eligible for treenails, axletrees and turnery; strength greater than that of the British oak, weight lighter (D. J. Browne). The natives used the wood for their bows. The tree is of rapid growth, and attains an age of several hundred years. A tree, raised in 1635, in the Paris Jardin des Plantes, is still alive. It may be planted closely for timber-belts and hedge-shelter on farm-lands. It is one of the best trees for renovating exhausted land and for improving poor soil. Also a bee-plant. Recommended as one of the easiest grown of all trees on bare sand, though standing in need of twice as much mineral aliment as Pinus silvestris and nearly as much as poplars. It pushes through shifting sand its spreading roots, which may attain a length of seventy feet. It will maintain its hold in hollows of drifts, where even poplars fail (Wessely). The roots are poisonous. The allied R. viscosa (Ventenat) attains a height of 40 feet.

Roccella tinctoria, De Candolle.

Canary-Islands, Azores, also in Western and Southern Europe and North-Africa. This lichen furnishes the litmus, orseille or orchil for dyes and chemical tests. It is a question of interest, whether it could be translocated and naturalized on the cliffs of our shores also. Other dye-lichens might perhaps still more easily be naturalized; for instance, Lecanora tartarea, L. parella, Pertusaria communis, Parmelia sordida, Isidium corallinum and some others, which furnish the Cudbear or Persio.

Rosa canina, Linné.

Europe, Northern and Middle Asia, North-Africa. This species attains a very great age; the famed and sacred rose at the cathedral of Hildesheim existed before that edifice was built, therefore before the ninth century (Langethal). In some of the German monasteries real roses of tree-size occur, which have also lived through several centuries and are regarded with veneration.

Rosa centifolia, Linné.

The Cabbage-Rose, Moss-Rose, Provence-Rose. Indigenous on the Caucasus and seemingly also in other parts of the Orient. It will endure the frosts of Norway as far north as lat. 70° (Schuebeler). Much grown in South-Europe and Southern Asia for the distillation of rose-water and oil or attar of roses. No pruning is resorted to, only the dead branches are removed; the harvest of flowers is from the middle of May till nearly the middle of June; the gathering takes place before sunrise (Simmonds). From 12,000 to 16,000 roses, or from 250 lbs. to 300 lbs. of rose-petals, are required according to some calculations for producing a single ounce of attar through ordinary distillation. The flowers require to be cut just before expansion;

the calyx is separated and rejected; the remaining portions of the flowers are then subjected to aqueous distillation, and the saturated rose-water so obtained is repeatedly used for renewed distillation, when on any cold place the oil separates from the overcharged water and floats on the surface, whence it can be collected after refrigeration by fine birds' feathers. Rose-oil consists of a hydro-carbon stearopten, which is scentless, and an elaeopten, which is the fragrant principle. But some other methods exist for producing the oil; for instance, it may be got by distilling the rosebuds without water at the heat of a salt water bath, or by merely passing steam through the still. The odor may also be withdrawn by alcoholic distillation from the roses, or be extracted by the "enfleurage" process. The latter is effected by placing the flowers, collected while the weather is warm, into shallow frames covered with a glass-plate, on the inner side of which a pure fatty substance has been thinly spread. The scent of the flowers is absorbed by the adipose or oleous substance, though the blossoms do not come in direct contact with it; fresh flowers are supplied daily for weeks. The scent is finally withdrawn from its matrix by maceration with pure alcohol. Purified eucalyptus-oil can be used for diluting rose-oil, when it is required for the preparation of scented soap. The essential oil of orange-peel might similarly be employed as a vehicle.

Rosa Damascena, Miller.

Orient. Allied to the preceding species, and also largely used for the production of essential oil of roses. The annual time of flowering extends over several months.

Rosa Gallica, Linné. (*R. provincialis,* Miller.)

The French or Dutch Rose. Middle and Southern Europe, Orient. Hardy to lat. 70° in Norway. The intensely colored buds of this species are particularly chosen for drying. These however may be got also from other kinds of roses.

Rosa Indica, Linné. (*R. Sinica,* L.; *R. Chinensis,* Jacquin.)

China, thence brought to India. The "Hybrid Perpetuals" are largely traceable to this plant. Flowering time of long duration annually. Some roses of the sweetest scent are derived from this species. R. fragrans (Redoute), the Tea-Rose, is a variety. The Noisette Rose is a cross of this and R. moschata.

Rosa laevigata, Michaux. (*R. Sinica,* Murray, *non* Linné.)

The Cherokee-Rose. China and Japan. Considered one of the best hedge-roses, and for that purpose much employed in North-America. It serves well also for bowers. Allied to the foregoing species.—Rosa rugosa (Thunberg) of Japan, a large-fruited and large-leaved rose, is exceedingly well adapted for garden-hedges also.

Rosa moschata, Miller.*

North-Africa and South-Asia, ascending the Indian mountains to 11,000 feet. Blooming all the year round in warm climes, but more profusely in the cool season. From the flowers of this extremely tall climbing species also essential oil is obtained. The attar thus derived from roses of not only different varieties, but even distinct species, must necessarily be of various qualities. In the Balkan-mountains, on basalt-slopes facing south, the most odorous roses are produced. At Kesanlik rose-distillation is the main - industry. Shoots of rose-bushes are placed in trenches 3 feet deep and 5 feet apart. Irrigation promotes the growth. The gathering commences in the third and lasts till about the fifteenth year (Simmonds). The pure oil as a European commodity is worth from £20 to £23 per pound. This is also the rose, according to Schlagintweit, used for attar-distillation in Tunis. Pure attar, valued at 30 shillings per ounce, is produced in Roumelia to the amount of £80,000 annually (Piesse).

Rosa sempervirens, Linné.

From South-Europe through Southern Asia to Japan. Hardy still at Christiania. One of the best rose-bushes for covering walls, fences and similar structures. The flowers of this species also can be utilized for rose-oil.

Rosa setigera, Michaux.

North-Eastern America, where it is the only climbing rose-bush. It deserves introduction on account of its extremely rapid growth,— 10 to 20 feet in a season. Its flowers however are nearly inodorous. Other original species of roses are worthy of our attention, Sir Joseph Hooker admitting about thirty, all from the northern hemi-sphere. But on the snow-clad unascended mountains of New Guinea and Africa south of the equator, perhaps new roses may yet be discovered, as they have been traced southward to Abyssinia already.

Rosa spinosissima, Linné.

Europe, North-Africa, Middle and Northern Asia. The Burnet-Rose. Adapted for holding coast-sands; unapproachable to pasture-animals, and not spreading into culture-land or pastures like the sweetbrier, R. rubiginosa, L.

Rosmarinus officinalis, Linné.

The Rosemary. Countries around the Mediterranean Sea, extending to Switzerland. This well-known bush is mentioned here as a medicinal plant. One of our best plants for large garden-edgings. The oil, distilled from its foliage, enters into certain compositions of per-fumery; one cwt. of fresh herb yields about 24 ounces of oil (Piesse). The flowers are much sought by bees. Vilmorin states that the seeds will keep for about four years; but the propagation from cuttings is easy also.

Rottboellia ophiuroides, Bentham.

Tropical East-Australia. A tall perennial grass, praised by Mr. Walter Hill for fodder. Hardy in regions free of frost.

Royenia Pseudebenus, E. Meyer.

South-Africa. Only a small tree, but its wood jet-black, hard and durable; in Capeland and Caffraria called ebony. R. pubescens (Willdenow), according to Dr. Pappe, furnishes there a wood adapted for xylography; this may give a clue to the adaptability of many other kinds of woods in the large order of Ebenaceæ as substitutes for the Turkish boxwood.

Rubia cordifolia, Linné. (*R. Mungista*, Roxburgh.)

From the Indian highlands through China and Siberia to Japan; also occurring in various parts of Africa, as far south as Caffraria and Natal. This perennial plant produces a kind of madder. Probably other species likewise yield dye-roots. The genus is represented widely over the globe, but as far as known not in Australia.

Rubia peregrina, Linné.

Middle and Southern Europe, South-Western Asia. This perennial species also yields madder-root. Several other kinds deserve comparative test-culture.

Rubia tinctorum, Linné.

The Madder. Countries at the Mediterranean Sea, extending to temperate Western Asia. Hardy still at Christiania. A perennial herb of extremely easy culture. Soil, fit for barley, is also suitable for madder. Its culture opens any deep subsoil and suffocates weeds, but requires much manure, leaving the land enriched however. Stagnant water in the soil must be avoided, if madder is to succeed. The harvest is in the second or third year. It can be raised from seeds, or planted from off-shoots. The roots merely dried and pounded form the dye. The chemical contents are numerous: in the herb: rubichloric and rubitannic acid; in the root: alizarin, purpurin, rubiacin, rubian, ruberythric acid and three distinct resins; also chlorogenin, xanthin and rubichloric acid. On the five first depend the pigments produced from the root. Madder is one of the requisites for alizarin-ink. Since the manufacture of artificial alizarin from anthracene, a constituent of coal-tar, was commenced, the cultivation of madder has declined. Still it remains a valuable root, handy for domestic dye. The root is also important as an emmenagogue.

Rubus acuminatus, Smith.

Indian mountains, at elevations between 4,000 and 7,000 feet. A scandent species with large fruits.

Rubus biflorus, Hamilton.

Indian mountains, at temperate altitudes between 7,000 and 10,000 feet. A rambling shrub, with sweet red or orange-colored fruit. Hardy in England. Another Himalayan species, R. macilentus (Cambessedes), has bright yellow fruits.

Rubus caesius, Linné.

The British Dewberry. Europe, Western and Northern Asia. Resists extreme frosts, protracted dryness and also heat of exceptional seasons. In this respect the most accommodating of all blackberry-bushes. In Russia the berries are boiled together with apples into a preserve, which is of particularly pleasant taste. This Rubus supplies fruit till late in the season. Easily naturalized on ground, subject to occasional inundations, and sheltered by bushy vegetation (Burmeister). Some regard R. caesius as one of the numerous forms of R. fruticosus.

Rubus Canadensis, Linné.*

The Dewberry of Eastern North-America. A shrub of trailing habit. Fruit large, black, of excellent taste, ripening earlier than that of R. villosus (Aiton). All the species can readily be raised from seeds; thus the naturalization of these plants in adapted localities is easy by mere dissemination. The astringent root is a popular remedy in dysentery and diarrhœa.

Rubus Chamæmorus, Linné.

The Cloudberry. North-Europe, North-Asia, North-America, particularly in the frigid zone. In Norway it will grow northward to lat. 71° 10′ (Schuebeler). A perennial but herbaceous plant; a pigmy amongst its congeners; nevertheless it is recommended for introduction to spongy, mossy, alpine moors, on account of its grateful amber-colored or red fruit. R. Arcticus (Linné), also with edible fruit, is usually its companion in the high north. A similar little herb, living for a great part of the year in snow—namely R. Gunnianus, Hooker, —occurs on the alpine heights of Tasmania, whence it might be easily transferred to snowy mountains of other countries. The fruit of R. Gunnianus is red and juicy, but not always well developed. R. calycinus (Wallich), occurring on the Indian mountains in regions between 4,000 and 9,000 feet, is also a dwarf herbaceous species, having a creeping stem, and scarlet fruits, usually however with but few fruitlets.

Rubus cuneifolius, Pursh.

The Sand-Blackberry. Eastern North-America. A dwarf shrub. The fruit is of agreeable taste.

Rubus deliciosus, Torrey.*

About the sources of the Missouri. An erect, exceedingly handsome shrub. Fruit raspberry-like, large and grateful.

Rubus ellipticus, Smith.* (*R. flavus,* Hamilton.)

On the mountains of India, from 4,000 to 7,000 feet elevation, also in Ceylon and Yunan. A large rather erect bush with yellow fruits, which are reckoned in flavor fully equal to the ordinary raspberry (C. B. Clarke).

Rubus fruticosus, Linné.*

The ordinary Bramble or Blackberry-bush. All Europe, North- and South-Africa, Middle and Northern Asia. Hardy in Norway to lat. 60° 24'. The shrub bears well in a temperate clime. In some countries it is a favorite plant for hedges. It likes above all calca- reous soil, though it is content with almost any, and deserves to be naturalized on the rivulets of any ranges. R. corylifolius (Smith), R. suberectus (Andrews) and R. leucostachys (Smith) are varieties like many other named kinds of European blackberries, or perhaps belong to the closely allied R. caesius; or in some instances hybrid- forms may have arisen from the two, although the generality of these various blackberry-bushes bear their fruit freely enough.

Rubus geoides, Smith.

Falkland-Islands, Fuegia, Patagonia and Chiloe. An herbaceous kind of raspberry-plant with greenish-yellow fruits, resembling the Cloudberry, and possessing a very agreeable taste. Best adapted for mountainous regions.

Rubus Havaiensis, A. Gray.

Sandwich-Islands. The fruit of this bramble-shrub is raspberry- like.

Rubus Idaeus, Linné.*

The ordinary Raspberry-bush. Europe and Northern Asia, east- ward to Japan. In Norway hardy to lat. 70° 22'. It is mentioned here to point out the desirability of naturalizing the plant on moun- tains and on river-banks. The fruits contain a stearopten. The leaves are sometimes used as a substitute for tea.

Rubus imperialis, Chamisso.

Brazil and Argentina. Furnishes superior fruits.

Rubus lasiocarpus, Smith.

India, reaching in the Himalayas an elevation of about 10,000 feet, in Ceylon of 7,000 feet, in Java of 6,000 feet. The fruit is very palatable. R. opulifolius (Bertoloni) is closely allied. R. lanatus (Wallich) affords also edible but rather insipid fruits in Upper India (Atkinson).

Rubus Moluccanus, Linné. (R. rugosus, Smith.)

India, continental as well as insular, there ascending to 7,000 feet, advancing southward through New Guinea and East-Australia to Gippsland, northward to China and eastward to the Philippine- Islands and Fiji. A very tall and variable species. Fruit red. A variety, R. reticulatus (Wallich), ascends the Indian mountains to 10,000 feet (Sir J. Hooker), and is remarkable for its large fruit. The plant proved hardy at Christiania. It ripens in warm climes its fruits all the year round. R. tiliaceus (Smith) is an allied congener from the same region.

Rubus nutans, Wallich.

Himalayan mountains, ascending to about 10,000 feet; growing on the ground like strawberry-plants, yielding fruits of very pleasant subacid taste (Atkinson), but not of large size (J. Hooker). A species easily spreading and probably improvable by culture.

Rubus occidentalis, Linné.*

The "Black Cap "-Raspberry or " Thimbleberry "-bush. North-America. A species with woody stems and nice fruits, the latter with a glaucous bloom, well flavored and large; it ripens early. To this bears near affinity R. leucodermis (Douglas) from California, Utah and Arizona; its fruit is yellowish-red, rather large and of agreeable flavor (A. Gray).

Rubus odoratus, Cornuti.*

North-America. A kind of raspberry-bush. Handsome on account of its large purple flowers. Berries edible. Hardy in Norway to lat. 67° 56'. Culture would doubtless enhance the value of the fruits of many of these Rubi. Hybridising might be tried. R. Nutkanus (Mocino) is the Salmon-Raspberry of Western North-America and closely allied to R. odoratus.

Rubus parvifolius, Linné.

East-Asia, Eastern and Southern Australia. It produces much finer fruits in the Alps of Australia than in the lowlands. It extends as a native to Japan, where according to Maximowicz 22 species of Rubus exist, many of them endemic, and probably some eligible for special fruit-culture.

Rubus phœnicolasius, Maximowicz.

Japan. A Raspberry-Bramble with fair-sized fruits.

Rubus rosifolius, Smith.

Tropical and sub-tropical regions of Africa and Asia, ascending the Himalayas to about 10,000 feet, also occurring throughout the litoral forests of East-Australia. In woody regions this shrub bears an abundance of fruits of large size, and these early and long in the season, though not so excellent as those of many other species.

Rubus strigosus, Michaux.*

Eastern North-America, extending to Canada. Closely allied to the European raspberry. Its fruits large, also of excellent taste.

Rubus tiliaceus, Smith.

Indian mountains, at altitudes between 3,000 and 8,000 feet. A rambling species. Fruit large, purplish-black, but rather insipid.

Rubus trivialis, Michaux.*

South-Eastern States of North-America. Another shrubby species with good edible fruits, which are large and black. The plant will thrive in dry sandy soil. Like many other congeners, this one has the bark rich in tannic acid.

Rubus ursinus, Chamisso and Schlechtendahl. (*R. macropetalus,* Douglas.)

California and Oregon. An unisexual shrub. Fruit black, oval-cylindric, particularly sweet. Readily rendered spontaneous. It would lead too far to enumerate other utilitarian kinds of Rubus, although altogether about one hundred genuine species do occur, which render the genus one of very wide dispersion over the globe.

Rubus villosus, Aiton.

Eastern North-America, reaching Canada, there the ordinary Blackberry-bush. Growth tall. Fruit large and pleasant (Asa Gray). The Rochelle- and Lawton-varieties are of this species (C. Koch). The root-bark is praised for medicinal purposes. .

Rumex Acetosa, Linné.

The Kitchen-Sorrel. Europe, Middle and Northern Asia to Japan, also in the frigid zone of North-America. Endures the frosts of Norway northward to lat. 71° 10' (Schuebeler). A perennial herb. The tender varieties, particularly the Spanish one (R. Hispanicus, Koch), also the alpine one (R. montanus, Desfontaines) serve as pleasant acidulous vegetables, but must be used in moderation, as their acidity, like that of the species of Oxalis (Wood-Sorrel), depends on binoxalate of potash. The South-African R. luxurians, L., serves likewise as culinary sorrel. Aquatic species of Rumex help to solidify embankments subject to floods.

Rumex hymenosepalus, Torrey.

Texas, New Mexico. This "Dock," vernacularly known as "Canaigré," has come into use in tanneries, the roots acting by their powerful tan-principle doubly as quick as oak-bark. This may prove a hint for testing other large species of the extensive genus Rumex in this respect. The root of R. hymenosepalus yields in a dry state 23½ per cent. rheo-tannic acid.

Rumex Patientia, Linné.

Middle and Southern Europe, South-Western Asia. Biennial. It is the R. sativus of Plinius according to Fraas. Bears the cold of Norwegian winters to lat. 70°. The young leaves furnish a palatable sorrel, like spinach. In cold climes it pushes forth its leaves, before the frost is hardly gone, and thus comes in as one of the first vegetables of the season.

Rumex scutatus, Linné.

The French Sorrel. Middle and Southern Europe, Northern Africa, Orient. Also perennial, and superior to the foregoing as a culinary plant. They are all of use against scurvy, and most easily reared. Dr. Rosenthal lauds even the common R. crispus (Linné) of Europe and Asia as a spinage-plant; further for culinary purposes, R. pulcher (Linné) and R. bucephalophorus (Linné) from the countries at or near the Mediterranean Sea.

Rumex vesicarius, Linné.

Southern Europe, Middle Asia, Northern Africa.   An annual herb of the same utility as other sorrels.

Ruscus aculeatus, Linné.

Middle and Southern Europe, Northern Africa, South-Western Asia.   This odd plant serves for forming garden-hedges.   The young shoots of this and a few allied plants are edible.

Russula vesca, Fries.

Europe.   One of the best of mushrooms for the kitchen within the genus Russula, which has representatives in most parts of the globe. Dr. Cooke mentions further as culinary R. lepida and R. virescens (Fries).   Professor Morren notes R. integra (Fries) as used among the Belgian champignons.   Under any circumstances, mushrooms should only be used when fresh collected or quickly dried.

Ruta graveolens, Linné.

The Rue.   Mediterranean countries and the Orient.   Hardy in Norway to lat. 63° 26'.   The foliage of this acrid and odorous shrub, simply dried, constitutes the rue-herb of medicine.   The allied R. sylvestris (Miller) is still more powerful in its effect.   These plants and others of the genus contain a peculiar volatile oil and a glycosid, the rutin.   Fresh they should be handled most cautiously, best with gloves.

Sabal Adansoni, Guernsent.

Dwarf Palmetto.   South-Carolina, Georgia and Florida.   A stemless Fan-palm, with the following congeners, Rhaphidophyllum Hystrix and Washingtonia filifera, attaining the most northerly positions of any American palms.   According to Count de Saporta it resists a temperature as low as 17° F.   Professor Ch. Naudin found it to endure the frosts in Southern France to 43° 20' north latitude. This palm does well in marshy places.

Sabal Palmetto, Loddiges.*

Extends from Florida to North-Carolina.   The stem attains a height of 40 feet.   This hardy palm delights on sandy coast-tracts.   Stems almost imperishable under water, not attacked by the teredo.

Sabal serrulata, Roemer and Schultes.   (*Serenaea serrulata,* J. Hooker.)

The Saw-Palmetto.   South-Carolina, Georgia and Florida, particularly well adapted for sea-coasts.   The stem grows to eight feet in height, but according to Mr. A. J. Cook may slimly creep along the ground for 20 feet, sending roots beneath for nourishment.   This is a grand honey-plant.   Mr. G. Damköhler mentions this Sabal as a tan-plant.   The leaves can be used for cabbage-tree hats, mats, baskets and other purposes, for which palm-leaves are sought.   The fibrous-spongy parts of the stem serve as brushes.

Sabal umbraculifera, Grisebach.

West-Indies. Attains a height of 80 feet, or occasionally even over 100 feet. Though naturally a tropical Savannah-palm, it has proved even hardier than the orange. A near relative is S. Blackburniana (Glazebrook), a native of Bermuda, where, according to Sir John Lefroy, it gains a stem-height of 50 feet, and where the leaves are extensively used for plat; the sweet pulp of the fruit is edible. At Hyères this palm withstood a temperature of 22° F. (Bonnet). Another equally tall Antillan palm is S. glaucescens, Loddiges.

Sabbatia angularis, Pursh.

North-Eastern America. This pretty biennial herb is lauded as a substitute for gentian by American physicians, and might with its congeners be grown in medicinal gardens, though its naturalization would not be desirable, as pastoral animals avoid the bitter gentian-aceous plants.

Saccharum officinarum, Linné.*

The Sugar-Cane. India, Cochin-China, South-Sea Islands, spontaneous; probably derived from one of the native South-Asiatic species of Saccharum, according to Loureiro indigenous in Cochin-China, an observation confirmed by Dr. Bretschneider. Sugar-cane having been cultivated in Spain and other countries on the Mediterranean Sea, it will be worthy of further trial. at what distance from the equator and at what elevations in tropical parts of the globe sugar from cane can be produced to advantage. In the United States the profitable culture of cane ceases at 32° north latitude; in Japan it is carried on with advantage to 36° north latitude and even further northward (General Capron); the average-yield of raw sugar even there is 3,300 lbs. per acre; in China this crop extends only to 30° north latitude. In South-Asia the culture of the sugar-cane dates from the remotest antiquity; from China we have a particular kind (S. Sinense, Roxburgh), which is hardier and bears the drought better than the ordinary cane; this kind needs renewal only every third year, and ripens in seven months, if planted early in spring; but if planted in autumn and left standing for fully a year the return of sugar is larger. Moderate proximity to the sea is favorable for the growth of canes. Prolific yields have been secured in East-Australia south to 29°. A very saccharine but hard cane is wild in New Guinea (Maclay). The multiplication of all sorts of sugar-cane is usually effected from top-cuttings; but this cannot be carried on from the same original stock for an indefinite period without deterioration; and as seeds fit to germinate do not ripen on cultivated canes, new plants must from time to time be brought from a distance. Thus, New Caledonia and Fiji have latterly supplied their almost wild-growing splendid varieties for replanting many sugar-fields in Mauritius and some other places. The Bourbon-variety is praised as one of the richest for sugar; the Batavian variety, S. violaceum (Tussac), is content with less fertile soil. Many other varieties are known. The

sugar-cane is one of the best of all plants of economic value, to keep
cleared ground in tropical forests free from weeds or the invasion of
other plants. Excessive rains produce a rank luxuriance of the canes
at the expense of the saccharine principle. Rich manuring is necessary to attain good crops, unless in the best of virgin soil. The lower
leaves of the stem must successively be removed, also superabundant
suckers, to promote the growth upwards, and to provide ventilation
and light. Out of the remnants of sugar-cane either molasses or rum
or taffia can be prepared. The average-yield of sugar varies from
1 ton 6 cwt. to 3 tons for the acre; but exceptionally as much as
6 tons per acre have even been obtained in the hardly tropical
Hawaian Islands. The world's production of cane-sugar in 1875
amounted to 2,140,000 tons (Boucheraux). Among some other works
for fuller information the valuable volume of Mr. A. McKay, " The
Sugar-Cane in Australia," should be consulted, particularly in the far
southern colonies. The stately S. spontaneum (Linné), which extends
from India to Egypt and New Guinea, is available for scenic culture.
It attains a height of 15 feet, and ascends in Java, according to Dr.
Junghuhn's observations; to a height of 17,000 feet. Other tall kinds
of Saccharum occur in South-Asia. For the conditions and prospects
of the cane-sugar industry in the Southern United States see the
special report of the Hon. General Le Duc, Washington, 1877: also
the " Sugar Industries of the United States," by H. W. Wiley, 1885.

Sagittaria lancifolia, Linné.

From Virginia to the Antilles. This very handsome aquatic
plant can doubtless be utilized like the following species. It attains
a height of five feet.

Sagittaria obtusa, Muehlenberg. (*S. latifolia,* Willdenow.)

North-America, where it replaces the closely-allied S. sagittifolia.
A few other conspicuous species are worthy of introduction. The
Tule or Wapatoo-root of California is derived from a species of
Sagittaria.

Sagittaria sagittifolia, Linné.

Europe, Northern and Middle Asia, east to Japan. One of the
most showy of all hardy water-plants; still not alone on that account
deserving naturalization, but also because its root is edible. If once
established, this plant maintains its ground well, and might occupy
submerged spots not otherwise utilized.

Salix alba, Linné.*

The Huntingdon or Silky Willow of Europe, originally of North-
Africa, Northern and Western Asia; according to Prof. Andersson
of exclusively Asiatic origin. It bears the frosts of Norway to lat.
63° 52'. It is positively known, that the Silky Willow will live to
an age of 150 years, and probably much longer. Available for wet
places not otherwise in cultivation. Height reaching to 80 feet, circumference of stem sometimes to 20 feet; of rapid growth. Foliage

silvery-pubescent. Wood smooth, soft and tough, bearing pounding and knocking better than that of any other British tree; eligible where lightness, pliancy and elasticity are required; hence in request for wheel-floats and shrouding of water-wheels, as it is not subject to splinter; for the sides and bottoms of carts and barrows, for break-blocks of trucks; also used for turnery, trays, fenders, shoe-lasts, light handles (Simmonds). Its weight is from 26 to 33 lbs. per cubic foot. Timber, according to Robb, the lightest and softest of all prominently utilitarian woods; available for bungs; it. is planed into chips for hat-boxes, baskets and woven bonnets; also worked up for cricket-bats, boxes and many utensils. The bark is particularly valued as a tan for certain kinds of glove-leather, to which it imparts an agreeable odor. Mr. Scaling records, that in rich ground on the banks of streams this willow will grow to a height of 24 feet in 5 years, with 2 feet basal girth of the stem; in 8 years he found it to grow 35 feet, with 33 inches girth at 1 foot from the ground. Loudon noticed the height to be 53 feet in 20 years, and the girth 7$\frac{1}{2}$ feet. In winterless countries the growth is still more rapid. To produce straight stems for timber, the cuttings must be planted very close, some of the trees to be removed from time to time. After 30 or 40 years the trees will deteriorate. Scaling estimates the value of an acre of willow-timber to be about £300. The Golden Osier, Salix vitellina, L., is a variety. The shoots are used for hoops and wicker-work. With other large willows and poplars one of the best scavengers for back-yards, where drainage cannot readily be applied; highly valuable also for forming lines along narrow watercourses or valleys in forests, to stay bush-fires. The charcoal excellent for gun-powder. The wood in demand for matches. All willows, as early flowering, are of particular importance to apiarists (Cook, Quinby). The extreme rapidity of growth of most willowtrees and poplars, particularly in mild climes, renders it quite feasible, to rear them purposely for providing wood as an adjunct to paper-material, particularly on ground not eligible as agrarian.

Salix Babylonica, Tournefort.

The Weeping Willow. Indigenous in North-China, sparingly wild, according to Stewart, in the Himalayas; probably likewise in Persia and Kurdistan. One of the most grateful of all trees for the facility of its culture and its fitness for embellishment; also as one of the quickest growing and most easily reared of all shade-trees. Fifty feet upward growth has been witnessed in five years. The tree is important for consolidating river-banks, and everywhere available for cemeteries. In frostless climes annually only for a few weeks without leaves. In Norway it will grow northward to lat. 58° 8′. A powerful. scavenger of back-yards, but apt to undermine masonry and to get into cisterns. Dr. C. Koch prefers Moench's name S. pendula, as the Weeping Willow is not a native of Babylon, and he distinguishes another Weeping Willow from Japan as S. elegantissima, which is still hardier than S. Babylonica; allied also is S. Japonica.

Salix Capensis, Thunberg. (*S. Gariepina,* Burchell.)

South-Africa. This willow might be introduced on account of its resemblance to the ordinary Weeping Willow. Prof. Harvey says of it, that it is one of the greatest ornaments of the banks of the Gariep-River.

Salix caprea, Linné.*

Europe, Northern and Middle Asia. The British Sallow or Hedge-Willow. In Norway it extends to lat. 70° 37′; in 65° 28′ Prof. Schuebeler found it to attain a height of nearly 70 feet. The Kilmarnock Weeping Willow is a form of this species. Wood used for handles and other implements, the shoots for hoops; it is also largely employed for gunpowder-coal. Bark available for tanning, particularly glove-leather. The flowers are eagerly sought by bees. It is one of the earliest flowering of willows, hence with S. daphnoides, as the harbinger of spring, particularly gladdening to bees, although all willows are honey-plants.

Salix cordata, Muehlenberg.

One of the Osiers of North-America, extending to Canada. Fit also to bind sand. One of the dwarf Californian willows has been found on the coast-sands to send out root-like stems to 120 feet in length.

Salix daphnoides, Villars.

Northern and Middle Europe and Northern Asia, eastward as far as the Amoor, ascending to 15,000 feet in the Himalayas, growing in Norway northward to lat. 62° 20′. A tree, rising to about 60 feet in height, rapid of growth, attaining 12 feet in four years. It is much chosen to fix the ground at railway-embankments, on sandy ridges and slopes, for which purposes its long-spreading and strong roots render it particularly fit. The twigs can be used for baskets, wicker-work and twig-bridges (Stewart and Brandis). The variety *pruinosa* is considered by Dr. Sonder to be as valuable as the Bedford-Willow. The foliage furnishes cattle-fodder. The tree is comparatively rich in salicin, like S. pentandra (Linné) and the following.

Salix fragilis, Linné.

The Crack-Willow or Withy. Indigenous in South-Western Asia; widely spontaneous also in Europe. Hardy in Norway to lat. 64° 5′. Height to 90 feet; stem reaching 20 feet in girth. According to Scaling next to S. alba the best of the European timber-willows, but the wood not quite so tough and the tree requiring more space for growth. Both species are recommended for shelter-plantations, on account of their rapidity of growth, uninflammability and easy propagation; the latter quality they share with most willows. A variety or hybrid of this species is the Bedford-Willow, also called Leicester-Willow, Salix Russelliana (Smith), which yields a light, elastic, tough timber, more tannin in its bark than oak, and more salicin (a substitute for quinine and most valuable as an anti-rheumatic remedy) than most of its congeners. According to Sir H. Davy the inner layers of the bark contain fully 16 per cent. tannin, the whole bark only about 7 per cent.

Salix Humboldtiana, Willdenow.

Through a great part of South-America, southward as far as Patagonia, there furnishing building timber for inside-structures. This willow is of pyramidal habit, attains a height of 50 feet and more, and is in moist ground of quick and ready growth. The wood is locally much in use for yokes and other implements. Many kinds of willows can be grown for consolidating shifting sand-ridges.

Salix Japonica, Thunberg.

Japan. Here mentioned, to draw attention to its distinctness from S. Babylonica, to which Prof. Andersson referred it as a variety.

Salix longifolia, Muehlenberg.

North-America from the Atlantic to the Pacific Ocean. Acquires finally a height of about 20 feet; likes to creep on sand and gravel along river-banks. It is one of the species forming long flexible wythes. S. petiolaris (Smith) and S. tristis (Aiton) are among the North-American species best fitted for binding sand.

Salix lucida, Muehlenberg.

One of the Osiers of North-America, reaching Canada. Likes river-banks for its habitation.

Salix nigra, Marshall. (*S. Purshiana*, Sprengel.)

The Black Willow of North-America. It attains a height of 30 feet. One of the willows used for basket-work, although it is surpassed in excellence by some other species, and is more important as a timber-willow. Mr. W. Scaling of Basford includes it among the sorts, which he recommends in his valuable publication, " The Willow," London 1871.

Salix purpurea, Linné.*

Of wide range in Europe, Northern and Western Asia, extending also to North-Africa, according to Sir. J. Hooker. The Bitter Willow; one of the Osiers. Hardy in Norway northward to lat. 67° 56'. In deep moist soil, not readily otherwise utilized, it will yield annually, four to five tons of the best of rods, qualified for the finest work. Impenetrable, not readily inflammable screens as much as 25 feet high can be reared from it in five years. In localities exposed to storms, willow-copses fully 40 feet high can be raised from this species. It forms also a variety with pendent branches. It is most valuable also for the reclamation of land along watercourses. Rich in salicin, which collaterally can be obtained from the peelings of the twigs, when the latter are prepared for basket-material. From Mr. Scaling's treatise on the Willow, resting on unrivalled experience, it will be observed, that he anew urges the adoption of the Bitter

Willow (also called the Rose-Willow or the Whipcord-Willow) for game-proof hedges, the species scarcely ever being touched by cattle, rabbits and other herbivorous animals. Not only for this reason, but also for its very rapid growth and remunerative yield of the very best of basket-material, he recommends it for field-hedges. Cuttings are planted only half a foot apart and must be entirely pushed into the ground. The annual produce from such a hedge is worth 4s. to 5s. for the chain. To obtain additional strength, the shoots can be interwoven. In rich bottoms the shoots will grow from 7 to 13 feet in a year. The supply of basket-material from this willow has fallen very far short of the demand in England. The plant grows vigorously on light soil or warp-land, but not on clay. It likes sandy loam, and will even do fairly well on gravelly soil, but it is not so easily reared as S. triandra. Mr. Scaling's renewed advocacy for the formation of willow-planta-tions comes with so much force, that his advice is here given, though condensed in a few words. Osier-plantations come into full bearing in the third year; they bear for about ten years and then slowly decline. The raw produce from an acre in a year averages 6 to 7½ tons, ranging from £2 10s. to £3 10s. for the ton (unpeeled). Although 7,000 acres are devoted in Britain to the culture of basket-willows (exclusive of spinneys and plantations for the farmer's own use), yet in 1866 there had to be imported from the Continent 4,400 tons of willow-branches, at an expense of £44,000, while besides the value of the made baskets imported that year was equal to that sum. In recent years the importation into the United States of willow-material for baskets, chairs and other utensils has, according to Simmonds, been estimated as approaching $1,000,000. Land, com-paratively unfit for root- or grain-crops, can be used very remunera-tively for osier-plantations. The soft-wooded willows like to grow in damper ground than the hard-wooded species. The best peeled willow-branches fetch as much ac £25 for the ton. Peeling is easiest effected by steam, by which means the material is also increased in durability. No basket-willow will thrive in stagnant water. Osier-plantations in humid places should therefore be drained. The cuttings are best taken from branches one or two years old, and are to be planted as close as one foot by one foot and a half. No part of the cutting must remain uncovered, in order that only straight shoots may be obtained; manuring and ploughing between the rows is thus also facilitated, after the crop has been gathered, and this, according to the approved Belgian method, must be done by cutting the shoots close to the ground after the fall of the leaves. The accidental introduction from abroad of destructive saw-flies (particularly Ne-matus ventralis), which prey also on currant- and gooseberry-bushes, should be guarded against.

Salix rubra, Hudson.*

Throughout Europe, also in West-Asia and North-Africa; much chosen for osier-beds. When cut down, it will make shoots eight feet long in a season. Dr. Porcher regards it as one of the most valuable species for work, in which unpeeled rods are used. It is also admirably adapted for hedges. The bark is one of the best for salicin. Considered by some as a hybrid between S. purpurea and. S. viminalis.

Salix tetrasperma, Roxburgh.

Mountains of India, from 2,000 to 7,000 feet. Height of tree reaching 40 feet. This . thick-stemmed willow is worthy of a place on banks of watercourses. The twigs can be worked into baskets, the wood serves for gunpowder, the foliage for cattle-fodder.

Salix triandra, Linné.* (*S. amygdalina*, Linné.)

The Almond-Willow. Through nearly all Europe and extra-tropical Asia. Height of tree at length 30 feet. It sheds its bark annually after the third year. Likes rich loamy soil; requires less space 'than S. viminalis, more than S. purpurea. It is a prominent representative of the hard-wooded basket-willows, and comprises some of the finest varieties in use by the manufacturers. Shoots are obtainable 9 feet long; they answer for hoops and white basket-work, being pliant and durable. The bark contains a good deal of salicin. For basket-purposes 20,000 to 30,000 cuttings can be planted on an acre, and 2,000 to 3,000 can be planted in a day by an expert; the second year's crop is already of considerable value; at five years it comes to its prime, the plantation holding good for 15 to 25 years. The rods for baskets should be cut as soon as the leaves have dropped. The annual value of a crop of basket-willows is in England from £25 to £35 per acre (Scaling). The cultivation of basket-willows along railroads has proved in cold countries one of the best protectives against snow-drifts. S. lanceolata (Smith) is a hybrid between S. triandra and S. viminalis, according to Prof. Andersson.

Salix viminalis, Linné.

The common Osier of Europe, North- and West-Asia. Height to 30 feet. The best of basket-willows for banks, subject to occasional inundations. It is a vigorous grower, very hardy (to lat. 67° 56' in Norway), likes to be fed by deposits of floods or by irrigation, and disposes readily of sewage (Scaling). One of the best for wicker-work and hoops; when cut, it shoots up to a length of 12 feet; distinguished by the basket-makers as the soft-wooded willow; it is best for rods requiring two years' age, but inferior to several other species for basket-manufacture. Prof. Wiesner mentions this species among those drawn into use for tanning purposes. S. Smithiana (Willdenow) is a hybrid of S. viminalis and S. caprea, and has proved one of the best willows for copses and hedges. Its growth is very quick and its foliage remarkably umbrageous. It would lead too far, to enumerate even all the more important willows on this occasion. Professor Andersson, of Stockholm, admits 158 species. Besides these, numerous hybrids exist. Many of the taller of these willows could be grown to rural or technologic advantage.

z

Salpichroma rhomboides, Miers,

Extra-tropical South-America, as far south as Magelhaen's Straits. A half-shrub, with good-sized berries of vinous taste (Lorentz).

Salvia Matico, Grisebach.

Sub-alpine Argentina. An important medicinal herb.

Salvia officinalis, Linné.

The Garden-Sage. South-Europe, ranging to Switzerland. Endures the climate of Norway to lat. 70°. A somewhat shrubby plant of medicinal value, pervaded by essential oil. Prefers calcareous soil. Counted also with honey-plants; enters into some condiments. S. Sclarea (Linné), which ranges from South-Europe to Persia, is similarly drawn into use, and was grown by Imperial order already in the gardens of Charles the Great (E. Meyer). Among nearly half a thousand species of this genus some are gorgeously ornamental.

Sambucus Australis, Chamisso and Schlechtendal.

Southern Brazil and La Plata-States. Resembles the ordinary elder, and is locally used for tall hedges (Dr. Lorentz).

Sambucus Canadensis, Linné.

North-Eastern America. The berries of this half-woody elder are used, like those of Phytolacca decandra, for coloring vinous liquids. Dr. Gibbons observes, that this species is recognised in the United States Pharmacopœia, and that S. Mexicana (Presl.) and S. racemosa (Linné) possess similar medicinal properties. The flowers are gently excitant and sudorific, the berries diaphoretic and aperient; a kind of wine is frequently manufactured from them; the inner bark in large doses acts as a hyrdagogue cathartic and as an emetic. S. xanthocarpa (F. v. Mueller) is a large elder-tree of extra-tropical East-Australia.

Sambucus nigra, Linné.

The ordinary Elder. Europe, Northern Africa, Western Asia. Endures the frosts of Norway northward to lat 66° 5′ (Schuebeler). Known to have exceptionally attained a height of 35 feet. The flowers are of medicinal value, and an essential oil can be obtained from them. The wood can be utilized for shoe-pegs and other purposes of artisans. The berries are used for coloring portwine and for other dyeing purposes. The roots of the elder possess highly valuable therapeutic properties, according to Dr. Al. Buettner.

Sanguinaria Canadensis, Linné.

North-Eastern America. " Blood-root." A perennial herb. Hardy to lat. 63° 26′ in Norway. The root important as a therapeutic agent. It contains also dye-principles. An excellent bee-plant (Cook).

Sanguisorba minor, Scopoli. (*Poterium Sanguisorba*, Linné.)

The Salad-Burnet. Europe, North-Africa, Northern and Western Asia. A perennial herb, easily disseminated and naturalized, particularly adapted for calcareous soils. Serves as salad and particularly as a sheep-fodder.

Sanseviera Zeilanica, Willdenow.

India. This thick-leaved liliaceous plant should not be passed in this enumeration, as it has proved hardy in temperate climes, free from frost. Four pounds of leaves give nearly one pound of fibre, which unites softness and silky lustre with extraordinary strength and tenacity, serving in its native country particularly for bow-strings. The plant might be left to itself for continued growth in rocky unutilized places. Several species, South-Asiatic as well as African, exist.

Santalum album, C. Bauhin.

India, ascending to the temperate elevations of Mysore. A small or middle-sized tree, famed for its fragrant wood and roots. In the drier and stony parts of ranges the greatest fragrance of the wood is generated. S. Freycinetianum (Gaudichaud) produces santal-wood on the mountains of Hawaia up to 3,000 feet. Several other species occur in Polynesia. The precious santal-oil is obtained by slow distillation from the heartwood and root, the yield being about two and a half per cent. It is worth about £3 per pound, and important not only for cosmetic but also for medicinal purposes. Santalum Austro-Caledonicum (Vieillard), from New Caledonia, furnishes there santal-wood, excellent for strength and agreeableness of odor (Simmonds).

Santalum cygnorum, Miquel.

South-Western Australia, where this small tree furnishes scented santal-wood. The wood on distillation yields nearly 2 per cent. oil (Seppelt).

Santalum Preissianum, Miquel. (*S. acuminatum,* A. de Candolle.)

The Quandong. Desert-country of extra-tropical Australia. The fruits of this small tree are called Native Peaches. As both the succulent outer part and kernel are edible, it is advisable to raise the plant in desert-tracts. Dromedaries, according to Mr. E. Giles, feed on the foliage.

Santalum Yasi, Seemann.

The Santal-tree of the Fiji-Islands, where it grows on dry and rocky hills. It is likely to prove hardy, and deserves, with a few other species from the South-Sea Islands yielding scented wood, test-culture in warm temperate regions.

Santolina Cyparissias, Linné.

Countries on the Mediterranean Sea. A very aromatic and handsome bush, of medicinal value. There are several allied species.

Saponaria officinalis, Linné.

The Soapwort or Fuller's Herb. Europe, Northern and Western Asia. Hardy in Norway to lat. 63° 26' (Schuebeler). A perennial herb of some technologic interest, as the root can be employed with advantage in some final processes of washing silk and wool, to which it imparts a peculiar gloss and dazzling whiteness, without injuring in the least any subsequent application of the most sensitive colors. The same purpose serves Saponaria Struthium (Loeffling) of Spain.

Sassafras officinale, Hayne.

The deciduous Sassafras-tree, indigenous from Canada to Florida, occurring in dry open woods. Height sometimes to 80 feet; the stem has been known to attain a girth of more than 19 feet at 3 feet from the ground. It furnishes the medicinal sassafras bark and wood, and from this again an essential oil is obtainable. The deciduous and often jagged leaves are remarkable among those of Lauraceæ; they are used as a condiment in cookery. The root-bark contains 58 per cent. tannin (Reinsch). The wood ranks also as a material for a lasting dye; for turnery it is easily worked, and proved of great resistance to the influence of water (Dr. C. Mohr).

Satureja hortensis, C. Bauhin.

The Summer-Savory. Countries around the Mediterranean Sea. An annual scent-herb, from which an essential aromatic oil can be distilled; it is used also as a condiment. The culture of this and allied plants is easy in the extreme.. This is one of about 100 kinds of plants, which Charles the Great already ordered to be grown on his domains or gardens (E. Meyer).

Satureja montana, Linné.

The Winter-Savory. On arid hilly places at and near the Mediterranean Sea. A perennial somewhat shrubby herb, frequently used as a culinary condiment along with or in place of the foregoing species, although it is scarcely equal to it in fragrance.

Satureja Thymbra, Linné.

Countries on or near the Mediterranean Sea. A small evergreen bush, with the flavor almost of thyme. The likewise odorous S. Graeca, and S. Juliana (Linné) have been transferred by Bentham to the closely cognate genus Micromeria; they have been in use since ·Dioscorides' time, though not representing, as long supposed, the Hyssop of that ancient physician, which according to Sprengel and Fraas was Origanum Smyrnaeum or some allied species.

Saussurea Lappa, Bentham. (*Haplotaxis Lappa*, Decaisne.)

Cashmere. The aromatic root of this perennial species is of medicinal value, and by some considered to be the Costus of the ancients.

Saxono-Gothæa conspicua, Lindley.

The Mahin of Southern Chili and Patagonia. A middle-sized tree, with fine-grained yellowish timber.

Scandix grandiflora, Linné.

Countries around the Mediterranean Sea. An annual herb, much liked there as a salad of pleasant aromatic taste.

Schima Wallichii, Choisy.

India, up to 5,000 feet. A tree attaining a height of about 100 feet. Timber highly valuable (C. B. Clarke).

Schinus Molle, Linné.

From Mexico to Chili, ascending the Andes to about 12,750 feet. A tree, thriving on dry and sandy soil, odorous in all its parts; the foliage in bouquets a good substitute for ferns and not quickly shrivelling; the jerking motion of leaf-fragments thrown into water very remarkable; the somewhat spicy small fruits serving as condiment. Rate of growth of tree at Port Phillip about 1 foot a year. S. terebinthifolia (Raddi) from Brazil proved a good promenade-tree in Victoria.

Schizostachyum Blumei, Nees.

Java, at an elevation of about 3,000 feet. A lofty Bamboo. A few other species, less elevated, occur in China, in the South-Sea and Philippine-Islands, also in Madagascar. The Bamboos being thus brought once more before us, it may be deemed advisable, to place together in one brief list all kinds, which are recorded either as very tall or as particularly hardy. Accordingly, from Major-General Munro's admirable monography (Transactions of the Linnean Society, 1868) the succeeding enumeration is compiled, and from that masterly essay, resting on very many years' close study of the richest collections, a few prefatory remarks are likewise offered, to vindicate the wish of the writer of seeing these noble and graceful forms of vegetation largely transferred to every part of Australia, and indeed to many other portions of the globe, where they would impress a grand tropical feature on the landscapes. Even in the far southern latitudes of Victoria, Tasmania and New Zealand, some Bamboos from the Indian lowlands have proved able to resist our occasional night-frosts of the low country; but in colder places the many sub-alpine species could be reared. Be it remembered, that Chusquea aristata advances to an elevation of 15,000 feet on the Andes of Quito, indeed to near the zone of perpetual ice. Arundinaria racemosa and A. spathiflora live on the Indian highlands, at a zone between 10,000 and 11,000 feet, where they are annually beaten down by snow. Forms of Bambusaceæ still occur, according to Grisebach, in the Kurilian archipelagus up to 46° N., and in Japan even to 51°. We may further recognise the great importance of these plants, when we reflect on their manifest industrial uses, when we consider their grandeur for picturesque scenery, when we observe their resistance to storms or heat, or when we watch the marvellous rapidity with which many develop. Their seeds, though generally produced only at long intervals, are valued in many instances higher than rice. The ordinary great Bamboo of India is known to grow 40 feet in forty days, when bathed in the moist heat of the jungles. Delchevalerie noticed the growth of some Indian Bamboos at Cairo to have been 10 inches in one night. Their power of growth is such, as to upset stone-walls or demolish substantial buildings. As shelter-plants for grazing animals these giant-reeds are most eligible. The Bourbon-Bamboo forms an impenetrable sub-alpine belt of extraordinary magnificence in that island. One of the Tenasserim-Bambusas rises to about 150 feet, with the mast-like cane sometimes measuring fully one foot in diameter. The great West-Indian Arthrostylidium is sometimes nearly as high and quite as columnar in its form, while the Dendrocalamus at Pulo-Genm is equally colossal. The Platonia-Bamboo of the highest wooded mountains of Panama sends forth leaves occasionally 15 feet in length and 1 foot in width. Arundinaria macrosperma, as far north as Philadelphia, still rises to a height of nearly 40 feet in favorable spots, and one of the Japan-Bamboos, according to Mr. Christy, gains the height of 60 feet even in those extra-tropical latitudes. Through perforating with artistic care the huge canes of various Bamboos, musical sounds can be melodiously produced, when the air wafts through the groves, and this singular fact may possibly be turned to practice for checking the devastations from birds on many a cultured spot. Altogether twenty genera, with one hundred and seventy well-marked species, are circumscribed by General Munro's consummate care ; but how may these treasures yet be enriched, when once the alpine mountains of New Guinea through Bamboo jungles have been scaled, or when the highlands on the sources of the Nile, which Ptolemæus and Julius Cæsar already longed to ascend, have become the territory also of phytologic researches, not to speak of many other tropical regions as yet left unexplored ! Europe possesses no Bamboo; Australia, as far as hitherto ascertained, only three. Almost all Bamboos are local, and there seems really no exception to the fact, that none are indigenous to both hemispheres, a remark which applies to Palms as well, with the sole exception of Cocos nucifera, the nuts of which indeed may have drifted from the western to the eastern world. All true Bambusas are Oriental. Observations on the growth of many Bamboos in Italy are recently offered by Chevalier Fenzi. The introduction of these exquisite plants is one of the easiest imaginable, either from seeds or the living roots. The consuls at distant ports, the missionaries, the mercantile and navigating gentlemen abroad, and particularly also many travellers could all easily aid in transferring the various Bamboos from one country to another— from hemisphere to hemisphere. Most plants of this kind, once well established in strength under glass, can be trusted out in mild temperate climes to permanent locations with perfect and lasting safety at the commencement of the warm season. Indeed, Bamboos are hardier than most intra-tropical plants, and the majority of them are not the denizens of the hottest lowlands, but delight in the cooler air of mountain-regions. Strong manuring brings some tardy flowering Bamboos early into bloom. In selecting the following array from General Munro's monograph, it must be noted, that it comprises only a limited number, and that among those, which are already to some extent known, several as yet cannot be defined with precision in their generic and specific relation; evidently some occur, which in elegance, grace and utility surpass even many of those now specially mentioned:—

Arthrostylidium excelsum, Grisebach. West-Indies. Height reaching at length 80 feet, stem-diameter 1 foot.

Arthrostylidium longiflorum, Munro. Venezuela; ascends to 6,000 feet.

Arthrostylidium racemiflorum, Steudel. Mexico; ascends to 7,500 feet. Height 30 feet.

Arthrostylidium Schomburgkii, Munro. Guiana; ascends to 6,000 feet. Height 60 feet.

Arundinaria acuminata, Munro. Mexico. Height 20 feet.

Arundinaria collosa, Munro. Himalaya; ascends to 6,000 feet. Height 12 feet.

Arundinaria debilis, Thwaites. Ceylon; ascends to 8,000 feet. A tall species.

Arundinaria Hookeriana, Munro. Sikkim; ascends to 7,000 feet. Height 15 feet.

Arundinaria Japonica, Siebold and Zuccarini. Japan. Height becoming 12 feet.

Arundinaria Khasiana, Munro. Himalaya; ascends to 6,000 feet. Height 12 feet.

Arundinaria spathiflora, Trinius. The true "Ringal." Himalaya. Height 30 feet. Often confounded with A. falcata.

Arundinaria suberecta, Munro. Himalaya; ascends to 4,500 feet. Height 15 feet.

Arundinaria tessellata, Munro. South-Africa; ascends to 6,500 feet. Height 20 feet.

Aulonemia Quexo, Goudot. New Granada, Venezuela, in cool regions. Tall, climbing.

Bambusa Balcooa, Roxburgh. Bengal to Assam. Height 70 feet.

Bambusa Beecheyana, Munro. China. Height 20 feet.

Bambusa Brandisii, Munro. Tenasserim; ascends to 4,000 feet. Height reaching 120 feet, stem-circumference finally 2 feet.

Bambusa marginata, Munro. Tenasserim; ascends to 5,000 feet. Tall, scandent.

Bambusa nutans, Wallich. Himalaya; ascends to 7,000 feet.

Bambusa pallida, Munro. Bengal to Khasia; ascends to 3,500 feet. Height 50 feet.

Bambusa polymorpha, Munro. Burmah, in the Teak-region. Height 80 feet.

Bambusa regia, Th. Thomson. Tenasserim. Height 40 feet.

Bambusa Tulda, Roxburgh. Bengal to Burmah. Height 70 feet.

Bambusa tuldoides, Munro. China, Hong Kong, Formosa.

Beesha capitata, Munro. Madagascar. Height 50 feet.

Beesha stridula, Munro. Ceylon.

Cephalostachyum capitatum, Munro. Himalaya; ascends to 6,000 feet. Height 30 feet.

Cephalostachyum pallidum, Munro. Himalaya; ascends to 5,000 feet. Rather tall.

Cephalostachyum pergracile, Munro. Burmah. Height 50 feet.

Cephalostachyum schizostachyoides, Kurz. Burmah. Height 30 feet.

Chusquea abietifolia, Grisebach. West-Indies. Tall, scandent.

Chusquea capituliflora, Trinius. South-Brazil. Very tall.

Chusquea Culcou, E. Desvaux. Chili. Height 20 feet. Straight.

Chusquea Dombeyana, Kunth. Peru; ascends to 6,000 feet. Height 10 feet.

Chusquea Fendleri, Munro. Central America; ascends to 12,000 feet.

Chusquea Galleottiana, Ruprecht. Mexico; ascends to 8,000 feet.

Chusquea Gaudichaudiana, Kunth. South-Brazil. Very tall.

Chusquea Lorentziana, Grisebach. Sub-tropic Argentina. Height 30 feet; not hollow. Useful for many kinds of utensils and structures.

Chusquea montana, Philippi. Chilian Andes. Height 10 feet.

Chusquea Muelleri, Munro. Mexico; ascends to 8,000 feet. Climbing.

Chusquea Quila, Kunth. Chili. Tall.

Chusquea scandens, Kunth. Colder Central America. Climbing, tall.

Chusquea simpliciflora, Munro. Panama. Height 80 feet. Scandent.

Chusquea tenuiflora, Philippi. Chili. Height 12 feet.

Chusquea uniflora, Steudel. Central America. Height 20 feet.

Dendrocalamus flagellifer, Munro. Malacca. Very tall.

Dendrocalamus Hamiltoni, Nees. Himalaya; ascends to 6,000 feet. Height 60 feet.

Dendrocalamus Hookeri, Munro. Himalaya; ascends to 6,000 feet. Height 50 feet.

Dendrocalamus sericeus, Munro. Behar; ascends to 4,000 feet. Tall.

Dinochloa Tjankorreh, Buese. Java, Philippines; ascends to 4,000 feet. Climbing.

Gigantochloa heterostachya, Munro. Malacca. Height 30 feet.

Guadua capitata, Munro. South-Brazil. Height 20 feet.

Guadua macrostachya, Ruprecht. Guiana to Brazil. Height 30 feet.

Guadua paniculata, Munro. Brazil. Height 30 feet.

Guadua refracta, Munro. Brazil. Height 30 feet.

Guadua Tagoara, Kunth. South-Brazil; ascends to 2,000 feet. Height 30 feet.

Guadua virgata, Ruprecht, South-Brazil. Height 25 feet.

Merostachys Clausseni, Munro. South-Brazil. Height 80 feet.

Merostachys Kunthii, Ruprecht. South-Brazil. Height 30 feet.

Merostachys ternata, Nees. South-Brazil. Height 20 feet.

Nastus Borbonicus, Gmelin. Bourbon; ascends to 4,000 feet. Height 50 feet.

Oxytenanthera Abyssinica, Munro. Abyssinia to Angola; ascends to 4,000 feet. Height 50 feet.

Oxytenanthera albo-ciliata, Munro. Pegu, Moulmein. Tall, scandent.

Phyllostachys bambusoides, Siebold and Zuccarini. Himalaya, China and Japan. Height 12 feet.

Phyllostachys nigra, Munro. China, Japan. Height 25 feet.

Platonia nobilis, Munro. New Granada, colder region.

Pseudostachyum polymorphum, Munro. Himalaya; ascends to 6,000 feet. Very tall.

Teinostachyum Griffithi, Munro. Tall and slender. Referred by Kurz to Cephalostachyum.

Besides these, various others are specially enumerated in the alphabetic sequence of this work. See index also.

Schizostachyum brachycladum, Kurz.

Sunda-Islands and Moluccas. Stems at length 40 feet high, very hollow. The short branches give to this Bamboo a peculiar habit. One variety has splendidly yellow stems.

Schizostachyum elegantissimum, Kurz.

Java; at elevations from 3,000 to 6,000 feet. Unlike all other Bamboos, this bears flowers at an age of three years, and is therefore of special importance for scenic effect. Height 25 feet; stems stout. It requires renewal after flowering, like many allied plants.

Schizostachyum Hasskarlianum, Kurz.

Java. This and S. serpentinum afford the best kinds of Bamboo-vegetables for cookery; the young shoots, when bursting out of the ground, being used for the purpose. Kurz mentions as culinary "Rebong"-Bamboos: Gigantochloa aspera, G. robusta, G. maxima, G. atter. For ornamental culture the same meritorious writer singles out: Schizostachyum brachycladum, the varieties of Bambusa vulgaris, with gaudy and glossy coloring of the stems, in contrast with the black-stemmed species of Phyllostachys from China and Japan.

Schizostachyum irratum, Steudel.

Sunda-Islands and Moluccas. Stems to 30 feet high, remarkably slender.

Schizostachyum Zollingeri, Steudel.

Hills of Java. Much cultivated. Height to 35 feet; stems slender.

Schkuhria abrotanoides, Roth.

From Peru to Argentina. This annual herb yields locally an insecticidal powder.

Schoenocaulon officinale, A. Gray. (*Asa-Graya officinalis*, Lindley; *Sabadilla officinalis*, Brandt and Dierbach.)

Mountains of Mexico. A bulbous-rooted herb with leafless stem, thus far specially distinct from any Veratrum. It furnishes the sabadilla-seeds, and yields two alkaloids, veratrin and sabadillin; further a resinous substance, helonin; also sabadillic and veratric acid. The generic names adopted for this plant by Gray and by Dierbach are coetaneous.

Sciadopitys verticillata, Siebold.

The curious "Kooya-maki" or Umbrella-Fir of Japan. Becoming 140 feet high; pyramidal in habit. Resists severe frosts. Wood white and compact.

Scilla esculenta, Ker. (*Camassia esculenta*, Lindley.)

The Quamash. In the western extra-tropical parts of North-America, on moist prairies. The onion-like bulbs in a roasted state form a considerable portion of the vegetable food, on which the aboriginal tribes of that part of the globe extensively live. This is a pretty plant, and might be naturalized on moist meadows.

Scilla Fraseri, A. Gray.

The Quamash of the Eastern States of North-America. Most prolific in the production of its bulbs, which taste somewhat like potatoes.

Scirpus nodosus, Rottboell.

South-Africa, extra-tropic South-America, New Zealand and extra-tropic Australia. A tufty sedge with widely creeping rhizomes, well fitted to bind driftsand on coasts even in saline wet places (Ch. French).

Scolymus Hispanicus, Linné.

Countries at or near the Mediterranean Sea. The young roots and tender shoots of this biennial herb serve as culinary vegetable; much like salsify; the aged root acts as a diuretic. S. maculatus (Linné) is a cognate annual species.

Scorzonera crocifolia, Sibthorp.

Greece. A perennial herb; the leaves, according to Dr. Heldreich, used there for a favorite salad and spinach.

Scorzonera deliciosa, Gussone.* ·

Sicily. One of the purple-flowered species; equal, if not superior, in its culinary use to the allied salsify.

Scorzonera Hispanica, Linné.* .

Middle and Southern Europe, South-Western Asia. In Norway hardy to lat. 63° 26′. The root of this yellow-flowered perennial herb fur-. nishes not only a wholesome and palatable food, but also serves as a therapeutic remedy much like dandelion. Long boiling destroys its medicinal value (B. Clark). Some other kinds of Scorzonera may perhaps be drawn into similar use, there being many Asiatic species; they should be cultivated as annuals. The leaves of some may be used as salad.

Scorzonera tuberosa, Pallas.

On the Volga and in Turkestan, in sandy desert country. This species also yields an edible root, and so perhaps the Chinese S. albicaulis (Bunge), the Persian S. Scowitzii (Candolle), the North-African S. undulata (Vahl), the Greek S. ramosa (Sibthorp), the Russian S. Astrachanica, the Turkish S. semicana (Candolle), the Iberian S. lanata (Bieberstein). At all events, careful culture may render some of them valuable esculents.

Scutia Indica, Brogniart.

Southern Asia. This, on Dr. Cleghorn's recommendation, might be utilized as a thorny hedge-shrub.

Sebæa ovata, R. Brown.

Extra-tropical Australia and New Zealand. This neat little annual herb can be utilized for its bitter tonic principle (Gentian-bitter). S. albidiflora (F. v. M.) is an allied species from somewhat saline ground. These plants get disseminated most readily, but are unacceptable to stock. S. crassulifolia (Chamisso) and Chironia baccifera (Linné) serve for the same therapeutic purposes in South-Africa (McOwan).

Secale cereale, Linné.

The Rye. Orient, but perhaps wild only in Afghanistan, and, as recently noted by Dr. A. von Regel, also in Turkestan. Mentioned as one of the hardiest of all grain-plants for sub-arctic and sub-alpine regions. In Norway it can be grown as far north as lat. 69° 30′ (Schuebeler). There are annual and biennial varieties, while a few allied species, hitherto not generally used for fodder or cereal culture, are perennial. The rye, though not so nutritious as wheat, furnishes a most wholesome well-flavored bread, which keeps for many days, and is most extensively used in Middle and Northern Europe and Asia. This cereal moreover can be reared in poor soil and cold climates, where wheat will no longer thrive. In produce of grain, rye is not inferior to wheat in colder countries, while the yield of straw is larger, and the culture less exhaustive. It is not readily subject to disease, and can be grown on some kinds of peaty or sandy or moory ground. The sowing must not be effected at a period of much wetness. Wide sand-tracts would be uninhabitable, if it were not for the ease of providing human sustenance from this grateful corn. It dislikes moist ground. Sandy soil gives the best grain. It is a very remarkable fact, that for ages in some tracts of Europe rye has been prolifically cultivated from year to year without interruption. In this respect rye stands favorably alone among alimentary plants. It also furnishes in cold countries the earliest green fodder, and the return is large. Dr. Sonder observed in cultivated turf-heaths with much humus, that the spikelets produce three or even four fertile florets, and thus each spike will yield as many as eighty beautiful seeds. Langethal recommends for argillaceous soils a mixture of early varieties of wheat and rye, the united crops furnishing grain for excellent bread. When the rye grains get attacked by Cordyceps purpurea (Fries) or similar species of fungs, it becomes dangerously unwholesome; but then also a very important medicinal substance—namely Ergot—is obtained. The biennial Wallachian variety of rye can be mown or depastured prior to the season of its forming grain. In alpine regions Wallachian rye is sown with pine-seeds, for shelter of the pine-seedlings in the first year. Rye-grain is also extensively used for the distillation of gin. Rye-straw serves as paper-material.

Sechium edule, Swartz.

Central America. The Chocho or Chayota. The large starchy root of this climber can be consumed as a culinary vegetable, while the good-sized fruits are also edible. The fruit often germinates before it drops. The plant bears even in the first year, and may ripen one hundred fruits in a year. It comes to perfection in the warmer parts of the temperate zone.

Selinum anesorrhizum, F. v. Mueller. (*Anesorrhiza Capensis*, Ch. and Schl.)

South-Africa. The root of this biennial herb is edible. A. montana (Ecklon and Zeyher), a closely allied plant, yields likewise an edible root; and so it is with a few other species of the section Anesorrhiza.

Selago leptostachya, E. Meyer.

South-Africa. There an excellent bush for sheep-pastures in the Karro-grounds, reproduced spontaneously with great readiness from dropping seeds, and maintaining itself also by the running stems. It is the "Waterfinder" of the Orange-river regions, indicating generally humidity beneath the ground (McOwan).

Selinum Monnieri, Linné.

Eastern-Asia, preferring moist places. An annual herb, praised by the Chinese as valuable for medicinal purposes.

Sequoia sempervirens, Endlicher.* (*Taxodium sempervirens*, Lambert.)

Red Wood or Bastard-Cedar of North-Western America, chiefly California. One of the most colossal trees of the globe, exceptionally becoming 360 feet high, occasionally with a stem-diameter of 55 feet. Likes humidity of soil, particularly in its early youth (Prof. Meehan). The wood is reddish, soft, easily split, very durable, but light and brittle. The timber of mission-buildings one hundred years old are still quite sound. The growth of this tree is about 32 feet in sixteen years. The tree is often found on metamorphic sandstone. It luxuriates in the cool dampness of sea-fogs. Shinn describes these Sequoias as rugged shafts, rising like huge monolithic columns, crowned with downward curving branches of shining green. Dr. Gibbons writes, that this tree forms forests along the coast-range for a distance of about 200 miles in a belt 20 miles wide. The wood is suitable for external as well as internal finish. It constitutes almost the sole material for weather-boarding along the Californian coast; and for fence-posts, foundations of buildings and railway-sleepers it is almost the only material used there. Is also susceptible of a splendid polish for furniture; is largely sawn into boards and shingles, furnishing in California the cheapest lumber. Stem bare for 100 feet or more; when cut, sending suckers from the root for renovation. Dr. Gibbons records as the stoutest stems some of 33 feet diameter at 3 feet from the ground. The foliage of this Sequoia is much like that of a Fir or Abies, while the foliage of the following species is more resembling that of a Spruce or Picea.

Sequoia Wellingtonia, Seemann.* (*Wellingtonia gigantea*, Lindley; *Sequoia gigantea*, Decaisne not Endlicher.)

Mammoth-tree. California, up to 8,000 feet above the sea. This, the biggest of all trees, attains a stem-length of 320 feet and a circumference of 112 feet, the age of the oldest trees being estimated at 1,100 years. The total height of a tree has been recorded as occasionally 450 feet, but such heights have never been confirmed by actual clinometric measurements of trees existing now. A stem broken at about 300 feet had yet a diameter of 18 feet. The wood is soft and white when felled; afterwards it turns red; it is very durable. Traditional accounts seem to have overrated the height of the Mammoth-tree. In the Calveras-grove two of the largest trees, which may have been the tallest of all, were destroyed; the two highest now existing there are respectively 325 and 319 feet high, with a circumference of 45 feet and 40 feet at 6 feet from the ground. At the Mariposa-grove the highest really measured trees are 272, 270 and 260 feet high; but one of these has the enormous circumference of 67 feet at 6 feet from the ground, while another, the height of which is not recorded, is 93 feet in girth at the ground, and

64 feet at 11 feet from it; the branches of this individual tree are as thick as the stems of large elms. The elevation of Calveras is about 4,760 feet above sea-level. A stump 33 feet in diameter is known at Yosemite. According to Dr. Gibbons this giant of the forest has a far wider range, than was formerly supposed, Mr. John Muir having shown, that it stretches over nearly 200 miles at an altitude of 5,000 to 8,000 feet. From the Calveras to the King-River it occurs in small and isolated groves, but from the latter point south to Deer-Creek, a distance of about 70 miles, there are almost unbroken forests of this noble tree. Growth of the tree about 2 feet a year under ordinary culture, much more in damp forest-glens. (Professor Schuebeler found it to endure the climate of Norway northward to ·lat. 61° 15'. Both Sequoias produce shoots from the root after the stem is cut away. The genus Sequoia can be reduced to Athrotaxis, as shown by Bentham and J. Hooker.

Sesamum Indicum, Linné.

The Gingili. Southern Asia, extending eastward to Japan, originating according to A. de Candolle in the Sunda-Islands. This annual herb is cultivated as far as 42° north latitude in Japan. The oil, fresh expressed from the seeds, is one of the best for table-use and free from any unpleasant taste; it congeals far less readily than olive-oil. There are varieties of this plant with white, red and black seeds; the latter is the earliest and richest, but gives a darker oil. Yield 45 to 50 per cent. oil. Nearly a million acres are under cultivation with this plant in the Madras-Presidency. The export of the oil from Bangkok in 1870 was valued, according to Simmonds, at £183,000; the market value is from 25s. to 35s. per cwt. The plant still succeeds at Malta and at Gaza, and is much grown in Turkey. Parched and pounded, the seeds make a rich soup. In Greece the seeds are often sprinkled over cakes. One of the advantages of the culture of this plant consists in its quick return of produce. The soot of the oil is used for China-ink.

Sesbania aculeata, Persoon.

The Danchi. Intra-tropical and sub-tropical Asia, Africa and Australia. This tall annual plant has proved adapted for desert-regions. It yields a tough fibre for ropes, nets and cordage, valued at from £30 to £40 per ton. Several congeneric plants can be equally well utilized.

Sesbania Ægyptiaca, Persoon.

Africa, Southern Asia, Northern and Central Australia, ranging to 33° north in Afghanistan and to 33° south on the Darling-River, ascending to 4,000 feet in the Himalayas. The foliage of this tall perennial herb and of the allied annual S. brachycarpa (F. v. M.) serves as fodder, which cattle are ravenously fond of. According to Mr. T. Gulliver, the green pods, as well as the seeds, are nutritious, wholesome and of pleasant taste.

Sesbania cannabina, Persoon.

South-Asia. An annual herb of easy growth in wet localities, requiring less attention in weeding and otherwise than the Jute-plant. The crop for fibre ripens in about five months. Regarded by Baker as a variety of S. aculeata, in which case S. cannabina would claim the right of priority for its specific name.

Sesbania grandiflora, Persoon.

North-Western Australia to the verge of the tropics, Indian Archipelagus. Called in Australia the Corkwood-tree; valuable for various utilitarian purposes. The red-flowered variety is grandly ornamental. Dr. Roxburgh recommends the leaves and young pods as an exquisite spinage. The plant is shy of frost.

Sesleria coerulea, Arduino.*

Many parts of Europe, but not in the far northern regions, though in Scotland ascending to 2,500 feet. Of this perennial grass Langethal observes, that it is for dry and loose limestone what Elymus arenarius is for loose sand. It stands depasturing by sheep well, and is one of the earliest grasses of the season.

Sesuvium Portulacastrum, Linné.

All round the globe on the shores of tropical and sub-tropical countries, occurring naturally as far south as Port Jackson. A perennial creeping herb, fit to fix the sandy silt on the edges of sea-coasts.

Shepherdia argentea, Nuttall.

The Buffalo-Berry. From the Missouri to Hudson's Bay. This bush bears red, acidulous, edible berries.

Shorea robusta, Gaertner.

The Sal-tree. India, up to about 3,000 feet. It attains as a maximum a height of 150 feet and a stem-girth of 25 feet. Foliage evergreen; leaves annual. One of the most famed of Indian timber trees, likes a rather humid clime (about 70 inches rain), thrives in annual extremes of temperature from the freezing point to about 100° F. Drs. Stewart and Brandis found it on sandstone, conglomerate, gravelly and shingly ground, where loose water-transmitting soils are mixed with a large portion of vegetable mould. The climatic conditions within a Sal-area may be expressed as—mean annual rainfall, 40 to 100 inches; mean temperature, in the cool season 55° to 77°, in the hot season 77° to 85° F. The heartwood is dark-brown; coarse-grained, hard, very heavy, strong, tough, with fibrous cross-structure, the fibres interlaced. It requires careful seasoning, otherwise it will rend and warp. For buildings, river-boats and railway-sleepers it is the most important timber of North-India. It exudes a pale, aromatic, dammar-like resin. The Tussa-silkworm derives food from this tree.

Shorea Talura, Roxburgh. (*S. laccifera*, Heyne.)

India, abounding in Mysore, where South-European fruits prosper. On this tree also the Lac-insect lives. It furnishes a peculiar dammar.

Sison Amomum, Linné.

Middle and Southern Europe. An herb of one or two years' duration. It grows best on soil rich in lime. The seeds can be used for condiment.

Smilax bona nox, Linné. (*S. tamnoides*, A. Gray.)

Southern States of North-America and Mexico. The young shoots of this, of S. laurifolia (L.) and some others are edible.

Smilax China, Linné.

Japan and China. Stems of medicinal value; in its native country the young shoots used for food.

Smilax glauca, Walter.

Southern States of North-America, extending into Mexico. One of the Sarsaparilla-plants, regarded by Dr. Porcher of undoubted medicinial efficacy, both sudorific and alterative, containing much smilacin; it likes rich soil.

Smilax medica, Chamisso and Schlechtendal.

Mexico. This plant produces mainly the Sarsaparilla-root of that country. It is one of the richest in smilacin.

Smilax officinalis, Humboldt.

New Granada and other parts of Central America. This climbing shrub produces at least a portion of the Columbian sarsaparilla, S. syphilitica (Willdenow) yields some Brazilian sarsaparilla, S. febrifuga (Kunth) the Purhampui-sarsaparilla of Peru.

Smilax papyracea, Duhamel.

Guiana to Brazil. The original of the principal supply of Brazilian sarsaparilla is ascribed to this species, although several others of this genus, largely represented in Brazil, may yield the medicinal root also, so S. Brasiliensis (Sprengel). In warm humid gullies of the temperate zone these plants would probably succeed in establishing themselves. Smilax Australis (R. Brown) extends from the tropical coast-parts of Australia to East-Gippsland. Neither this, nor the East-Australian S. glycyphylla (Smith), nor the New Zealand Ripogonum scandens (Forster), has ever been subjected to accurate therapeutic tests, and the same may be said of numerous other Smilaces, scattered through the warmer countries of the globe. Even the Italian sarsaparilla, which is derived from the Mediterranean S. aspera (Linné), has been introduced into medicine.

Smilax Pseudo-China, Linné.

South-Eastern States of North-America. This climber likes swampy banks of streams for its habitation. Serves as sarsaparilla locally. The round root is starchy, and when boiled edible; the young shoots serve as asparagus (Dr. Porcher).

Smilax rotundifolia, Linné.

Eastern States of North-America, extending to Canada. A prickly climber with deciduous foliage. An immense local use is made of the roots for the bowls of tobacco-pipes. It is estimated, that nearly three millions of these briar-root pipes are now made a year. The reed-portion of these pipes is generally prepared from Alnus serrulata (Willdenow) according to Professor Meehan.

Smilax rubiginosa, Grisebach.

South-Brazil and La Plata-States. The roots of this and S. campestris are there employed as sarsaparilla (Prof. Hieronymus).

Smyrnium Olusatrum, Linné.

The Alisander. Southern Europe, Northern Africa, Western Asia. A biennial herb, which raw or boiled can be utilized in the manner of celery. The roots and fruitlets serve medicinal purposes.

Solanum Aculeastrum, Dunal.

South-Africa. Recommended for hedges as one of the tallest species of this genus, and as armed with the most formidable prickles.

Solanum Æthiopicum, Linné.

Tropical Africa. Cultivated there and elsewhere on account of its edible berries, which are large, red, globular and uneven. The plant is annual.

Solanum betaceum, Cavanilles. (*Cyphomandra betacea*, Sendtner.)

Central America. This shrub is cultivated as far south as Buenos Ayres and Valparaiso, also on the Mediterranean Sea, for the sake of its tomato-like berries. Bears about five months each year, and fruits in the second season already. The berries can be eaten raw, when quite ripe, and are of pleasant taste (Morris).

Solanum cardiophyllum, Lindley.

Central Mexico. A Potato-Solanum with naturally large tubers.

Solanum Commersonii, Dunal.

La Plata-States, in rocky arid situations at a low level. Also a Potato-Solanum.

Solanum Dulcamara, Linné.

Europe, Northern Africa, Middle Asia, indigenous in Norway to lat. 66° 32'. A trailing half-shrub, with deciduous leaves. The stems are used in medicine, and contain two alkaloids: dulcamarin and solanin. Concerning these ample information is given by Husemann and Hilger in their large work "Pflanzenstoffe" (1884).

2 A

Solanum edule, Schumacher and Thonning.

Guinea. The berry is of the size of an apple, yellow and edible. How far this species is hardy remains to be ascertained.

Solanum Gilo, Raddi.

Tropical America ; much cultivated there for the sake of its large, spherical, orange-colored berries, which are eatable.

Solanum Guinense, Lamarck.

Within the tropics of both hemispheres. The berries of this shrub serve as a dye of various shades, particularly violet, for silk.

Solanum indigoferum, St. Hilaire.

Southern Brazil. A dye-shrub, deserving trial-culture.

Solanum Jamesii, Torrey.

Mountains of the South-Western States of North-America and of Mexico. A potato-bearing species, with small tubers, on which however continued cultural care may improve. See Mr. J. G. Baker's review of the tuber-bearing species of Solanum in the Journal of the Linnean Society 1884.

Solanum Lycopersicum, Linné.* (*Lycopersicum esculentum,* Miller.)

The Tomato. South-America, particularly Peru. Annual. Several varieties exist, differing in shape and color of the berries. It is one of the most eligible plants with esculent fruits for naturalization in desert-countries. As well known, the tomato is adapted for various culinary purposes. Export of berries from Bermuda alone in 1871 as much as 672 tons (Sir J. Lefroy). Tomato-foliage may be placed round fruit-trees, like the equally poisonous potato-leaves, to prevent the access of insects, and an infusion of the herb serves also as an insecticide for syringing, as first adopted for Mr. Sircy.

Solanum macrocarpum, Linné.

Mauritius and Madagascar. A perennial herb. The berries are of the size of an apple, globular and yellow. S. Thonningi (F. Jacquin) from Guinea, is a nearly related plant. S. calycinum (Mocino and Sessé) from Mexico is also allied.

Solanum Maglia, Molina.

From Chili to the Chonos-Archipelagus along the shores. A Potato-Solanum with naturally large tubers.

Solanum Melongena, Linné. (*S. ovigerum,* Dunal; *S. esculentum,* Dunal.)

The Egg-Plant. Arabia. Ripening still at Christiania like the tomato. A perennial plant, usually renewed in cultivation like an annual. The egg-shaped large berries are known under the name of Aubergines, Bringals or Begoons as culinary esculents. Prepared in France like an omelette. The seeds will keep for several years. Allied plants are S. insanum (L.), S. longum (Roxb.), S. serpentinum (Desf.), S. undatum (Lam.), S. ferox (L.), S. pseudo-saponaceum

(Blume), S. album (Dour.), which all bear large berries, considered harmless; but all may not represent well-marked species. Absolute ripeness of all such kinds of fruit is an unavoidable requisite, as otherwise even wholesome sorts may prove acrid or even poisonous. Probably many other of the exceedingly numerous species of the genus Solanum may offer yet good-sized edible berries. Seeds keep 7 years.

Solanum muricatum, L'Heritier.

The Pepino of Peru. A shrubby species, yielding egg-shaped edible berries, which are white with purple spots, and attain a length of 6 inches.

Solanum oxycarpum, Schiede.

Central Mexico. A species with very small tubers only in its wild state, which culture may however further develop.

Solanum Quitoense, Lamarck.

Ecuador, Peru. A shrubby plant. The berries resemble small oranges in size, color and even somewhat in taste, and are of a peculiar fragrance. To this S. Plumierii (Dunal) from the West-Indian Islands is also cognate, as well as S. Topiro (Kunth) from the Orinoco-region.

Solanum torvum, Swartz.

From the West-Indies to Peru. A shrubby species with yellow spherical berries of good size, which seem also wholesome. Other species from tropical America have shown themselves sufficiently hardy, to induce us to recommend the test-culture of such kinds of plants. Many of them are highly curious and ornamental. S. sisymbrifolium (Lamarck) of South-America, where it becomes extra-tropical, also yields edible berries.

Solanum tuberosum, Linné.*

The Potato. Andes of South America, particularly of Chili and Peru, but not absolutely trans-equatorial, as it extends into Bolivia, Ecuador and Columbia. It is also wild in the Argentine territory, and extends northward into the United States and Mexico, in its variety boreale (S. Fendleri, Gray). In Norway it can be grown as far north as 71° 7' (Schuebeler). From some varieties of potatoes three crops may be obtained within a year in regions free from frost. In rich coast-lands of Victoria as much as 14 tons of potatoes have been taken from an acre in a single harvest. As a starch-plant, the potato interests us on this occasion particularly. Considering its prolific yield in rich soil, we possess as yet too few factories for potato-starch. The average yield is 10 per cent. Great Britain imported in 1884 about 14,000 tons of dry yeast, for the manufacture of which potato-starch is largely used, at a value of £777,000. The starch, by being heated with mineral acids or malt, can be converted into dextrin and dextro-glucose for many purposes of the arts. Dextrin, as a substitute for gum, is also obtainable by subjecting potato-starch in a dry state to a heat of 400° F. Alcohol may be largely produced from the tubers. The berries and shoots contain solanin. Baron von Liebig remarks, " So far as its foliage is concerned, it is a lime-plant; as regards its tuber a potash-plant." Langethal says, " It surpasses in easy range of cultivation all other root-crops; its culture suppresses weeds and opens up the soil, besides preparing the land for cereals." Seeds of the potato-berries should be sown in adapted places by explorers of new countries. The most formidable potato-disease of the last thirty years, from the Peronospora infestans, seems to have originated from the use of objectionable kinds of guano, with the introduction of which the murrain was contemporaneous; along with this fung the Fucisporium Solani works almost constantly its mischief also ; to destroy their hibernating spores, all vegetable remnants on potato-fields must be burnt (W. G. Smith). The foliage of potato-plants, when thickly placed under trees or shrubs infected by blights, checks materially the spread of insects, which cause the disease. The most destructive potato-grub is Lita Solanella. The Colorado-beetle, injurious to the potato-crop in North-America, is Doryphora decemlineata. See in reference to nativity Sir Joseph Hooker's notes on the wild forms of the potato-plant in the Flora Antarctica, II., 329–332. The plant in one of its wild states (S. Fendleri) in New Mexico and Arizona is said to endure a temperature of zero.

Solanum Uporo, Dunal.

In many of the islands of the Pacific Ocean. The large red spherical berries of this shrub can be used like tomatoes. Proves hardy at Port Phillip. The leaves, macerated in cold water, render it quickly mucilaginous, and the liquid thus obtained is regarded as valuable against dysentery (Rob. MacDougall).

Solanum vescum, F. v. Mueller.

The Gunyang. South-Eastern Australia, extending to Tasmania. A shrub, yielding edible berries, which need however to be fully ripe for securing absence of deleterious properties.

Sophora Japonica, Linné.*

A deciduous tree of China and Japan, resembling the Laburnum, at length 60 feet high. Hardy in Middle Europe. Wood hard and compact, valued for turners' work. All parts of the plant purgative; the flowers rich in yellow dye, used for silk. The variety pendula, desirable for cemeteries, when trained as a creeper, has few rivals in handsomeness.

Sophora tetraptera, J. Miller.

New Zealand, Lord Howe's Island, Juan Fernandez Island, Chili, Patagonia. The " Pelu " of the latter countries. A small tree with exceedingly hard and durable wood, which can be used for cog-wheels and other select structures. Trunk exceptionally attaining a diameter of three feet. The wood differs much from that of S. Tomairo of the Easter-Island (Dr. Philippi).

Spartina cynosuroides, Willdenow.

Eastern part of North-America, there often called Prairie-grass. A perennial grass of fresh-water swamps; it can be utilized for fodder, and its value as paper-material seems almost equal to that of Esparto. Emits shoots copiously, hence is recommended by Bouché for binding maritime driftsands, covering the ground densely with its persistent rigid foliage.

Spartina juncea, Willdenow.

Salt-marshes of North-America. A grass with creeping roots; it can be utilized to bind moist sand on coasts. A tough fibre can readily be obtained from the leaves. S. polystachya (Willdenow) is a stately grass, adapted for saline soil; it is also a North-American grass. S. juncea and S. cynorsuroides extend to Canada (J. Macoun).

Spartina stricta, Roth.

The Twin-spiked Cord-grass. Countries on the Mediterranean Sea, extending to Britain and also to North-America. A rigid perennial with creeping roots, recommended for fixing and rendering solid any mud-flats on low shores and at the mouths of rivers; only suitable for brackish ground.

Spartium junceum, Linné.

Countries around the Mediterranean Sea, thence to the Canary Islands, Styria and Persia. The flowers of this bush provide a yellow dye. A textile fibre can be separated from the branches.

Spergula arvensis, Linné.

All Europe, North-Africa, West-Asia. This annual herb, though easily becoming a troublesome weed, is here mentioned for the desirable completeness of this enumeration. The tall variety with large seeds (S. maxima, Weihe) can be chosen with advantage for the commencement of tillage on any sandy soil too poor for barley. It takes up the land only for about two months, if grown for green fodder, and as such much increases the yield of milk. It serves also for admixture to hay (Langethal). It is one of the earliest of fodder-plants, and imparts a particularly pleasant taste to butter.

Spigelia Marylandica, Linné.

The "Pinkroot" of North-Eastern America, north to Pennsylvania and Wisconsin. A perennial handsome herb, requiring cautious administration as a vermifuge. S. anthelmia, L., is an annual pretty plant of tropical America, and possesses similar medicinal properties, in which probably some others of the many species likewise share.

Spilanthes oleracea, N. Jacquin.

South-America. Inaptly called the "Para-Cress." An annual herb of considerable pungency, used as a medicinal salad.

Spinacia oleracea, ·Linné.

Siberia. The ordinary Spinach, an agreeable culinary annual of rapid growth. Can be grown in Norway to lat. 70° 4' (Schuebeler). It has a mild aperient property, like several species of Chenopodium. Two varieties are distinguished, the Summer- and the Winter-Spinach, the former less inclined to run into seed, but also less hardy. The seeds are found by Burbridge and Vilmorin to retain their vitality for about five years.

Spinacea tetrandra, Steven.

The "Schamum." From the Caucasus and Persia to Turkestan and Afghanistan. Also an annual and unisexual plant like the preceding, with which it has equal value, though it is less known. A. de Candolle surmises, that it may be the original parent of the Spinage-plant.

Spinifex hirsutus, La Billardière.

On the whole coast of extra-tropical Australia. Highly valuable for binding coast-sand with its long creeping roots; deserving transfer to other parts of the world.

Spinifex longifolius, R. Brown.

On the tropical and western extra-tropical coast of Australia. Available like the former.

Spinifex paradoxus, Bentham.

Central Australia. Not unimportant as a large perennial fodder-grass on sand-ridges, and remarkable for its endurance of protracted drought and extremely high temperature (Flierl).

Spinifex squarrosus, Linné.

India. Useful for binding sand. Tennant remarks, that the radiating heads become detached when the seed is matured, and are carried by the wind along the sand, over the surface of which they are impelled by their elastic spines, dropping their seeds as they roll along. The heads are so buoyant as to float lightly on water, and while the uppermost spiny rays are acting as sails, they are carried across narrow estuaries, to continue the process of embanking beyond on any newly formed sandbars.

Spondias dulcis, G. Forster.

Polynesia, perhaps also Philippine Islands. This noble tree is introduced into this indicative list, with a view of showing the desirability, that trials should be instituted as regards the culture of the various good fruit-bearing species of this genus, one of which, S. Solandri (Bentham), crosses the tropical circle in East-Australia.

The lamented Dr. Seemann saw S. dulcis 60 feet high, and describes it as laden with fruits of agreeable apple-flavor, called " Wi," some attaining over 1 lb. weight.

Spondias mangifera, Willdenow.

Continental and insular India, ascending the Himalayas to about 5,000 feet. A small tree with deciduous foliage and edible fruit, the latter of the size of a small mango.

Sporobolus Virginicus, R. Brown.

Warmer regions of both hemispheres. A perennial grass, which will luxuriate even in sandy maritime places, and keep perfectly green after three or four months' drought. In Jamaica horses become rapidly and astonishingly fat while feeding upon this grass (Jenman).' S. Indicus, S. purpurascens (Hamilton) and S. Jacquemonti (Kunth) are also highly spoken of as pasture-grasses in the West-Indian Islands. Several other congeners deserve attention, but S. elongatus (Brown), though a very resisting grass, is rather too hard for fodder-purposes.

Stenopetalum nutans, F. v. Mueller.

Central Australia. An excellent annual herb for sheep-pastures, disseminating itself over the ground readily (Rev. H. Kempe). The naturalization of other species, all Australian, might be effected in arid hot sandy deserts.

Stenotaphrum Americanum, Schranck.* (*S. glabrum,* Trinius.)

South-Asia, North-Australia, Africa, warmer countries of America; not known as really indigenous from any part of Europe. Here called the Buffalo-grass. It is perennial, creeping, and admirably adapted for binding sea-sand and river-banks, also for forming garden-edges, and for establishing a grass-sward on lawns much subjected to traffic; it is besides of some value on pastures, and is one of the best of shade-grasses also, though not particularly nutritious. It kept alive in the hottest and driest regions of Central Australia, where it was first introduced by the writer of this work; it endures also some frost, even the tender portions of its blade, and has shown itself adapted for recently reclaimed swamp-land. The chemical analysis, instituted late in spring, gave the following results:—Water, 80·25; albumen, 0·50; gluten, 5·44; starch, 0·08; gum, 1·60; sugar, 1·60; fibre, 10·53 (F. v. Mueller and L. Rummel). It consolidates rolling sands into a firm pasture-turf. It was this grass, which Mr. John C. Bell reared with so much advantage for fodder on the bare rocks of the Island of Ascension; and it was there, where Australian Acacias took the lead, to establish wood-vegetation and to secure permanency of drinking water. S. subulatum (Trinius) is a closely allied grass of New Guinea and some of the adjoining islands.

Sterculia Carthaginensis, Cavanilles. (*S. Chicka,* St. Hilaire.)

South-Brazil. This and some other South-American species furnish seeds of almond-like taste.

Sterculia nobilis, Smith.

From India to China. A middle-sized spreading tree. The rather large seeds can be used as chestnuts in a roasted state.

Sterculia quadrifida, R. Brown.

Eastern and Northern Australia. This tree might be naturalized in rich and humid forest-regions within mild climes. It is the "Calool" of the natives. The black seeds are of filbert-taste, like those of some other *Sterculiæ.* As many as eleven of the brilliantly scarlet large fruitlets may occur in a cluster, and each of them may contain as many as ten or eleven seeds (Ch. Fawcett). The fruit is almost alike to that of, S. nobilis in size and color.

Sterculia urens, Roxburgh.

India, extending to the north-western provinces, to Assam and Ceylon. A tree with deciduous foliage; likes dry, rocky, hilly situations. This and S..urceolata (Smith), from the Moluccas and Sunda-Islands, produce edible seeds, and may prove hardy in mild extra-tropical regions. The same may be said of S. foetida (Linné), which extends from India to tropical Africa and North-Australia.

Stilbocarpa polaris, Decaisne and Planchon.

Auckland's and Campbell's Islands, also in the southern extremity of New Zealand, and also in Macquarie's Island, luxuriating in a frigid zone and in exposed, boisterous localities. An herbaceous plant with long roots, which are saccharine, and served some wrecked people for a lengthened period as sustenance. The plant is recommended here for further attention, as it may prove through culture a valuable addition to the stock of culinary vegetables of cold countries. Herbage liked by some pasture-animals.

Stipa aristiglumis, F. v. Mueller.

South-Eastern Australia, in the dry inland-regions. Graziers consider this perennial grass as very fattening and as yielding a large quantity of feed. Its celerity of growth is such that, when it springs up, it will grow at the rate of 6 inches in a fortnight. Horses, cattle and sheep are extremely fond of it. It ripens seeds in little more than two months, should the season be favorable.

Stipa tenacissima, Linné.* (*Macrochloa tenacissima,* Kunth.)

The Esparto or Atocha. Spain, Portugal, Greece, North-Africa, ascending the Sierra Nevada to 4,000 feet. This grass has become celebrated for some years, having already afforded a vast quantity of material for British paper-mills. In 1884 Great Britain imported of Esparto and other vegetable fibres for paper-mills 184,000 tons, representing a value of £1,128,000. It is tall and perennial, and would prove a valuable acquisition anywhere, inasmuch as it lives on any kind of poor soil, occurring naturally on sand and gravel, as well as on clayey, calcareous or gypseous soil, and even on the very brink of the coast. Possibly the value of some Australian grasses, allied to the Atocha, may in a like manner become commercially established; and mainly with this view paper-samples of several grass-kinds were prepared by the writer. (*Vide* " Report Industrial Exhibition, Melbourne, 1867.") Even in the scorching heat and the forbidding sands of the Sahara-region the Atocha maintains itself, and it may thus yet be destined, to play an important part in the introduced vegetation of any arid places of desert-tracts, particularly where lime and gypsum exist. The very tenacious fibre resists decay, and is much employed for the manufacture of ropes, also for baskets, mats, hats and other articles. During 1870 the export of Esparto-ropes into England was 18,500 tons, while the raw material to the extent of about 130,000 tons was imported. Extensive culture of this grass has commenced in the south of France. It is pulled once a year, in the earlier part of the summer. The propagation can be effected from seeds, but is done usually by division of the root. 10 tons of dry Esparto, worth from £4 to £5 each, can be obtained from an acre under favorable circumstances. The supply has fallen short of the demand. Good writing-paper is made from Esparto without admixture; the process is similar to that for rags, but cleaner. The price of Esparto-paper ranges from £40 to £50 for the ton. Stipa arenaria (Brotero) is a closely allied and still taller species, confined to Spain and Portugal. Consul W. P. Marks deserves great praise for having brought the Atocha into commercial and manufactural recognition. In some places it passes under the name "Alfa." Mr. Christy notes, that half its weight consists of fibre, fit for paper-mills. Stipa pennata, S. capillata and S. elegantissima will grow in pure sand, are pretty for bouquets, must however be kept away from pasture-lands.

Streblus asper, Loureiro.

South-Asia. This bears a good recommendation for live fences, in being a shrub of remarkable closeness of branches.

Stylosanthes elatior, Swartz.

South-Eastern North-America, West-Indies. A perennial herb, possessing valuable medicinal properties (Dr. E. Sell).

Styrax officinalis, Linné.

Countries on the Mediterranean Sea. A tall bush or small tree. The fragrant solid storax-resin exudes from this plant, or is particularly obtained by pressure of the bark.

Swertia Chirata, Bentham. (*Ophelia Chirata*, Grisebach.)

Widely dispersed over the higher mountain-regions of India. A perennial herb, considered as one of the best tonics; it possesses also febrifugal and antarthritic properties. Its administration in the form of an infusion, prepared with cold water, is the best. Besides S. elegans (Wallich) and some of the other Upper Indian, Chinese and Japanese species probably deserve equal attention. Hanbury and Flückiger mention as chiratas or chirettas of Indian bazaars: S.

angustifolia (D. Don), S. densifolia (Grisebach) and S. multiflora (Dalzell), besides species of Exacum and Andrographis. All come from the cooler uplands,

Swietenia Mahagoni, Linné.*

The Mahogany-tree of the West-Indies, extending naturally to Florida and Mexico. The degree of endurance of this famous tree seems not sufficiently ascertained. In its native mountains it ascends to 3,000 feet. It reaches a considerable height, and the stem a diameter of 6 or 7 feet, indicating a very great age. Sir William Hooker counted 200 wood-rings in a block of 4 feet width, which may not however indicate as many years of age. The importation of Mahogany-wood into the United Kingdom in 1884 amounted to 62,000 tons at a value of £562,000.

Symphytum officinale, Linne.*

The Comfrey. Europe, Western Asia. A perennial herb. The root is utilized chiefly in veterinary practice.

Symphytum peregrinum, Ledebour.*

The Prickly Comfrey. Caucasus and Persia. The growth of this hardy plant may be recommended as an adjunct to lupine-culture. The Hon. Arthur Holroyd of Sydney has recently devoted a special publication to this plant. He quotes on seemingly good authority the return of foliage even in the first year as 20 tons to the acre, in the second year 50 tons, and every year after the astonishing quantity of 80 to even 120 tons on richly manured land. Anyhow this herb yields a nutritive and relished foliage in rapid and continuous reproduction. It is likewise recommended for green manure. Dr. Curl observed it to grow well in the moist climate of New Zealand during the hottest and driest as well as coldest weather. Fit especially for sub-alpine country. Dr. Voelcker found much mucilage but little sugar in this plant. The massive root, known to penetrate to a great depth, sometimes to fully 9 feet, sustains the plant in vigor, allowing it to be cut almost throughout the year. The propagation is easy from root-cuttings, difficult from seeds; 4,000 of the former to an acre; it will thrive even in sand and tough clay, but prefers moist and even boggy land. In tropical countries cattle have a predilection for it; there it likes shade. The likewise borragineous Cynoglossum Morrisonii (De Candolle), of North America, yields three cuttings annually. Horses and cattle relish it. It ought to be naturalized along swamps, lagoons and river-banks. It can be dried for hay. Finally it is recommended as a plant for game. It ought to suit well for ensilage.

Symplocos ramosissima, Wallich.

Himalayas, up to 7,500 feet. In Sikkim, according to Dr. Stewart, the yellow silkworm is reared on the leaves of this tree. Two allied species occur spontaneously in the forests of East-Australia far south, . many in Southern Asia, several in tropical America.

Syncarpia laurifolia, Tenore. (*Metrosideros glomulifera*, Smith.)

Queensland and New South Wales. Vernacular name, Turpentine-tree; attains a height of 200 feet, with a stem of great thickness, to 30 feet in girth; rather of quick growth; well adapted to shade road-sides. The wood is very durable, mostly used for flooring and cabinet-making, as it takes a high polish (Hartmann); extremely durable, of excellent qualities (A. McDowall).

Synoon glandulosum, A. de Jussieu.

New South Wales and South-Queensland. This tall evergreen tree deserves cultivation in sheltered warm forest-valleys on account of its rose-colored easily worked wood. Some species of Dysoxylon of East-Australia also produce rose-wood, for instance, D. Fraseranum (Bentham), of New South Wales, a tree reaching 100 feet in height and 3 feet in stem-diameter, with fragrant wood, and D. Muelleri (Bentham), 80 feet high; the wood of the last mentioned species is of a rich color, valuable for cabinet-work, and fetches in Brisbane £3 to £4 per 1,000 feet (W. Hill).

Tacca pinnatifida, G. Forster.

From India to tropical Australia and Polynesia, also in Madagascar. This perennial plant will live even on sand-shores. From the tubers of this herb the main supply of the Fiji-arrowroot is prepared. It is not unlikely that this plant will endure a temperate clime. The Tacca-starch is much valued in medicine, and particularly used in cases of dysentery and diarrhœa. Its characteristics are readily recognized under the microscope. Several other kinds of Tacca are distinguished, but their specific limits are not yet well ascertained. From the leaves and flower-stalks light kinds of bonnets are plaited. A Tacca, occurring in Hawaia, yields a large quantity of the so-called arrowroot exported thence. Other species (including those of Ataccia) occur in India, Guinea, Guiana and North-Brazil, all deserving tests in reference to their hardiness and their value as starch-plants.

Tagetes glanduligera, Schranck.

South-America. This vigorous annual plant is said by Dr. Prentice to be pulicifugous.

Talinum patens, Willdenow.

From Mexico to Argentina. A perennial succulent herb, which might easily be naturalized on coast- and river-rocks. It furnishes the "Puchero" vegetable. The Talinum-species can all be placed well enough into the genus Claytonia.

Tamarindus Indica, Linné.

Tropical Asia and Africa. This magnificent, large, expansive tree ranges northward to Egypt, and was found in North-Western Australia by the writer of this list; reaches a very great age. Final stem-girth of 25 feet not rare. Never leafless. Varieties occur, according to Brandis, with sweetish red pulp. It is indicated here not without hesitation, to suggest trials of its acclimation in regions of the temperate zone with a warm, humid and equable temperature. The acid pulp of the pods forms the medicinal tamarind, rich in formic and butyric acid, irrespective of its other contents.

Tamarix dioica, Roxburgh.

India, up to 2,500 feet. An important shrub for binding newly formed river-banks, even in saline soil.

Tamarix Gallica, Linné.*

Southern Europe, Northern and tropical Africa, Southern and Eastern Asia, ascending the Himalayas to 11,000 feet; hardy still at Christiania. Attains a height of 30 feet in Algeria, according to Cosson. This shrub or small tree adapts itself in the most extraordinary manner to the most different localities. It will grow alike in water and the driest soil, also in salty ground, and is one of the most grateful and tractable plants in culture; it is readily multiplied from cuttings, which strike root as easily as a willow, and push forth stems with unusual vigor. Hence it is one of the most eligible bushes for planting on coast-sand, to stay its movements, or for lining embankments. Furnishes material for a superior charcoal (W. H. Colvill) and various implements (Brandis). Planted much in cemeteries. In Australia first largely sent out by the writer.

Tamarix Germanica, Linné.

Europe and Western Asia, ascending to about 15,000 feet in the Himalayas; hardy in Norway to lat. 70° 20′ (Schuebeler). Likewise available for arresting the ingress of shifting sand, particularly in moist places, also for solidifying precipitous river-banks. The allied T. elegans (Myricaria elegans, Royle) attains a height of 20 feet.

Tamarix orientalis, Forskael. (*T. articulata*, Vahl.)

Northern and Middle Africa, Southern Asia. A fast-growing tree, attaining a height of 60 feet, the trunk occasionally enlarging to a circumference of 12 feet. Springs up readily from seeds, and is also easily propagated from cuttings. Coppices well. The wood serves for ploughs, wheels and many implements (Stewart and Brandis). With T. Gallica it grows with sufficient rapidity, to be reared in India for fuel. Dye-galls and a kind of manna are also produced by this tree. The same, or an allied species, extends to Japan.

Tanacetum vulgare, Linné.

The Tansy. Northern and Middle Europe, Northern Asia, North-Western America. A perennial herb of well-known medicinal value, which mainly depends on its volatile oil. One of over 100 kinds of plants, ordered to be grown in the gardens of Charles the Great already.

Taraxacum officinale, Weber.

Dispersed over most of the temperate and cold parts of the globe, but apparently not a native of Australia. It lives in Norway northward to lat. 71° 10′ (Schuebeler). This well-known plant is mentioned, as it can be brought under regular cultivation, to obtain the medicinal extract from its roots. It is also considered wholesome to grazing animals. The young leaves furnish a medicinal salad. It is also an important honey-plant and flowers early in the season.

Tarchonanthus camphoratus, Linné.

South-Africa. This bush deserves attention, being of medicinal value. As an odorous garden-plant it is also very acceptable.

Taxodium distichum, Richard.[*]

Virginian Swamp- or Bald-Cypress. In swampy places of Eastern North-America, extending from 38° to 47° north latitude. Thought to attain occasionally an age of 2,000 years; wood-rings to the number of about 4,000 have been counted, but perhaps these not all fully annual. A valuable tree, 100 feet high or more, sometimes with a stem-circumference of 40 feet above the conical base; of rapid growth; foliage deciduous, like that of the Larch and Ginkgo. Important as antimalarian for wet fever-regions. It is found fossil in the miocene formation of many parts of Europe. The wood is fine-grained, hard, strong, light, elastic and very durable, splits well; it is much used for shingles, rails, cabinet-work and planks; it is almost indestructible in water. The tree requires a rich soil, a well-sheltered site, with much moisture but also good drainage (Lawson). It yields a superior kind of turpentine, and thus also much oil on distillation. Useful for avenues on swampy margins of lakes or river-banks. Dr. Porcher says, "This tree, lifting its giant-form above the others, gives a striking feature to many of the swamps of Carolina and Georgia; they seem like watch-towers for the feathered race."

Taxodium mucronatum, Tenore.

The famed Montezuma-Cypress of Mexico; to 120 feet high, with a trunk reaching 44 feet in circumference; it forms extensive forests between Chapultepec and Tescuco.

Taxus baccata, Linné.

Yew. Europe, North-Africa and Asia, in the Himalayas up to 11,000 feet elevation. In Norway it extends northward to lat. 67° 30′ (indigenous); Professor Schuebeler found it to attain still a height of 45 feet and a circumference of 4 feet in lat 59° 26′. Generally a shrub, finally a tree as many as 100 feet high; it furnishes a yellow or brown wood, which is exceedingly tough, elastic and durable, and much esteemed by turners; one of the best of all woods for bows. Simmonds observed, that "a post of yew will outlast a post of iron." Much valued also for pumps, piles and water-pipes, as more lasting than any other wood; also for particular musical instruments, the strongest axletrees and select implements. The tree is of very slow growth, but attains a great age, perhaps three thousand years; some ancient ones are known with a stem of 50 feet in girth; wood-rings to the extent of 2,880 have been counted. In the " Garden " it is stated, that a yew, still existing at the Fountain-Abbey in Yorkshire, was already in 1132 a large tree, when this monastery was founded. It should be kept out of the reach of grazing animals, as leaves and fruit are deadly poisonous. T. cuspidata (Siebold) is a closely cognate Japan-Yew.

Taxus brevifolia, Nuttall. (*T. Lindleyana*, Lawson.)

Western Yew. North-Western America. A stately tree, finally to 75 feet high, with a stem to 5 feet in circumference. Wood beautifully whitish or slightly yellow, or rose-colored, tough, very hard, but remarkably elastic; as fine and close-grained as that of the European yew. The Indians use it for their bows. Sir Joseph Hooker regards this as well as the Japanese and some other yews all as forms of one species.

Tectona grandis, Linné fil.*

The "Teak" of South-Asia. This superb timber-tree has its northern limit in Bandalkhand, at elevations of 3,000 feet; it ascends to 4,000 feet, but is then not of tall size; it extends to the Sunda-Islands and New Guinea; likes rather open forest-land. In Western India, according to Stewart and Brandis, frost is not uncommon in the teak-districts. The leaves drop annually. Price now in London £15 for 50 cubic feet. Teak-wood is held in the highest esteem by ship-builders; for the backing of ironclad men-of-war preferred to any other wood; also used for the panels of coaches, and for various other select purposes unsurpassed. It scarcely shrinks.

Tectona Hamiltoniana, Wallich.

Lower India. Yields the Burma-wood, which is heavy, close-grained, streaked and susceptible to a high polish. In habit and size it is similar to the ordinary teak (Kurz), but perhaps not so hardy.

Teinostachyum attenuatum, Munro.

One of the hardier Bamboos of Ceylon, there growing on the mountains at elevations between 4,000 and 6,000 feet. It attains a height of about 25 feet. Three species of this genus from New Caledonia have been described as Greslanias. Doubtless Mr. Th. Christy's use of " Thiolyte " (value £8 10s. per ton) would be particularly applicable also for converting the foliage of many kinds of Bamboos into paper-pulp; it has a most remarkable effect on fibres for separating and cleansing, and it prevents oxydation, when green parts of plants are boiled under high pressure.

Telfairia pedata, Hooker.

Mozambique. A cucurbitaceous climber with perennial stems, attaining sometimes a length of fully 100 feet, bearing fringed lilac flowers of extraordinary beauty, and fruits attaining occasionally a weight of 60 lbs. and a length of 3 feet, containing at times as many as 500 large seeds. The latter in a boiled state are edible, or a large quanticy of oil, fit for the table, can be pressed from them. The root is fleshy.. A second huge species of similar use, T. occidentalis (J. Hooker), occurs in Guinea. The genus-name Joliffia is coetaneous, if not anterior. These superb plants may not be absolutely restricted to hot tropical lowlands, and therefore possibly endure a warm temperate clime.

Terfezia leonis, Tulasne. (*Cheiromyces leonis*, Tul.)

Southern Europe, Northern Africa. This edible truffle, together with other species of this and other genera, is deserving of wide naturalization.

Terminalia Buceras, J. Hooker. (*Bucida Buceras*, Linné.)

From the Antilles to Brazil. One of the Mangrove-trees, living in salt-water. Possibly hardy and calculated to consolidate mud-shores. The Tussa-silkworm inhabits, among other trees, several Terminalias.

Terminalia Catappa, Linné.

India, continental as well as insular, ascending only lower mountain-regions, also North-Eastern Australia. Few trees, as stated by Roxburgh, surpass this in elegance and beauty. We have yet to learn, whether it can be naturalized in temperate climes, which it especially deserves for its nuts. The seeds are almond-like, of filbert taste, and wholesome. The astringent fruits of several other species constitute an article of trade, sought for a lasting black dye. T. parviflora ('Thwaites) forms a large tree in Ceylon, at elevations up to 4,000 feet. Several of their congeners reach extra-tropic latitudes in Eastern Australia.

Terminalia Chebula, Retzius.

On the drier mountains of India, ascending to 5,000 feet. A tree rising to about 100 feet. The seeds are of hazel taste; the galls of the leaves and also the young fruits, the latter known as Myrobalans, serve for superior dye and tanning material. Some of its congeners answer the same purpose, among them T. Belerica (Roxburgh).

Tetragonia expansa, Murray.

The New Zealand Spinach, occurring also on many places of the coast and in the desert-interior of Australia. Known also from New Caledonia, China, Japan and Valdivia. An annual herb, useful as a culinary vegetable, also for binding drift-sand. It produces its succulent branchlets and leaves also in the hottest weather and driest localities (Rhind). Rapid in growth. The seeds will keep fully five years (Vilmorin).

Tetragonia implexicoma, J. Hooker.

Extra-tropical Australia, New Zealand, Chatham-Island. A frutescent, widely expanding plant, forming often large natural festoons, or trailing and climbing over rocks and sand, never far away from the coast. As a spinach-plant it is of not less value than the preceding species. It is well adapted for the formation of bowers even in arid places; it also helps to bind sand. T. trigyna (Banks and Solander) seems identical.

Teucrium Marum, Linné.

Countries on the Mediterranean Sea. A small somewhat shrubby plant, in use for the sake of its scent, containing a peculiar stearopten. T. Scordium, from Europe and Middle Asia, T. Chamaedrys, T. Polium and T. Creticum, L., from South-Europe, are occasionally drawn into medical use. All these, together with many other species from various countries, are pleasantly odorous.

Thapsia edulis, Bentham. (*Monizia edulis*, Lowe.)

On the Island of Deserte Grande, near Madeira, where it is called the Carrot-tree. It might be of some use, to bring this almost shrubby umbellate to the cliffs of other shores; though the root is inferior to a carrot, perhaps cultivation would improve it. T. decipiens, Bentham (Melanoselinum decipiens, Lowe) from Madeira, is of palm-like habit and desirable for scenic effects in plant-grouping. T. Silphium (Viviani) is the Silphion of the ancients.

Theligonum Cynocrambe, Linné.

Countries around the Mediterranean Sea, extending to the Canary-Islands. An annual spinach-plant of somewhat aperient effect.

Thouarea sarmentosa, Persoon.

Tropical shores of the eastern hemisphere. This curious and tender grass might be easily introduced, to help in binding the sand on sea-beaches.

Thrinax parviflora, Swartz.

South-Florida, West-Indies and also on the continent of Central America. The stem of this Fan-palm attains a height, according to Dr. Chapman, of about 40 feet, but is extremely slender. Belongs to the sand-tracts of the coast; hardy in the South of France to 43° 32′ N. lat. (Naudin). The fibre of this palm forms material for ropes. T argentea, Loddiges, is an allied palm. The few other species of the genus from the West-Indies also deserve trial-culture.

Thuya dolabrata, Linné fil. (*Thuyopsis dolabrata*, Siebold and Zuccarini.)

The "Akeki" of Japan, ascending to an elevation of 8,000 feet. A majestic tree, of conical shape and drooping habit, growing to 50 feet high, attaining a stem-diameter of 3 feet. It delights in shaded and rather moist situations, and is used in China and Japan for avenues. Hardy in Norway to lat. 58° 27′ (Schuebeler). It furnishes an excellent hard timber of a red color.

Thuya gigantea, Nuttall.

North-Western America. The Yellow Cypress of the colonists, also known as Oregon Red or White Cedar. A straight tree, in some instances known to have attained a height of 325 feet, with a stem 22 feet in diameter; it furnishes a valuable building timber of a pale or light-yellow color, susceptible of high polish. It is light, soft, smooth and durable, and makes the finest sashes, doors, mouldings and similar articles (Vasey); also used for shingles (Dawson). Canoes carrying 4 tons have been obtained out of one stem. The bast can be converted into ropes and mats. The tree can be trained into hedges and bowers. It endures the climate of Christiania. To Thuya are referred by Bentham and J. Hooker all the cypresses of the sections Chamæcyparis and Retinospora.

Thuya Japonica, Maximowicz.

Japan. Closely allied to T. gigantea. Dr. Masters has pointed out the characteristic differences between the two in an essay on Conifers of Japan, published in the Journal of the Linnean Society, 1881.

Thuya occidentalis, Linné.

North-America, extending from Carolina to Canada. Northern White Cedar or Arbor Vitæ. A fine tree, to 70 feet high. Bears the frosts of Norway northward to lat. 63° 52′. The wood is reddish or yellowish, fine-grained, very tough and resinous, light, soft, durable, and well fit for building, especially for water-work and railway-ties, also for turnery and machinery. Michaux mentions, that posts of this wood last forty years; a house built of it was found perfectly sound after sixty years. The tree prefers moist soil; it is valuable for copses; it can also be trained into garden-bowers. Dr. Porcher says, that it makes the finest ornamental hedge or screen in the United States, attaining any required height and being very compact and beautiful; such hedges indeed were observed by the writer himself many years ago in Rio de Janeiro. The shoots and also an essential oil from this tree are used in medicine; the bast can be converted into ropes; the branches serve for brooms.

Thuya orientalis, Linné. (*Biotia orientalis,* Endlicher.)

China and Japan. The Chinese "Arbor-Vitæ" of gardens. Though seldom exceeding 20 feet in height, this common garden-plant is mentioned here, as it will admit of clipping for hedge-growth, and as the "Fi-Moro" variety should on account of its elongated slender and pendent branches be chosen extensively for cemeteries. `

Thymelæa tinctoria, Endlicher. (*Passerina tinctoria,* Pourret.)

Portugal, Spain, South-France. A small shrub, which yields a yellow dye. Cursorily it may be noted here, that some of the Australian Pimeleæ contain a blue pigment, which has not yet been fully tested. Their bark produces more or less of daphnin and of the volatile acrid principle, for which the bark of Daphne Mezereum (Linné) is used; these are remarkably developed in the South-Eastern Australian Pimelea stricta (Meissner). The bark of many is also pervaded by a tough fibre, that of the tall Pimelea clavata (Labillardière), a West-Australian bush, being hence particularly tenacious, and used for whips.

Thymus capitatus, Hoffmannsegg and Link. (*Satureja capitata*, Linné.)

Around the whole Mediterranean Sea. Since the time of Hippocrates, Theophrastos and Galenus this small scented shrub has been employed in medicine.

Thymus mastichina, Linné.

Spain, Portugal, Morocco. A half-shrub of agreeable scent, used also occasionally in medicine.

Thymus Serpillum, Linné.

Europe, Western Asia, North-Eastern Africa. A perennial herb of some medicinal value. It would live on the highest alps. An essential oil can be obtained from it. One particular variety is lemon-scented.

Thymus vulgaris, Linné.

The Garden-Thyme. South-Europe, particularly westward. Both this and the preceding species can be grown in Norway up to lat. 70° 22′ (Schuebeler). This small shrubby plant is available for scent and for condiments; further as a honey-plant. It is also well adapted for forming garden-edges. The essential oil of this plant can be separated into the crystalline thymol and the liquid thymen and cymol. T. æstivus (Reuter) and T. hiemalis (Lange) are closely cognate plants. Several other species with aromatic scent occur at or near the Mediterranean Sea.

Tilia alba, Aiton. (*T. heterophylla*, Ventenat.)

The "Silver-Linden" of the Eastern and Middle States of North-America.

Tilia Americana, Linné.

The Basswood-Tree or North-American Linden-Tree, growing there to 52° north latitude in the eastern regions. In Norway it is hardy as far as Christiania (59° 55′) as well as the following (Schuebeler). Height of tree reaching about 80 feet, diameter of stem 4 feet. The wood is close-grained and firm, as soft as deal; used in the construction of musical instruments, particularly pianofortes. Specially valuable for the cutting-boards of curriers and shoemakers, bowls, pails, shovels, panelling of carriages (Robb). As the wood is free from knots, it is particularly eligible for turnery and carving and certain portions of machinery. The tree is highly valued for street-planting in its native land, where it also furnishes linden-bast or bass. This is one of the principal honey-plants in many parts of the United States. Quinby saw 25 lbs. of honey gathered by a single swarm in one day from Basswood-flowers, bees travelling gradually several miles for them if necessary. For profusion and quality of honey the flowers are unsurpassed (A. J. Cook). Rate of stem-growth in Nebraska 35 inches girth at 2 feet from the ground in fourteen years (Furnas).

Tilia argentea, Desfontaines.*

The Silver-Linden of South-Eastern Europe and Asia Minor. The wood is not attacked by boring insects. The flowers are deliciously fragrant and yield a precious oil. The oldest specific name, according to Prof. Koch, is T. tomentosa (Moench).

Tilia Europæa, Linné.

The common Linden-tree of Europe, extending naturally to Japan; the large-leaved variety of South-European origin. It lives to a great age. A very hardy tree, living in Norway as far north as lat. 67° 56' (Schueboler). A weeping variety is known. Height sometimes to 120 feet; stem exceptionally to 50 feet in girth. One of the best of promenade-trees in climes not too dry. The wood pale, soft and close-grained; sought for turnery, piano-keys, carving; used by shoe-makers, saddlers and glovers, to cut leather on; also for toys (Simmonds). The flowers yield a highly aromatic honey. The bast excellent for mats.

Tilia Mandschurica, Ruprecht.

Amoor and Mandschuria. Not so tall as T. tomentosa, to which it bears close alliance (Maximowicz, Koch).

Tillandsia usneoides, Linné.

Black Moss, Long Moss, Florida-Moss. From Carolina and Florida to Uruguay and Chili, on trees. Might be naturalized in forests of countries with mild climes. In its native country a favorite material for upholsterers' work.

Tinguarra Sicula, Parlatore.

Southern Italy. The root is edible and celery-like.

Todea Africana, Willdenow. (*Osmunda barbara,* Thunberg.)

South-Africa, South-Eastern Australia, New Zealand. Most important for scenic effects in wet places; as an export article the very aged stems of this fern are now much sought, and have endured wide transits, which were initiated by the writer. Stems have been found bearing from 500 to 600 fronds. A gigantic specimen was got in the Dandenong forests, the trunk of which weighed 4,368 lbs., after the fronds were cut away, the extreme dimensions of the stem being about 6 feet in height, breadth and width. Supplies of this massive fern in the gullies ought to be maintained for future generations by the artificial dispersion of the dust-like spores.

Torreya Californica, Torrey. (*T. myristica,* Hooker.)

California, extending from the coast to the Sierras. A symmetrical tree, becoming about 100 feet high, with a clear straight trunk to 30 feet in length and 6 feet in diameter (Dr. Gibbons). The wood is hard and firm.

Torreya grandis, Fortune.

The "Kaya" of China. A tree at length 60 feet high, with an umbrella-shaped crown; it produces good timber.

Torreya nucifera, Siebold. (*Caryotaxus nucifera,* Zuccarini.)

Japan. Height of tree about 30 feet. From the nuts the Japanese press an oil, used as an article of food. The wood is highly valued in Japan by coopers, also by turners; it resembles to some extent boxwood (Dupont).

Torreya taxifolia, Arnott.

Florida. A tree, reaching about 50 feet in height. Wood firm, close-grained, light, durable, of a reddish color; very lasting also underground. Prostrated trees did not decay in half a century. Timber slightly more yellow than that of the white pine (P. J. White). The tree yields a reddish turpentine (Hoopes).

Touchardia latifolia, Gaudichaud.

In the Hawaian Islands. A shrub, allied to Boehmeria nivea, yielding a tough and easily separable fibre, as shown by Dr. Hillebrand. Probably best adapted to humid warm gullies.

Tragopogon porrifolius, Linné.

The "Salsify." Middle and Southern Europe, Northern Africa, South-Western Asia; hardy to lat. 70° in Norway. Biennial. The root of this herb is well known as a useful culinary vegetable ; it is obtainable even in cold countries through the winter; the young leaves supply a very good salad (Vilmorin).

Trapa bicornis, Linné fil.*

The Leng, Ling or Links of China. The nuts of this water-plant are extensively brought to market in that country. The horns of the fruit are blunt. The kernel, like that of the two following species, is of an excellent taste. The plant is regularly cultivated in the lakes and ponds of China.

Trapa bispinosa, Roxburgh.*

Middle and Southern Asia, where it is called " Singhara," extending to Ceylon and Japan ; found also in Africa as far south as the Zambezi. The nuts are often worked for starch. They can be converted into most palatable cakes or porridge, and may be stored for food, even for several years. The produce is copious and quite maintained by spontaneous dissemination. In some countries, for instance in Cashmere, the nuts in a raw or cooked state form an important staple of food to the population. To this species probably belong T. Cochin-Chinensis (Loureiro) and T. incisa (Siebold and Zuccarini).

Trapa natans, Linné.*

The ordinary Waternut. Middle and Southern Europe, Middle Asia, Northern and Central Africa. Recorded as an annual. T. quadrispinosa (Roxburgh) from Sylhet, is a mere variety.

Tremella mesenterica, Retzius.

From Europe to Australia. Arrayed by Dr. L. Planchon with the fungs fit for human food.

Trichodesma Zeylanicum, Brown. (*Pollichia Zeylanica,* F. v. M.)

From Abyssinia and Southern Asia to extra-tropical Australia. An annual herb, perhaps available for green manure. The dromedaries show an extraordinary predilection for the herb (Giles). Several other species deserve trial for fodder-growth.

Trichostema lanatum, Bentham.

California, where it is called the "Black Sage." A half-shrub, recorded by Mr. A. J. Cook in his "Bee-keepers' Guide" among important honey-plants.

Trifolium agrarium, Dodoens. (*T. aureum,* Pollich.)

The Perennial Yellow Clover or Hop-Clover. All Europe, Northern Africa, Western Asia; wild in Norway northward to lat. 63° 26′ (Schuebeler). Of considerable value in sandy soil as a fodder-herb. It is easily naturalized.

Trifolium Alexandrinum, Linné.*

The Bersin-Clover. North-Eastern Africa, South-Western Asia, South-Europe. Much grown for forage in Egypt, where it is used as the main-fodder. On the Nile it gives three green crops during the season, each up to 2 feet high. Seeds of this and other clovers must be sifted, to free them from any of the destructive Dodder-plants or Cuscutas. About 20 lbs. of seed are required for an acre (Morton). Recorded as annual.

Trifolium alpestre, Linné.

Europe, Western-Asia. Perennial. Content with lighter soil than that needed for most clovers, but the constituents must be fairly marly or limy. This clover is early out and very palatable to herds and flocks (Langethal).

Trifolium fragiferum, Vaillant.

The Strawberry-Clover. Europe, North-Africa, Middle and Northern-Asia. Indigenous in Norway to lat. 59° 55′. A perennial species, well adapted for clay-soils. Foliage closer and more tender than that of the white clover, but its vegetation later (Langethal). Morton recommends it for moist sandy soil. It delights in ground much wetter than suits most other clovers; it spreads over humid pastures most readily, with a growth more luxuriant than that of white clover, consequently stands the summer-heat better, smothering most other plants and covering the ground with a thick and close herbage. Cattle are very fond of it, and fatten well on it (Geo. Black).

Trifolium furcatum, Lindley.

California. A stout and somewhat succulent species, with large flower-heads. Affords good pasturage (A. Gray), and gets disseminated readily; but it is annual. Several other native clovers occur in Western North-America; 25 are described from California.

Trifolium hybridum, Linné.*

The Alsike-Clover. Europe, Northern Africa, Western Asia. Wild in Norway to lat. 63° 50'. A valuable perennial pasture-herb, particularly for swampy localities. It succeeds, where the ground becomes too sandy for lucerne and too wet for red clover, but does not withstand drought so well, while it produces a heavier bulk of forage than white clover, and maintains its ground, when the soil has become too much exhausted for other clovers. The seed being very small, less than half the quantity is required for the same area as of red clover. Much frequented by bees for honey.

Trifolium incarnatum, Linné.

The Carnation-Clover, also called Crimson or Italian Clover. In Norway it can be grown to lat. 70° 22'. Southern Europe, extending naturally to Switzerland. Though annual only, or sometimes biennial, it is valued in some of the systems of rotations of crops. In the south of England it is much sown on harrowed stubble-fields, to obtain an early feed of great fattening value. It forms particularly a good fodder for sheep, and is recommended especially for gypsum-regions. A white-flowering variety exists. Bees are very fond of this clover (Darwin).

Trifolium medium, Linné.

The Red Zigzag-Clover. Europe, Northern and Middle Asia. Indigenous in Norway to lat 63° 26' (Schuebeler). A deep-rooting, wide-creeping perennial herb, much better adapted for dry sandy places than. T. pratense. It would also endure the inclemency of the clime of alpine heights, if disseminated there; also one of the best of clovers for forest-regions. For regular culture it needs lime, like most plants of its class. More hardy than T. hybridum, less productive than T. pratense (Langethal). It ought not to be omitted among mixed clovers and grasses. According to Morton it is not so much sought and relished by grazing-animals as many other clovers. T. Quartiniánum (A. Richard) is an allied plant from Abyssinia, where several endemic species exist.

Trifolium montanum, Linné.

Europe, Western Asia. Perennial. Not without importance for limy or marly ground. It is indigenous northward to Christiania.

Trifolium ochroleucum, Linné.

Pale-yellow Clover. Middle and Southern Europe, Western Asia. Perennial. This species is much cultivated in Upper Italy; its value is that of T. medium (Langethal).

Trifolium Pannonicum, Linné.

The Hungarian Clover. Southern Europe. Perennial. Earlier in the season than red clover, to which it is allied, but less tender in foliage (Morton).

Trifolium pratense, Camerarius.*

The ordinary Red Clover. All Europe, North-Africa, Northern and Middle Asia. It is found wild as far north as 69° 20′ in Norway (Schuebeler). A biennial, or under certain circumstances also a perennial herb, of special importance for stable-fodder. The perennial variety passes under the name of Cow-clover, by which name also T. medium is sometimes designated. Highly recommendable for permanent pastures, particularly in cool humid climes, as it continues to grow year after year, and produces a large amount of herbage (Dr. Curl). It prefers rich ground and particularly soil, which is not devoid of lime; gypsum-dressings are recommendable for the fields. It enters into the rotation-system of crops very advantageously. This species lives also in alpine regions, where it much enriches the pastures. The nectar of the flowers is sucked by bumble bees, which tends to facilitate the production of seeds.

Trifolium reflexum, Linné.*

The Pennsylvania- or Buffalo-Clover. Eastern North-America, Annual or biennial; flower-heads larger than those of the red clover; likes alluvial flats.

Trifolium repens, Rivin.*

The ordinary White Clover, called also Dutch Clover. Europe, North-Africa, Northern and Middle Asia, sub-arctic America. In Norway indigenous to lat. 70° 57′. Perennial. Most valuable as a fodder-plant on grazing land. It has a predilection for moist soil, but also springs again from dry spots after rain. It likes soil containing lime, prospers on poorer ground than red clover, is more nourishing and better digested, and less exhaustive to the soil. Dressing with gypsum vastly enhances the value and productiveness of any clover-field. Important as a bee-plant.

Trifolium resupinatum, Linné.

The annual Strawberry-Clover. From South-Europe and North-Africa to Persia; also in the Canary-Islands and Azores. Admitted here, though annual, as this clover is cultivated with predilection in Upper India; it is of tall growth and succulent foliage.

Trifolium spadiceum, Linné.

Brown Clover. Europe, Western Asia. Perennial. This has been recommended for wet sandy moorland, on which it gets disseminated with readiness.

Trifolium subrotundum, Hochstetter.

The Mayad-Clover. Northern and Middle Africa, ascending to 9,000 feet. A perennial species, in its native countries utilized with advantage for clover-culture.

This by no means closes the list of the clovers variously desirable for introduction, inasmuch as about 150 well-marked species are recognized, many doubtless of value for pasture. But the notes of rural observers on any of these kinds are so sparingly extant, that much uncertainty about the yield and nutritive value of various kinds continues to prevail. Most clovers come from the temperate zone of Europe and Asia; only two are indigenous to the eastern of the United States of North-America, none occur in Australia, a few are found in South-Africa, a good number in California and the adjoining countries, several also in Chili, no species is peculiar to Japan.

Trigonella Fœnum Græcum, Linné.

Countries on the Mediterranean Sea, Western and Central Asia. The seeds of this annual herb find their use in veterinary medicine.

Trigonella suavissima, Lindley.

Interior of Australia, from the Murray-River and its tributaries to the vicinity of Shark-Bay. This perennial, fragrant, clover-like plant proved a good pasture-herb. A lithogram, illustrating this plant, occurs in the work on the "Plants indigenous to Victoria." Some of the many European, Asiatic and African species of this genus deserve local tests.

Trillium erectum, Linné.

"The Birthroot." Eastern North-America. This liliaceous plant has found its way into the materia medica.

Triodia exigua, Kirk.

South-Island of New Zealand, at 1,200 to 3,000 feet elevation. Forms naturally almost even plots, often many square-yards in extent; the leaves are hard, short and shining; the compact growth of the turf or sward prevents weeds and other grasses to encroach. It is particularly to be recommended for croquet-lawns, never requiring mowing (Prof. Th. Kirk). Should prove especially valuable in colder countries for lawns, and may hold its ground also in hotter climes through some irrigation.

Triphasia Aurantiola, Loureiro.

South-Eastern Asia. This shrub is worth cultivation for the exquisite fragrance of its flowers. The fruits, though small, are of pleasant sweetness. The plant may also prove well adapted for hedges. Glycosmis citrifolia, Lindley, and Claussena punctata, Oliver, also East-Asiatic fruit-shrubs, may possibly show themselves hardy in sheltered forest-regions of warm temperate climes.

Tripsacum dactyloides, Linné.

Central and Northern America; known popularly as a Gama-grass. A reedy perennial grass, more ornamental than utilitarian. It is the original Buffalo-grass, and attains a height of 7 feet, assuming the aspect of maize. It seems of inferior value for feed, but serves for binding sand. Prof. C. Mohr however regards it as a valuable fodder-grass. The seeds are available for food. Howard, speaking in Carolina of this grass, contends, that it may be cut three

. or four times in a season, that it makes a coarse but nutritious hay, and that the quantity of forage, which can be made from it, is enormous; cattle and horses are fond of it, and the hay can be harvested easily.

Trisetum antarcticum, Trinius.

New Zealand. Ascending to sub-alpine elevations. A perennial lustrous grass, particularly fit for cool climes. According to Mr. John Buchanan it keeps its ground well, becoming an important element locally in the pasture-vegetation. The abundant natal growth indicates, how easily the grass by dissemination could be naturalized elsewhere.

Tristania conferta, R. Brown.

New South Wales and Queensland. A noble shady tree, attaining a height of about 150 feet. It is not only eligible as an avenue-tree, but also as producing select timber; ribs of vessels from this tree have lasted unimpaired thirty years and more. Growth in height 20–30 feet at Port Phillip in twenty years.

Trithrinax Acanthocoma, Drude.

Rio Grande do Sul, in dry elevations. A dwarf Fan-palm for window- or table-decoration, attaining only a height of about 6 feet; foliage not leathery. '

Trithrinax Brasiliensis, Martius.

Rio Grande do Sul and Parana, Uruguay and Paraguay. A very hardy palm, not tall.

Trithrinax campestris, Drude.*

Argentina, as far south as 32° 40'. Height reaching about 30 feet. One of the most southern of all palms. Content with even less humidity than Chamærops humilis. The leaves are almost of a woody hardness and stiffer than those of any other palm (Drude). Germination from seeds easy (Lorentz and Hieronymus). Another species occurs in Southern Bolivia.

Triticum junceum, Linné. (*Agropyrum junceum*, Beauvois.)

Europe and North-Africa. A rigid coast-grass, with pungent leaves and extensively creeping roots, requiring sea-sand for its permanent growth. One of the best grasses, to keep rolling sand-ridges together, and particularly eligible, where cattle and other domestic animals cannot readily be prevented from getting access.

Triticum vulgare, Villars.*

The Wheat. Indigenous to the Euphrates-regions, according to A. de Candolle. Traced back more than 5,000 years as an Egyptian and Chinese culture-plant; indeed the earliest lacustrine people in Switzerland reared wheat in a stone-age (Heer). In many intra-tropical countries, not too wet, wheat and barley can be grown as winter-crop. In Japan wheat is of extraordinary precocity (Lartigne), and it is greatly recommended there as a forage-plant. The Punjab-Wheat with a few other varieties is rust-proof. This is not the place, to enter into details about a plant universally known, unless we may allude to the much overlooked fact, that a light beer can be brewed from wheat; it may therefore suffice merely to mention, that three primary varieties must be distinguished among the very numerous sorts of cultivated wheat: 1. *Var.* muticum, T. hybernum, L., the Winter-Wheat or Unbearded Wheat; 2. *Var.* aristatum, T. æstivum, L., the Summer-Wheat or Bearded-Wheat; 3. *Var.* adhærens, T. Spelta, L., Wheat with fragile axis and adherent grain. Metzger enumerates as distinct kinds of cultivated wheat:—

T. vulgare, Villars, which includes among other varieties the ordinary Spring-Wheat, the Fox-Wheat and the Kentish Wheat. It comprises also the best Italian sorts for plaiting straw-bonnets and straw-hats, for which only the upper part of the stem is used, collected before the ripening of the grain, and bleached through exposure to the sun while kept moistened.

T. turgidum, Linné, comprising some varieties of White and Red Wheat, also the Clock-Wheat and the Revet-Wheat.

T. durum, Desfontaines, which contains some sorts of the Bearded Wheat.

T. Polonicum, Linné, the Polish Wheat, some kinds of which are well adapted for peeled Wheat.

T. Spelta, Linné, the Spelt-Corn or Dinkel-Wheat, a kind not readily subject to disease, succeeding on soil of very limited fertility, not easily attacked by birds, furnishing a flour of excellence for cakes, also yielding a superior grain for peeled wheat. For preparing the latter it is necessary to collect the spikes while yet somewhat green, and to dry them in baking-houses.

T. dicoccum, Schrank, (*T. amyleum*, Seringe). The Emmer-Wheat. Its varieties are content with and prolific on poor soil, produce excellent starch, are most hardy and not subject to diseases. To this belongs the Arras-Wheat of Abyssinia, where a few other peculiar sorts of wheat are to be found. A large-grained variety of wheat is baked in Persia like rice (Colvill).

T. monococcum, Linné. St. Peter's Corn, which is hardier than most other wheats; exists in the poorest soils, but produces grains less adapted for flour than for peeled wheat. Indigenous to Serbia, Greece and Turkey, if derived from T. Baeoticum (Boissier). The Champlain-Wheat, recently here introduced by me, yields about 40 fold and seems quite rust- and smut-proof; the crop is heavy; but this variety is preferable for green fodder and hay, the grain carrying too much bran (Hermiston). Dr. Bancroft's experiments in Southern Queensland showed the common Indian Bearded Wheat to be exempt from rust, as well as two beardless varieties from the same part of Asia. On this subject see also the print of my lecture before the Agricultural Society of Bendigo, "on rust in wheat," 1865. According to the report of the Commissioner of Agriculture in 1882, 37 millions of àcres were under wheat-culture in the United States. In various parts of the world the prodigious quantity of 60 bushels on an acre is sometimes obtained on rich and new land.

Tropæolum majus, Linné.

Peru. This showy perennial climber passes with impropriety under the name of Nasturtium. The herbage and flowers serve as cress, and are also considered antiscorbutic. The plant can be grown in Norway northward to lat. 70° 22′ (Schuebeler). A smaller species, T. minus, L., from Peru, can likewise be chosen for a cress-salad; both besides furnish in their flower-buds and young fruits a substitute for capers. A volatile oil of burning taste can be distilled from the foliage of both, and this is more acrid even than the distilled oil of mustard-seeds. In colder countries these plants are only of one year's duration. Numerous other species, all highly ornamental, occur in South-America and a few also in Mexico. The seeds will keep for several years.

Tropæolum sessilifolium, Pœppig.

Chili. Among the species of this genus one of the most eligible for its tubers, which can be consumed even in a raw state, and are larger than those of most other Tropæolums, while the stems are short and procumbent (Prof. Philippi).

Tropæolum tuberosum, Ruiz and Pavon.

Peru. The tuberous root serves as an esculent; some frost improves it.

Trophis Americana, Linné.

West-Indian Archipelagus. The foliage of this milky tree has been recommended as food for the silk-insect. In Cuba and Jamaica it is used as provender for cattle and sheep.

Tuber æstivum, Micheli.

Middle and Southern Europe. The truffle most frequent in the markets of England. The White British Truffle, Chairomyces meandriformis, though large, is valued less. In the Department Vaucluse alone about 60,000 lbs. of truffles are collected annually, at a value of about £4,000. Many other kinds of truffles are in use. The Australian truffle, Mylitta Australis (Berkeley) or Notiohydnum Australe, sometimes attains the size of the cocoa-nut, and is also a fair esculent. It seems quite feasible, to naturalize the best edible fungs of these and other genera, although such may not be amenable to regular culture; thus efforts should be made for the introduction of all the superior kinds of truffles, as an insight into the manner, in which vegetables of the fungus-species might be transferred to wide distances, has gradually been obtained. The total value of the export of truffles from France in 1877 amounted to considerably over half a million pounds sterling, the total production in that year being valued at about £800,000. The annual revenue of the truffle-ground of Carpentras is, according to Simmonds, £80,000. The great White North-American Truffle (Tuber album) is as white as snow and as tender as curds (Millington).

Tuber albidum, Cesalpini.

Occurs with T. æstivum, but is smaller and less agreeable in taste.

Tuber cibarium, Sibthorp.

.The Black Truffle. Middle and Southern Europe. Like all others growing underground, and generally found in forest-soil of limestoue-formation. It attains a weight of over one pound. Experiments for naturalization may be effected with every prospect of success by conveying the truffle in its native soil and locating it in calcareous places of forest-regions. As a condiment or merely in a roasted state, it affords an aromatic food. The famous Quercy- or Perigold-Truffle is derived from this species. T. melanosporum (Vittadini) from France, Germany and Italy, is of a still more exquisite taste than T. cibarium —indeed, of strawberry-flavor.

Tuber magnatum, Pico.

Grey Truffle. South-Europe. One of the most esteemed of all truffles, with some garlic-flavor. Hymenogaster Bulliardi (Vittadini) and Melanogaster variegatus (Tulasne) of South-Europe are also excellent truffles.

Tuber rufum, Pico.

Red Truffle, especially in vineyards. Much used for food, but smaller than Terfezia-Truffles.

Typha latifolia, Linné.

The Cattail, large Reedmace or Bulrush. Widely distributed over the northern hemisphere—in Norway to lat. 60° 41'. Worthy of being encouraged in its growth on rivers and around lakes, and of being transferred to unutilized waters, as the very light and soft foliage can be converted into material for mattresses, which in the Royal Navy of Italy have come into universal use as additional means of saving human life in the event of shipwreck. These mattresses continue to float for a very long time and bear a great weight; thus one mattress is capable of supporting several persons in water (Marquis Toverena and Captain Romano). The large rootstocks are rich in nourishing starch. The closely allied T. angustifolia extends to Australia.

Ulex Europæus, Linné.

The Whin, Gorse or Furze. Western and Southern Europe, Azores, Canary-Islauds; hardy in Norway to lat. 58° 58'. A bush, important for covering quickly drift-sands on coasts, not readily approached by pastoral animals. Too apt to stray as a hedge-plant. Prof. C. Koch recommends a thornless variety for sheep-pastures.

Ullucus tuberosus, Lozano. (*Melloca tuberosa*, Lindley.)

Andes of New Granada and Peru, up to an elevation of 9,000 feet. A perennial herb, the tubers of which are edible; they are of about the size of hens' eggs. Can also be propagated from cuttings, and will endure some frost (Watson). A mean temperature of about 50° F. is favorable for the production of tubers of this plant (Vilmorin). Shablee found the tubers in a dried state to contain 3 per cent. fat, 4 per cent. gum, 19 per cent. grape-sugar, 33 per cent. starch, 12 per cent. albumen.

Ulmus alata, Michaux.

The Whahoo-Elm of North-America, extending to Newfoundland and Texas. Of quick growth. Height of tree reaching about 40 feet. Wood fine-grained, heavier and stronger than that of the White Elm, of a dull-red color, unwedgeable, used by wheelwrights, but like that of U. Americana not equal to the wood of the European elm.

Ulmus Americana, Linné.*

The White Elm of Eastern North-America, also called Rock- or Swamp-Elm. A tree of longevity, fond of moist river-banks, becoming fully a hundred feet high; trunk to 60 feet and as much as 5 feet in diameter. The tree is found hardy in Norway at least to lat. 59° 55'. Manning mentions that trees have been known to attain a circumference of 27 feet at 3 feet from the ground, and of 13 feet where the branches burst forth. It is highly prized for street-planting in North-America. Can be propagated from suckers like the European elm, irrespective of multiplication from cuttings or seeds. Almost indifferent to soil. The timber is light, used for wheelwrights' work, for tubes, water-pipes; bears driving bolts well (Robb); it is durable, if either kept quite dry or permanently submerged in water. U. floridana (Chapman) is a variety. Rate of growth in Nebraska: stem-circumference, 63 inches in 24 years (Furnas) 2 feet above ground.

Ulmus campestris, Linné.*

The ordinary Elm, indigenous to Europe and temperate Asia, as far east as Japan. Several marked varieties, such as the Cork Elm and Wych-Elm, exist, also a weeping variety. The elm in attaining an age of several centuries becomes finally of enormous size. Sir Joseph Hooker records the height of a tree at 125 feet, with a stem-circumference of 50 feet. In Britain it has been occasionally attacked by Scolytus destructor, and irrespective of this beetle, also by the Goat-moth, Cossus ligniperda, both boring into the stem. The wood is tough, hard, fine-grained and remarkably durable, if constantly under water. Next to the yew it is the best of European woods, where great elacticity is required, as for archery-bows. It is also used for keels, blocks, wheels, piles, pumps, gun- and railway-carriages, gunwales, various tools and implements.   The Wych-Elm (U. montana, Withering) grows still further north than the Cork-Elm, in Norway to lat. 66° 59'; even in lat. 59° 45' Professor Schuebeler found a tree still over 100 feet high, with a stem 4 feet in diameter. The wood of the Wych-Elm is preferred for bending purposes (Eassie).   The bast is tough.   The average growth at Port Phillip proved 40 feet in 25 years.   De Candolle estimated a particular aged elm in France to be 335 years old then.

Ulmus crassifolia, Nuttall. .

The Evergreen Elm of Mexico, Arkansas and Texas. A tree fully 90 feet high and 2 feet in stem-diameter.

Ulmus fulva, Michaux.

The Slippery or Red Elm of Eastern North-America.   Reaching a height of about 60 feet.   Splendid for tree-planting.   There is a pendent-branched variety.   Wood red, tenacious, useful for wagon-hubs and wheels (Vasey).   Regarded as the best North-American wood for blocks of rigging, according to Simmonds. The leaves seem available as food for the silkmoth; the bark is employed in medicine. Rate of growth, little more than half that of the White Elm (Furnas).

Ulmus Mexicana, Planchon.

Cordilleras of South-Western North-America.   This elm attains a height of 60 feet or perhaps more.   Many of these elms are available as quick-growing avenue-trees for shade-lines.   .

Ulmus parvifolia, Jacquin.

The Evergreen Elm of China, Japan, Upper India and Burmah. A similar tree is found on the Himalayan mountains. Well eligible for big hedges also.

Ulmus pedunculata, Fougeraux. · (*U. ciliata*, Ehrhart.)

Europe and Asia, through the middle zone.   A fine avenue-tree.

Ulmus racemosa, Thomas. *

The Cork-Elm of North-America, also called Western Rock-Elm Wood as valuable as that of U. Americana, but much heavier; it is fine-grained and compact, tough, flexible, not liable to split, holds bolts better than most timber, and is extremely durable when constantly wet; deserves unqualified praise as a furniture-wood for hardness, strength, beauty and buff-reddish tint; largely also employed for piles, pumps, naves, tackle-blocks, keels, heavy agricultural implements, such as mowing and threshing machines, ploughs, gunwales (Robb, Sargent).   .

Ulmus Wallichiana, Planchon.

Himalayan Elm. In the mountains of India from 3,500 feet to 10,000 feet. A tree sometimes to 90 feet high, the stem attaining a girth of 24 feet. Bark very tough; foliage locally lopped off for cattle-fodder (Brandis).

Umbellularia Californica, Nuttall. (*Oreodaphne Californica,* Nees; *Tetranthera Californica,* Hooker and Arnott.)

Oregon and California, where it is called the "Mountain-Laurel" or "Bay-tree." Tree becoming 100 feet high; throughout pervaded by a somewhat camphoric odor. Wood most valuable for cabinet-work, also for the best of flooring; that of the root splendid for turnery; it is hard, close-grained, durable, easily worked, susceptible of high polish (Dr. Behr and Prof. Bolander).

Uniola gracilis, Michaux.

North-America. A perennial pasture-grass of considerable value, content with sandy soil, and liking the vicinity of the sea. Root creeping.

Uniola latifolia, Michaux.

North-America. This rather tall perennial grass forms large tufts, and affords valuable fodder; it is best adapted for shady woodlands (C. Mohr).

Uniola paniculata, Linné.

North-Eastern America. This tall maritime grass can be chosen on account of its creeping roots to bind rolling coast-sands.

Urena lobata, Linné.

Intra-tropic girdle around the globe. This perennial herb has recently been enumerated among plants with comparatively tenacious fibre; it can be reared far beyond the tropics. Some congeneric plants can similarly be utilized.

Urginea Scilla, Steinheil. (*Scilla maritima,* Linné.)

The medicinal Squill. Countries around the Mediterranean Sea, Canary-Islands. Already ordered by Charles the Great to be grown in the imperial gardens. This coast-plant needs no regular cultivation; but settlers living near the sea might encourage its dissemination, and thus obtain the bulbs as drugs from natural localities. Its peculiar bitter principle is called scillitin. The bulb contains 24 per cent. tannin. U. altissima (Baker) serves in South-Africa as squill.

Uvularia sessilifolia, Linné.

North-America, in forests. This pretty herb is mentioned as yielding a good substitute for asparagus.

Vaccinium alatum, Dombey. (*Thibaudia alata,* Dunal.)

Frigid regions of the Andes of Peru. A tall evergreen shrub, with pink berries of the size of a cherry. This highly ornamental plant could be grown in sub-alpine regions.

Vaccinium Arctostaphylos, Linné.

From Greece to the Caucasus. The leaves, dried and slightly heated, furnish the Broussa-tea, the material for a fairly palatable beverage (G. Maw).

Vaccinium bicolor, F. v. Mueller. (*Thibaudia bicolor*, Ruiz and Pavon.)

Cold zones of Peruvian Andes. A high evergreen bush, with red berries of about the size of a hazel-nut. All Thibaudias seem best to form a section in the genus Vaccinium, some species of the latter— for instance, Vaccinium Imrayi (Hooker) from Dominica—mediating the transit. The species of the section Thibaudia, as a rule, produce red berries of acidulous grateful taste. Many others may therefore deserve culture or naturalization in forest-ravines or on sub-alpine heights. They occur from Peru to Mexico, also in the West-Indies. One species, Vaccinium melliflorum (Thibaudia melliflora, Ruiz and Pavon), has its flowers particularly rich in honey-nectar.

Vaccinium caespitosum, Michaux.

Labrador, Canada and North-Eastern States of the American Union. A deciduous-leaved small bush, with bluish edible berries. V. ovali-folium (Smith) is an allied species.

Vaccinium Canadense, Kalm.*

From the Middle States of North-America northwards. A dwarf shrub in swampy ground of woodlands. Yields, like V. Pennsyl-vanicum, to which it is allied, edible blueberries or huckleberries. Mr. Marity calls the berries delicious, fetching a high price—up to 11 dollars a bushel, never lower than 5 dollars, in New York. One bush yields from a pint to a quart of berries. It thrives through all grades of soil and exposure. The berries are rather large and aro-matic; for cooking and preserves they locally take precedence to any other kind of berry; they are easily dried, and retain their full delicious flavor. The bush grows occasionally to a height of 15 feet.

Vaccinium corymbosum, Linné.

The Swamp-Blueberry or Blue Huckleberry. Canada and Eastern States of North-America. A good sized shrub, reaching a height of 15 feet, with deciduous foliage. Berries bluish-black, rather large, aromatic, of sweetish taste, ripening late in the season. The fre-quency of this bush in its native countries induces the anticipation, that it could readily be disseminated elsewhere in apt climes and soils.

Vaccinium erythrocarpum, Michaux. (*Oxycoccus erectus*, Pursh.)

Carolina and Virginia, on high mountains. An upright bush, a few feet in height, with deciduous leaves. The transparent scarlet berries, according to Pursh, are of excellent taste.

Vaccinium grandiflorum, Dombey. (*Ceratostemma grandiflorum*, Ruiz and Pavon).

Andes of Peru. A tall evergreen shrub. The berries of a pleasant acidulous taste.

Vaccinium humifusum, Graham.

North-Western America, particularly on the Rocky Mountains. Berries of this bush well flavored.

Vaccinium Leschenaultii, Wight. (*Agapetes arborea,* Dunal.)

Southern India, extending to Ceylon, at elevations from 4,000 to 8,000 feet. This evergreen species attains the size of a small tree, flowering and fruiting throughout the year. The fruits resemble cranberries.

Vaccinium leucanthum, Chamisso.

Mountains of Mexico. An arborescent species. The blackish berries are edible.

Vaccinium macrocarpon, Aiton.* (*Oxycoccus macrocarpus,* Persoon.)

The large Cranberry. From Canada to Virginia and Carolina, particularly in sandy and peaty bogs, and in cold mossy swamps, Hardy to Christiania. A trailing evergreen bush, with stems attaining a length of 3 feet. In sunny places more fruitful than in shady localities. It is this species, which has become so extensively cultivated in the eastern parts of the United States, where on moory land, often not otherwise to be utilized, enormous quantities of this fruit have been produced by regular culture at a highly profitable scale. The berries are of acid taste, pleasant aroma and the scarlet brightness of the British cranberry, but considerably larger. The plant is rooting also along its depressed stem, from which it can be readily multiplied.

Vaccinium meridionale, Swartz.

Jamaica, from the summits of the highest ranges down to the coffee-regions. It attains a height of 30 feet, and is evergreen. The small berries are of the taste and color of those of V. Vitis Idæa.

Vaccinium Mortinía, Bentham.

Mountains of Columhia. A shrub, several feet high. The fruits resemble those of V. Myrtillus, but are more acid. They come to the Quito-market under the name Mortina.

Vaccinium myrtilloides, Michaux.

Michigan, Canada, Newfoundland, Labrador. The large edible berries are called Bluets. This little bush is adapted for alpine country.

Vaccinium Myrtillus, Linné.*

The British Whortleberry or Bilberry. Throughout Europe, Northern and Middle Asia, remotest North-America, extending to the Californian Sierra Nevada; in heathy and turfy forest-land. In Norway it is found wild to lat. 71° 10′ (Schuebeler). A shrub, a few feet high or less, deciduous, erect, of great value for its copious supply of berries. They are, as well known, black with a bluish-grey hue, of exceedingly grateful taste and very wholesome. The naturalization of this plant on alpine ranges and in cooler woodlands would prove a boon. The berries can be utilized also for their dye. The whole bush contains quina-acid.

Vaccinium ovalifolium, Smith.

North-Western America from Mendocino to Oregon. This shrub bears large edible berries (Dr. Gibbons).

Vaccinium ovatum, Pursh.

Common throughout California, also in British Columbia, at altitudes from 1,000 to 2,000 feet, attaining a height of about 8 feet. It bears its fruit in densely crowded racemes, the dark-blue but small berries being of good flav... This species would doubtlessly form a valuable accession among cultivated fruits (Gibbons).

Vaccinium Oxycoccus, Linné. (*Oxycoccus palustris*, Persoon.)

The British Cranberry. Throughout Europe, Northern and Middle Asia, North-America; on turf-moss in moory heaths. A creeping evergreen shrub of particular neatness. The berries give a most agreeable preserve, and are of antiscorbutic value. This species is particularly eligible for the spongy, mossy bogs of alpine mountains. Indigenous in Norway northward to lat. 70° 45′.

Vaccinium parvifolium, Smith.

North-Western America, from Mendocino to Sitka. A tall shrub. The berries are excellent for preserves.

Vaccinium penduliflorum, Gaudichaud.

Hawaia, where it is called the "Ohelo." The acidulous berries of this bush are edible.

Vaccinium Pennsylvanicum, Lamarck.* (*V. angustifolium*, Aiton.)

The early Blueberry or Blue Huckleberry. North-America, on dry woody hills. A dwarf bush with deciduous foliage, producing fruit in abundance and early in the season. The berries are large, bluish-black and of sweet taste. V. Canadense (Kalm), according to Dr. Asa Gray, is closely allied.

Vaccinium præstans, Rudolphi.

Kamtschatka. A minute plant, but with large delicious fruits. It might perhaps easily be disseminated on any alpine mountains.

Vaccinium uliginosum, Linné.

British Bog-Bilberry. Europe, Northern and Middle Asia, North-America. A deciduous bush, with blackish berries, similar to those of V. Myrtillus, but hardly of equal excellence. Wild to lat. 78° north in Greenland.

Vaccinium vacillans, Solander.

Eastern North-America, in sandy forest-lands. A deciduous small bush, with its blue berries coming later into season than V. Pennsylvanicum (Dr. A. Gray).

Vaccinium Vitis Idæa, Linné.

Europe, Northern and Middle Asia, North-America. Extends in Greenland to 76° N. L. (Nathorst); therefore fit for subglacier-regions. A dwarf shrub with evergreen leaves. The purplish-red berries are sought for jellies and other preserves. It is as yet impossible to say, how many other species of Vaccinium produce good-sized and well-flavored fruits. The genus ranges also in tropical species from Continental Asia to the Indian Archipelagus, and has a wide extension likewise in South-America, occupying in hot countries higher mountain-regions; but few reliable notes on the tropical species are extant as far as their fruits are concerned.

Vahea florida, F. v. Mueller. (*Landolphia florida,* Bentham.)

West-Africa, up to 2,500 feet. This may prove hardy in mild extra-tropic regions. Dr. Welwitsch describes the Aboh-fruit of this species as sweet and acidulous, but was not less gratified with the beauty and marvellous abundance of its large snow-white and jasmin-scented flowers. V. florida also yields caoutchouc, like V. Heudelotii (*Landolphia Heudelotii,* D.C.) from the Senegal-regions. The genus Vahea was fully established by Lamarck as early as 1791. The excellent work on the caoutchoucs of commerce, by James Collins, may be consulted as regards the sources of various kinds of India-rubbers. Prof. Wiesner (Rohstoffe des Pflanzenreichs, 1873) enumerates at p. 154-156 the various plants then known to yield caoutchouc, giving also a chemical account of these substances.

Vahea Owariensis, F. v. Mueller. (*Landolphia Owariensis,* Beauvois.)

Tropical West-Africa, but ascending to the highlands of Angola, according to Dr. Welwitsch. This climber, with several other Vaheas, yields the West-African caoutchouc; others furnish the Madagascar-sort, particularly V. gummifera (Lamarck), now cultivated also in India. Prof. Wiesner of Vienna enumerates 47 species of various genera, which yield either rubber, gutta percha, or balata. It is said, that the addition of ammonia to the sap improves the rubber. V. Owariensis produces edible fruits as large as middle-sized oranges, with sweet and slightly acid pulp.

Valeriana Celtica, Linné.

Alps of Europe; hardy at Christiania. The "Speik." The root of this perennial herb is particularly aromatic.

Valeriana edulis, Nuttall.

North-Western America, from Oregon to the Rocky Mountains. The thick spindle-shaped root of this herb affords food to the natives of that part of the globe. When baked, the root proves agreeable and wholesome. When we consider the wild state of the plants, from which many of our important root-crops arose, this Valeriana and several other plants, suggestively mentioned in these pages, may well be admitted for trial-culture.

Valeriana officinalis, Linné.

Europe, Northern and Middle Asia, in swampy grass-land, with a predilection for forests and river-banks. In Norway it extends northward to lat. 70° 22′ (Prof. Schuebeler). This perennial herb would do particularly well on higher mountains. It is the only one among numerous congeners of Europe, Asia and America, which is drawn to a considerable extent into medicinal use. The root and herb contain valerianic acid and a peculiar tannic acid; the root furnishes also an essential oil, which again resolves itself into valerol (70 per cent), valeren, barneol and valerianic acid. Concerning these see Husemann and Hilger's Pflanzenstoffe 1884. The order of Valerianeæ is not represented by any native plant in Australia.

Valerianella olitoria, Moench.

Lamb's Lettuce. Europe, North-Africa, Northern and Middle Asia. Northward to lat. 59° 16′ in Norway. A fair and early salad-plant. It is an annual, and has several congeners in Europe and Asia. V. eriocarpa (Desvaux) is similarly utilitarian. With still more force this may be said of the co-ordinal Fedia cornucopiæ (Gaertner). The seeds will keep about five years (Vilmorin).

Vangueria infausta, Burchell.

From tropical Africa to Natal and Caffraria. The fruit of this shrub or small tree is medlar-like, but superior in taste. Worth test-cultivation with a view of improving the fruit. V. edulis (Vahl) from the warmer regions of Africa and from Madagascar proved hardy as far south as Port Jackson, and yields esculent rather small fruits.

Veratrum album, Linné.

Europe, Northern and Middle Asia, extending eastward to Japan. Hardy still at Christiania. It delights particularly in sub-alpine localities. The root furnishes veratrin, jervin and sabadillic acid. The root is used in medicine particularly for external applications.

Veratrum viride, Aiton.

Canada and United States of North-America. A near relative of the former plant. Professor Schuebeler found it hardy in Norway to lat. 71°. Its root has recently come into medicinal use; especially as an arterial sedative (Porcher).

Verbascum Thapsus, Linné.

The Mullein. A biennial herb of some use in medicine, but adapted also for scenic cultural effects.

Veronica Virginica, Linné.

Eastern North-America. A perennial herb, which for medicinal use furnishes the "Culver's-root," from which again the Leptandrin as a chologogue is prepared. The showy shrubby species, such as V. speciosa (R. Cunningham) of New Zealand and their hybrids can easily be multiplied from cuttings in the open air; they are grateful in culture, and afford good material for table-bouquets.

Viburnum Tinus, Linné.

The Lauristine. Countries around the Mediterranean Sea. An evergreen shrub, one of the earliest flowering of the season; well adapted for ornamental hedges. Hardy in the south of England. An excellent plant as a standard of comparison for floral calendars.

Vicia Cracca, Linné.

Europe, North-Africa, Northern and Middle Asia, North-America; in Norway it extends to lat. 71° 10'. Perennial. Recommendable for naturalization as a fodder-plant in sylvan and sub-alpine lands. It yields in shade a return three times larger than in open places (Langethal). Lauded as most nourishing to cattle by Dr. Plot of Staffordshire. The cognate V. Cassubica and V. biennis, Linné, serve also for field-culture.

Vicia Ervilia, Willdenow. (*Ervum Ervilia,* L.)

South-Europe, North-Africa, South-Western Asia. An annual. herb, praised as a valuable fodder-plant particularly fit for dry calcareous soil. Cultivated already at Troja (Virchow, Wittmack).

Vicia Faba, Linné.

The Straight Bean, called also Common Field-Bean. Orient, particularly on the Caspian Sea. Professor Schuebeler found it to bear seeds still in lat. 67° 17'. Was cultivated already at Troja (Virchow, Wittmack). This productive annual herb not only affords its seeds for table-use, as Broad-Bean and Windsor-Bean, but provides also a particularly fattening stable-food, in its common form the Horse-Bean. The seeds contain about 33 per cent. starch. V. Narbonensis, L., from South-Europe and South-Western Asia, is preferable for the table, because its seeds contain less bitter principle, though they are smaller. They will retain their vitality for six years or more.

Vicia peregrina, Linné.

South-Europe. Annual. In Italy preferred to the ordinary Tare for sandy soil; recommends itself also for its close growth.

Vicia sativa, Linné.* (*V. angustifolia,* Roth.)

The ordinary Vetch or Tare. Europe, North-Africa, Western and Northern Asia, extending to Japan. According to Professor Schuebeler it will grow in Norway to lat. 70°; it perfected its seeds there still in 63° 26'. One of the best fodder-plants, but only of one or two years' duration. Praised particularly for dairy-cattle by G. Don. This plant according to Middleton has yielded as large a crop as 12 tons on an acre, cut green. Horses thrive remarkably on it. Important also for green manure and as a companion of clovers. The allied V. cordata, Wulfen, and V. globosa, Retzius, are similarly cultivated in Italy (Langethal). Many of the other European and Asiatic species of Vicia are deserving of our attention.

Vicia sepium, Rivinus.

Europe, Western and Northern Asia. A perennial Vetch, enduring an alpine clime; indigenous in Norway northward to lat. 69° 40'. It might with advantage be naturalized in forests and on high mountains, but it can also readily be subjected to field-culture, the yield being large and nutritious in regions with humid air, though the soil might be poor. This vetch can be kept continually on the same field for about fifteen years (Langethal). V. Pannonica, Jacquin, is an allied but annual species.

Vicia Sitchensis, Bongard. (*V. gigantea,* Hooker.)

From California to Sitka. Dr. Asa Gray remarks, that the young seeds of this tall vetch are eatable like green peas.

Vicia sylvatica, Linné.

The Wood-Vetch. Europe, Northern Asia. Indigenous in Norway to lat. 67° 56'. Perennial. Recommendable to culturists settling in new forest-land; available also for growth in sub-alpine copses. Pasture-animals have a predilection for this vetch; its yield is large and very nourishing. In limestone-soil of forests V. pisiformis and V. dumetorum, Linné, can best be selected for introduction.

Vicia tetrasperma, Koch. (*Ervum tetraspermum,* Linné.)

The Lentil-Tare. Europe, Western Asia, North-Africa. Annual. According to Langethal this species is preferable to the ordinary tare for sandy soil. It is also less hard as fodder and very palatable. Lime in the sand enlarges the yield. V. monantha and V. hirsuta, Koch, serve nearly as well.

Vigna lanceolata, Bentham.

Tropical and sub-tropical Australia. Mr. P. O'Shanesy observes, that this twiner produces, along with the ordinary cylindrical pods, others underground from buried flowers, and these somewhat resemble the fruit of Arachis. The plant might be rendered perhaps available for culinary purposes.

Vigna Sinensis, Endlicher.* (*Dolichos Sinensis,* Rumph.)

Tropical Asia and Africa. The cultivation of this twining annual pulse-herb extends to Southern Europe, the United States and many other countries with a temperate clime. One of the many vernaculars of this plant is the " Cow-Pea." The pods are remarkable for their great length, and used like French-beans, dry as well as preferentially also green. This plant bears plentifully even in seasons of severe drought in Central Australia (Rev. H. Kempe). V. Catjang, V. unguiculata, V. sesquipedalis and V. melanophthalma are varieties of this species. In fair soil the produce is about forty fold. The Laubich-grains of Egypt are from a variety of this species (Sir J. Hooker). This Vigna is satisfied with comparatively poor soil and stands also dryness well.

Villebrunia integrifolia, Gaudichaud.

India, ascending the Himalayan mountains to 5,000 feet. A small tree, allied to the Ramie-plant (Boehmeria nivea). Mr. C. B. Clarke regards the fibre as one of the strongest available in India, it being used for bow-strings. Other Villebrunias—for instance, V. frutescens, and also some species of Debregeasia, particularly D. velutina—likewise deserve regular culture, for the sake of their fibre. Moist forest-tracts seem particularly adapted for these plants, because V. integrifolia grows in Sikkim at an elevation, where the rainfall ranges from 100 to 200 inches. This fibre is much more easily separable than that of Maoutia Puya, according to Dr. G. King's observations.

Viola odorata, Renealm.

The Violet. Middle and Southern Europe, North-Africa, Western and Middle Asia. Passingly alluded to here, as this modest though lovely plant should be extensively naturalized in forest-glens; it furnishes its delicate scent by enfleurage for various compositions of perfumery. It flowers in the southern regions of Australia through the whole of our almost six-monthly spring. The annual produce of flowers from violets obtained at Nice and Cannes alone amounts to about 50,000 lbs. Violets are there often grown as an extra-crop under lemon- and orange-trees; the kind chiefly cultivated for perfumery is the "Double Parma" (Piesse). Varieties specially cultivated for bouquets, are: Lee's Victoria, the Czar and the Neapolitan and Semperflorens. Their culture proves quite remunerative.

Vitis acetosa, F. v. Mueller.

Carpentaria and Arnhem's Land. Stems rather herbaceous than shrubby, erect. The whole plant is pervaded with acidity, and proved valuable in cases of scurvy. The berries are edible. This species, if planted in countries with a mild temperate clime, would probably spring afresh from the roots annually.

Vitis æstivalis, Michaux.[1]

The Summer-Grape of the Middle and Eastern States of North-America. Flowers fragrant. The berries are deep blue, of pleasant taste, and ripen late in the season, but are generally rather small and in some kinds somewhat sour. Among the varieties derived from this species, the Jacques, Herbemont, Norton's Virginia, Elsinburg, Cunningham, Rulander and Pauline are the best known; all resist the attacks of the Phylloxera vastatrix, as has been fully demonstrated by experience in the United States as well as in the south of France. Several of these give an excellent produce; Jacques and Norton's Virginia gained a first prize in competition with the wines of Southern France, at an exhibition held in Montpellier. The Jacques-variety especially is much esteemed in the Provence for its resistance to Phylloxera, also for its luxuriant growth, great fertility and excellent wine of rich color. The whole group of Vitis æstivalis is however rather difficult to propagate, and is for this reason not so valuable for stock of the European vine as V. riparia. As these vines are of larger growth than V. vinifera, they should be planted further apart; a distance of 8 or 10 feet, and 6 feet between the rows is considered the most suitable. In Europe the flowering season is at the end of June, about a fortnight later than that of the European vine. The following method has been recommended for propagating these American vines in districts infested by the Phylloxera. Cut the best old stocks of European vines down to six or eight inches underground, graft upon them American scions having at most three eyes, fasten with clay and cover the graft with soil, preferably with sand. To obtain then a number of American vines, cut off any European shoots which may have sprouted, leave all the best American shoots, make furrows about four inches deep, radiating from the stock, in which layer the shoots, fixing them down with pegs, and cover them with sand. It is to be observed, that in very poor dry soil, where the European vine still yields a fair crop, American vines do not succeed (Planchon, Vignes Américaines).

Vitis Baudiniana, F. v. Mueller. (*Cissus Antarctica,* Ventenat.)

East-Australia. With V. hypoglauca the most southern of all species of grapes, none extending to New Zealand. It is evergreen, and a vigorous plant for bowers, but suffers even from slight frosts. The berries are freely produced and edible, though not large.

Vitis candicans, Engelmann.

The Mustang-Grape of Texas, extending to Florida. Suited for warm dry climes. Climbs to a maximum height of 80 feet, and gets finally a stem of nearly 2 feet diameter. Bears abundantly. Mr. Buckley obtained from a plant 8 years old, 54 gallons of juice; but the wine obtained is inferior to that from some other American species. The variety Solonis is derived from crosses between V. riparia, V. rupestris and V. candicans (Prof. Millardet).

Vitis cinerea, Engelmann.

Valley of the middle and lower Mississippi. Of near affinity to V. æstivalis. A large vine. Resists Phylloxera. Some hybrids from this serve well for stock to graft on (Millardet).

Vitis cordifolia, Michaux.*

The Winter-Grape or Frost-Grape. From Canada to Florida. A very large deciduous vine. The scent of the flowers reminds of Reseda. The berries are small, either blackish or amber-colored and very acid. They can be used for preserves, and are only fully matured when touched by frosts. A succession of seedlings may give us a superior vine, with the recommendation of particular hardiness; this species developes. however also well in rather warm climes and bears also considerable dryness. Resists the attacks of Phylloxera very well, and seems also safe against mildew (Professor Millardet).

Vitis hypoglauca, F. v. Mueller.

East-Australia, as far south as Gippsland. An evergreen climber of enormous length, forming a very stout stem in age. The black berries attain the size of small cherries. This species also may perhaps be vastly changed in its fruit by continued culture. Bears slight frost; but it is best in cool climes to keep seedlings for two or three years under shelter, so that sufficient increment and induration of the woody stem takes place for its resisting subsequently some frost, a remark applying to many other kinds of plants to be acclimatised.

Vitis Indica, Linné.

On the mountains of various parts of India, ascending to an altitude of 3,000 feet in Ceylon. The small berries are edible. The plant should be subjected to horticultural experiments. This is an apt opportunity, to draw attention to some of the various Indian species of Vitis with large edible berries—for instance, V. lævigata (Blume), V. thyrsiflora (Miquel), V. mutabilis (Blume), V. Blumeana (Steudel), all from the mountains of Java, and all producing berries as large as cherries, those of V. Blumeana being particularly sweet. Further may here be inserted V. imperialis (Miquel), from Borneo, V. auriculata and V. elongata (Wallich), the latter two from the mountainous mainland of Coromandel, and all producing very large juicy berries; even in the jungle-wilderness. V. quadrangularis (Linné) stretches from Arabia to India and Central Africa, and has also edible fruits. Many such plants may be far more eligible for grape-culture in hot wet climates than the ordinary vine. About 250 species of Vitis are already known, mostly from intra-tropical latitudes, and mostly evergreen; but in regard to their elevation above the ocean and to the nature of their fruits we are almost utterly without data. An herbaceous species of a tuberous vine, occurring in Soudan, is recommended by Mr. Lécard; another tuberous species is noted by Mr. J. B. Martin as wild in Cochin-China, the herbaceous stems being reproduced annually from the roots; both kinds bear excellent grapes; the species from Cochin-China forms long shoots, sometimes to a length of 60 and exceptionally 150 feet, bearing grapes all along the branches. Occasionally more than a cwt. of grapes are obtained from one plant, according to General Haldeman. It would be a grand acquisition to tropical countries; its ripe grapes are produced successively through fully three months; the berries are very large.

Vitis Labrusca. Linné.

The Isabella-Grape. North-America, from Canada to Texas and Florida, also in Japan. The Schuylkill-Grape is derived from this species. A pale-fruited variety furnishes the Bland's Grape; another yields the American Alexander-Grape (Torrey and Gray). The Concord, Catawba, Isabella, Martha, Ives-Seedling, Hartford-Prolific and a number of other less known varieties are also derived from this species. Among these the Concord takes the first rank as well for wine as for dessert-grapes in the Eastern United States, where it is cultivated more than all the other varieties put together, although it has a strong so-called foxy taste. It is not quite proof against the attacks of the Phylloxera vastatrix, but suffers less than most other varieties of this species (Planchon, Vignes Américaines). Many good and fertile crosses between V. Labrusca and V. vinifera occur in North-American cultivation; the Delaware-Grape is a hybrid from V. Labrusca according to Bush and Meisner, and has in its turn given rise to many òther good crosses. The berries of V. Labrusca are large among American kinds, and are of pleasant taste. Flowers fragrant. It is the only species which thrives well and bears largely in the clime of Brisbane, according to Dr. Bancroft. This and the other hardy North-American vines seem never to be attacked by the Oïdium-disease. Dr. Regel unites the South-Asiatic V. lanata (Roxburgh) with this.

Vitis riparia, Michaux.* (*V. cordifolia* var. *riparia*, A. Gray.)

From the Northern and Central United States to the Rocky Mountains of Colorado. To this species belong the Clinton, Franklin, Taylor and some other varieties, probably also Vitis Solonis, which seem more particularly destined to revive viticulture in Southern France and other countries, where the Phylloxera vastatrix has annihilated such a vast extent of vineyards. They serve as grafting stock for the European vine, the majority of them showing a sufficient if not a complete resistance to this pest, while they are for the most part not difficult to propagate. The experiments hitherto made in the Provence and elsewhere have given good results, and the produce of the European vine on American stock has been found as good as if grown on its own root. Professor Planchon places the varieties in the following order of merit: Vitis Solonis, Clinton-Vialla or Franklin, wild Vitis riparia, Taylor, Clinton. The York-Madeira, which may be a hybrid between V. riparia and V. Labrusca, is by some growers placed next to Vitis Solonis and answers well for grafting. The seedlings of V. Solonis retain the typical characteristics of the parent-plant —which the other varieties do not. To raise vines from seeds, the pips may be taken either before or after fermentation of the grape; the essential point is, not to let them get dry; they should be kept in a cool place and mixed with sand, to prevent mould. For transmission to great distances they should be sent dried in the peel and pulp to ensure the preservation of their vitality. Several French cultivators recommend grafting " by approach." For this purpose an American and an European vine are planted side by side; early in spring, when the shoots are about the size of a small goose-quill, two from the different stocks are brought together and in the most convenient place a slice is taken out of the bark and the outer portion of the wood of each, about half an inch in length, care being taken that the two surfaces exactly fit each other; they have only to be tied together, the sap which is then at the height of its flow soon closing up the wound; the American shoot is pinched off when it has made 3 or 4 leaves, the following winter the root of V. vinifera is cut off. Phylloxera-galls are frequently found on the leaves of V. riparia as well as of V. æstivalis, but the roots are not so often attacked; if the latter happens, the wounds inflicted by the insect are superficial and soon heal up (Planchon,

Vignes Américaines). Professor A. Millardet of Bordeaux has in
1885 issued an excellent illustrated work on the principal Varieties
and Species of Vines of American Origin resisting the Phylloxera.
At present in the Department Herault already 170,000 acres are
planted with American vines. Unfortunately the mildew, which has
attacked so much the European vine is equally hurtful to the American species, unless V. rubra and V. cordifolia be proof (Planchon).
The Phylloxera has now found its way to Algeria, Smyrna, and
New South Wales, so that all the five great parts of the globe are
invaded.

Vitis rubra, Michaux.

The Cat-Vine. Illinois and adjoining country, on river-banks.
May climb to half a hundred feet height. Proof against Phylloxera
and mildew. Promises to become of value for hybridisation
(Millardet).

Vitis rupestris, Scheele.

The Sand-Grape or Sugar-Grape. From the Missouri to Texas.
Likes naturally gravelly borders of torrents, along which elsewhere
this species might be naturalized. Hybridises easily; also well
adapted for grafting on it the European vine (Prof. Millardet).

Vitis Schimperiana, Hochstetter.

From Abyssinia to Guinea. This vine may perhaps become
valuable, with many other Central African kinds, for tropical culture,
and may show itself hardy also in extra-tropical countries. Barter
compares the edible berries to clusters of Frontignac-Grape.

Vitis vinifera, C. Bauhin.*

The Grape-Vine. Greece, Turkey, Persia, Tartary; probably also
in the Himalayas. One of the most thankful of plants over a wide
cultural range. Praised already by Homer; cultivated in Italy as
early as the bronze-age, in Armenia since Noah's time. This is not
the place, to discuss at length the great industrial questions concerning this highly important plant, even had these not already
engaged the attention of a large number of colonists for many years.
A large territory of West- and South-Australia, also of Victoria
and New South Wales stretches essentially through the Vine-zone,
and thus most kinds of vine can be produced here, either on the lowlands or the less elevated mountains in various climatic regions and in
different geological formations. The best grapes with us are produced
mainly between the 30th and 38th degree of latitude. Cultivation
for wine advances on the Rhine to 50° north; on trellis it extends to
52° or 53° N., in Norway even to 61° 17'. In Italy vines are often
trained high up over maples, willows and elms, since Pliny's time;
in the Caucasus they sometimes grow on Pterocarya. Vines attain
an age of centuries and get stems 3 feet in diameter. The doors of
the dome of the Ravenna-Cathedral are of vine-wood (Soderim).
Tozetti saw a vine with branches extending diametrically, as a whole,.

over 3,000 feet at Montebamboli. Rezier notes a plant, bearing about 4,000 bunches of grapes annually at Besançon (Regel). A single plant of "Black Hamburg" under glass at Rockhampton, England, bore annually 900–1,000 lbs. of grapes (Davis). A vine of enormous dimensions at Hampton-Court has also gained wide celebrity. In Italy the establishing of vine-plantations on ordinary culture-land is regarded as enhancing the value of the latter four or five fold, and elsewhere often even more (whereas cereal-land is apt to deteriorate), provided that vine-diseases can be kept off. The imports of wine into the United Kingdom in 1884 amounted to about 15 million gallons, worth more than £5,000,000, of which only a very small proportion came from British colonies.

The Corinthian variety, producing the currants of commerce, also thrives well in some districts of extra-tropic Australia, where with raisins its fruit may become a staple-article of export beyond home-consumption. The Sultana-variety is not to be much pruned; the bunches.when gathered are dipped in an alkaline liquid obtained from wood-ashes, to which a little olive oil is added, to expedite drying, which is effected in about a week (G. Maw). The produce of Sultana-raisins fluctuates from 7 to 30 cwt. per acre. The plant is best reared on limestone-formations. In Greece the average-yield of ordinary raisins is about 2,000 lbs. per acre (Simmonds). Great Britain imported in 1884 about 60,000 tons of currants and 25,000 tons of raisins, nearly all for home-consumption. Dr. W. Hamm, of Vienna, has issued a Vine-map of Europe, indicating the distribution of the different varieties and the principal sources of the various sorts of wine. The writer would now merely add, that the preservation of the grapes in a fresh state, according to M. Charmeux's method, and the sundry modes of effecting the transit of ripe grapes to long distances, ought to be turned to industrial advantage. The pigment of the dark wine-berries is known as racemic acid. The juice contains along with tartaric acid also grape-acid. All these chemically defined substances have uses of their own in art and science. It might be worthy of a trial, how far the Grape-vine can be grafted on such other species, not American, of the extensive genus Vitis, as may not be attacked by the destructive Pemphigus or Phylloxera. Irrespective of sulphur, borax has also latterly been recommended against the Oïdium-disease. Professor Monnier, of Geneva, has introduced the very expansive sulphurous anhydrous acid gas against the Phylloxera. The cultivation of insecticidal herbs to check the ingress of Phylloxera should be more extensively tried, as such plants might ward off the insect at all events in its wingless state. Dr. Herman Behr suggests for the mitigation of this plague the ignition of wood near vineyards, when the insect is on its wings, as all such insects seek fires, and succumb in them largely, the attraction to the fiery light being greatest when the sky is overcast, or when the nights are without moonlight. Mr. Leacock, in Madeira, applies a coating of a sticky solution of resin in oil of turpentine advantageously to the roots of Vines affected by Phylloxera. None of the remedies hitherto

suggested however seem to have proved really effective, or they are not
of sufficiently easy and cheap application, and the Phylloxera-pest is
still rapidly on the increase in Europe; according to the latest
accounts one-third of all the vineyards of France are affected, and the
disease is also spreading in Italy and Spain. Inundation to the depth
of a few inches for about a month, where that is practicable, completely suffocates the Phylloxera, but renders the vine for a while
much less productive. In sandy soil this dreadful insect is retarded
in its development, action and progress. Bisulphide of carbon has
proved an efficient remedy; this expansive fluid is introduced into
the soil by a peculiar injector, or through porous subtances (wood,
earth), saturated with the bisulphide, the cost of this operation being,
in France, £3 10s.–£4 per acre annually. (Planchou, David, Marion,
Robart. See also translations by K. Staiger and A. K. Findlay.)
Dressing with sulpho-carbonate of potassium is still more efficacious
and less dangerous, but involves an annual expenditure of about £8
per acre (W. T. Dyer). Sand might be dug in at the roots of vines,
which may be in imminent danger of becoming a prey of Phylloxera.
Recently it has been insisted on by Mr. Bauer, of San Francisco, that
it would be best to put minute quantities of mercury, triturated with
chalk, near the roots of vines affected with Phylloxera, a measure
which deserves every consideration, as the particles of quicksilver
would only very gradually become dissolved. and long remain
stationary; and we know that metal in its solutions to be the most
powerful antiseptic, a dilution of one part of bichloride of mercury in
5,000 parts of water proving strong enough for surgical purposes. It
is reported from California, likewise, that there cereals seem also
attacked by Phylloxera. Little's soluble Phenyl is among the remedies, recommended by the chief viticultural officer in San Francisco
against the insect. Wetmore urges the use of sulphate of iron against
the mildew of vines. Travellers through new temperate regions might
include carefully kept vine-seeds among those to be disseminated.

Vitis vulpina, Linné.* (*Vitis rotundifolia,* Michaux.)

The Muscadine- or Fox-Grape. South-Eastern States of North-
America; extends also to Japan, Manchuria and the Himalayas.
This species includes as varieties the Bullace, the Mustang, the Bull-
ate-Grape and both kinds of the Scuppernongs. The berries are of
pleasant taste, but in some instances of strong flavor; they are the
largest among Amerian grapes. All the varieties derived from Vitis
vulpina are perfectly proof against the attacks of Phylloxera vastatrix. Although in infected districts a few insects may sometimes be
found on it, yet no ill effects are ever manifested. The flowering
season is about six weeks later than that of the European vine. This
species is not easily propagated from cuttings, but must be raised
from seeds or by layering. As this is a very large species, the vines
should be planted 20 to 30 feet apart, and grown in bower-fashion or
on trellises. It does not bear pruning, but some of the superfluous
wood may be trimmed off during summer. It is only suited for mild

climates; even in the latitude of Washington it succumbs to the cold, being thus not hardy like most other North-American species in Northern Germany. The bunches contain generally only from 4 to 10 large berries, but are produced abundantly all over the plant. The berries are of a brownish-yellow color with a bronze-tinge when ripe; the peel is coriaceous, the juice vinous, of delicate perfume, resembling muscat. The grapes do not ripen together, but successively during about a month, and drop off the stalk when ripe. To gather them a sheet is generally spread under the vine and the latter shaken. The Muscadine vine grows sometimes to an extraordinary size, rising to the top of the tallest trees. A Scuppernong, planted on the island of Roanoke, covers an area of more than 40 acres; another is mentioned by M. Labinux as extending still further. Vitis vulpina is not suited for stock, on which to graft the European vine (Planchon). Hybrids of this species with the European and with other American vines are but little fertile, but by further crossing the first hybrids can furnish fertile sorts, whereas crosses between Vitis vinifera, V. æstivalis, V. cordifolia, V. riparia and V. Labrusca in any way are hardly less fertile than the original species (Bush and Meisner). V. candicans, the Mustang-grape of Texas, is recommended by Professor Millardet for grape-culture. Dr. Regel refers to V. vulpina also V. parviflora, Roxburgh. The important memoirs " Les Vignes Américaines," published by Planchon since 1875, should be consulted in reference to American vines.

Voandzeia subterranea, Thouars.

Madagascar and various parts of Africa, as far south as Natal. This Earth-Pea is annual, and pushes its pods underground for maturation in the manner of Arachis hypogæa. The pods are edible and consumed in some tropical countries.

Wallichia caryotoides, Roxburgh. (*Harina caryotoides,* Hamilton.)

India, up to 4,000 feet elevation (Kurz). A dwarf tufted palm, eligible for scenic group-planting.

Wallichia densiflora, Martius. (*W. oblongifolia,* Griffith.)

Himalaya as far as 27° north. There one of the hardiest of all palms. It is not a tall one, yet a graceful and useful object for cultural industries.

Washingtonia filifera, H. Wendland. (*Pritchardia filifera,* Linden.)

From South-California to Arizona and Colorado. One of the most northern and therefore most hardy of American palms. This species attains a height of about 50 feet. In gardens it passes often under the name Brahea filamentosa. W. robusta (H. Wendland) occurs on the Sacramento-River, and will endure long continued drought as well as a few degrees of frost (Prof. Naudin).

Wettinia augusta, Poeppig.

Peru, on mountains several thousand feet high. This palm is therefore likely to endure mild, temperate climes.

Wettinia Maynensis, Spruce.

Cordilleras of Peru. Like the foregoing, it attains a height of about 40 feet, and advances to elevations of 3,000 to 4,000 feet.

Before finally parting from the American palms, it may be appropriate to allude briefly to some of the hardier kinds, which were left unnoticed in the course of this compilation. From Dr. Spruce's important essay on the Palms of the Amazon-River may be learned that, besides other species as yet imperfectly known from the sources of this great river, the following kinds are comparatively hardy, and hence might find places for cultivation or even naturalization within the limits of extra-tropical countries: Geonoma undata (Klotzsch), Iriartea deltoidea (Ruiz and Pavon), Iriartea ventricosa (Martius), which latter rises in its magnificence to fully 100 feet; Iriartea exorrhiza (Martius); this, with the two other Iriarteas, ascends the Andes to 5,000 feet. Oenocarpus multicaulis (Spruce) ascends to 4,000 feet; from six to ten stems are developed from the same root, each from 15 to 30 feet high. Of Euterpe two species occur in a zone between 3,000 and 6,000 feet. Phytelephas microcarpa (Ruiz and Pavon) ascends to 3,000 feet on the eastern slope of the Peru-Andes. Phytelephas macrocarpa, R. & P., grows also on the eastern side of the Andes, up to 4,000 feet; it is this superb species, which yields by its seeds much of the vegetable ivory. Phytelephas æquatorialis, Spruce, occurs on the western slope of the Peruvian Andes, up to 5,000 feet; this palm is one of the grandest objects in the whole vegetable creation, its leaves attaining a length of 30 feet ! The stem rises to 20 feet. Palm-ivory is also largely secured from this plant. Though equinoctial, it lives only in the milder regions of the mountains. Carludovica palmata (R. & P.), on the eastern side of the Andes of Peru and Ecuador ascends to about 4,000 feet ; the fan-shaped leaves from cultivated specimens furnish the main-material for the best Panama-hats. Count de Castelnau saw many palms on the borders of Paraguay during his great Brazilian expedition. Most of these, together with the palms of Uruguay and the wide Argentine territory, would probably prove adapted for acclimation in mild temperate latitudes ; but hitherto the limited access to those countries has left us largely unacquainted with their vegetable treasures also in this direction. Von Martius demonstrated so early as 1850 the occurrence of the following palms in extra-tropical South-America : Juania australis (H. Wendland), on high mountains in Juan Fernandez, at 30° south latitude ; Jubœa spectabilis (Humboldt), in Chili; at 40° south latitude ; Trithrinax Brasiliana (Mart.), at 31° south latitude ; Copernicia cerifera, (Mart.), at 29° south latitude ; Acrocomia Totai (Mart.), at 28° south latitude ; Cocos Australis (Mart.), at 34° south latitude; Cocos Yatai (Mart.), at 32° south latitude ; Cocos Romanzoffiana (Chamisso), at 28° south latitude ; Diplothemium littorale (Mart.), at 30° south latitude. All the last-mentioned palms occur in Brazil, the Acrocomia and Trithrinax extending to Paraguay, and Cocos Australis to Uruguay and the La Plata-States.

While some palms, as indicated, descend to cooler latitudes, others--
ascend to temperate and even cold mountain-regions. Among the
American species are prominent in this respect—Euterpe andicola
(Brongniart), E Haenkeana (Brogn.), E. longivaginata (Mart.),
Diplothemium Porallyi (Mart.) and Ceroxylon pithyrophyllum
(Mart.), all occurring on the Bolivian Andes at an elevation of about
8,000 feet. Ceroxylon andicola (Humboldt), Kunthia montana
(Humb.), Oreodoxa frigida (Humb.)' and Geonoma densa (Linden),
also reach on the Andes of New Granada an elevation of 8,000 feet.
Ceroxylon Klopstockia (Mart.) advances on the Andes of Venezuela
fully to a zone of 7,500 feet altitude, where Karsten saw stems 200
feet high, with leaves 24 feet long. There also occur Syagrus
Sancona (Karsten) and Platenia Chiragua (Karsten), at elevations of
5,000 feet, both very lofty palms, and both recently reduced by Sir
Joseph Hooker to the genus Cocos. From the temperate mountain-
regions of sub-tropical Mexico are known, among others, Chamædorea
concolor (Mart.), Copernicia Pumos (Humb.), C. nana (Kunth) and
Brahea dulcis (Mart.), at elevations of from 7,000 to 8,000 feet.

Wissadula rostrata, Planchon.

Tropical Africa and America. A perennial somewhat shrubby
plant, easily naturalized in frostless regions. The bark abounds in
serviceable fibre ; and as the plant shoots quickly into long simple
twigs, if cut near the root, fibre of good length is easily produced
(Dr. Roxburgh).

Wistaria Chinensis, De Candolle.

The "Fuji" of Japan and China ; hardy still at Christiauia.
Lives through a century and more. The stem is carried up straight,
and the branches are trained on horizontal trellises at Japanese
dwellings, affording shade for seats beneath. One Wistaria tree will
thus cover readily a square of 50 feet by 50 feet, the delightfully
odorous trusses of flowers pendent through the trellis overhead
(Christy). Fortune tells us of a tree of great age, which measured
at 3 feet from the ground 7 feet in circumference, and covered
a space of trellis-work 60 feet by 100 feet. At Sunningdale
(England) a single plant covers a wall 9 feet high for a length of
340 feet (J. B. Torry). Flowers probably available for scent-
distillation.

Wistaria frutescens, Candolle. (*W. speciosa,* Nuttall.)

South-Eastern States of North-America. A woody tall-climbing
plant, of grand value, with the preceding species, for bees.

Withania coagulans, Dunal.

Mountains of India. A somewhat shrubby plant. With the fruit
milk can be coagulated into curd for cheese, as with rennet ; the
active principle, according to Mr. Sheridan Lee, is best extracted by
a weak aqueous solution of kitchen-salt.

Withania somnifera, Dunal.

Countries around the Mediterranean Sea, thence to South-Asia and South-Africa. A half-shrub. The root, according to Professor McOwan, acts much like that of Podophyllum medicinally.

Xanthorrhiza apiifolia, L'Heritier.

Eastern North-America. A perennial, almost shrubby plant, of medicinal value. The root produces a yellow pigment, similar to that of Hydrastis Canadensis. Both also contain berberin.

Xanthorrhœa Tatei, F. v. Mueller.

Kangaroo-Island. One of the largest of the so-called "Grass-trees," and one of the best for furnishing the fragrant resin of this genus of plants, it being in demand for particular sorts of varnishes, for the manufacture of sealing-wax, for picric acid, which it yields in large percentage, for coloring walls as an admixture to lime and for some other technologic purposes. Approximate London-price now £8 for the ton, according to Mr. Will. Somerville. Resin is also commercially exported from X. australis (R. Brown) of Tasmania and Victoria, from X. resinosa (Persoon) of N. S. Wales and Queensland, from X. quadrangulata (F. v. M.) of South-Australia, from X. Preissii (Endlicher) of West-Australia and from X. hastilis of New South Wales. Mechanical redissemination should be effected, wherever the plants largely become sacrificed for obtaining the resin. For technologic and geographic notes on various Xanthorrhœas see also Zeitschrift des oesterreich. Apotheker-Vereins xxiii., 293–295 (1885).

Xanthosoma sagittifolium, Schott.

West-Indies. The tubers are largely cultivated there, and used as an esculent like those of Colocasia. The plant may be as hardy as the latter.

Xanthoxylon piperitum, De Candolle.

Used as a condiment in China and Japan. Fruit-capsules remarkably fragrant.

Ximenia Americana, Linné.

Tropical-Asia, Africa and America, passing the tropics however in Queensland, and gaining also an indigenous position in Florida. This bush may therefore accommodate itself to cooler climes in localities free from frost. The fruits are edible, resembling yellow plums in appearance; their taste is agreeable. The wood is scented. In Mexico called "Alvarillo del campo." Mr. P. O'Shanesy recommended this shrub for hedges.

Xylia dolabriformis, Bentham.

The "Pyengadu" of India, extending to China and the Philippine-Islands, ascending mountains to 3,000 feet. An Acacia-like tree, attaining a height of about 120 feet, the stem often clear up

2 D

to about 80 feet and of very considerable girth. Foliage deciduous. The wood is reddish-brown, close-grained, and pervaded when fresh by an oily glutinous clamminess. The heartwood is of greater durability than even teak, and of a marvellous resistance to shocks through its extreme hardness. It is used for gun-carriages, crooks of ships, railway-sleepers, tools, gauges, ploughs, house- and bridge-posts (Laslett). It is as indestructible as iron, hence locally called iron-wood; a rifle shot at 20 yards distance will scarcely cause any penetration into it (Colonel Blake). Neither the teredo nor termites will touch the heartwood (J. Hooker). It can only be sawn up in a fresh state. The stem exudes a red gum-resin (Kurz). This tree yields also saponin.

Yucca aloifolia, Linné.

Carolina, Florida, West-India, Mexico, in coast-sand. Stem to 20 feet high. With its congeners a fibre-plant.

Yucca angustifolia, Pursh.

From Missouri and Iowa to Colorado, Arizona and New Mexico. Height according to Mr. Greene to about 15 feet. One of the hardiest of all.

Yucca baccata, Torrey.

Colorado, Texas, Southern California, Utah, Northern Mexico. In its ordinary state not tall; but the variety Y. filifera (Chabaud) will sometimes produce a stem half a hundred feet high with a diameter to 3 feet. The leaves are singularly short (S. Watson). This furnishes the Tambico-fibre for cordage, ropes, rugs and other fabrics.

Yucca brevifolia, Engelmann.

Southern California, Arizona and Utah, in the deserts, ascending to 4,000 feet. Attains a height of 30 feet. The whole plant can be converted into paper (Vasey, Baker).

Yucca filamentosa, Linné.

The Adam's Needle. From Maryland to Florida. An almost stemless species. It would hardly be right, to omit the plants of this genus altogether here, as they furnish a fibre of great strength, similar to that of the Agaves. Moreover, all these plants are decorative, and live in the poorest soil, even in drifting coast-sand. They are also not hurt, as is the case with the Fourcroyas, by slight frosts.

Yucca gloriosa, Linné.

Carolina and Florida, along the sandy coast-tracts. Stem not tall, but leaves very numerous. The fibre of the leaves furnishes much material for rope to supply the wants for ships and boats locally. Yucca-ropes are lighter, stronger and more durable than those of hemp (H. M. Brakenridge). At Edinburgh it bore a temperature of 0° F. with impunity (Gorlie).

Yucca Guatemalensis, Baker.

Mexico and Guatemala. Acquires finally a height of about 20 feet. Regarding the specific characteristics of the various Yuccas see particularly Baker's descriptions in the journal of the Linn. Soc. 1880.

Yucca Treculiana, Carrière,

From Texas to Mexico. Stem to about 50 feet high, branched only near the summit. Grand in aspect and also most showy on account of its vast number of white flowers of porcelain-lustre. The fruit tastes like that of the Papaw (Lindheimer).

Yucca Yucatana, Engelmann.

Mexico. This species attains a height of about 25 feet, branching from the base. Y. canaliculata (Hooker) ranges from Texas to North-Mexico, and has a stem up to 25 feet high, with very long leaves.

Zalacca secunda, Griffith.

Assam, as far north as 28°. A stemless palm with large feathery leaves, exquisitely adapted for decorative purposes. Before we quit the Asiatic palms, we may learn from Von Martius' great work, how many extra-tropical members of this princely order were already known in 1850, when that masterly publication was concluded. Martius enumerates as belonging to the boreal extra-tropical zone in Asia; *From Silhet at 24° north latitude:* Calamus erectus, Roxb.; C. extensus, Roxb.; C. quinquenervius, Roxb.; — *from Garo at 26° north latitude:* Wallichia caryotoides, Roxb.; Ptychosperma gracilis, Miq.; Caryota urens, L.; Calamus leptospadix, Griff.; — *from Khasya, in 26° north latitude:* Calamus acanthospathus, Griff.; C. macrospathus, Griff.; Plectocomia Khasyana, Griff.; — *from Assam, about 27° north latitude:* Areca Nagensis, Griff.; A. triandra, Roxb.; Livistona Jenkinsii, Griff.; Daemonorops nutantiflorus, Griff.; D. Jenkinsii, Griff.; D. Guruba, Mart.; Plectocomia Assamica, Griff.; Calamus tenuis, Roxb.; C. Flagellum, Griff.; C. Heliotropium, Hamilt.; C. floribundus, Griff.; Phœnix Ouseloyana, Griff.; — *from Upper Assam, between 28° and 29° north latitude:* Caryota obtusa, Griff.; Zalacca secunda, Griff.; Calamus Mishmelensis, Griff.; — *from Darjiling, at 27° north latitude:* Wallichia obtusifolia, Griff.; Licuala peltata, Roxb.; Plectocomia Himalaiana, Griff.; Calamus schizospathus, Griff.; — *from Nepal, between 28° and 29° north latitude:* Chamærops Martiana, Wall.; — *from Guhrvall, in 30° north latitude:* Calamus Royleanus, Griff.; — *from Saharanpoor, in 30° north latitude:* Borassus flabelliformis, L.; — *from Duab, in 31° north latitude:* Phœnix sylvestris, Roxb.; — *from Kheree, in 30° north latitude:* Phœnix humilis, Royle; — *from Dekan:* Bentinckia Coddapanna, Berry, at an elevation of 4,000 feet. Miquel mentions as palms of Japan (entirely extra-tropical): Rhapis flabelliformis, Aiton; R. humilis, Blume; Chamærops excelsa, Thunb.; Livistona Chinensis, Br. and Arenga saccharifera, Labill., or a species closely allied to that palm.

Zea Mays, Linné.*

The Maize or Indian Corn. Indigenous to the warmer parts of South-America. St. Hilaire mentions as its native country Paraguay. A. de Candolle believes it to have come originally from New Granada. Found—as cultivated—in Central America already by Columbus. This conspicuous, though annual, cereal grass interests us on this occasion as being applicable to far more uses than those, for which it has been employed in most parts of the globe. In North-America, for instance, maize is converted into a variety of dishes for the daily table, being thus boiled in an immature state, as "green corn." Mixed with other flour it furnishes good bread. For some kinds of cakes it is solely used, also for maizena, macaroni and polenta. Several varieties exist, the Inca-Maize of Peru being remarkable for its gigantic size and large grains; the variety named is very hardy, having matured seeds in Norway as far north as 63° 15′ according to Professor Schuebeler. Some varieties in wet tropical countries ripen grain within six weeks from the time of sowing. Maize is not readily subject to the ordinary corn-diseases, but to prosper it requires fair access to potash and lime. Good writing and printing papers can be prepared from maize-straw. Meyen calculated, that the return from maize under most favorable circumstances in tropical countries would be eight hundred fold, and under almost any circumstances it is the largest yielder among cereals in warm countries. Acosta counted on some cobs of the Inca-Maize as many as 700 grains, and says that it is not uncommon to harvest of this variety 300 fold the seeds sown; it grows to a height of 15 feet in rich soil and under careful cultivation, by which means the grains will become 4 or 5 times as large as the ordinary kind. In Peru it can be grown up to an altitude of 8,000 feet. Mr. Buchanan, of Lindenau, obtained 150 bushels of ordinary maize from an acre in Gippsland-flats, colony Victoria. Even in the very dry clime of the Murray-River districts maize, but under irrigation, has yielded 80 bushels per acre (D. Cormack). According to the Report of the Commissioner of Agriculture in Washington, the maize culture extended over 68,804,685 acres in the United States in 1882, that being over one-third of all the land under tillage in the Union. From the stalks of ordinary maize, after the ripened grains have been plucked, sugar at the rate of 900 lbs. per acre is still obtainable (Department of Agriculture, Washington). Maize has also come into extensive use for alcoholic distillation. In 1879 already the United States produced 1,547,900,000 bushels of maize on 53 millions of acres, to the value of 580 million dollars, or about £140,000,000. In 1882 the maize-produce there was 1,617,000,000 bushels, realizing monetarily 783,867,000 dollars, equal to £188,128,000. Maize-grain will retain its power of germination for two years with certainty. As a fattening saccharine green-fodder, maize is justly and universally in warm countries appreciated. In Middle Europe the Horse-tooth variety is frequently grown for this purpose and attains occasionally a height of fully 12 feet, although the seeds do not come to perfection there. Any ergot from it is used, like that of rye, for medicinal purposes. Maize-corn contains about 75 per cent. of starch. Dierbach recommends mellago or treacle from maize instead of that prepared from the roots of Triticum repens, L., and the molasses so obtained serves also for culinary uses. Sugar and treacle are now made on a large scale from maize-stems in the manner indicated under Andropogon saccharatus. Exposure to extreme and protracted cold—four years in Polaris-Bay, Smith-Sound, 81° 38' north latitude—did not destroy the vitality of .wheat- and maize-grains (R. J. Lynch). The elongated threadlike styles have come recently into medicinal use.

Zelkova acuminata, Planchon. (*Planera acuminata,* Lindley; *P. Japonica,* . Miquel; *Zelkovia Keaki,* Savatier and Franchet.)

The "Keaki," considered one of the best timber-trees of Japan; it proved of rapid growth and valuable as a shade-tree also at Melbourne. The wood never cracks, and is hence most extensively used for turnery, also much for furniture (Rein). Stems occasionally to 20 feet in girth. For out-door work the most valued wood in Japan (Christie).

Zelkova crenata, Spach. (*Planera Richardi,* Michaux.)

South-Western Asia, ascending to 5,000 feet. In favorable localities a good-sized tree, with qualities resembling those of the elms. The allied Z. Cretica (Spach) is restricted to South-Europe.

Zingiber officinale, Roscoe.

The Ginger. India and China. Possibly this plant may be productive also in the warmer temperate zone, and give satisfactory results. The multiplication is effected by division of the root. For candied ginger only the young succulent roots are used, which are peeled and scalded prior to immersion into the saccharine liquid. Great Britain imported in 1884 about 56,000 cwt. of ginger, valued at £124,000.

Zizania aquatica, Linné.* (*Hydropyrum esculentum,* Link.)

The Canada-Rice. In shallow streams and around ponds and lakes, from Canada to Florida. This grass might be readily naturalized. Annual. It attains a height of 9 feet. Although its grain can be utilized for bread-corn, we would wish to possess the plant chiefly, to obtain additional food of a superior kind for water-birds.

Zizania fluitans, Michaux. (*Hydrochloa Carolinensis,* Beauvois.)

Southern-States . of North-America. This grass, floating in shallow streams, or creeping on muddy banks of rivers or swamps, is praised by Prof. C. Mohr as valuable for fodder, lasting throughout the year.

Zizania latifolia, Turczaninow. (*Hydropyrum latifolium,* Grisebach.)

The Kau-sun of China. In lakes of Amur, Manchuria, China and Japan. Regarded by Bentham as conspecific with Z. aquatica.

From Dr. Hance we know, that the solid base of the stem forms a very choice vegetable, largely used in China, where this tall water-grass undergoes regular cultivation like the Trapa.

Zizania miliacea, Michaux.*

Southern part of North-America, West-Indies. Tall and perennial, but more restricted to the tide-water meadows and ditches, according to Pursh; while according to Chapman's note it is generally distributed like Z. aquatica, with which it has similar use. In Southern Brazil occurs a similar if not identical grass—namely Z. microstachya (Nees).

Zizyphus Joazeiro, Martius.

Brazil. Recommended as yielding edible fruit in arid regions.

Zizyphus Jujuba, Lamarck.

From India to China and East-Australia, extending also to tropical Africa, ascending the Himalayas to 4,500 feet. This shrub or tree can only be expected to bear its pleasant fruits within the temperate zone in warm regions. The fruit is red or yellow and of the size of a cherry. The Tussa-silkworm, which according to Dr. Forbes Watson is the most important and widely distributed of the wild silk-insects of India, feeds on Z. Jujuba, but also on Terminalias, Shorea, Bombax heptaphyllum (Cavanilles) and some other trees. Often the cocoons are merely collected in the forests.

Zizyphus Lotus, Lamarck.

Countries around the Mediterranean Sea. The fruits are small and less sweet than those of Z. vulgaris; nevertheless they are largely used for food in the native country of this bush, and are quite a staple-product for the local fruit-markets there (Dr. Shaw). Z. nummularia (Wight and Arnott) is an allied species from the mountains of India, ascending to about 3,000 feet. It is much used for garden-hedges. The fruit is sweet and acidulous and of a pleasant flavor (Brandis).

Zizyphus Mistal, Grisebach.

Argentina. A fine tree with edible fruits.

Zizyphus rugosa, Lamarck.

Nepal and other mountainous parts of India. A small tree, hardier than Z. Lotus. The drupe of this is also edible, and the same may be said of a few other Indian species.

Zizyphus Sinensis, Lamarck.

China and Japan. Similar in use to the last.

Zizyphus Spina Christi, Willdenow.

Middle and North-Africa, South-Western Asia. Rather a hedge-plant than a fruit-bush.

Zizyphus vulgaris, Lamarck.

Orient, particularly Syria, extending to China; in the Himalayas up to 6,500 feet, A small tree, adapted for a mild temperate clime. Fruits scarlet, about an inch long, with edible pulp; they are known as South-European Jujubes. The allied Z. oxyphylla (Edgeworth) has a very acid fruit.

Zoysia pungens, Willdenow.

Eastern and Southern Asia, East-Australia. This creeping grass, although not large, is important for binding coast-sands; it will live on saline soil, and can also be utilized as a lawn-grass.

Number of plants primarily recorded	... ...	2,279
Number of plants secondarily mentioned	... ...	1,347
Total	...	3,626

TABLE OF AVERAGES AND EXTREMES OF TEMPERATURE OF AIR IN VICTORIA.

Furnished by the Melbourne-Observatory.

Stations.	Yearly Mean Temperature.		Extreme Maximum Temperature.		Extreme Minimum Temperature.	
	Years.	Fahr.	Years.	Fahr.	Years.	Fahr.
		°		°		°
Cape Otway	22	55·1	20	108·0	16	30·0
Portland	8	56·8	21	108·0	17	27·0
Melbourne	27	57·3	27	111·2	27	27·0
Cape Schanck	2	56·7	2	98·0	2	33·0
Wilson's Promontory ...	9	56·2	7	101·0	8	30·0
Gabo Island	16	58·2	6	93·0	7	38·0
Ballarat	25	54·3	24	114·0	24	22·0
Birregurra	5	55·1	5	103·0	5	29·0
Macedon	11	53·0	...	...	...	...
Romsey	3	54·2	3	105·5	3	25·0
Sunbury	5	55·7	5	107·7	5	28·0
Berwick	18	57·0	...	...	...	...
Stratford	6	56·6	6	108·0	6	22·0
Stawell	16	57·7	17	120·0	19	25·0
Dimboola	4	57·8	4	116·0	4	20·0
Sandhurst	22	58·6	23	117·4	22	27·5
Clunes	5	55·0	5	113·0	4	23·0
Cashel	3	59·1	3	111·0	4	25·0
Beechworth	8	55·4	8	111·0	7	26·5
Echuca	3	58·6	3	111·0	3	23·0
Yarrawonga	5	60·5	5	112·0	5	25·0
Omeo	3	53·7	3	107·0	3	19·0

TABLE OF AVERAGE ANNUAL RAINFALL IN VICTORIA.

Furnished by the Melbourne-Observatory.

District or Basin.	Names of Stations.	Amount.	Years.
		inches	
West Coast	Cape Otway	35·09	21
	Portland	32·65	24
	Warrnambool	28·15	5
	Glenhuntly, Macarthur	31·74	5
South Coast	Geelong	18·36	10
	Wyndham	18·58	10
	Melbourne	25·76	27
	Tyabb, Hastings	36·42	10
	Cape Schanck	27·96	5
East Coast	Wilson's Promontory	42·79	10
	Port Albert	24·92	19
	Gabo Island	36·87	16
Glenelg and Wannon	Hamilton	24·90	10
	Glenthomson	26·32	7
	Retreat, Casterton	24·96	5
Hopkins River and Mount Emu Creek	Ercildoune, Burrumbeet	25·00	8
	Ararat	21·14	10
	Yalla-y-Poora, Streatham	20·95	6
	Wickliffe	21·88	5
	Keilambete, Terang	29·92	6
Mount Elephant and Lake Corangamite	Camperdown	29·41	10
	Rokewood	20·73	10
	Poligolet, Derrinallum	23·08	6
	Mount Bute, Lintons	24·79	7
	Pirron Yalloak	27·40	6
Moorabool and Barwon rivers	Buninyong	29·10	10
	Ballarat	26·74	25
	Birregurra	34·16	10
	Stony Creek Reservoir	24·51	8
	Murdeduke, Winchelsea	20·50	5
	Lovely Banks Reservoir	17·44	8
	Mount Gellibrand	23·95	6
Saltwater and Werribee rivers	Macedon	35·20	10
	Romsey	28·04	6
	Blackwood	40·97	6
	Sunbury	20·68	6
	Ballan	28·75	5
	Bacchus Marsh	18·25	5
Yarra River and Dandenong Creek	Yan Yean Reservoir	23·70	10
	Warrandyte	34·64	10
	Berwick	37·17	10
	Ferntree Gully	43·83	10
	Kew	28·43	10
	Gembrook	46·17	6
	Dandenong	30·11	6
	Yering	32·18	4
	Beenak	65·45	6
	Warburton	60·26	6

TABLE OF AVERAGE RAINFALL—*continued.*

District or Basin.	Names of Stations.	Amount.	Years.
		inches	
	Sale	24·68	10
	Stratford	27·46	7
	Rosedale	26·88	7
Mitchell and La Trobe-rivers	Warragul	43·45	6
	Maffra	25·96	6
	Grant	41·33	5
	Bairnsdale	27·90	5
Lowan Shire and Mallee ...	Werracknabeal	13·96	7
	Neuarpur, Apsley	19·91	5
	Stawell	19·01	10
Wimmera-River	Horsham	15·78	10
	Hall's Gap, Grampians ...	34·78	10
	Dimboola	13·44	6
	Banyenong, Donald	14·34	8
Avon and Richardson-rivers	Wallaloo, Glenorchy	15·22	5
	Warranooke, Glenorchy ...	16·34	7
	Daylesford	34·45	8
	Sandhurst	21·66	23
	Castlemaine	23·75	8
	Newlyn	29·91	8
	Clunes Weir	21·85	8
	Barker's Creek Reservoir ...	24·89	8
	Expedition Pass Reservoir ...	24·88	8
Loddon-River	Crusoe Reservoir	21·98	8
	Maryborough	18·44	7
	Maldon·	23·32	6
	Clunes	21·03	6
	Tragowell Station	12·46	5
	Lake Meran Station	12·68	4
	Durham Ox	12·03	3
	Kerang	11·44	5
	Kyneton	30·88	10
	Malmsbury Reservoir ...	26·62	8
Campaspe-River and Lake	Sutton Grange Reservoir ...	28·18	8
Corop	Trentham	40·22	7
	Whroo ·... ... ·...	21·36	7
	Metcalfe	25·00.	6
	Kaarimba, north of Shepparton	18·40	10
	Shepparton	17·22	7
Goulburn and Broken-rivers	Alexandra	25·65	6
	Cashel	19·27	5
	Edenvale, Strathbogie Ranges	39·79	5
Ovens-River	Beechworth	30·09	9
	Wangaratta	21·05	6
Mitta Mitta and Kiewa-rivers	Omeo	23·55	5
	Wahgunyah	19·77	9
	Piangil	13·14	10
	Coonanga	20·30	6
	Pannoobamawm	16·15	9
Murray-River	Chiltern	21·48	6
	Picola	14·26	4
	Echuca ... ·,· ...	14·90	4
	Yarrawonga	17·32	5

GENERA INDICATING :

Alimentary Plants—.

1. YIELDING HERBAGE (culinary)—

   Agriophyllum, Allium, Amarantus, Anthriscus, Apium, Aralia, Atriplex, Barbarea, Basella, Beta, Bongardia, Borrago, Brassica, Chenopodium, Claytonia, Corchorus, Crambe, Cynara, Eremurus, Euchlæna, Fagopyrum, Gigantochloa, Gunnera, Hibiscus, Lactuca, Lepidium, Musa, Œnanthe, Pharnaceum, Pringlea, Pugionium, Rheum, Rumex, Sanguisorba, Scandix, Schizostachyum, Scorzonera, Spinacia, Talinum, Tetragonia, Theligonum, Tropæolum, Valerianella, Zizauia.

2. YIELDINC ROOTS (culinary)—

   Allium, Apios, Aponogeton, Arracacha, Asparagus, Bassowia, Beta, Boussingaultia, Brassica, Butomus, Carum, Chærophyllum, Cichorium, Colocasia, Conopodium, Cordyline, Crambe, Cymopterus, Cyperus, Daucus, Dendrocalamus, Dioscorea, Diposis, Eustrephus, Ferula, Flemingia, Flucggea, Geitonoplesium, Gigantochloa, Gladiolus, Heleocharis, Helianthus, Hypochœris, Ipomœa, Iris, Manihot, Microseris, Nelumbo, Nepeta, Ophiopogon, Oxalis, Pachyrrhizus, Peucedanum, Pimpinella, Pouzolzia, Priva, Psoralea, Pueraria, Raphanus, Rhaponticum, Ruscus, Scilla, Scolymus, Scorzonera, Sechium, Selinum, Solanum, Stilbocarpa, Thapsia, Tinguarra, Tragopogon, Tropæolum, Ullucus, Uvularia, Valeriana, Xanthosoma.

3. YIELDING CEREAL GRAIN—

   Andropogon, Avena, Eleusine, Hordeum, Oryza, Panicum, Pennisetum, Poa, Secale, Triticum, Zea, Zizania.

4. YIELDING TABLE-PULSE—

   Cajanus, Caragana, Cicer, Cyamopsis, Dolichos, Ervum, Lupinus, Mucuna, Phaseolus, Pisum, Psophocarpus, Vicia, Vigna.

5. YIELDING VARIOUS ESCULENT FRUITS—

   Aberia, Acanthosicyos, Achras, Adenostemon, Agriophyllum, Albizzia, Alibertia, Amarantus, Amelanchier, Anona, Arachis, Araucaria, Aristotelia, Artocarpus, Atalantia, Averrhoa, Benincasa, Berberis, Borassus, Brabejum, Canavalia, Carica, Carissa, Carya, Casimiroa, Castanea, Castanopsis, Celtis, Ceratonia, Cereus, Cervantesia, Citrus, Coccoloba, Condalia, Corynocarpus, Coryuosicyos, Cratægus, Cucumis, Cucurbita, Cudrania, Cupania, Cynara, Debregeasia, Diospyros, Euclea, Eugenia, Fagopyrum, Ficus, Fragaria, Fuchsia, Gaultiera, Gaylussacia, Gingko, Gourliaea, Guevina, Hibiscus, Hovenia, Hymenæa, Illipe, Juglans, Juniperus, Lapageria, Limonia, Macadamia, Maclura, Mangifera, Marlea, Marliera,

GENERA INDICATING—*continued.*

Melicocca, Mesembrianthemum, Moringa, Morus, Musa, Myrica, Myrtus, Nageia, Nelumbo, Nephelium, Nicmeyera, Nuphar, Nyssa, Opuntia, Pappea, Parinarium, Passiflora, Peireskia, Persea, Peumus, Phœnix, Photinia, Phyllanthus, Physalis, Pinus, Pistacia, Prunus, Psidium, Punica, Pyrularia, Pyrus, Quercus, Ribes, Rubus, Salpichroma, Sambucus, Santalum, Sechium, Shepherdia, Solanum, Spondias, Sterculia, Tamarindus, Telfairia, Terminalia, Trapa, Triphasia, Vaccinium, Vahea, Vangueria, Vitis, Voandzeia, Ximenia, Zizyphus.

6. TRUFFLES, MUSHROOMS AND OTHER FUNGS—

Agaricus, Boletus, Cantharellus, Clavaria, Coprinus, Cortinarius, Fistulina, Helvella, Hydnum, Morchella, Pachyma, Peziza, Polygaster, Polyporus, Rhizopogon, Russula, Terfezia, Tremella, Tuber.

Avenue-Plants (partly also for street-planting)—

Acer, Æsculus, Castanea, Corylus, Cupressus, Eucalyptus, Ficus, Fraxinus, Gleditschia, Grevillea, Jubæa, Juglans, Melia, Oreodoxa, Pinus, Pircunia, Pistacia, Planera, Platanus, Populus, Prunus, Pyrus, Quercus, Robinia, Salix, Sequoia, Thespesia, Tilia, Ulmus, Zelkova.

Bamboo-Plants—

Arundinaria, (Arundo), Bambusa, Beesha, Dendrocalamus, Gigantochloa, Guadua, Melocalamus, Melocanna, Phyllostachys, Schizostachyum (many other genera mentioned under Schizostachyum), Teinostachyum, Thamnocalamus.

Camphor-Plant—

Cinnamomum.

Coffee-Plant.

Coffea.

Condiment-Plants—

Acorus, Allium, Apium, Archangelica, Artemisia, Asperula, Benincasa, Borrago, Brassica, Calamintha, Calyptranthes, Capparis, Capsicum, Carum, Chærophyllum, Cinnamomum, Citrus, Cochlearia, Coriandrum, Crithmum, Cuminum, Fœniculum, Glycine, Illicium, Laserpitium, Laurus, Lepidium, Lindera, Mentha, Meriandra, Monarda, Monodora, Myrrhis, Nyssa, Ocimum, Olea, Origanum, Peucedanum, Pimpinella, Prunus, Pycnanthemum, Salvia, Satureja, Sison, Smyrnium, Spilanthes, Tropæolum, Thymus, Tuber, Valerianella, Xanthoxylon, Zingiber.

Cork-Plants—

Quercus.—Substitutes: Æschynomene, Agave, Nyssa.

GENERA INDICATING—*continued.*

Dye-Plants—

Acacia, Acer, Albizzia, Aleurites, Alkanna, Alnus, Anthemis, Baloghia, Baptisia, Cæsalpinia, Carthamus, Carya, Chlorogalum, Cladrastris, Coccoloba, Coprinus, Crocus, Crozophora, Cytisus, Dracæna, Excæcaria, Fagopyrum, Fraxinus, Garcinia, Gunnera, Hedera, Helianthus, Heterothalamus, Indigofera, Isatis, Juglans, Lawsonia, Lithospermum, Lyperia, Maclura, Maharanga, Mallotus, Onosma, Opuntia, Peireskia, Peltophorum, Perilla, Peumus, Phyllocladus, Pinus, Polygonum, Quercus, Reseda, Rhamnus, Rhus, Roccella, Rubia, Sambucus, Saponaria, Solanum, Sophora, Spartium, Terminalia, Thymelæa, Vaccinium, Xanthorrhiza.

Edging-Plants—

Anthemis, Buxus, Rosa, Stenotaphrum, Thymus.

Fibre-Plants—

Agave, Apocynum, Beschorneria, Bœhmeria, Broussonetia, Camelina, Cannabis, Caryota, Chlorogalum, Copernicia, Corchorus, Cordyline, Crotalaria, Cyperus, Debregeasia, Eryngium, Fitzroya, Fourcroya, Gossypium, Hardwickia, Helianthus, Hibiscus, Humulus, Lardizabala, Lavatera, Linum, Malachra, Maoutia, Musa, Pachyrrhizus, Phormium, Pipturus, Poa, Sanseviera, Sesbania, Spartina, Spartium, Thuya, Tillandsia, Touchardia, Urena, Villebrunia, Yucca. (See also Paper-plants).

Fullers' Plant—

Dipsacus.

Fodder-Plants—

1. GRASSES—

Agrostis, Aira, Alopecurus, Andropogon, Anthistiria, Anthoxanthum, Aristida, Arundinaria, Arundinella, Avena, Bouteloua, Bromus, Buchloa, Carex, Chionachne, Chloris, Cinna, Cynodon, Cynosurus, Dactylis, Danthonia, Ehrharta, Eleusine, Euchlæna, Erianthus, Eriochloa, Festuca, Hemarthria, Hierochloa, Holcus, Hordeum, Kœleria, Leersia, Lolium, Melica, Milium, Muehlenbergia, Neurachne, Oryzopsis, Panicum, Pappophorum, Paspalum, Pennisetum, Phalaris, Phleum, Poa, Rottboellia, Sclerachne, Secale, Sesleria, Spartina, Spinifex, Stenotaphrum, Tricholæna, Tripsacum, Trisetum, Triticum, Uniola, Zizania.

2. HERBAGE—

Achillea, Alchemilla, Anthyllis, Arachis, Astragalus, Atriplex, Brassica, Cichorium, Conospermum, Crotalaria, Desmodium, Erodium, Ervum, Exomis, Heracleum, Hippocrepis, Jacksonia, Kochia, Lespedeza, Lotus, Lupinus, Medicago, Pentzia, Peucedanum, Phymaspermum, Plantago, Portulacaria, Prangos, Sanguisorba, Selago, Sesbania, Spergula, Stenopetalum, Symphytum, Trichodesma, Trifolium, Trophis.

3. STABLE PULSE (Pods and Herbs)—

Cicer, Dolichos, Hedysarum, Lathyrus, Lupinus, Medicago, Melilotus, Onobrychis, Ornithopus, Oxytropis, Pisum, Trifolium, Trigonella, Vicia.

4. OTHER FRUITS—

Argania, Carya, Castanea, Ceratonia, Helianthus, Prosopis, Quercus.

Garland-Plants—

Baccharis, Cupressus, Helichrysum, Laurus, Lycopodium, Melaleuca, Pinus, Quercus.

Grave-Plants—

Acacia, Agonis, Boronia, Cupressus, Dacrydium, Fraxinus, Helichrysum, Helipterum, Lycopodium, Salix, Tamarix, Thuya, Viola.

Gum-Plants—

Acacia, Albizzia, Astragalus, Bambusa, Brachychiton, Caragana, Diospyros, Olea, Piptadenia, Prosopis, Xylia.

Hedge-Plants—

Aberia, Acacia, Acer, Agave, Albizzia, Alnus, Azima, Baccharis, Bambusa, Berberis, Buddleya, Cæsalpinia, Capparis, Carissa, Ceanothus, Celtis, Citrus, Cratægus, Cupressus, Cytisus, Euonymus, Elæagnus, Flacourtia, Gleditschia, Guilandina, Hymenanthera, Justicia, Lawsonia, Ligustrum, Lycium, Maclura, Mimosa, Opuntia, Paliurus, Parkinsonia, Peireskia, Pisonia, Pistacia, Pittosporum, Plectronia, Prosopis, Prunus, Punica, Pyrus, Rhamnus, Rhus, Rosa, Rubus, Ruscus, Salix, Scutia, Streblus, Thuya, Viburnum, Zizyphus.

Honey-Plants—

Acacia, Acer, Æsculus, Agave, Angophora, Audibertia, Avicennia, Barbarea, Borago, Brassica, Catalpa, Cephalanthus, Cerinthe, Citrus, Crocus, Cucurbita, Diospyros, Dipsacus, Echium, Ervum, Eucalyptus, Eucryphia, Eupatorium, Fagopyrum, Gleditschia, Gossypium, Grevillea, Hedera, Helianthus, Lavandula, Liriodendron, Lupinus. Marrubium, Medicago, Melilotus, Melaleuca, Melianthus, Melissa, Mentha, Monarda, Musa, Nepeta, Ocimum, Origanum, Ornithopus, Onobrychis, Prenanthes, Protea, Prunus, Pyrus, Reseda, Rhamnus, Rhus, Robinia, Rosa, Rosmarinus, Sabal, Salix, Salvia, Taraxacum, Thymus, Tilia, Trichostema, Trifolium, Tropæolum, Viola.

Hop-Plant—

Humulus.

Insecticidal Plants—

Ailantus, Artemisia, Eucalyptus, Chrysanthemum, Cannabis, Gymnocladus, Melia, Ricinus, Schkuhria, Solanum, Tagetes.

Medicinal Plants—

1. YIELDING HERBAGE OR FLOWERS—

Achillea, Aconitum, Agave, Aletris, Aloe, Althæa, Anemone, Anthemis, Arctostaphylos, Aristolochia, Arnica, Artemisia, Atropa, Baptisia, Barosma, Cannabis, Carica, Cassia, Catha, Chelidonium, Chenopodium, Chrysanthemum, Cochlearia, Conium, Convallaria, Crocus, Cytisus, Digitalis, Duboisia, Erythroxylon, Eupatorium, Garulcum, Grindelia, Hagenia, Hedeoma, Hyoscyamus, Ilex, Justicia, Lactuca, Leonotis, Leyssera, Lippia, Marrubium, Matricaria, Melia, Melianthus, Mentha, Menyanthes, Nepeta, Osmitopsis, Papaver, Parthenium, Pilocarpus, Polygala, Prunus, Rafnia, Ricinus, Rosmarinus, Ruta, Salvia, Sambucus, Santolina, Schkuhria, Sebæa, Selinum, Solanum, Sophora, Spigelia, Spilanthes, Swertia, Tanacetum, Tarchonanthus, Teucrium, Thuya, Thymus.

2. YIELDING BARK—

Achras, Alstonia, Aspidosperma, Cinchona, Juglans, Melia, Pilocarpus, Salix.

3. YIELDING ROOTS—

Acorus, Actæa, Althæa, Anacyclus, Apocynum, Archangelica, Aristolochia, Arnica, Atropa, Carex, Cephælis, Cimicifuga, Colchicum, Convolvulus, Euryangium, Gentiana, Glycyrrhiza, Helleborus, Hydrastis, Inula, Ipomœa, Krameria, Leontice, Nardostachys, Panax, Periandra, Peucedanum, Pimpinella, Piscidia, Podophyllum, Polygala, Punica, Rafnia, Rheum, Rubus, Sabbatia, Sanguinaria, Saponaria, Sassafras, Saussurea, Schœnocaulon, Scorzonera, Smilax, Smyrnium, Stylosanthes, Symphytum, Taraxacum, Urginia, Valeriana, Veratrum, Withania, Xanthorrhiza.

4. YIELDING FRUITS (or only Seeds)—

Carica, Cassia, Cucumis, Cuminum, Ecballion, Fœniculum, Illicium, Mallotus, Œnanthe, Persea, Punica, Rhamnus, Rheum, Ricinus, Schœnocaulon, Smyrnium, Tamarindus, Trigonella.

Oil-Plants—

Aleurites, Arachis, Argania, Brassica, Camelina, Camellia, Cannabis, Carya, Combretum, Cucurbita, Cyperus, Eruca, Excœcaria, Ginkgo, Gossypium, Guizotia, Helianthus, Juglans, Linum, Litsea, Olea, Papaver, Prunus, Pyrularia, Ricinus, Sesamum, Telfairia.

Palm-Plants—

Acanthophœnix, Acrocomia, Bactris, Bacularia, Borassus, Brahea, Calamus, Caryota, Ceroxylon, Chamædora, Chamærops, Cocos, Copernicia, Diplothemium, Euterpe, Geonoma, Hyphæne, Jubæa, Kentia, Livistona, Mauritia, Oncosperma, Oreodoxa, Phœnix, Plectocomia, Prestoa, Pritchardia, Ptychosperma, Rhapidophyllum, Rhapis, Sabal, Thrinax, Trithrinax, Wallichia, Wettinia, Zalacca, (many other American genera under Wettinia, many other Asiatic genera under Zalacca).

Paper-Plants—

Arundo, Broussonetia, Cyperus, Fatsia, Lepidosperma, Lygeum. Phormium, Pinus, Populus, Psamma, Salix, Spartina, Stipa, Zea. (See also Fibre-plants.)

Resin-Plants—

Achras, Adesmia, Balsamodendron, Belis, Boswellia, Bursera, Butea, Cajanus, Callitris, Ceroxylon, Chloroxylon, Cistus, Croton, Dammara, Dichopsis, Dorema, Ferula, Ficus, Frenela, Garcinia, Hancornia, Hymenæa, Isonandra, Juniperus, Liquidambar, Manihot, Melanorrhœa, Myrica, Pinus, Pistacia, Pterocarpus, Rhus, Shorea, Styrax, Vahea, Xanthorrhœa.

Saline Plants—

Agrostis, Albizzia, Alopecurus, Atriplex, Avicennia, Batis, Cæsalpinia, Casuarina, Cynodon, Cyperus, Kochia, Leptospermum, Melaleuca, Myoporum, Paspalum, Phœnix, Phormium, Poa, Salicornia, Sesuvium, Spartina, Tamarix, Zoysia.

Sandcoast-Plants—

Acacia, Agrostis, Ailantus, Alkanna, Aloe, Andropogon, Apium, Asparagus, Atriplex, Baccharis, Beta, Cæsalpinia, Cakile, Calamagrostis, Callitris, Carex, Casuarina, Ceanothus, Coccoloba, Crambe, Crithmum, Cupressus, Cynodon, Cytisus, Dactylis, Distichlis, Ehrharta, Elegia, Elymus, Festuca, Genista, Glaucium, Imperata, Launæa, Lavandula, Lepidosperma, Leptospermum, Lupinus, Medicago, Melaleuca, Mesembrianthemum, Myoporum, Myrica, Opuntia, Ornithopus, Oxytropis, Panicum, Paspalum, Phormium, Pinus, Plantago, Poa, Populus, Prunus, Psamma, Quercus, Remirea, Rhagodia, Robinia, Sabal, Salix, Scirpus, Sesuvium, Spartina, Spinifex, Stenotaphrum, Stipa, Tamarix, Tetragonia, Thouarea, Thrinax, Tripsacum, Triticum, Ulex, Uniola, Urginea, Yucca, Zoysia.

Scenic Plants (other than Palms or Bamboos)—

Agave, Ailantus, Aloe, Andropogon, Angelica, Arundo, Asplenium, Berberis, Bœhmeria, Canna, Cereus, Colocasia, Cordyline, Cycas, Cynara, Cyperus, Datura, Dicksonia, Dirca, Dracæna,

GENERA INDICATING—*continued.*

Elegia, Encephalartos, Euchlæna, Eustrephus, Fatsia, Ferula, Festuca, Fœniculum, Fourcroya, Gunnera, Helianthus, Heracleum, Inula, Lavatera, Leucadendron, Melianthus, Musa, Opuntia, Pandanus, Paulownia, Phormium, Pipturus, Podachænium, Rheum, Richardia, Ricinus, Todea, Touchardia, Watsonia, Yucca, Zea.

Scent-Plants—

Acacia, Adesmia, Aloexylon, Andropogon, Anthoxanthum, Aquilaria, Backhousia, Boronia, Calamintha, Cedronella, Citrus, Convolvulus, Dracocephalum, Eucalyptus, Gelsemium, Lavandula, Liatris, Lippia, Liquidambar, Melia, Melissa, Mentha, Monarda, Murraya, Myrtus, Nyctanthes, Ocimum, Origauum, Osmanthus, Pelargonium, Pittosporum, Pogostemon, Polianthes, Prunus, Pycnanthemum, Reseda, Rosa, Rosmarinus, Santalum, Satureja, Styrax, Synoon, Teucrium, Thymus, Tilia, Triphasia, Viola, Wistaria.

Silk-Plants—

Ailantus, Cajanus, Castanea, Cudrania, Liquidambar, Maclura, Morus, Quercus, Ricinus, Shorea, Symplocos, Terminalia, Trophis, Ulmus, Zizyphus.

Starch-Plants—

Alstrœmeria, Canna, Caryota, Colocasia, Copernicia, Cycas, Fagopyrum, Hordeum, Levisia, Manihot, Maranta, Musa, Oreodoxa, Oryza, Secale, Solanum, Tacca, Triticum, Zea.

Sugar-Plants—

Acer, Andropogon, Beta, Borassus, Caryota, Copernicia, Cucumis, Euchlæna, Phœnix, Saccharum, Zea.

Tan-Plants—

Acacia, Æsculus, Alnus, Albizzia, Angophora, Aspidosperma, Banksia, Butea, Cæsalpinia, Cedrela, Coccoloba, Comptonia, Cytisus, Dacrydium, Duvaua, Elephanthorrhiza, Eucalyptus, Eugenia, Gordonia, Gunnera, Osiris, Pinus, Piptadenia, Populus, Prosopis, Prunus, Pterocarpus, Quercus, Rhus, Rumex, Salix, Terminalia.

Tea-Plants—

Achillea, Andropogon, Astartea, Camellia, Catha, Ceanothus, Hydrangea, Ilex, Vaccinium.

Tide-Plants—

Ægiceras, Avicennia, Batis, Cyperus, Melaleuca, Myoporum, Salicornia, Spartina, Terminalia.

Timber-Plants—

1. TREES, CONIFEROUS—

   a. Evergreen—

   Araucaria, Belis, Callitris, Cephalotaxus, Cryptomeria, Cupressus, Dacrydium, Dammara, Fitzroya, Frenela, Juniperus, Libocedrus, Nageia, Phyllocladus, Pinus, Saxono-Gothæa, Sciadopitys, Sequoia, Taxus, Thuya, Torreya.

   b. Deciduous—

   Ginkgo, Glyptostrobus, Pinus, Taxodium.

2. TREES, NOT CONIFEROUS—

   a. Evergreen—

   Acacia, Adenostemon, Albizzia, Angophora, Castanopsis, Casuarina, Cedrela, Cercocarpus, Chloroxylon, Corynocarpus, Dalbergia, Diospyros, Embothrium, Eucalyptus, Eucryphia, Fagus, Flindersia, Gmelina, Gourliæa, Grevillea, Harpullia, Hymenæa, Jacaranda, Knightia, Laurelia, Maba, Machilus, Magnolia, Marlea, Maytenus, Metrosideros, Myrtus, Owenia, Peltophorum, Persea, Peumus, Psychotria, Quercus, Rhus, Royenia, Santalum, Shorea, Swietenia, Syncarpia, Tetranthera, Tristania.

   b. Deciduous—

   Acer, Æsculus, Ailantus, Alnus, Betula, Butea, Carpinus, Carya, Castanea, Catalpa, Celtis, Corylus, Diospyros, Engelhardtia, Excæcaria, Fagus, Fraxinus, Gleditschia, Gymnocladus, Holoptelea, Juglans, Liriodendron, Magnolia, Melia, Ostrya, Pircunia, Planera, Platanus, Populus, Pterocarpus, Pterocarya, Quercus, Robinia, Salix, Sophora, Tectona, Tilia, Ulmus, Umbellularia, Xylia, Zelkova.

Tobacco-Plant—

Nicotiana.

Water-Plants—

Acorus, Æschynomene, Aponogeton, Butomus, Cyperus, Euryale, Menyanthes, Nelumbo, Nuphar, Nyssa, Oryza, Poa, Richardia, Sagittaria, Trapa, Zizania.

Wicker-Plants—

Cyperus, Parrotia, Salix (also genera mentioned under Bamboo Plants).

Wood-engravers' Plants—

Aspidosperma, Buxus, Dacrydium, Camellia, Cratægus, Eucalyptus, Gonioma, Ilex, Pittosporum, Prunus, Pyrus, Rhododendron, Royenia, Torreya.

SYSTEMATIC INDEX TO GENERA.

DICOTYLEDONEÆ.

Ranunculaceæ.
Aconitum.
Actæa.
Anemone.
Cimicifuga.
Helleborus.
Hydrastis.
Xanthorrhiza.

Nymphæaceæ.
Nelumbo.
Nuphar.

Piperaceæ.
Piper.

Aristolochieæ.
Aristolochia.

Canellaceæ.
Canella.

Magnoliaceæ.
Drimys.
Illicium.
Liriodendron.
Magnolia.
Michelia.

Calycantheæ.
Calycanthus.

Anonaceæ.
Anona.
Monodora.

Laurineæ.
Adenostemum.
Cinnamomum.
Laurus.
Lindera.
Persea.
Sassafras.
Umbellularia.

Monimieæ.
Laurelia.
Peumus.

Berberideæ.
Berberis.
Bongardia.

Decaisnea.
Lardizabala.
Leontice.
Podophyllum.

Papaveraceæ.
Chelidonium.
Glaucium.
Papaver.
Sanguinaria.

Cruciferæ.
Barbarea.
Brassica.
Cakile.
Camelina.
Cochlearia.
Crambe.
Eruca.
Isatis.
Lepidium.
Pringlea.
Raphanus.

Capparideæ.
Capparis.

Violaceæ.
Hymenanthera.
Viola.

Moringaceæ.
Moringa.

Bixaceæ.
Aberia.

Cistaceæ.
Cistus.

Resedaceæ.
Reseda.

Pittosporeæ.
Pittosporum.

Polygalaceæ.
Krameria.
Polygala.

Guttiferæ.
Garcinia.

Camelliaceæ.
Camellia.
Gordonia.
Schima.

Dipterocarpeæ.
Shorea.

Linaceæ.
Erythroxylon.
Linum.

Geraniaceæ.
Averrhoa.
Erodium.
Oxalis.
Pelargonium.
Tropæolum.

Malvaceæ.
Althæa.
Gossypium.
Hibiscus.
Urena.
Wissadula.

Sterculiaceæ.
Brachychiton.
Sterculia.

Tiliaceæ.
Aristotelia.
Corchorus.
Tilia.

Euphorbiaceæ.
Aleurites.
Baloghia.
Buxus.
Croton.
Crozophora.
Mallotus.
Manihot.
Phyllanthus.
Ricinus.

Urticaceæ.
Artocarpus.
Bœhmeria.
Broussonetia.
Cannabis.
Celtis.
Cudrania.
Debregeasia.
Holoptelea.
Humulus.
Maclura.
Maoutia.
Morus.
Planera.
Pipturus.
Pouzolzia.
Streblus.
Touchardia.
Trophis.
Ulmus.
Villebrunia.
Zelkova.

Juglandeæ.
Carya.
Engelhardtia.
Juglans.

Amentaceæ.
Alnus.
Betula.
Carpinus.
Castanea.
Castanopsis.
Comptonia.
Corylus.
Myrica.
Ostrya.
Platanus.
Populus.
Quercus.
Salix.

Casuarineæ.
Casuarina.

Rutaceæ.
Atalantia.
Barosma.
Boronia.
Calodendron.
Casimiroa.
Citrus.
Limonia.
Murraya.
Pilocarpus.
Ruta.
Triphasia.
Xanthoxylon.

Simarubeæ.
Ailantus.

Anacardiaceæ.
Corynocarpus.
Mangifera.
Melanorrhœa.
Odina.
Pistacia.
Rhus.
Schinus.
Spondias.

Burseraceæ.
Amyris.
Balsamodendron.
Boswellia.
Bursera.

Meliaceæ.
Cedrela.
Chloroxylon.
Flindersia.
Melia.
Owenia.
Synoon.
Swietenia.

Sapindaceæ.
Acer.
Æsculus.
Blighia.
Cupania.
Harpullia.
Melianthus.
Melicocca.
Nephelium.
Pappea.

Celastrinæ,
Catha.
Euonymus.
Maytenus.

Rhamnaceæ.
Ceanothus.
Colletia.
Condalia.
Hovenia.
Paliurus.
Rhamnus.
Scutia.
Zizyphus.

Aquifoliaceæ.
Ilex.

Datisceæ.
Datisca.

Tamariscinæ.
Tamarix.

Cacteæ.
Cereus.
Echinocactus.
Opuntia.
Peireskia.

Ficoideæ.
Mesembrianthemum.
Sesuvium.
Tetragonia.

Caryophylleæ.
Saponaria.
Spergula.

Portulaceæ.
Claytonia.
Lewisia.
Talinum.

Amarantaceæ.
Amarantus.

Salsolaceæ.
Agriophyllum.
Atriplex.
Basella.
Beta.
Boussingaultia.
Chenopodium.
Exomis.
Rhagodia.
Spinacia.
Theligonum.
Ullucus.

Polygonaceæ.
Calligonum.
Coccoloba.
Polygonum.
Rheum.
Rumex.

Nyctagineæ.
Pisonia.

Phytolacceæ.
Pircunia.

Halorageæ.
Batis.
Gunnera.

SYSTEMATIC INDEX TO GENERA—*continued.*

Rosaceæ.

Alchemilla.
Amelanchier.
Cercocarpus.
Cratægus.
Fragaria.
Hagenia.
Parinarium.
Prunus.
Pyrus.
Quillaja.
Rosa.
Rubus.
Sanguisorba.

Leguminosæ.

Acacia.
Adesmia.
Æschynomene.
Albizzia.
Aloexylon.
Anthyllis.
Apios.
Arachis.
Astralagus.
Butea.
Cæsalpinia.
Cajanus.
Canavalia.
Caragana.
Cassia.
Ceratonia.
Cercis.
Cicer.
Cladrastis.
Crotalaria.
Cyamopsis.
Cytisus.
Dalbergia.
Desmodium.
Dolichos.
Elephanthorrhiza.
Ervum.
Genista.
Gleditschia.
Glycine.
Glycyrrhiza.
Gymnocladus.
Hardwickia.
Hedysarum.
Hippocrepis.
Hymenæa.
Indigofera.
Jacksonia.
Lathyrus.
Lespedeza.
Lippia.
Lotus.
Lupinus.

Medicago.
Melilotus.
Mimosa.
Onobrychis.
Ornithopus.
Oxytropis.
Pachyrrhizus.
Parkinsonia.
Peltophorum.
Periandra.
Phaseolus.
Piptadenia.
Piscidia.
Pisum.
Prosopis.
Psophocarpus.
Psoralea.
Pterocarpus.
Pueraria.
Rafnia.
Robinia.
Sesbania.
Sophora.
Spartium.
Stylosanthes.
Tamarindus.
Trifolium.
Trigonella.
Ulex.
Vicia.
Vigna.
Voandzeia.
Wistaria.
Xylia.

Saxifrageæ.

Eucryphia.
Hydrangea.
Ribes.

Hamamelideæ.

Liquidambar.
Parrotia.

Myrtaceæ.

Agonis.
Angophora.
Astartea.
Backhousia.
Calyptranthes.
Eucalyptus.
Eugenia.
Leptospermum.
Marliera.
Melaleuca.
Metrosideros.
Myrtus.
Psidium.
Tristania.

Combretaceæ.

Combretum.
Terminalia.

Onagreæ.

Fuchsia.
Trapa.

Lythraceæ.

Lawsonia.
Punica.

Passifloreæ.

Carica.
Passiflora.

Cucurbitaceæ.

Acanthosicyos.
Benincasa.
Corynosicyos.
Cucumis.
Cucurbita.
Ecballion.
Sechium.
Telfairia.

Olacinæ.

Ximenia.

Santalaceæ.

Cervantesia.
Pyrularia.
Santalum.

Proteaceæ.

Brabejum.
Conospermum.
Embothrium.
Grevillea.
Guevina.
Leucadendron.
Macadamia.
Protea.

Thymeleæ.

Aquilaria.
Dirca.
Thymelæa.

Elæagneæ.

Elæagnus.
Shepherdia.

Cornaceæ.

Cornus.
Marlea.
Nyssa.

Viniferæ.

Vitis.

Umbelliferæ.

Anthriscus.
Apium.
Arracacha.
Aralia.
Archangelica.
Carum.
Chærophyllum.
Conium.
Conopodium.
Coriandrum.
Crithmum.
Cuminum.
Cymopterus.
Daucus.
Diposis.
Dorema.
Eryngium.
Heracleum.
Laserpitium.
Myrrhis.
Œnanthe.
Panax.
Peucedanum.
Pimpinella.
Prangos.
Scandix.
Selinum.
Sison.
Smyrnium.
Stilbocarpa.
Thapsia.
Tinguarra.

Rubiaceæ.

Alibertia.
Cephælis.
Cephalanthus.
Cinchona.
Coffea.
Plectronia.
Psychotria.
Rubia.
Vangueria.

Caprifoliaceæ.

Sambucus.

Valerianeæ.

Nardostachys.
Valeriana.
Valerianella.

Dipsaceæ.

Dipsacus.

Compositæ.

Achilléa.
Anacyclus.
Anthemis.
Arnica.
Artemisia.
Baccharis.
Carthamus.
Cichorium.
Chrysanthemum.
Crepis.
Cynara.
Garuleum.
Grindelia.
Guizotia.
Helianthus.
Helichrysum.
Heterothalamus.
Hypochœris.
Inula.
Launæa.
Lactuca.
Leyssera.
Liatris.
Matricaria.
Microseris.
Osmitopsis.
Parthenium.
Pentzia.
Podachænium.
Rhaponticum.
Santolina.
Saussurea.
Scorzonera.
Schkuhria.
Scolymus.
Spilanthes.
Tagetes.
Tanacetum.
Taraxacum.
Tarchonanthus.
Tragopogon.

Ericaceæ.

Arbutus.
Arctostaphylos.
Gaultiera.
Gaylussacia.
Rhododendron.
Vaccinium.

Styraceæ.

Styrax.
Symplocos.

Ebenaceæ.

Diospyros.
Euclea.
Maba.
Royenia.

Sapotaceæ.

Achras.
Argania.
Dichopsis.
Illipe.
Niemeyera.

Myrsinaceæ.

Ægiceras.

Jasmineæ.

Azima.
Fraxinus,
Jasminum.
Ligustrum.
Nyctanthes.
Olea.
Osmanthus.

Apocyneæ.

Alstonia.
Apocynum.
Aspidosperma.
Carissa.
Gonioma.
Hancornia,
Vahea.

Loganiaceæ.

Buddlea.
Gelsemium.
Spigelia.

Plantagineæ.

Plantago.

Gentianeæ.

Gentiana.
Menyanthes.
Sabbatia.
Sebæa.
Swertia.

Convolvulaceæ.

Convolvulus.
Ipomœa.

2 G

SYSTEMATIC INDEX TO GENERA—*continued.*

Solanaceæ.
Atropa.
Bassovia.
Capsicum.
Duboisia.
Hyoscyamus.
Lycium.
Nicotiana.
Physalis.
Salpichroma.
Solanum.

Scrophularinæ.
Chelone.
Digitalis.
Lyperia.
Verbascum.
Veronica.

Acanthaceæ.
Justicia.

Bignoniaceæ.
Catalpa.
Jaracanda.

Pedalinæ.
Sesamum.

Asperifoliæ.
Alkanna.
Borrago.
Cerinthe.
Echium.

Musaceæ.
Musa.

Scitamineæ.
Canna.
Maranta.
Zingiber.

Bromeliaceæ.
Ananas.
Tillandsia.

Taccaceæ.
Tacca.

Dioscorideæ.
Dioscorea.

Irideæ.
Crocus.

Amaryllideæ.
Agave.
Aletris.

Heliotropium.
Lithospermum.
Onoma.
Symphytum.
Trichodesma.

Labiatæ.
Audibertia.
Calamintha.
Cedronella.
Dracocephalum.
Hedeoma.
Hyssopus.
Lavandula.
Leonotis.
Marrubium.
Melissa.
Mentha.
Meriandra.
Monarda.
Ocimum.
Origanum.
Pogostemon.
Perilla.
Prostanthera.
Pycnanthemum.
Rosmarinus.
Salvia.
Satureja.
Teucrium.
Thymus.
Trichostema.

Verbenaceæ.
Avicennia.
Gmelina.

MONOCOTYLEDONEÆ.

Alstrœmeria.
Beschorneria.
Fourcroya.
Polianthes.

Liliaceæ.
Aloe.
Allium.
Asparagus.
Chlorogalum.
Colchicum.
Convallaria.
Cordyline.
Dracæna.
Eremurus.
Geitonoplesium.
Lapageria.
Ophipogon.
Phormium.
Ruscus.
Sanseviera.
Schœnocaulon.

Lippia.
Priva.
Tectona.

Myoporineæ.
Eremophila.
Myoporum.

Coniferæ.
Araucaria.
Belis.
Callitris.
Cephalotaxus.
Cryptomeria.
Cupressus.
Dacrydium.
Dammara.
Fitzroya.
Ginkgo.
Glyptostrobus.
Juniperus.
Libocedrus.
Nageia.
Phyllocladus.
Pinus.
Saxono-Gothæa.
Sciadopitys.
Sequoia.
Taxodium.
Taxus.
Thuya.
Torreya.

Cycadeæ.
Cycas.
Encephalartos.

Scilla.
Smilax.
Trillium.
Urginia.
Uvularia.
Veratrum.
Xanthorrhœa.
Yucca.

Alismaceæ.
Aponogeton.
Butomus.
Sagittaria.

Aroideæ.
Acorus.
Colocasia.
Richardia.
Xanthosma.

Pandanaceæ.
Pandanus.

Palmaceæ.
Acanthophœnix.
Acrocomia.
Bactris.
Bacularia.
Brahea.
Borassus.
Calamus.
Calyptronoma.
Caryota.
Ceroxylon.
Chamædora.
Chamærops.
Cocos.
Copernicia.
Diplothemium.
Dypsis.
Geonoma.
Hyphæne.
Jubæa.
Kentia.
Livistona.
Mauritia.
Oncosperma.
Oreodoxa.
Phœnix.
Plectocomia.
Prestoa.
Ptychosperma.
Rhapis.
Rhapidophyllum.
Sabal.
Thrinax.
Trithrinax.
Wallichia.
Wettinia.
Zalacca.

Restiaceæ.
Elegia.

Cyperaceæ.
Carex.
Cyperus.

Heleocharis.
Lepidosperma.
Lepironia.
Scirpus.

Gramineæ.
Agrostis.
Aira.
Alopecurus.
Andropogon.
Anthistiria.
Anthoxanthum.
Aristida.
Arundinaria.
Arundinella.
Arundo.
Avena.
Bambusa.
Beesha.
Bouteloua.
Bromus.
Buchloa.
Calamagrostis.
Chionachne.
Chloris.
Chusquea.
Cinna.
Cynodon.
Cynosurus.
Dactylis.
Danthonia.
Dendrocalamus.
Dimochloa.
Ectrosia.
Ehrharta.
Eleusine.
Elymus.
Erianthus.
Eriochloa.
Euchlæna.
Festuca.
Gigantochloa.
Guadua.
Hemarthria.
Hierochloa.

Holcus.
Hordeum.
Imperata.
Kœleria.
Leersia.
Lolium.
Lygeum.
Melica.
Melocalamus.
Melocanna.
Milium.
Muehlenbergia.
Nastus.
Neurachne.
Oryza.
Oryzopsis.
Oxytenanthera.
Panicum.
Pappophorum.
Paspalum.
Pennisetum.
Phalaris.
Phleum.
Phyllostachys.
Poa.
Rottboellia.
Saccharum.
Schizostachyum.
Secale.
Sesleria.
Spartina.
Spinifex.
Stenotaphrum.
Stipa.
Thamnocalamus.
Teinostachyum.
Thouarea.
Triodia.
Tripsacum.
Trisetum.
Triticum.
Uniola.
Zea.
Zizania.
Zoysia.

ACOTYLEDONEÆ.

Filices.
Cyathea.
Dicksonia.
Lycopodium.
Todea.

Lichenes.
Cetraria.
Roccella.

Fungaceæ.
Agaricus.
Boletus.

Cantharellus.
Clavaria.
Coprinus.
Cortinarius.
Exidia.
Fistulina.
Helvella.
Hydnum.
Morchella.
Pachyma.
Peziza.
Polygaster.

Polyporus.
Rhizopogon.
Russula.
Terfezia.
Tremella.
Tuber.

Algæ.
Chondrus.
Porphyra.

GENERIC SYNONYMS, REFERRED RESPECTIVELY TO THE ADOPTED GENERA.

(In many cases the name of the genus changeable only for some of its species.)

Abbevillea	...	Psidium.	Cotinus ...	... Rhus.
Abies ...	...	Pinus.	Cunninghamia	... Belis.
Acacia ...	...	Albizzia, Pipta-denia.	Cydonia ...	... Pyrus.
			Dactylis ...	... Poa.
Achyrophorus	...	Hypochæris.	Danthonia	... Astrebla.
Adenachæna	...	Phymaspermum.	Diosma ...	... Barosma.
Aeluropus...	...	Dactylis.	Dolichos ..	... Vigna.
Agapetes ...	...	Vaccinium.	Eragrostis...	... Poa.
Agropyrum	...	Triticum.	Eriobotrya	... Photinia.
Alocasia ...	...	Colocasia.	Ervum ...	... Vicia.
Amarantus	...	Euxolus.	Euryangium	... Ferula.
Amphibromus	...	Danthonia.	Fabricia ...	... Leptospermum.
Amygdalus	...	Prunus.	Ferdinanda	... Podachænium.
Anesorrhiza	...	Selinum.	Festuca ...	... Distichlis.
Anethum ...	...	Carum, Peuceda-num.	Flueggea ...	... Ophiopogon.
			Frenela ...	... Callitris.
Apium	...	Carum.	Glechoma ...	... Nepeta.
Aralia	...	Fatsia.	Glyceria ...	... Poa.
Areca	...	Bacularia, Pty-chosperma.	Grumilia ...	... Psychotria.
			Guilandina	... Cæsalpinia.
Armeniaca	...	Prunus.	Guilielma ...	... Bactris.
Arrhenatherum	...	Avena.	Gynerium...	... Arundo.
Asa-Graya	...	Schoenocaulon.	Haplotaxis	... Saussurea.
Astragalus	...	Oxytropis.	Harina ...	... Wallichia.
Azadirachta	...	Melia.	Hedyscepe	... Kentia.
Balsamocarpon	...	Cæsalpinia.	Hirneola ...	... Exidia.
Bambusa ...	...	Gigantochloa, Guadua, Phyl-lostachys.	Howea ...	... Kentia.
			Hydrochloa	... Zizania.
			Hydropyrum	... Zizania.
Bassia ...	...	Illipe.	Hyospathe	... Prestoa.
Batatas ...	...	Ipomœa.	Imperatoria	... Peucedanum.
Beesha ...	...	Melocanna.	Isachne ...	... Panicum.
Biotia ...	...	Thuya.	Isonandra...	... Dichopsis.
Blighia ...	...	Cupania.	Juania ...	... Ceroxylon.
Blitum ...	...	Chenopodium.	Kentia ...	... Clinostigma.
Boaria ...	...	Maytenus.	Landolphia	... Vahea.
Brahea ...	...	Washingtonia.	Larix ...	... Pinus.
Brayera ...	...	Hagenia.	Lens ...	... Ervum.
Bunium ...	...	Carum.	Lithræa ...	... Rhus.
Cæsalpinia	...	Peltophorum.	Lucuma ...	... Niemeyera.
Calamagrostis	...	Psamma.	Lycopersicum	... Solanum.
Camassia ...	...	Scilla.	Lysiloma ...	... Albizzia.
Campomanesia	...	Psidium.	Macranthus	... Mucuna.
Caryotaxus	...	Torreya.	Macrochloa	... Stipa.
Catabrosa...	...	Poa.	Macrozamia	... Encephalartos.
Caulophyllum	...	Leontice.	Maharanga	... Onosma.
Cedrus ...	...	Pinus.	Malabanga...	... Peucedanum.
Ceratostemma	...	Vaccinium.	Melloca ...	... Ullucus.
Chamæcyparis	...	Cupressus, Thuya.	Mespilus ...	... Cratægus, Pyrus.
Chamærops	...	Rhapidophyllum.	Metrosideros	... Syncarpia.
Cibotium ...	...	Dicksonia.	Microlæna	... Ehrharta.
Cicca ...	...	Phyllanthus.	Monizia ...	... Thapsia.
Citrullus ...	...	Cucumis.	Myrcianthes	... Myrtus.
Cladosicyos	...	Corynosicyos.	Nannorrhops	... Chamærops.
Codiæum ...	...	Baloghia.	Nasturtium	... Cochlearia.
Commiphora	...	Balsamodendron.	Negundo ...	... Acer.

GENERIC SYNONYMS—*continued.*

Nelumbium	...	Nelumbo.	Rubachia ...	...	Marliera.
Nengella ...	...	Kentia.	Sabadilla ...	...	Schœnocaulon.
Ophelia ...	...	Swertia.	Salisburia...	...	Ginkgo.
Oplismenus	...	Panicum.	Sapota ...	...	Achras.
Oreodaphne	...	Umbellularia.	Satureja ...	...	Thymus.
Orobus ...	...	Lathyrus.	Scilla ...	...	Urginea.
Osmunda ...	...	Todea.	Scorodosma	...	Ferula.
Oxycoccos...	...	Vaccinium.	Seaforthia...	...	Ptychosperma.
Oxytenanthera	...	Gigantochloa.	Serenaea ...	...	Sabal.
Panax ...	...	Aralia.	Setaria ...	...	Panicum.
Panicum ...	...	Pennisetum.	Sinapis ...	...	Brassica.
Paspalum ...	...	Panicum.	Sium ...	...	Pimpinella.
Passerina ...	...	Thymelæa.	Slackea ...	...	Decaisnea.
Pastinaca ...	...	Peucedanum.	Soja ...	...	Glycine.
Penicillaria	...	Pennisetum.	Sorghum ...	...	Andropogon.
Picea ...	...	Pinus.	Spartium ...	...	Cytisus.
Piptochætium	...	Oryzopsis.	Stillingia ...	...	Excæcaria.
Pithecolobium	...	Albizzia.	Taxodium...	...	Glyptostrobus,
Planera ...	...	Zelkova.			Sequoia.
Poa ...	...	Festuca.	Tetranthera	...	Umbellularia.
Podocarpus	...	Nageia.	Thamnocalamus	...	Arundinaria.
Poinciana ...	...	Cæsalpinia.	Thea ...	...	Camellia.
Pollichia ...	...	Trichodesma.	Thibaudia...	...	Vaccinium.
Poterium ...	...	Sanguisorba.	Thuyopsis...	...	Thuya.
Prinos ...	...	Ilex.	Trachycarpus	...	Chamærops.
Pritchardia	...	Washingtonia.	Tricholæna	...	Panicum.
Prumnopithys	...	Nageia.	Tricuspis ...	...	Festuca.
Pyrethrum	...	Chrysanthemum.	Trisetum ...	...	Avena.
Quadria ...	...	Guevina.	Ulmus ...	...	Holoptelea.
Reana ...	...	Euchlæna.	Uralepis ...	...	Festuca.
Retinospora	...	Cupressus.	Wellingtonia	...	Sequoia.
Rhopalostylis	...	Kentia.	Widdringtonia	...	Callitris.
Rottlera ...	...	Mallotus.	Witheringia	...	Bassowia.

GEOGRAPHIC INDEX.

NORTHERN AND MIDDLE EUROPE.

Acer campestre, A. platanoides, A. Pseudo-platanus, Achillea atrata, A. Millefolium, A. moschata, A. nana, Aconitum Napellus, Acorus Calamus, Actæa spicata, Agaricus Auricula, A. cæsareus, A. campestris, A. Cardarella, A. decorus, A. deliciosus, A. eryngii, A. esculentus, A. extinctorius, A. flammeus, A. fusipes, A. gambosus, A. giganteus, A. Marzuolus, A. melleus, A. Mouzeron, A. odorus, A. oreades, A. ostreatus, A. procerus, A. scorodonius, A. socialis, A. splendens, A. sylvaticus, A. virgineus, A. volemus, Agrostis alba, A. rubra, A. vulgaris, Aira cæspitosa, Alchemilla alpina, A. vulgaris, Allium Ampeloprasum, A. Schœnoprasum, A. Scorodoprasum, Alnus glutinosa, A. incana, Alopecurus bulbosus, A. geniculatus, A. pratensis, Althœa officinalis, Anemone Pulsatilla, Anthemis nobilis, A. tinctoria, Anthriscus Cerefolium, Archangelica officinalis, Arctostaphylos uva-ursi, Arnica montana, Artemisia Absinthium, A. Mutellina, A. Pontica, Asparagus officinalis, Asperula odorata, Astragalus arenarius, A. glycyphyllos, A. hypoglottis, Atropa Belladonna, Avena elatior, A. fatua, A. flavescens, A. pratensis, A. pubescens, A. sativa, Barbarea vulgaris, Beta vulgaris, Betula alba, Boletus bovinus, B. circinans, B. edulis, B. luteus, B. sapidus, B. scaber, B. subtomentosus, B. variegatus, Brassica alba, B. Napus, B. nigra, B. oleracea, B. Rapa, Bromus asper, Butomus umbellatus, Buxus sempervirens, Cakile maritima, Calamintha officinalis, Camelina sativa, Cantharellus edulis, Carex arenaria, Carpinus Betulus, Carum Bulbocastanum, C. Carui, C. segetum, Cetraria Islandica, Chæromyces meandriformis, Chærophyllum bulbosum, C. sativum, Chenopodium Bonus Henricus, Chondrus crispus, Cichorium Intybus, Clavaria aurea, C. botrytis, C. brevipes, C. coralloides, C. crispa, C. flava, C. formosa, C. grisea, C. muscoides, C. palmata, Cochlearia Armoracia, C. officinalis, Colchicum autumnale, Conium maculatum, Convallaria majalis, Coprinus comatus, Cornus màs, Cortinarius cinnamomeus, Corylus Avellana, Crambe maritima, Cratægus Oxyacantha, Crepis biennis, Cynosurus cristatus, Cytisus scoparius, Dactylis glomerata, Daucus Carota, Digitalis purpurea, Dipsacus fullonum, Elymus arenarius, Euxolus viridis, Exidia auricula Judœa, Fagus sylvatica, Festuca arundinacea, F. drymeia, F. duriuscula, F. elatior, F. gigantea, F. heterophylla, F. loliacea, F. ovina, F. pratensis, F. rubra, F. silvatica, Fistulina hepatica, Fragaria collina, F. vesca, Fraxinus excelsior, Genista tinctoria, Gentiana lutea, Geum urbanum, Helleborus niger, Helvella esculenta, H. Gigas, H. infula, Heracleum Sibiricum, Holcus lanatus, H. mollis, Hordeum nodosum, H. secalinum, Humulus Lupulus, Hydnum album, H. auriscalpium, H. Caput Medusæ, H. coralloides, H. diversidens, H.

erinaceum, H. fuligineo-album, H. graveolens, H. repandum, H. suaveolens, H. hystrix, H. imbricatum, H. infundibulum, H. lævigatum, H. subsquamosum, H. violascens, Hyoscyamus niger, Ilex Aquifolium, Inula Helenium, Juniperus communis, Lactuca virosa, Laserpitium aquilegium, Lathyrus macrorrhizus,˙L. pratensis, L. sativus, Lavatera arborea, Leersia oryzoides, Lolium perenne, Lotus corniculatus, L. majòr, Marrubium vulgare, Matricaria Chamomilla, Medicago falcata, M. sativa, Melica altissima, M. ciliata, M. nutans, M. uniflora, Melilotus alba, M. officinalis, Mentha arvensis, M. citrata, M. crispa, M. piperita, M. Pulegium, M. rotundifolia, M. sylvestris, M. viridis, Menyanthes trifoliata, Milium effusum, Morchella conica, M. deliciosa, M. esculenta, M. Gigas, M. patula, Nepeta Cataria, N. Glechoma, Œnanthe Phellandrium, Onòbrychis sativa, Origanum vulgare, Panicum Germanicum, Peucedanum officinale, P. Ostruthrium, P. sativum, Peziza macropus, Phleum alpinum, P. pratense, Physalis Alkekengi, Pimpinella saxifraga, Pinus Abies, P. Cembra, P. Larix, P. montana, P. obovata, P. picea, P. silvestris, Plantago lanceolata, Poa airoides, P. alpina, P. angustifolia, P. aquatica, P. distans, P. fertilis, P. fluitans, P. maritima, P. nemoralis, P. palustris, P. pratensis, P. trivialis, Polyporus citrinus, P. frondosus, P. giganteus, P. ovinus, P. tuberaster, Populus alba, P. canescens, P. dilatata, P. fastigiata, P. nigra, P. tremula, Porphyra vulgaris, Prunus Mahaleb, P. Padus, P. spinosa, Psamma arenaria, P. Baltica, Pyrus aucuparia, P. Germanica, P. nivalis, Quercus Robur, Reseda Luteola, Rhamnus catharticus, R. Frangula, Rhizopogon magnatum, R. rubescens, Ribes Grossularia, R. nigrum, R. rubrum, Rosa Gallica, R. spinosissima, Rubia peregrina, Rubus cæsius, R. Chamæmorus, R. fruticosus, R. Idæus, Rumex Acetosa, R. scutatus, Ruscus aculeatus, Russula vesca, Salix alba, S. caprea, S.·daphnoides, S. fragilis, S. lanceolata, S. purpurea, S. rubra, S. triandra, S. viminalis, Sambucus nigra, Sanguisorba minor, Saponaria officinalis, Scorzonera Hispanica, Sesleria cœrulea, Sison Amomum, Smyrnium Olusatrum, Solanum Dulcamara, Spartina stricta, Spergula arvensis, Tanacetum vulgare, Taraxacum officinale, Tilia Europæa, Tragopogon porrifolius, Trapa natans, Tremella mesentorica, Trifolium agrarium, T. alpestre, T. fragiferum, T. hybridum, T. incarnatum, T. medium, T. montanum, T. ochroleucum, T. Pannonicum, T. pratense, T. repens, T. spadiceum, Triticum junceum, Tuber æstivum, T. albidum, T. cibarium, T. magnatum, T. melanosporum, Typha latifolia, Ulex Europæus, Ulmus campestris, U. pedunculata, Vaccinium Myrtillus. V. Oxycoccos, V. uliginosum, V. Vitis Idæa, Valeriana Celtica, Valerianella olitoria, Veratrum album, Verbascum Thapsus, Vicia sativa, V. sepium, V. sylvatica, Viola odorata.

COUNTRIES ON OR NEAR THE MEDITERANEAN SEA.

Acacia Arabica, A. gummifera, A Seyal, A. tortilis, A. Verek, Acer campestre, A. Creticum, A. Pseudo-Platanus, Achillea fragrantissima, Ægilops ovata, Æsculus Hippocastanum, Agaricus cæsareus, A. flammeus, Agrostis alba, A. vulgaris, Aira caespitosa,. Albizza Lebbeck, Alchemilla vulgaris, Alkanna tictoria, Allium Ampe-- loprasum, A. Ascallonicum, A. Cepa, A. Neapolitanum, A. Porrum, A. roseum, A. sativum, A. Scorodoprasum, Aloe vulgaris, Alopecurus, bulbosus, A. geniculatus, A. pratensis, Althæa officinalis, Amarantus Blitum, Anacyclus Pyrethrum, Andropogon Gryllos, A. Haleppensis, A. Ischæmum, A. provincialis, A. Schœnanthus, Anthemis nobilis, A. tinctoria, Anthoxanthum odoratum, Anthyllis vulneraria, Apium graveolens, Argania Sideroxylon, Artemisia Abrotanum, A. Absin- thium, A. Pontica, Arundo Ampelodesmos, A. Donax, A. Pliniana, Asparagus acutifolius, A. albus, A. aphyllus, A. horridus, A. officinalis, Astragalus adscendens, A. arenarius, A. brachycalyx, A. Cephalonicus,. A. Cicer, A. Creticus, A. glycyphyllos, A. gummifer, A. microcephalus,. A. Parnassi, A. strobiliferus, A, stromatodes, A. venosus, A. verus, Atriplex rosea, Atropa Belladonna, Avena elatior, A. fatua, A.. flavescens, A. pubescens, A. sativa, Balsamodendron Mukul, B. Myrrha,. B. Opobalsamum, Beta vulgaris, Betula alba, Bongardia Rauwolfii, Borassus Æthiopicus, Borrago officinalis, Brassica alba, B. campestris, B. Cretica, B. juncea, B. Napus, B. nigra, B. oleracea, B. Rapa, Bromus erectus, Buxus Balearica, B. longifolia, B. sempervirens, Cajanus Indicus, Cakile maritima, Calamintha Nepeta, C. officinalis, Callitris quadrivalvis, Camelina sativa, Cannabis sativa, Capparis spinosa, Carpinus Betulus, Carthamus tinctorius, Carum Bulbocastanum, C. Carui, C. ferulifolium, C. Petroselinum, C. segetum, Cassia acutifolia, C. angustifolia, C. obovata, Castanea sativa, Catha edulis, Cedronella triphylla, Celtis Australis, Ceratonia Siliqua, Cerinthe major, Chamæ-- rops. humilis, Chelidonium majus, Chenopodium Blitum, Chœrophyl- lum bulbosum, Choiromyces Leonis, Chrysanthemum carneum, C. parthenium, C. roseum, Cicer arietinum, Cichorium Endivia, C. Intybus,. Cistus Creticus, C. Cyprius, Cochlearia Armoracia, Coffea Arabica,. Colchicum autumnale, Colocasia antiquorum, Conium maculatum, Cono- podium denudatum, Convallaria majalis, Convolvulus floridus, C. Scammonia, C. scoparius, Coprinus comatus, Cornus mas, Cortinarius cinnamomeus, Coriandum sativum, Corylus Avellana, C. Colurna, C.. maxima, C. Pontica, Corynosicyos edulis, Crambe cordifolia, C. Kots- chyana, C. maritima, C. Tartaria, Cratægus Azarolus, C. Oxyacantha, C. Pyracantha, Crepis biennis, Crithmum maritimum, Crocus sativus, C. serotinus, Crozophora tinctoria, Cucumis Citrullus, C. Colocynthis, C. Melo, C. sativus, Cucurbita maxima, C. Melopepo, C. moschata, C. Pepo, Cuminum Cyminum, C. Hispanicum, Cupressus sempervirens, . Cynara Cardunculus, C. Scolymus, Cynodon Dactylon, Cynosurus cristatus, Cyperus esculentus, C. Papyrus, C. proliferus, C. Syriacus, Cytisus proliferus, C. scoparius, C. spinosus, Dactylis glomerata, D. litoralis, Daphne Mezereum, Datisca cannabina, Daucus Carota,. Digitalis purpurea, Diospyros Lotus, Dipsacus fullonum, Dolichos Lablab, D. uniflorus, Dorema Ammoniacum, Dracæna Draco, D. schizantha, Dracocephalum Moldavica, Ecballion Elaterium, Echium candicans, Elæagnus hortensis, Eleusine flagelligera, E. Tocussa,

Elymus arenarius, Eruca sativa, Ervum Lens, Eryngium maritimum, Euxolus viridis, Exidia auricula Judæ, Fagopyrum esculentum, F. Tataricum, Fagus sylvatica, Ferula galbaniflua, F. longifolia, Festuca elatior, F. gigantea, F. sylvatica, Ficus Carica, F. Sycomorus, Fistulina hepatica, Fœniculum officinale, Fragaria collina, F. pratensis, F. vesca, Fraxinus excelsior, F. Ornus, Genista monosperma, G. sphærocarpa, G. tinctoria, Gentiana lutea, Geum urbanum, Glaucium luteum, Glycyrrhiza echinata, G. glabra, Gossypium arboreum, Guilandina Bonduc, G. Bonducella, Hedysarum coronarium, Helichrysum orientale, Helleborus niger, Hippocrepis comosa, Holcus lanatus, H. mollis, Hordeum deficiens, H. distichon, H. hexastichon, H. macrolepis, H. nodosum, H. vulgare, H. zeocriton, Humulus Lupulus, Hydnum imbricatum, Hyoscyamus niger, Hyphæne Argun, H. coriacea, Hyssopus officinalis, Imperata arundinacea, Indigofera argentea, Inula Helenium, Iris Florentina, I. juncea, Isatis tinctoria, Jasminum odoratissimum, J. officinale, Juglans regia, Juniperus brevifolia, J. Cedrus, J. drupacea, J. excelsa, J. fœtidissima, J. Phœnicea, J. procera, Koeleria cristata, K. glauca, Lactuca virosa, Lathyrus Cicera, L. pratensis, L. tuberosus, Laserpitium aquilegium, Laurus nobilis, Lavandula angustifolia, L. latifolia, L. Stœchas, Lavatera arborea, Lawsonia alba, Leersia oryzoides, Lepidium latifolium, L. sativum, Linum usitatissimum, Liquidambar Altingia, L. orientalis, Lolium Italicum, L. perenne, Lotus corniculatus, L. major, L. siliquosus, L. tetragonolobus, Lupinus albus, L. angustifolius, L. luteus, L. varius, Lycium Afrum, L. barbarum, L. Europæum, Lygeum Spartum, Marrubium vulgare, Matricaria Chamomilla, Medicago arborea, M. lupulina, M. media, M. orbicularis, M. sativa, M. scutellata, Melica ciliata, M. nutans, M. uniflora, Melilotus alba, M. cœrulea, M. macrorrhiza, M. officinalis, Melissa officinalis, Mentha arvensis, M. citrata, M. crispa, M. piperita, M. Pulegium, M. rotundifolia, M. sylvestris, M. viridis, Menyanthes trifoliata, Meriandra Abyssinica, Milium effusum, Morchella deliciosa, M. esculenta, M. conica, Moringa aptera, Morus nigra, Musa Ensete, Myrica Faya, Myrrhis odorata, Myrtus communis, Nelumbo nucifera, Nepeta raphanorrhiza, N. Cataria, N. Glechoma, Nicotiana Persica, Ocimum basilicum, O. sanctum, O. suave, Œnanthe Phellandrium, Olea Europæa, Onobrychis sativa, Origanum Dictamnus. O. hirtum, O. Majorana, O. Maru, O. normale, O. Onites, O. virens, O. vulgare, Ornithopus sativus, Ostrya carpinifolia, Oxytenanthera Abyssinica, Oxytropis pilosa, Paliurus Spina Christi, Panicum brizanthum, P. Crus-Galli, P. glabrum, P. maximum, P. prostratum, P. repens, P. sanguinale, P. spectabile, P. turgidum, Papaver somniferum, Pennisetum typhoideum, Persea Teneriffæ, Peucedanum cachrydifolium, P. graveolens, P. officinale, P. Sekakul, Phalaris aquatica, P. brachystachys, P. Canariensis, P. minor, P. truncata, Phaseolus coccineus, Phleum alpinum, P. pratense, Phœnix dactylifera, Physalis Alkekengi, P. angulata, Pimpinella Anisum, P. magna, P. nigra, P. saxifraga, P. Sisarum, Pinus Abies, P. Canariensis, P. Cedrus, P. Cembra, P. Cilicica, P. Haleppensis, P. Laricio, P. Larix, P. montana, P. orientalis, P.

Pinaster, P. Pinea, P. Pinsapo, P. Pyrenaica, Pistacia Atlantica, P. Lentiscus, P. Terebinthus, P. vera, Peucedanum sativum, Plantago lanceolata, P. Psyllium, Platanus orientalis, Poa Abyssinica, P. airoides, P. angustifolia, P. aquatica, P. cynosuroides, P. distans, P. fluitans, P. maritima, P. nemoralis, P. palustris, P. trivialis, Populus alba, P. canescens, P. dilatata, P. Euphratica, P. fastigiata, P. nigra, P. tremula, Prosopis Stephaniana, Prunus Amygdalus, P. Armeniaca, P. avium, P. cerasifera, P. Cerâsus, P. domestica, P. insititia, P. Lauro-Cerasus, P. Lusitanica, P. Mahaleb, P. Padus, P. Persica, P. spinosa, Psamma arenaria, Pugionium cornutum, Punica Granatum, Pyrus aucuparia, P. communis, P. Cydonia, P. Germanica, P. Malus, P. nivalis, P. salicifolia, Quercus Ægilops, Q. Cerris, Q. coccifera, Q. Ilex, Q. infectoria, Q. macrolepis, Q. Robur, Q. Suber, Q. Toza, Reseda luteola, R. odorata, Rhamnus Alaternus, R. amygdalinus, R. catharticus, R. Frangula, R. Græcus, R. infectorius, R. oleoides, R. prunifolius, R. saxatalis, Rhaponticum acaule, Rheum Rhaponticum, Rhus Coriaria, R. Cotinus, Ribes Grossularia, R. nigrum, R. orientale, R. rubrum, Richardia Africana, Ricinus communis, Roccella tinctoria, Rosa centifolia, R. Damascena, R. Gallica, R. moschata, R. sempervirens, R. spinosissima, Rosamarinus officinalis, Rubia peregrina, R. tinctoria, Rubus fruticosus, R. Idæus, Rumex Acetosa, R. scutatus, R. vesicarius, Ruscus aculeatus, Russula vesca, Ruta graveolens, R. sylvestris, Sagittaria sagittifolia, Salix alba, S. Babylonica, S. daphnoides, S. fragilis, S. purpurea, S. rubra, S. viminalis, Salvia officinalis, Sambucus nigra, Sanguisorba minor, Santolina Cyparissias, Saponaria officinalis, Satureja Græca, S. hortensis, S. Juliana, S. montana, S. Thymbra, Saussurea Lappa, Scandix grandiflora, Scolymus Hispanicus, Scorzonera Astrachanica, S. crocifolia, S. deliciosa, S. Hispanica, S. lanata, S. ramosa, S. Scowitzii, S. semicana, S. tuberosa, S. undulata, Secale cereale, S. Creticum, Sesbania Ægyptica, Sesuvium Portulacastrum, Sison Amomum, Smilax aspera, Smyrnium Olusatrum, Solanum Æthiopicum, S. Dulcamara, S. edule, Spartina stricta, Spartium junceum, Spergula arvensis, Spinacia tetrandra, Stipa arenaria, S. tenacissima, Styrax officinalis, Symphytum peregrinum, S. officinale, Tamarindus Indica, Tamarix articulata, T. Gallica, T. Germanica, T. orientalis, Tanacetum vulgare, Taraxacum officinale, Taxus baccata, Terfezia Leonis, Teucrium Chamædrys, T. Creticum, T. Marum, T. Polium, T. Scordium, Thapsia edulis, Theligonum Cynocrambe, Thouarea sarmentosa, Thymelæa tinctoria, Thymus æstivus, T. capitatus, T. hiemalis, T. Mastichina, T. Serpillum, T. vulgaris, Tilia argentea, T. Europæa, Tinguarra Sicula, Tragopogon porrifolius, Trapa natans, Tremella mesenterica, Trichodesma Zeylanicum, Trifolium agrarium, T. Alexandrinum, T. alpestre, T. fragiferum, T. hybridum, T. incarnatum, T. medium, T. montanum, T. ochroleucum, T. pratense, T. Quartinianum, T. repens, T. resupinatum, T. spadiceum, T. subrotundum, Trigonella Fœnum Græcum, Triticum junceum, T. vulgare, Tuber æstivum, T. albidum, T. cibarium, T. magnatum, Ulex Europæus, Ulmus campestris, U. pedunculata, Urginia Scilla, Vaccinum Arctostaphylos, V. Myrtillus,

V. Oxycoccos, V. uliginosum, V. Vitis-Idæa, Valeriana officinalis;
Valerianella olitoria, Veratrum album, Verbascum Thapsus, Viburnum
Tinus, Vicia Cracca, V. Ervilia, V. Faba, V. peregrina, V. sativa, V.
sepium, V. sylvatica, V. tetrasperma, Viola odorata, Vitis Schimperiana,
V. vinifera, Withania somnifera, Zelkova crenata, Z. Cretica, Zizyphus
Lotus, Z. Spina Christi, Z. vulgaris.

NORTHERN AND TEMPERATE EASTERN ASIA.

Acer palmatum, A. pictum, Æsculus turbinata, Agaricus flammeus,
Agriophyllum Gobicum, Agrostis alba, A. vulgaris, Ailanthus glan-
dulosa, Albizzia Julibrissin, Aleurites cordata, Allium Cepa, A. fistulo-
sum, A. sativum, A. Schœnoprasum, Alopecurus geniculatus, Andro-
pogon involutus, Aralia cordata, Arenga saccharifera, Aristolochia
recurvilabra, Artemisia Cina, A. Dracunculus, Arundinaria Japonica,
Atriplex hortensis, Avena elatior, A. fatua, A. flavescens, A. pubescens,
Balsamodendron Mukul, Bambusa Beechyana, B. flexuosa, B. Senaensis,
B. tuldoides (under Schizostachyum), Barbarea vulgaris, Basella rubra,
Betula alba, Bœhmeria nivea, Brassica alba, B. Chinensis, B. juncea,
B. nigra, Bromus asper, Broussonetia papyrifera, Butomus umbellatus,
Buxus microphylla, B. sempervirens, Cæsalpinia sepiaria, Camellia
Japonica, C. Thea, Cannabis sativa, Caragana arborescens, Carissa
Carandas, Carpinus cordata, C. erosa, C. Japonica, C. laxiflora, Carum
Bulbocastanum, C. Carui, Catalpa Kæmpferi, Cedrela Sinensis, Cepha-
lotaxus Fortunei, C. drupacea, Cetraria Islandica, Chamærops excelsa,
C. Fortunei, Chenopodium Bonus Henricus, Cinnamomum Camphora,
Citrus Japonica, C. trifoliata, Convallaria majalis, Coprinus comatus,
Corchorus capsularis, Cordyline terminalis, Cornus mas, Cortinarius
cinnamomeus, Corylus heterophylla, Cryptomeria Japonica, Cucumis
Melo, Cudrania triloba, Cupressus funebris, C. obtusa, C. pisifera, Cycas
revoluta, Daucus Carota, Debregeasia edulis, Dendrocalamus strictus,
Dioscorea Japonica, D. oppositifolia, D. quinqueloba, D. sativa, Diospyros Kaki, D. Lotus, Ehrharta caudata, Elæagnus hortensis, E. parvifolius, E. umbellatus, Eleusine Coracana, Erianthus Japonicus, Eruca
sativa, Euonymus Japonicus, Euryale ferox, Euxolus viridis, Excæcaria
sebifera, Fagopyrum cymosum, F. emarginatum, F. esculentum, F.
Tataricum, Fagus Sieboldii, Fatsia papyrifera, Ferula Sambul. Fistulina
hepatica, Fraxinus Chinensis, Genista tinctoria, Geum urbanum, Ginkgo
biloba, Gleditschia horrida, Glycine hispida, G. Soya, Glyptostrobus
heterophyllus, Heleocharis tuberosa, Heracleum Sibiricum, Hordeum
secalinum, Hovenia dulcis, Hydrangea Thunbergi, Ilex crenata, Illicium
anisatum, Imperata arundinacea, Isatis indigotica, I. tinctoria, Jasminum
grandiflorum, J. officinale, J. Sambac, Juglans cordiformis, J. Mandschurica. J. Sieboldiana, J. stenocarpa, Juniperus Chinensis, J. sphærica,
Lathyrus macrorrhizus, Lepidium latifolium, Lespedeza striata,
Ligustrum Japonicum, Liquidambar Formosana, Livistona Chinensis,
Magnolia hypoleuca, M. Yulan, Melia Azadirachta, Melica altissima,
Mentha arvensis, Morchella conica, Morus alba, Mucuna Cochin-

· chinensis, Musa Cavendishii, Myrica rubra, Myrtus tomentosa, Nageia cupressina, Nephelium Litchi, N. Longanum, Œnanthe Phellandrium, Ophiopogon Japonicus, Osmanthus fragrans, Pachyma Hœlen, Paliurus ramosissimus, Panax Shinsing, Paulownia imperialis, Pennisetum cereale, Perilla arguta, Phœnix pusilla, Photinia Eriobotrya, Phyllostachys bambusoides, P. nigra, Physalis Alkekengi, P. angulata, Pinus Alcockiana, P. densiflora, P. firma, P. Fortunei, P. Jezoensis, P. Kæmpferi, P. Koraiensis, P. leptolepis, P. Massoniana, P. obovata, P. selenolepis, P. Thunbergi, P. parviflora, P. polita, P. Sibirica, P. stenolepis, P. Tsuga, Pisum sativum, Planera Japonica, Poa airoides, P. alpina, P. fertilis, P. palustris, Polygaster Sampadarius, Polygonum tinctorium, Populus nigra, P. tremula, Prangos pabularia, Porphyra vulgaris, Prunus Armeniaca, P. domestica, P. Padus, P. Persica, P. Pseudo-cerasus, P. tomentosa, Pterocarpus Indicus, Pterocarya fraxinifolia, P. stenoptera, Pueraria Thunbergiana, Pugionium cornutum, Pyrus aucuparia, P. Japonica, Quercus Chinensis, Q. cornea, Q. cuspidata, Q. dentata, Q. glabra, Q. glauca, Q. Mongolica, Q. serrata, Rhamnus chlorophorus, R. Frangula, R. utilis, Rhapis flabelliformis, R. humilis, Rheum officinale, R. palmatum, R. Rhaponticum, R. Tartaricum, R. undulatum, Rhus semialata, R. succedanea, R. vernicifera, Rosa Indica, R. lævigata, R. moschata, R. sempervirens, R. spinosissima, Rubia cordifolia, Rubus parvifolius, R. phœnicolasius, Rumex acetosa, R. Patientia, R. vesicarius, Saccharum officinarum, S. Sinense, Sagittaria sagittifolia, Salix Babylonica, S. Japonica, Sanguisorba minor, Sciadopitys verticillata, Scorzonera albicaulis, Selinum Monnieri, Smilax China, Sophora Japonica, Spergula arvensis, Spinacia oleracea, Sterculia nobilis, Tetragonia expansa, Tetranthera Japonica, Tilia Europæa, T. Manchurica, Thuya dolabrata, T. Japonica, T. orientalis, Torreya grandis, T. nucifera, Trapa bicornis, T. bispinosa, Trifolium pratense, Triphasia Aurantiola, Ulmus campestris, U. parvifolia, Vaccinium præstans, Veratrum album, Vicia Cracca, V. sepium, V. sylvatica, Vigna Sinensis, Vitis Labrusca. V. vulpina, Wistaria Chinensis, Xanthoxylon piperitum, Zelkova acuminata, Zizania latifolia, Zizyphus Jujuba, Z. Sinensis, Zoysia pungens.

SOUTHERN ASIA.

Acacia Arabica, A. Catechu, A. concinna, A. Farnesiana, A. latronum, A. Sundra, Acer Campbelli, A. lævigatum, A. niveum, A. sterculiaceum, A. villosum, Aconitum ferox, Ægiceras majus, Æschynomene aspera, Æsculus Indica, Albizzia bigemina, A. Lebbeck, A. micrantha, A. stipulata, Aleurites cordata, A. triloba, Allium rubellum, Alnus Nepalensis, Aloe socotrina, Aloxylon Agallochum, Amarantus paniculatus, Andropogon annulatus, A. Calamus, A. cernuus, A. falcatus, A. Gryllos, A. involutus, A. Ischæmum, A. montanus, A. muricatus, A. Nardus, A. pertusus, A. saccharatus, A. Schœnanthus, A. sericeus, A. Sorghum, Anthistiria ciliata, Aponogeton crispus, Aquilaria Agallocha, Aralia Ginseng, Areca Nagensis, A. triandra (under Zalacca), Aristolochia Indica, Artocarpus Bengalensis, A. integrifolia, Arundinaria collosa, A. debilis, A. falcata, A. Falconeri, A. Hookeriana, A. Japonica, A. Khasiana, A. macrosperma, A. spathiflora, A. suberecta, A. tecta (partly under Schizostachyum), Arundinella Nepalensis, Arundo Bengalensis, A. Karka, Averrhoa Bilimbi, A. Carambola, Avicennia officinalis, Azima tetracantha, Bambusa arundinacea, B. aspera, B. attenuata, B. Balcooa, B. Blumeana, B. Brandisii, B. elegantissima, B. flexuosa, B. marginata, B. monadelpha, B. nutans, B. pallida, B. polymorpha, B. regia, B. spinosa, B. stricta, B. Tulda (under Schizostachyum). B. verticillata, B. vulgaris, Basella lucida, B. rubra, Beesha elegantissima, B. Rheedei, B. stridula, B. Travancorica (under Schizostachyum), Belis jaculifolia, Benincasa cerifera, Bentinckia Coddapanna (under Zalacca), Berberis aristata, B. Asiatica, B. Lycium, B. Nepalensis, Betula acuminata, Bœhmeria nivea, Borassus flabelliformis, Boswellia serrata, Brassica juncea, Buddlea Asiatica, B. Colvillei, B. macrostachya, B. paniculata, Butea frondosa, Buxus Wallichiana, Cæsalpinia Sappan, C. sepiaria, Cajanus Indicus, Calamus montanus, C. acanthospathus, C. erectus, C. extensus, C. Flagellum, C. floribundus, C. leptospadix, C. macrospathus, C. Mishmelensis, C. quinquenervius, C. Royleanus, C. schizospathus, C. tenuis (under Zalacca), Camellia Thea, Canavalia gladiata, Capparis aphylla, C. horrida, C. Roxburghi, C. sepiaria, Carex Moorcroftiana, Carissa Carandas, Carpinus viminea, Carthamus tinctorius, Carum Ajowan, C. gracile, C. nigrum, C. Roxburghianum, Caryota obtusa (under Zalacca), C. urens, Cassia fistula, Castanopsis argentea, C. Indica, Casuarina equisetifolia, Cedrela febrifuga, C. Taona, Cephalostachyum capitatum, C. pallidum, C. pergracile (under Schizostachyum), Chamærops Khasyana, C. Martiana, C. Richieana, Chloroxylon Swietenia, Chrysanthemum roseum, Cinnamomum Cassia, Citrus Aurantium, C. medica, Colocasia antiquorum, C. Indica, Corchorus acutangulus, C. capsularis, C. olitorius, Cordyline terminalis, Crambe cordifolia, Crotalaria Burhia, C. juncea, C. retusa, Croton lacciferus, Cucumis cicatrisatus, C. Colocynthis, C. Momordica, C. utilissimus, Cupressus torulosa, Cyamopsis psoraloides, Cynodon Dactylon, Cyperus corymbosus, C. tegetum, Dæmonorops Guruba, D. Jenkinsii, D. nutantiflorus (under Zalacca), Dalbergia latifolia, D. Sissoo, Dammara alba, Debregeasia dichotoma, D. hypoleuca, D. velutina, D. Wallichiana, Decaisnea insignis, Dandrocalamus flagelllifer, D. giganteus, D. Hamiltoni, D. Hookeri, D. longispathus, D. sericeus, D. strictus (under Schizostachyum), Desmodium triflorum, Dichopsis Gutta, Dimochloa Andamanica, Dioscorea aculeata, D. alata, D. deltoidea, D. glabra, D. globosa, D. nummularia, D. oppositifolia, D. pentaphylla, D. purpurea, D. sativa, D. spicata, D. tomentosa, D. triphylla, Diospyros Chloroxylon, D. Ebenum, D. Melanoxylon, D. oppositifolia, D. quæsita, Dolichos uniflorus, Eleusine Coracana, E. stricta, Engelhardtia spicata, Eriochloa annulata, Eriophorum comosum, Eugenia cordifolia, E. Jambolana, E. Jambos, E. maboides, E. Malaccensis, E. revoluta, E. rotundifolia, Euryale ferox, Fagopyrum cymosum, F. emarginatum, F. rotun-

datum, F. triangulare, Ficus elastica, F. Indica, F. infectoria, F. laccifera
Flacourtia cataphracta, F. Ramontchi, Flemingia tuberosa, Fraxinus
floribunda, Garcinia Travancorica, Gigantochloa apus, G. aspera,
G. atter, G. heterostachya, G. maxima (under Schizostachyum), G.
nigro-ciliata, G. robusta, G. verticillata, Glycine hispida, Gossypium
arboreum, G. herbaceum, Guilandina Bonduc, Guizotia oleifera, Gunnera macrophylla, Hardwickia binata, Heleocharis fistulosa, H. plantaginea, Hemarthria compressa, Hibiscus cannabinus, H. Sabdariffa,
Holoptelea integrifolia, Hydnum coralloides, Illipe butyracea, I. latifolia,
Indigofera argentea, I. tinctoria, Ipomœa mammosa, I. paniculata, I.
pes capræ, Jasminum grandiflorum, J. Sambac, Juniperus recurva, J.
Wallichiana, Justicia Adhatoda, Kentia Moluccana, Lactuca sativa,
Lagerstrœmia Indica, Launæa pinnatifida, Lawsonia alba, Lepironia
mucronata, Licuala peltata (under Zalacca), Limonia acidissima, Liquidambar Altingia, Litsea Wightiana, Livistona Jenkinsii (under Zalacca),
Maba Ebenus, Machilus odoratissima, Magnolia Campbelli, M. sphærocarpa, Maharanga Emodi, Mallotus Philippinensis, Malvastrum spicatum,
Mangifera Indica, Maoutia Puya, Melaleuca Leucadendron, Melia Azadirachta, M. Azedarach, Melocanna bambusoides, M. humilis, M. Travancorica, Melanorrhœa usitata, Melocalamus compactiflorus, Michelia excelsa, Mimosa rubicaulis, Moringa pterygosperma, Morus atropurpurea,
Mucuna Cochinchinensis, Murraya exotica, Musa coccinea, M. cornicu-
lata, M. paradisiaca, M. sapientum, M. simiarum, M. textilis, M. trog-
lodytarum, Myrica sapida, Myrtus tomentosa, Nageia amara, N.
bracteata, N. cupressina, Nardostachys grandiflora, N. Jatamansi,
Nastus Borbonicus, Nephelium lappaceum, N. Longanum, Nyctanthes
Abortritis, Ocimum Basilicum, O. canum, O. gratissimum, O. sanctum,
Oncospermum fasciculatum, Oryza sativa, Oxytenanthera albo-ciliata,
O. nigro-ciliata, O. Thwaitesii (under Schizostachyum), Onosma Emodi,
Pandanus furcatus, Panicum atro-virens, P. brizanthum, P. coloratum,
P. compositum, P. flavidum, P. fluitans, P. foliosum, P. frumentaceum,
P. Italicum, P. Koenigii, P. miliaceum, P. molle, P. Myurus, P. pro-
stratum, P. repens, P. sarmentosum, P. semialatum, P. tenuiflorum, P.
virgatum, Parrotia Jacquemontiana, Paspalum distichum, P. scrobiculatum, Pelargonium odoratissimum, Pennisetum thyphoideum, Perilla
ocimoides, Peucedanum Sowa, Phaseolus aconitifolius, P. adenanthus,
P. lunatus, P. Max, Phœnix humilis, P. Hanceana, P. Kasya, P. Mer-
kusii, P. Ouseloyana (under Zalacca), P. paludosa, P. pusilla, P.
sylvestris, Phyllostachys bambusoides, Phyllanthus Cicca, Pinus Brunoniana, P. Cedrus, P. excelsa, P. Gerardiana, P. Griffithii, P. longifolia,
P. Pindrow, P. Smithiana, P. Webbiana, Pipturus propinquus, P.
velutinus, Plectocomia Assamica, P. Himalayana, P. Khasyana (under
Zalacca), P. macrostachya, Poa Chinensis, P. parviflora, P. cynosuroides,
Podophyllum Emodi, Pogostemon Heyneanus, P. parviflorus, P.
Patchouli, Polygala crotalaroides, Polygaster sampadarius, Populus
ciliata, P. Euphratica, Pouzolzia tuberosa, Prosopis spicifera, P. Ste-
phaniana, Pseudostachyum polymorphum (under Schizostachyum),
Pterocarpus Indicus, P. Marsupium, P. santalinus, Ptychosperma

disticha, P. Muschenbrockiana, Pueraria tuberosa, Pyrularia edulis, Quercus annulata, Q. dilatata, Q. incana, Q. lancifolia, Q. semicarpifolia, Q. squamata, Q. Sundaica, Raphanus caudatus, R. sativus, Remirea maritima, Rheum Australe, R. cfficinale, Rhododendron Falconeri, Rhus vernicifera, Ribes glaciale, R. Griffithii, R. laciniatum, R. villosum, Ricinus communis, Rosa Indica, R. moschata, R. sempervirens, Rubia cordifolia, Rubus acuminatus, R. biflorus, R. ellipticus, R. lasiocarpus, R. Moluccanus, R. nutans, R. tiliaceus, Saccharum officinarum, S. spontaneum, S. violaceum, Salix tetrasperma, Sanseviera Zeylanica, Santalum album, Schima Wallichii, Schizostachyum elegantissimum, S. Blumei, S. brachycladum, S. Hasskarlianum, S. irratum, S. Zollingeri, Scutia Indica, Sesamum Indicum, Sesbania aculeata, S. Ægyptiaca, S. cannabina, S. grandiflora, Sesuvium Portulacastrum, Shorea robusta, S. Talura, Solanum album, S. ferox, S. Guineense, S. insanum, S. longum, S. Melongena, S. pseudo-saponaceum, S. undatum, Spinifex squarrosus, Spondias mangifera, Stenotaphrum Americanum, Sterculia monosperma, S. urceolata, S. urens, Streblus asper, Swertia Chirata, S. elegans, Symplocos ramosissima, Tamarindus Indica, Tamarix articulata, T. dioica, T. Gallica, T. orientalis, Tectona grandis, T. Hamiltoniana, Teinostachyum attenuatum, T. Griffithii (under Schizostachyum), Terminalia Catappa, T. Chebula, T. parviflora, Tetranthera calophylla, T. laurifolia, Thouarea sarmentosa, Trapa bispinosa, T. Cochinchinensis, T. incisa, T. quadrispinosa, Trichodesma Zeylanicum, Triphasia Aurantiola, Ulmus Wallichiana, Urena lobata, Vaccinium Leschenaulti, Vigna Sinensis, Villebrunea frutescens, V. integrifolia, Vitis auriculata, V. Blumeana, V. elongata, V. imperialis, V. Indica, V. Labrusca, V. lævigata, V. mutabilis, V. quadrangularis, V. thyrsiflora, V. vulpina, Wallichia caryotoides, W. densiflora, Withania coagulans, W. somnifera, Ximenia Americana, Xylia dolabriformis, Zalacca secunda, Zingiber officinale, Zizyphus Jujuba, Z. rugosa, Zoysia pungens.

WESTERN SOUTH-AMERICA.

Acacia Cavenia, A. macracantha, Achras Balata, Adenostemum nitidum, Adesmia balsamica, Alchemilla pinnata, Alstrœmeria pallida, Ananas sativa, Andropogon argenteus, Anona Cherimolia, Apium Chilense, A. prostratum, Arachis hypogæa, Araucaria imbricata, Aristotelia Macqui, Arracacha xanthorriza, Bassovia solanacea, Berberis buxifolia, B. Darwinii, Boussingaultia baselloides, Buddlea globosa, Cæsalpinia brevifolia, C. tinctoria, Canna edulis, Carica Candamarcensis, C. Papaya, Cereus Quixo, Ceroxylon andicola, C. Australe, C. pithyrophyllum (under Wettinia), Cervantesia tomentosa, Chenopodium Quinoa, Chusquea Culcou, C. Dombeyana, C. montana, C. Quila, C. tenuiflora (under Schizostachyum), Cinchona Calisaya, C. cordifolia, C. micrantha, C. nitida, C. officinalis, C. succirubra, Condalia microphylla, Dactylis cæspitosa, Datura arborea, Dioscorea piperifolia, Diplothemium Porallys (under Wettinia), Diposis Bulbocastanum, Drimys Winteri, Elymus condensatus, Embothrium coccineum, E. emarginatum, E. lanceolatum,

Erythroxylon Coca, Eucryphia cordifolia, Eugenia Hallii, Euterpe andicola, E. Hœnkena, E. longivaginata (under Wettinia), Fagus Dombeyi, F. obliqua, F. procera, Fagus betuloides, Festuca Coiron, F. Magellanica, Fitzroya Patagonica, Fragaria Chiloensis, Fuschia racemosa, Geonoma densa (under Wettinia), Gossypium religiosum, Guadua angustifolia, G. latifolia, Guevina Avellana, Gunnera Chilensis, Helianthus annuus, H. tuberosus, Heliotropium Peruvianum, Hibiscus esculentus, Hypochœris apargioides, H. Scorzoneræ, Ipomœa Batatas, I. pes caprae, Jubæa spectabilis, Krameria triandra, Lapageria rosea, Lardizabala biternata, Laurelia aromatica, L. serrata, Libocedrus Chilensis, L. tetragona, Lippia citriodora, Manihot Aipi, Maranta arundinacea, Mauritia flexuosa, Maytenus Boaria, Melicocca bijuga, Mesembrianthemum æquilaterale, Morus celtidifolia, M. insignis, Myrtus Luma, M. Meli, M. nummularia, M. Ugni, Nageia andina, N. Chilina, N. nubigena, Opuntia vulgaris, Oreodoxa frigida (under Wettinia), Oryza latifolia, Oryzopsis cuspidata, O. panicoides, Oxalis crassicaulis, O. crenata, O. enneaphylla, O. succulenta, O. tuberosa, Pacchyrrhizus angulatus, Panicum pilosum, Paspalum ciliatum, P. dilatatum, Passiflora alata, P. tilifolia, P. ligularis, P. macrocarpa, Persea gratissima, Peumus Boldus, Phaseolus vulgaris, Physalis Peruviana, Phytelephas æquatorialis (under Wettinia), Piptadenia rigida, Prosopis horrida, P. juliflora, P. Siliquastrum, Priva lævis, Quillaja saponaria, Rhus caustica, Rubus geoides, Salix Humboldtiana, Saxono-Gothæa conspicua, Schkuhria abrotanoides, Schinus Molle, Scirpus nodosus, Sesuvium Portulacastrum, Smilax officinalis, Solanum Gilo, S. Maglia, S. Guinense, S. Lycopersicum. S. muricatum, S. Quitoense, S. tuberosum, S. torvum, Sophora tetraptera, Spilanthes oleracea, Tagetes glanduligera, Tetragonia expansa, Tillandsia usneoides, Trithrinax campestris, Tropæolum majus, T. minus, T. sessilifolium, T. tuberosum, Ullucus tuberosus, Vaccinium alatum, V. bicolor, V. grandiflorum, V. melliflorum, Wettinia augusta, W. Maynensis, Zea Mays, Zizyphus Joazeiro, Z. Mistal.

WESTERN NORTH-AMERICA.

Acer circinnatum, A. macrophyllum, Æsculus Californica, Arbutus Menziesii, Audibertia polystachya, Baccharis consanguinea, B. pilularis, Baptisia tinctoria, Barbarea vulgaris, Beschorneria yuccoides, Bouteloua polystachya, Carica Papaya, Carum Gairdneri, Castaneopsis chrysophylla, Ceanothus prostratus, C. rigidus, C. thyrsiflorus, Cercocarpus ledifolius, C. parvifolius, Cereus Engelmanni, C. Thurberi, Chamædora elatior, Chlorogalum pomeridianum, Claytonia perfoliata, Cornus Nuttallii, Cupressus fragrans, C. Lawsoniana, C. macrocarpa, C. Nutkaensis, Cymopterus glomeratus, Fragaria Californica, Fraxinus Oregana, Gaultiera Myrsinites, Geum urbanum, Juglans rupestris, Juniperus occidentalis, Libocedrus decurrens, Lupinus Douglasii, Myrica Californica, Myrrhis occidentalis, Nicotiana multivalvis, Nuphar multisepalum, Nyssa aquatica, Parkinsonia aculeata, P. microphylla, Pinus albicaulis, P. amabilis, P. Arizonica, P. bracteata, P. Chihua-huana, P. concolor, P.

contorta, P. Coulteri, P. Douglasii, P. edulis, P. flexilis, P. grandis, P. Hookeriana, P. Jeffreyi, P. Lambertiana, P. Menziesii, P. Merteusiana, P. monophylla, P. monticola, P. muricata, P. nobilis, P. Nuttallii, P. Pattoniana, P. ponderosa, P. radiata, P. reflexa, P. resiuosa, P. Sabiniana, P. Williamsonii, Platanus racemosa, Populus Fremontii, P. tremuloides, P. trichocarpa, Pritchardia filamentosa, Prosopis pubescens, Prunus demissa, P. ilicifolia, Pyrus rivularis, Quercus agrifolia, Q. chrysolepis, Q. densiflora, Q. Douglasii, Q. Garryana, Q. lobata, Ribes aureum, R. divaricatum, R. niveum, R. tenuiflorum, R. villosum, Rubus leucodermis, R. macropetalus, R. ursinus, Salix longifolia, Schinus Molle, Scilla esculenta, Sequoia sempervirens, S. Wellingtonia, Solanum Fendleri, S. tuberosum, Tetranthera Californica, Torreya Californica, Trichostema lanatum, Umbellularia Californica, Vaccinium humifusum, V. ovalifolium, V. ovatum, Valeriana edulis, Washingtonia filifera, Yucca augustifolia, Y. baccata, Y. brevifolia, Y. Sitchensis, Y. Treculiana.

EASTERN NORTH-AMERICA.

Acer dasycarpum, A. Negundo, A. rubrum, A. sacchariuum, Achillea Millefolium, Achras Sapota, Acorus Calamus, Acrocomia Mexicana, Actæa alba, A. spicata, Æsculus lutea, Agave Americana, A. Mexicana, Agrostis alba, A. rubra, A. scabra, A. vulgaris, Alchemilla alpina, A. vulgaris, Aletris farinosa, Allium Canadense, A. Schœnoprasum, Amelanchier Botryapium, Andropogon avenaceus, A. nutans, A. scoparius, Apios tuberosa, Apocynum cannabinum, Arctostaphylos uva ursi, Aristolochia anguicida, A. ovalifolia, A. serpentaria, Arundinaria macrosperma, A. tecta, Astragalus hypoglottis, Barbarea vulgaris, Betula lenta, B. lutea, B. nigra, B. papyracea, Bouteloua barbata, Brahea dulcis, B. edulis, Bromus ciliatus, B. marginatus, Buchloa dactyloides, Bursera elemifera, Cæsalpinia Bonduc, Cakile maritima, Calamagrostis longifolia, Canella alba, Canna flaccida, Carya alba, C. amara, C. glabra, C. microcarpa, C. oliviformis, C. sulcata, C. tomentosa, Carpinus Americana, Cassia Marylandica, Catalpa bignonioides, C. speciosa, Cedronella cordata, Celtis occidentalis, Cephalantus occidentalis, Cetraria Islandica, Chamædora concolor, Chelone glabra, Chondrus crispus, Cimicifuga racemosa, Cinna arundinacea, Cladastris tinctoria, Cochlearia officinalis, Comptonia asplenifolia, Cornus florida, Corylus Americana, Cratægus æstivalis, C. Mexicana, C. apiifolia, C. coccinea, C. cordata, C. Crus-Galli, C. parvifolia, C. tomentosa, Cupressus Benthami, C. Lindleyi, C. thurifera, C. thuyoides, Desmodium acuminatum, Diospyros Virginiana, Dirca palustris, Echinocactus Fendleri, Elymus mollis, E. Virginicus, Euonymus atropurpureus, Eupatorium purpureum, Fagus ferruginea, Festuca flava, F. purpurea, Fragaria Chiloensis, F. grandiflora, F. Illinoensis, F. vesca, F. Virginiana, Fraxinus Americana, F. platycarpa, F. pubescens, F. quadrangulata, F. sambucifolia, F. viridis, Gaultieria Shallon, Gaylussacia frondosa, G. resinosa, Gelsemium nitidum, Geum urbanum, Gleditschia monosperma, G. triacanthos, Gordonia

Robinia Pseudo-Acacia, Rosa setigera, Rubus arcticus, R. Canadensis, R. Chamæmorus, R. cuneifolius, R. deliciosus, R. occidentalis, R. odoratus, R. strigosus, R. trivialis, R. villosus, Rumex acetosa, R. hymenosepalus, Sabal Adansoni, S. Palmetto, S. serrulata, Sabbatia angularis, Sagittaria lancifolia, S. obtusa, Salix cordata, S. longifolia, S. lucida, S. nigra, S. petiolaris, S. tristis, Sambucus Canadensis, Sanguinaria Canadensis, Sassafras officinale, Schœnocaulon officinale, Scilla Fraseri, Shepherdia argentea, Smilax bona nox, S. glauca, S. medica, S. Pseudo-China, S. rotundifolia, Solanum calycinum, S. cardiophyllum, S. Fendleri, S. Jamesii, S. oxycarpum, S. tuberosum, Spartina cynosuroides, S. juncea, S. polystachya, S. stricta, Spigelia Marylandica, Stenotaphrum Americanum, Stylosanthes elatior, Tanacetum vulgare, Taraxacum officinale, Taxodium distichum, T. mucronatum, Taxus brevifolia, Thuya gigantea, T. occidentalis, Tilia alba, T. Americana, Tillandsia usneoides, Torreya taxifolia, Trifolium reflexum, T. repens, Trillium erectum, Tripsacum dactyloides, Typha latifolia, Ulmus alata, U. Americana, U. crassifolia, U. fulva, U. Mexicana, U. racemosa, Uniola gracilis, U. latifolia, U. paniculata, Uvularia sessilifolia, Vaccinium Canadense, V. cæspitosum, V. corymbosum, V. erythrocarpum, V. leucanthum, V. macrocarpum, V. myrtilloides, V. Myrtillus, V. ovalifolium, V. ovatum, V. Oxycoccos, V. parvifolium, V. Pennsylvanicum, V. uliginosum, V. vacillans, V. Vitis Idæa, Valeriana edulis, Veratrum viride, Vicia Cracca, V. Sitchensis, Vitis æstivalis, V. candicans, V. cinerea. V. cordifolia, V. Labrusca, V. riparia, V. rubra, V. rupestris, V. vulpina, Wistaria frutescens, Xanthorriza apiifolia, Ximenia Americana, Yucca aloifolia, Y. angustifolia, Y. filamentosa, Y. gloriosa, Zizania aquatica, Z. fluitans, Z. miliacea.

CENTRAL AMERICA.

Acacia macracantha, Achras Sapota, Acrocomia Mexicana, Agave Americana, A. inæquidens, A. rigida, Albizzia dulcis, A. latisiliqua, A. Saman, Aleurites triloba, Amarantus paniculatus, Andropogon avenaceus, Anona muricata, A. squamosa, Arracacha xanthorrhiza, Arthrostylidium excelsum, A. longiflorum, A. racemiferum (under Schizostachyum), Arundinaria acuminata, Aulonemia Quexo (under Schizostachyum), Bactris Gasipæs, Batis maritima, Beschorneria yuccoides, Brahea dulcis, Bursera elemifera, Buxus acuminata, B. citrifolia, B. Cubana, B. glomerata, B. gonoclada, B. lævigata, B. Purdieana, B. retusa, B. subcolumnaris, B. Vahlii, B. Wrightii, Cæsalpinia crista, C. vesicaria, Cakile maritima, Calyptronoma Swartzii, Canavalia gladiata, Canna coccinea, C. glauca, Canella alba, Carica Papaya, Carludovica palmata (under Wettinia), Cæsalpinia Bonduc, Casimiroa edulis, Celtis Tala, Ceroxylon andicola, C. Klopstockia (under Wettinia), Cestrum nocturnum, Chusquea abietifolia, C. Fendleri, C. Galeottiana, C. Muelleri, C. scandens, C. simpliciflora, C. uniflora (under Schizostachyum), Claytonia perfoliata, Coccoloba uvifera, Cocos regia, Copernicia nana, C. Pumos,

Cyperus giganteus, Dioscorea Cajennensis, D. esurientum, D. trifida, Eriochloa annulata, Euchlæna luxurians, Eupatorium triplinerve, Fourcroya Cubensis, F. gigantea, F. longæva, Geonoma vaga, Gossypium Barbadense, G. hirsutum, G. religiosum, Hibiscus esculentus, Indigofera Anil, Ipomœa Batatilla, Juniperus Bermudiana, Kunthia montana, Malvastrum spicatum, Maranta arundinacea, Melicocca bijuga, Mimusops globosa, M. Sieberi, Morus celtidifolia, Nageia coriacea, N. Purdieana, Opuntia coccinellifera, O. Dillenii, O. elatior, O. Hernandezii, O. spinosissima, O. Tuna, Oreodoxa frigida, O. oleracea, O. regia, Pachyrrhizus angulatus, Panicum altissimum, P. divaricatum, P. molle, P. Myurus, P. obtusum, P. striatum, Paspalum stoloniferum, Passiflora laurifolia, P. ligularis, P. maliformis, P. pedata, P. serrata, Peireskia aculeata, Persea gratissima, Pinus Cubensis, Piscidia erythrina, Platenia Chiragua (under Wettinia), Podachænium alatum, Polianthes tuberosa, Prestoa pubigera, Psidium acidum, P. Araca, P. cordatum, P. Guayava, P. polycarpum, Quercus agrifolia, Q. Castanea, Q. Skinneri, Remirea maritima, Richardsonia scabra, Sabal umbraculifera, Schinus Molle, Sechium edule, Sesuvium Portulacastrum, Smilax officinalis, S. papyracea, Solanum betaceum, S. Guineense, S. Plumieri, S. Topiro, S. torvum, Sporobolus Virginicus, Stylosanthes elatior, Swietenia Mahagoni, Talinum patens, Terminalia Buceras, Thrinax argentea, T. parviflora, Tillandsia usneoides, Trophis Americana, Urena lobata, Vaccinium meridionale, V. Mortinia, Wissadula rostrata, Xanthosoma sagittifolium, Yucca aloifolia, Y. Guatemalensis, Y. Yucatana, Zizania miliacea.

EASTERN SOUTH-AMERICA.

Acacia Cebil, A. macracantha, A. moniliformis, Acrocomia Totai (under Wettinia), Alibertia edulis, Ananas sativa, Apium prostratum, Araucaria Brasiliensis, Arundinaria verticillata (under Schizostachyum), Arundo saccharoides, A. Sellowiana, Aspidosperma Quebracho, Bactris Gasipaes, Boussingaultia baselloides, Bromus unioloides, Cæsalpinia coriaria, C. echinata, C. Gilliesii, Calyptranthes aromatica, Canna Achiras, Capsicum annuum, C. baccatum, C. frutescens, C. longum, C. microcarpum, Cedrela Brasiliensis, C. Velloziana, Celtis Sellowiana, C. Tala, Cephælis Ipecacuanha, Ceroxylon Klopstockia, Chenopodium ambrosioides, Chusquea capituliflora, C. Culeon, C. Gaudichaudiana (under Schizostachyum), C. Lorentziana, Cocos Australis, C. flexuosa, C. plumosa, C. Romanzoffiana, C. Yatay (under Wettinia), Copernicia cerifera, Condalia microphylla, Cyperus giganteus, Dalbergia nigra, D. Miscolobium, Desmodium triflorum, Dioscorea conferta, D. tuberosa, Diplothemium campestre, D. littorale (under Wettinia), Duvaua longifolia, Eryngium pandanifolium, Eugenia Nhanica, E. pyriformis, E. supra-axillaris, E. uniflora, Eupatorium tinctorium, Geonoma vaga, Gourliaea decorticans, Guadua augustifolia, G. capitata, G. latifolia, G. macrostachya, G. paniculata, G. refracta, G. Tagoara, G. virgata, Hancornia speciosa, Heterothalamus brunioides, Hordeum andicola, Hymenæa

Courbaril, Ilex Paraguensis, Indigofera Anil, Ipomœa Batatas, I. Batatilla, I. Megapotamica, I. operculata, I. paniculata, I. pes capræ, Iriartea deltoidea, I. exorrhiza, I. ventricosa (under Wettinia), Jacaranda mimosifolia, Lippia citriodora, Loxopterygium Lorentzii, Lupinus arboreus, Maclura Mora, Malvastrum spicatum, Manihot Aipi, M. Glazioni, M. utilissima, Maliera glomerata, M. tomentosa, Melica sarmentosa, Merostachys Claussenii, M. Kunthii, M. ternata (under Schizostachyum), Myrtus edulis, M. incana, M. mucronata, Nageia Lamberti, Nicotiana rustica, N. glauca, N. Tabacum, Ocimum gratissimum, Œnocarpus multicaulis (under Wettinia), Opuntia vulgaris, Oryza latifolia, Oryzopsis panicoides, Oxalis carnosa, O. conorrhiza, Pachyrrhizus angulatus, Panicum altissimum, P. barbinode, P. divaricatum, P. latissimum, P. molle, P. Myurus, Parkinsonia aculeata, Paspalum notatum, P. ciliatum, P. dilatatum, P. undulatum, Passiflora alata, P. coccinea, P. cœrulea, P. edulis, P. filamentosa, P. laurifolia, P. maliformis, P. mucronata, P. pedata, P. .quadrangularis, P. serrata, P. suberosa, Paullinia sorbilis, Peireskia aculeata, P. Bleo, P. portulacifolia, Peltophorum Linnei, Pennisetum latifolium, Periandra dulcis, Persea gratissima, Phaseolus adenanthus, P. lunatus, Physalis angulata, P. Peruviana, P. pubescens, Phytelephas macrocarpa, P. microcarpa (under Wettinia), Pilocarpus pinnatifolius, Piptadenia Cebil, P. rigida, Pircunia dioica, Poa Bergii, P. Forsteri, Prosopis alba, P. dulcis, P. Siliquastrum, Psidium Araca, P. arboreum, P. Cattleyanum, P. chrysophyllum, P. cinereum, P. cuneatum, P. grandifolium, P. Guayava, P. incanescens, P. lineatifolium, P. malifolium, P. polycarpon, P. rufum, Rubus imperialis, Salix Humboldtiana, Salpichroma rhomboidea, Salvia Matico, Sambucus Australis, Schinus Molle, Scirpus nodosus, Sesuvium Portulacastrum, Smilax papyracea, S. rubiginosa, Solanum Commersonii, S. Gilo, S. Guineense, S. indigoferum, S. Lycopersicum, S. torvum, S. tuberosum, Spilanthes oleracea, Sporobolus Indicus, Syagrus Sancona, Sterculia Carthaginensis, Tagetes glanduligera, Talinum patens, Tillandsia usneoides, Terminalia Buceras, Trithrinax Acanthocoma, T. Brasiliensis (under Wettinia), Trophis Americana, Ullucus tuberosus, Wissadula rostrata, Zea Mays, Zizania microstachya, Ziziphus Mistal.

MIDDLE AFRICA (AND MADAGASCAR).

Acacia Arabica, A. Catechu, A. stenocarpa, A. Verek, Acanthophœnix rubra, Acanthosicyos horrida, Æschynomene aspera, Aloe Perryi, Andropogon annulatus, Aristida prodigiosa, Arundinella Nepalensis, Asplenium Nidus, Astragalus venosus, Bacularia Arfakiana, Beesha capitata, Buddleya Madagascariensis, Buxus Madagascarica, Canavalia gladiata, Casuarina equisetifolia, Coffea Liberica, Corchorus acutangulus, Corynosicyos edulis, Cucumis Anguria, Cudrania Javensis, Cupania sapida, Dalbergia melanoxylon, Dypsis pinnatifrons, Eriochloa annulata, Hagenia Abyssinica, Hibiscus Sabdariffa, Hyphæne Thebaica, Launea pinnatifida, Lepironia mucronata, Maclura excelsa, Malvastrum spicatum, Monodora Angolensis, M. Myristica, Musa

Livingstoniana, Panicum coloratum, P. compositum, P. fluitans, P. molle, Pennisetum villosum, Pharnaceum acidum, Phœnix spinosa, Psophocarpus tetragonolobus, Pterolobium lacerans, Remirea maritima, Rubus rosifolius, Solanum edule, S. Æthiopicum, S. macrocarpum, S. Thonningi, Tamarix orientalis, Telfairia occidentalis, T. pedata, Trichodesma Zeylanicum, Urena lobata, Vahea florida, V. Owariensis, Vigna Sinensis, Vitis Schimperiana, Wissadula rostrata.

SOUTHERN AFRICA.

Aberia Caffra, A. tristis, A. Zeyheri, Acacia Giraffæ, A. horrida, Alchemilla Capensis, A. elongata, Aloe dichotoma, A. ferox, A. linguiformis, A. plicatilis, A. purpurascens, A. spicata, A. vera, A. Zeyheri, Andropogon Caffrorum, Anthistiria ciliata, Aponogeton distachyos, Arundinaria tesselata, Arundinella Nepalensis, Asparagus laricinus, Atriplex albicans, Avicennia officinalis, Azima tetracantha, Barosma serratifolia, Brabejum stellatifolium, Callitris arborea, Calodendron Capense, Cannamois cephalotes, Carissa Arduina, C. ferox, C. grandiflora, Carum Capense, Combretum butyraceum, Ehrharta longiflora, Elegia nuda, Elephanthorrhiza Burchelli, Euclea myrtina, E. undulata, E. Pseudebenus, Eugenia Zeyheri, Exomis axyrioides, Garuleum bipinnatum, Gladiolus edulis, Gonioma Kamassia, Hemarthria compressa, Hibiscus Ludwigii, Hyphæne ventricosa, Kochia pubescens, Lasiocorys Capensis, Leonotis Leonurus, Leucadendron argenteum, Leyssera gnaphalioides, Lyperia crocea, Matricaria glabrata, Melianthus major, Mesembrianthemum acinaciforme, M. capitatum, M. crystallium, M. edule, M. floribundum, Myrica cordifolia, M. quercifolia, M. serrata, Nageia elongata, N. Thunbergi, Nastus Borbonicus, Osmitopsis asteriscoides, Osyris compressa, Panicum coloratum, P. compositum, P. roseum, Pappea Capensis, Parkinsonia Africana, Pentzia virgata, Phœnix reclinata, Phymaspermum parvifolium, Plectronia ciliata, P. spinosa, P. ventosa, Portulacaria Afra, Protea mellifera, Psychotria Eckloniana, Rafnia amplexicaulis, R. perfoliata, Rhus lucida, Royenia Pseudebenus, R. pubescens, Rubus fruticosus, Salix Capensis, Scirpus nodosus, Selago leptostachya, Selinum anesorrhizum, S. montanum, Solanum Aculeastrum, Tarchonanthus camphoratus, Todea Africana, Voandzeia subterranea, Vangueria infausta, Withania somnifera.

WESTERN AUSTRALIA.

Acacia acuminata, A. aneura, A. armata, A. heteroclita, A. leiophylla, A. microbotrya, A. Sentis, Agonis flexuosa, Albizzia lophantha, Astartea fascicularis, Atriplex Muelleri, Avicennia officinalis, Boronia megastigma, Cassia artemisioides, Casuarina Decaisneana, C. distyla, C. Fraseriana, C. Huegeliana, C. trichodon, Conospermum Stoechadis, Danthonia bipartita, Dioscorea hastifolia, Duboisia Hopwoodii, Encephalartos Preissii, Eriantbus fulvus, Eucalyptus calophylla, E. cornuta, E. diversicolor, E. Doratoxylon, E. ficifolia, E. gom-

phocephala, E. loxophleba, E. marginata, E. oleosa, E. patens, E.
redunca, E. rudis, E. salmonophloia, E. salubris, Grevillea annulifera,
Helichrysum, lucidum, H. Manglesii, Jacksonia cupulifera, Kochia
villosa, Lepidosperma gladiatum, Livistona Mariæ, Oryza sativa,
Panicum flavidum, P. semialatum, Phaseolus vulgaris, Pimelea clavata,
Santalum cygnorum, S. Preissianum, Scirpus nodosus, Sesbania
Ægyptiaca, S. grandiflora, Spinifex hirsutus, S. longifolius, Strychnos
Nux vomica, Tamarindus Indica.

EASTERN AUSTRALIA (INCLUDING TASMANIA).

Acacia aneura, A. armata, A. binervata, A. dealbata, A. decurrens,
A. estrophiolata, A. excelsa, A. falcata, A. Farnesiana, A. fasciculifera,
A. glaucescens, A. harpophylla, A. homalophylla, A. implexa, A.
longifolia, A. Melanoxylon, A. pendula, A. penninervis, A. pycnantha,
A. retinodes, A. salicina, A. Sentis, A. stenophylla, A. supporosa,
Ægiceras majus, Agrostis Solandri, Aira cæspitosa, Albizzia basaltica,
Alchemilla vulgaris, Aleurites triloba, Alstonia constricta, Andropogon
annulatus, A. Australis, A. bombycinus, A. erianthoides, A. falcatus, A.
Gryllos, A. pertusus, A, refractus, A. sericeus, Angophora intermedia,
A. lanceolata, A. subvelutina, Anthistiria avenacea, A. ciliata, A.
membranacea, Apium prostratum, Aponogeton crispus, Araucaria Bidwilli, A. Cunninghami, Aristolochia Indica, Astrebla pectinata, A.
triticoides, Atalantia glauca, Atriplex crystallinum, A. halimoides, A.
holocarpum, A. Muelleri, A. nummularium, A. semibaccatum, A. spongiosum, A. versicarium, Avicennia officinalis, Backhousia citriodora,
Bacularia monostachya, Barbarea vulgaris, Baloghia lucida, Brachychiton acerifolius, Cakile maritima, Callitris calcarata, C. columellaris, C. Endlicherii, C. Macleayana, C. Parlatorei, C. verrucosa, Carissa
. Brownii, Cassia artemisioides, Casuarina distyla, C. equisetifolia, C.
glauca, C. quadrivalvis, C. suberosa, C. torulosa, Cedrela Australis,
Ceratopetalum apetalum, Chenopodium auricomum, C. nitrariaceum,
Chionachne cyathopoda, Chloris scariosa, C. truncata, Citrus Austral-
asica, C. Planchoni, Colocasia Indica, Corchorus acutangulus, C.
Cunninghami, C. olitorius, Cordyline terminalis, Crotalaria juncea, C.
retusa, Cudrania Javanensis, Cycas angulata, C. Normanbyana, Cynodon
Dactylon, Cyperus textilis, Dacrydium Frauklini, Dammara robusta,
Danthonia bipartita, D. nervosa, D. penicillata, D. robusta, Dicksonia
Billardieri, Dioscorea sativa, D. transversa, Distichlis maritima,
Duboisia Hopwoodii, D. myoporoides, Ectrosia Gulliverii, Ehrharta
stipoides, Embothrium Wickhami, Encephalartos Denisonii, E. spiralis,
Eremophila longifolia, Erianthus fulvus, Eriochloa annulata, Erodium
cygnorum, Eucalyptus alpina, E. amygdalina, E. Baileyana, E.
botryoides, E. capitellata, E. citriodora, E. coccifera, E. cordata, E.
corymbosa, E. corynocalyx, E. crebra, E. drepanophylla, E. eugenioides,
E. Globulus, E. goniocalyx, E. Gunnii, E. hæmastoma, E. hemiphloia,
E. Howittiana, E. largiflorens, E. leptophleba, E. Leucoxylon, E.
longifolia, E. macrorrhyncha, E. maculata, E. melanophloia, E. melliodora,

E. microcorys, E. microtheca, E. miniata, E. obliqua, E. ochrophloia, E. oleosa, E. paniculata, E. pauciflora, E. phœnicea, E. pilularis, E. Planchoniana, E. platyphylla, E. polyanthema, E. populifolia, E. punctata, E. Raveretiana, E. resinifera, E. robusta, E. rostrata, E. saligna, E. siderophloia, E. Sieberiana, E. Staigeriana, E. Stuartiana, E. tereticornis, E. terminalis, E. tesselaris, E. trachyphloia, E. triantha, E. urnigera, E. vernicosa, E. viminalis, Eucryphia Billardieri, E. Moorei, Eugenia Jambolana, E. Australis, E. Smithii, Eustrephus Brownii, Exidia auricula Judæ, Fagus Cunninghami, F. Moorei, Festuca Billardieri, F. Hookeriana, F. litoralis, F. dives, Ficus colossea, F. columnaris, F. Cunninghami, F. eugenioides, F. macrophylla, F. rubiginosa, Fistulina hepatica, Flindersia Australis, F. Bennettiana, F. Oxleyana, Geitonoplesium cymosum, Geum urbanum, Gmelina Leichhardtii, Grevillea robusta, Harpullia Hillii, Heleocharis sphacelata, Helichrysum lucidum, Hemarthria compressa, Hibiscus cannabinus, Hierochloa redolens, Hymenanthera Banksii, Imperata arundinacea, Ipomœa Calobra, I. costata, I. graminea, I. paniculata, I. pes capræ, Jasminum calcareum, J. didymum, J. lineare, J. racemosum, J. simplicifolium, J. suavissimum, Kentia Belmoreana, K. Canterburyana, K. Mooreana, Kochia eriantha, K. villosa, Lagerstroemia Indica, Leersia hexandra, Lepidosperma gladiatum, Lepironia mucronata, Leptospermum lævigatum, L. lanigerum, Livistona Australis, L. Leichhardtii, L. Mariæ, Lycopodium clavatum, L. densum, L. laterale, L. varium, Maba fasciculosa, M. geminata, Macadamia ternifolia, Mallotus Philippinensis, Malvastrum spicatum, Marlea Vitiensis, Melaleuca ericifolia, M. genistifolia, M. Leucadendron, M. linarifolia, M. parviflora, M. styphelioides, Melia Azedarach, Mentha Australis, M. gracilis, M. laxiflora, M. saturejoides, Mesembrianthemum æquilaterale, Microseris Forsteri, Morchella conica, Murraya exotica, Mylitta Australis, Myoporum insulare, Myrtus acmenoides, Nageia elata, Neurachne Mitchelliana, Niemeyera prunifera, Ocimum sanctum, Owenia acidula, O. venosa, Oryza sativa, Pandanus Forsteri, P. pedunculatus, Panicum atro-virens, P. bicolor, P. cœnicolum, P. coloratum, P. compositum, P. decompositum, P. divaricatissimum, P. flavidum, P. foliosum, P. marginatum, P. melanthum, P. Myurus, P. Italicum, P. Kœnigii, P. miliaceum, P. parvifolium, P. prolutum, P. prostratum, P. pygmæum, P. repens, P. sanguinale, P. semialatum, P. tenuiflorum, P. virgatum, Pappophorum commune, Parinarium Nonda, Paspalum distichum, P. scrobiculatum, Peltophorum ferrugineum, Phaseolus adenanthus, P. Max, Phyllocladus rhomboidalis, Pimelea stricta, Pipturus propinquus, Pisonia aculeata, Pittosporum undulatum, Poa Australis, P. Billardieri, P. Brownii, P. cæspitosa, P. Chinensis, P. digitata, Prostanthera lasiantha, Ptychosperma Alexandræ, P. Cunninghami, P. elegans, Rhagodia Billardieri, R. nutans, Rhus rhodanthema, Rottbœllia ophiuroides, Rubus Gunnianus, R. parvifolius, R. rosifolius, Santalum Preissianum, Scirpus nodosus, Sclerachne cyathopoda, Sebæa albidiflora, S. ovata, Selaginella uliginosa, Sesbania aculeata, S. Ægyptiaca, Sesuvium Portulacastrum, Smilax Australis, S. glycyphylla, Solanum vescum, Spinifex hirsutus, S. paradoxus, Spondias Solandri, Stenocarpus sinuosus, Stenopetalum nutans, Sterculia quadrifida, Stipa artistiglumis, Syncarpia laurifolia, Synoou glandulosum, Tacca pinnatifida, Tetragonia expansa, T. implexicoma, Tetranthera laurifolia, Todea Africana, Tremella mesenterica, Trichodesma Zeylanicum, Trigonella suavissima, Tristania conferta, Vigna lanceolata, Vitis acetosa, V. Baudiniana, V. hypoglauca, Xanthorrhœa Tatei, Ximenia Americana, Zizyphus Jujuba, Zoysia pungens.

NEW ZEALAND.

Agrostis Solandri, Apium prostratum, Arundo conspicua, Avicennia officinalis, Cordyline Banksii, C. indivisa, C. superbiens, Corynocarpus lævigata, Dacrydium Colensoi, D. cupressinum, D. Kirkii, Dammara Australis, Danthonia Cunninghami, Dicksonia Billardieri, Ehrharta Diplax, E. stipoides, Fagus cliffortioides, F. fusca, F. Menziesii, F. Solandri, Festuca litoralis, Fuchsia excorticata, Heleocharis sphacelata, Hierochloa redolens, Hymenanthera Banksii, Kentia sapida, Knightia excelsa, Libocedrus Doniana, Metrosideros florida, M. lucida, M. robusta, M. tomentosa, Myoporum lætum, Nageia dacrydioides, N. ferruginea, N. spicata, N. Totara, Panicum atro-virens, Phormium tenax, Phyllocladus trichomanoides, Pittosporum eugenioides, P. tenuifolium, Poa cæspitosa, P. foliosa, Ripogonum scandens, Scirpus nodosus, Sebæa ovata, Stilbocarpa polaris, Tetragonia expansa, T. implexicoma, Triodia exigua, Trisetum antarcticum.

POLYNESIA,

Acacia Koa, Ægiceras majus, Aleurites triloba, Andropogon refractus, Araucaria Cookii, A. excelsa, A. Rulei, Aristolochia Indica, Artocarpus communis, Bacularia Arfakiana, Batis maritima, Broussonetia papyrifera, Casuarina equisetifolia, Colocasia antiquorum, C. Indica, Cordyline Baueri, C. terminalis, Cyrtosperma edule, Dammara macrophylla, D. Moorei, D. obtusa, D. ovata, D. Vitiensis, Dioscorea aculeata, D. alata, D. nummularia, D. pentaphylla, D. sativa, Eugenia Jambolana, Exidia auricula Judæ, Gossypium Taitense, G. tomentosum, Heleocharis sphacelata, Ipomœa paniculata, Kentia Baueri, K. Beccarii, Lagerstroemia Indica, Musa Troglodytarum, Ocimum gratissimum, Pipturus propinquus, Pringlea antiscorbutica, Ptychosperma Arfakiana, Rubus Hawaiensis, Saccharum officinarum, Santalum Freycinetianum, S. Yasi, Solanum Uporo, Spondias dulcis, Tacca pinnatifida, Tetragonia expansa, Touchardia latifolia, Vaccinium penduliflorum.

I.

IMPORTANT CULTURAL PLANTS, YIELDING A RETURN IN ONE YEAR, BUT REQUIRING RENEWAL.

(Annuals, including a few Biennials.)

Allium Ampeloprasum, A. Ascalonicum, A. Cepa, A. fistulosum, A. Porrum, A. sativum, A. Schœnoprasum, A. Scorodoprasum, Andropogon saccharatus, A. Sorghum, Apium graveolens, Arachis hypogæa, Avena sativa, Beta vulgaris, Brassica alba, B. nigra, B. oleracea, Cannabis sativa, Capsicum annuum, Carum Petroselinum, Chærophyllum bulbosum, Cicer arietinum, Cichorium Endivia, Corchorus capsularis, C. olitorius, Crotalaria juncea, Cucumis Citrullus, C. Melo, C. sativus, Cucurbita Pepo, Daucus Carota, Dioscorea aculeata, D. alata, D. Batatas, D. nummularia, D. pentaphylla, D. sativa, Dolichos Lablab, Ervum Lens, Euchlæna luxurians, Fagopyrum emarginatum, F. esculentum, F. Tataricum, F. triangulare, Glycine hispida, Helianthus annuus, Hibiscus esculentus, Hordeum deficiens, H. distichon, H. hexastichon, H. vulgare, H. zeocriton, Ipomœa Batatas, Lactuca sativa, Linum usitatissimum, Lupinus albus, L. angustifolius, L. luteus, L. varius, Medicago orbicularis, M. scutellata, Nicotiana Tabacum, Oryza sativa, Panicum Crus Galli, P. frumentaceum, P. Italicum, P. miliaceum, P. sanguinale, Papaver somniferum, Pennisetum typhoideum, Peucedanum sativum, Phalaris Canariensis, Phaseolus derasus, P. lunatus, P. Max, P. vulgaris, Physalis pubescens, Pisum sativum, Raphanus sativus, Rumex vesicarius, Secale cereale, Sesamum Indicum, Sesbania cannabina, Solanum Lycopersicum, S. tuberosum, Spinacia oleracea, Tragopogon porrifolius, Trapa natans, Trifolium Alexandrinum, T. furcatum, T. incarnatum, T. pratense, T. reflexum, T. resupinatum, T. spadiceum, Triticum vulgare, Vicia Faba, V. sativa, Vigna Sinensis, Zea Mais.

II.

IMPORTANT CULTURAL PLANTS, YIELDING A RETURN IN THE FIRST OR SECOND SEASON, AND ALSO FOR SOME YEARS AFTERWARDS:

(Perennials and some Shrubs.)

Æschynomene aspera, Agrostis alba, A. rubra, Aloe ferox, A. linguiformis, A. Perryi, A. purpurascens, A. vera, A. vulgaris, Alopecurus pratensis, Ananas sativa, Andropogon Haleppensis, Anthistiria ciliata, Artemisia Dracunculus, Arundinaria spathiflora, Asparagus

officinalis, Astrebla pectinata, A. triticoides, Atriplex halimoides, A.
nummularium, A. vesicarium, Bambusa arundinacea, B. Brandisii, B.
Balcooa, B. spinosa, B. vulgaris, Bœhmeria nivea, Bromus unioloides,
Cajanus Indicus, Canavalia gladiata, Canna Achiras, C. coccinea, C.
edulis, C. glauca, Capparis spinosa, Capsicum frutescens, Cedronella
triphylla, Chenopodium auricomum, Chrysanthemum cinerarifolium, C.
coronopifolium, C. roseum, Cichorium Intybus, Cochlearia Armoracia,
Crambe cordifolia, C. maritima, C. Tataria, Cynodon Dactylon, Cyperus
esculentus, Cytisus scoparius, Dactylis glomerata, Danthonia penicil-
lata, Dendrocalamus giganteus, D. strictus, Desmodium triflorum,
Elymus arenarius, Fagopyrum cymosum, Festuca elatior, F. ovina,
Fragaria Californica, F. Chiloensis, F. collina, F. grandiflora, F.
Illinoensis, F. pratensis, F. vesca, F. Virginiana, Gigantochloa Apus,
G. atter, G. maxima, G. robusta, Glycyrrhiza glabra, Gossypium
arboreum, G. Barbadense, G. herbaceum, G. hirsutum, G. religiosum,
Guadua angustifolia, G. latifolia, Hedysarum coronarium, Helianthus
tuberosus, Humulus Lupulus, Indigofera Anil, I. tinctoria, Jasminum
grandiflorum, J. odoratissimum, J. officinale, J. Sambac, Kochia villosa,
Lavandula angustifolia, L. latifolia, L. Stœchas, Lippia citriodora,
Lolium perenne, Lotus corniculatus, Lupinus arboreus, L. Douglasii,
Manihot Aipi, M. utilissima, Medicago sativa, Mentha piperita, Morus
alba, Musa Cavendishii, M. paradisiaca, M. simiarum, Nelumbo lutea,
N. nucifera, Ocimum gratissimum, Onobrychis sativa, Origanum
Majorana, Panicum decompositum, P. maximum, P. spectabile, Pap-
pophorum commune, Paspalum distichum, Passiflora alata, P. edulis, P.
ligularis, P. macrocarpa, P. quadrangularis, Pelargonium capitatum,
P. odoratissimum, P. Radula, Phaseolus coccineus, Phleum pratense,
Phormium tenax, Phyllostachys bambusoides, P. nigra, Physalis
Peruviana, Pimpinella Sisarum, Poa araninifera, P. Brownii, P. Forsteri,
P. pratensis, P. trivialis, Portulacaria Afra, Psamma arenaria,
Rheum australe, R. officinale, R. palmatum, R. Rhaponticum, Ribes
floridum, R. Griffithii, R. Grossularia, R. hirtellum, R. nigrum, R.
rubrum, Ricinus communis, Rosa centifolia, R. Damascena, R. moschata,
R. sempervirens, Rubia tinctorum, Rubus cæsius, R. Canadensis, R.
deliciosus, R. ellipticus, R. fruticosus, R. geoides, R. Idæus, R. imperialis, R. lasiocarpus, R. nutans, R. occidentalis, R. phœnicolasius, R.
rugosus, R. strigosus, R. trivialis, R. ursinus, R. villosus, Rumex
Acetosa, R. Patientia, R. scutatus, Salix purpurea, S. rubra, S.
triandra, S. viminalis, Scorzonera crocifolia, S. deliciosa, S. Hispanica,
S. tuberosa, Sechium edule, Sesbania Ægyptiaca, Sesleria cœrulea,
Solanum betaceum, Stenotaphrum Americanum, Symphytum peregrinum, Tinguarra Spicula, Trifolium agrarium, T. alpestre, T.
fragiferum, T. hybridum, T. medium, T. montanum, T. ochroleucum,
T. Pannonicum, T. repens, Vaccinium cæspitosum, V. Canadense, V.
corymbosum, V. erythrocarpum, V. humifusum, V. macrocarpon, V.
myrtilloides, V. Myrtillus, V. ovalifolium, V. ovatum, V. Oxycoccos, V.
Pennsylvanicum, V. vacillans, Zingiber officinale.

III.

IMPORTANT CULTURAL PLANTS, YIELDING A RETURN
IN THE THIRD OR FOURTH SEASON AND FOR SOME
OR MANY YEARS AFTERWARDS.

(Shrubs and some small Trees.)

Aberia Caffra, Agave Americana, A. inæquidens, A. rigida, Alibertia
edulis, Aloe dichotoma, A. plicatilis, A. spicata, Astragalus adscendens,
A. brachycalyx, A. Cephalonicus, A. Creticus, A. gummifer, A. mi-
crocephalus, A. Parnassi, A. verus, Camellia Thea, Carica Candamar-
censis, C. Papaya, Citrus Aurantium, C. medica, Coffea Arabica,
Erythroxylon Coca, Gaylussacia frondosa, G. resinosa, Ilex Paraguensis,
Laurus nobilis, Myrtus Ugni, Olea Europæa, Opuntia coccinellifera,
O. Hernandezii, O. Tuna, Photinia eriobotrya, Pilocarpus pinnatifolius,
Pistacia Lentiscus, P. Terebinthus, Prunus Amygdalus, P. Armenica, P.
Cerasus, P. domestica, P. Persica, Psidium Cattleyanum, P. Guayava,
P. polycarpon, Pyrus communis, P. Cydonia, P. Malus, Rhamnus
catharticus, Rhus coriaria, R. Cotinus, R. glabra, R. typhina, Vaccinium
Leschenaultii, V. leucanthum, V. meridionale, V. Mortinia, V. parvi-
folium, V. penduliflorum, Vitis æstivalis, V. candicans, V. cinerea, V.
cordifolia, V. Labrusca, V. riparia, V. rupestris, V. Schimperiana, V.
vinifera, V. vulpina, Yucca aloifolia, Y. brevifolia, Y. filamentosa, Y.
gloriosa, Y. Yucatana, Zizyphus Jujuba.

IV.

IMPORTANT CULTURAL PLANTS, YIELDING A RETURN
AFTER SEVERAL YEARS, AND OFTEN FOR MANY
SUBSEQUENT YEARS ALSO.

(Trees, mostly large.)

Acacia aneura, A. Arabica, A. Catechu, A. decurrens, A. Koa, A.
leiophylla, A. Melanoxylon, A. microbotrya, A. pycnantha, A. Verek,
Acer saccharinum, Achras Sapota, Albizzia Saman, Amelanchier
Botryapium, Anona Cherimolia, Argania Sideroxylon, Callitris
cupressiformis, C. quadrivalvis, C. verrucosa, Carya alba, C. amara,
C. glabra, C. microcarpa, C. oliviformis, C. sulcata, C. tomentosa,
Casimiroa edulis, Castanea sativa, Cedrela australis, C. Sinensis, C.
Taona, Ceratonia Siliqua, Ceroxylon andicola, Cinchona Calisaya,
C. cordifolia, C. micrantha, C. nitida, C. officinalis, C. succirubra,
Copernicia cerifera, Corylus Avellana, C. Colurna, C. maxima, C.
Pontica, C. rostrata, Dammara australis, D. robusta, Dichopsis Gutta,

IMPORTANT CULTURAL PLANTS—*continued.*

Diospyros Ebenum, D. Kaki, D. Virginiana, Eucalyptus citriodora, E. crebra, E. Globulus, E. gomphocephala, E. goniocalyx, E. Leucoxylon, E. melliodora, E. rostrata, E. siderophloia, Ficus Carica, F. elastica, Fraxinus Americana, F. excelsior, F. Ornus, F. quadrangulata, Ginkgo biloba, Guevina Avellana, Hovenia dulcis, Hymenæa Courbaril, Jubœa spectabilis, Juglans cinerea, J. nigra, J. regia, Juniperus Bermudiana, J. Chinensis, J. drupacea, J. Virginiana, Liquidambar Altingia, L. orientalis, L. styraciflua, Macadamia ternifolia, Mangifera Indica, Morus nigra, M. rubra, Nephelium Litchi, Persea gratissima, Phœnix dactylifera, Pinus Abies, P. amabilis, P. australis, P. balsamea, P. Cedrus, P. cembroides, P. Coulteri, P. Douglasii, P. edulis, P. excelsa, P. Fraseri, P. Gerardiana, P. Haleppensis, P. Hartwegii, P. Kæmpferi, P. Lambertiana, P. Laricio, P. Larix, P. longifolia, P. mitis, P. monticola, P. nigra, P. Picea, P. Pinaster, P. Pinea, P. ponderosa, P. radiata, P. rigida, P. Sabiniana, P. silvestris, P. Sitkensis, P. Strobus, P. Webbiana, Pistacia vera, Populus alba, P. monilifera, P. nigra, Quercus Ægilops, Q. alba, Q. Cerris, Q. chrysolepis, Q. coccinea, Q. Douglasii, Q. Garryana, Q. Ilex, Q. incana, Q. lyrata, Q. macrocarpa, Q. macrolepis, Q. Phellos, Q. Robur, Q. serrata, Q. Skinneri, Q. Suber, Q. virens, Sequoia sempervirens. S. Wellingtonia, Spondias dulcis, Swietenia Mahagoni, Taxodium distichum, Thuya gigantea, Tilia Americana, T. Europæa, Ulmus campestris, U. fulva, U. racemosa.

All these plants fit for extra-tropical countries, but many only for particular climatic regions; for information in respect to the latter, the geographic index is to some extent indicative.

456   *Select Plants for Iudustrial Culture*

INDEX TO VERNACULAR NAMES.

Name	PAGE	Name	PAGE
Abele	295	Baboot-bark	2
Aboh-fruit	393	Babur	1
Acacia, Locust	328	Badjong	8
Acajou-wood	80	Bajree	256
Adam's Needle	408	Balata	14, 218, 393
Adeira	68	Balm-herb	215
Agallochum	35	Balmony	85
Agath-Dammar	116	Balsam	50
Aggur	35	Bamboo reed	42
Akamatsou	272	Bamboos	417
Akeki	374	Bananas	222
Alder	24	Bandakai	178
Aleppo-grass	29	Baneberry	14
Alerce	161	Bangalay	132
Alexandra-Palm	306	Banyan-tree	158, 160
Alfalfa	211	Barley	179
Algaroba	82, 182, 299	Barnyard-grass	245
Algoborillo	63	Basil	233
Alisander	359	Basswood-tree	376
Alkanna	22, 202	Bastard Mahogany	132
Alkanet	202	Battari	30
Almond-tree	300	Bay-berry	225
Aloe, gigantic	18	Bay-tree	389
„ ordinary	24	„ Sweet	195
Aloe-wood	26	Beach-Plum	302
Aloja	299	Bean, Broad	395
Alvarillo	407	„ Field	395
Angelica	38	„ French	260
Angico-gum	287	„ Haricot	260
Anise	266	„ Horse	395
Apple	308	„ Kidney	260
Apple-gumtree	147	„ Straight	395
Apple-haw	103	„ Windsor	395
Apricot	300	Beebalm-Tea	280
Aracacha	243	Beeches	153
Aracua	14	Beet	53
Arbor vitæ	307	Begoon	360
Argan-tree	38	Bembil	143
Arhar	64	Bent-grass	18, 304
Aroche	46	Berberry	53
Arrowroot	68, 209, 369	Bermuda-grass	112
Artichoke	111	Bhaib-grass	29
„ Jerusalem	176	Bilberry	391
Aru-root	209	Birch	55
Ash	163	Birdsfoot Clover	241
Asparagus	43	Birdsfoot Trefoil	203
Aspen	297	Bitternut-tree	74
Assafetida	155	Birthroot	382
Atocha	366	Blackbutt-tree	142
Aubergine	360	Black Gumtree	233
Avens	167	Blackberry	333
Avocado-pear	257	Blackthorn	303

INDEX TO VERNACULAR NAMES—*continued.*

INDEX TO VERNACULAR NAMES—*continued.*

By Authority: JOHN FERRES, Government Printer, Melbourne.

www.ingramcontent.com/pod-product-compliance
Lightning Source LLC
Chambersburg PA
CBHW020902210326

41598CB00018B/1754